S0-AAA-854

THIRD EDITION

COMPANY OFFICER

THIRD EDITION

COMPANY OFFICER

DELMAR
CENGAGE Learning

Australia • Brazil • Japan • Korea • Mexico • Singapore • Spain • United Kingdom • United States

DELMAR
CENGAGE Learning™

Company Officer, Third Edition

Vice President, Career and Professional
  Editorial: Dave Garza

Director of Learning Solutions: Sandy Clark

Product Development Manager: Janet Maker

Managing Editor: Larry Main

Senior Product Manager: Jennifer Starr

Associate Product Manager: Meaghan O'Brien

Editorial Assistant: Amy Wetsel

Vice President, Career and Professional
  Marketing: Jennifer Baker

Marketing Director: Deborah S. Yarnell

Senior Marketing Manager: Erin Coffin

Marketing Coordinator: Shanna Gibbs

Production Director: Wendy Troeger

Production Manager: Mark Bernard

Senior Content Project Manager:
  Jennifer Hanley

Art Director: Benj Gleeksman

Cover photo courtesy of Andy Thomas

© 2010 Delmar, Cengage Learning

ALL RIGHTS RESERVED. No part of this work covered by the copyright herein may be reproduced, transmitted, stored or used in any form or by any means graphic, electronic, or mechanical, including but not limited to photocopying, recording, scanning, digitizing, taping, Web distribution, information networks, or information storage and retrieval systems, except as permitted under Section 107 or 108 of the 1976 United States Copyright Act, without the prior written permission of the publisher.

For product information and technology assistance, contact us at
Cengage Learning Customer & Sales Support, 1-800-354-9706
For permission to use material from this text or product,
submit all requests online at **www.cengage.com/permissions**.
Further permissions questions can be e-mailed to
permissionrequest@cengage.com.

Library of Congress Control Number: 2009929187

ISBN-13: 978-1-4354-2725-9

ISBN-10: 1-4354-2725-4

Delmar
Executive Woods
5 Maxwell Drive
Clifton Park, NY 12065
USA

Cengage Learning is a leading provider of customized learning solutions with office locations around the globe, including Singapore, the United Kingdom, Australia, Mexico, Brazil, and Japan. Locate your local office at **www.cengage.com/global**

Cengage Learning products are represented in Canada by Nelson Education, Ltd.

To learn more about Delmar, visit **www.cengage.com/delmar**

Purchase any of our products at your local bookstore or at our preferred online store **www.cengagebrain.com**

Notice to the Reader

Publisher does not warrant or guarantee any of the products described herein or perform any independent analysis in connection with any of the product information contained herein. Publisher does not assume, and expressly disclaims, any obligation to obtain and include information other than that provided to it by the manufacturer. The reader is expressly warned to consider and adopt all safety precautions that might be indicated by the activities described herein and to avoid all potential hazards. By following the instructions contained herein, the reader willingly assumes all risks in connection with such instructions. The publisher makes no representations or warranties of any kind, including but not limited to, the warranties of fitness for particular purpose or merchantability, nor are any such representations implied with respect to the material set forth herein, and the publisher takes no responsibility with respect to such material. The publisher shall not be liable for any special, consequential, or exemplary damages resulting, in whole or part, from the readers' use of, or reliance upon, this material.

Printed in the United States of America
3 4 5 6 7   19 18 17 16 15

This book is dedicated to firefighters everywhere.

The sculpture "To Lift A Nation" is based on the now-famous photograph taken by Thomas E. Franklin of *The Record* of Bergen County, New Jersey, in 2001, and it was produced by Atlas Bronze Casting in Cooperation with the North Jersey Media Group Disaster Relief Fund and The Bravest Fund (Sculptor: Stan Watts).

The inscription on the base reads:

Three firemen raised a flag at ground zero in silent tribute to those brave firefighters who answered the call. This noble flag is raised permanently in honor of those heroes and all who serve this great nation. May God continue to bless America.

Photo courtesy of Mark A. Whitney

This photograph was taken at the National Emergency Training Center, home of the National Fire Academy, in Emmitsburg, Maryland, by Mark Whitney. Mr. Whitney is a Fire Programs Specialist at the National Fire Data Center, United States Fire Administration.

Contents

1

The Company Officer's Role—Challenges and Opportunities

2

The Company Officer's Role in Effective Communications

3

The Company Officer's Role in the Organization

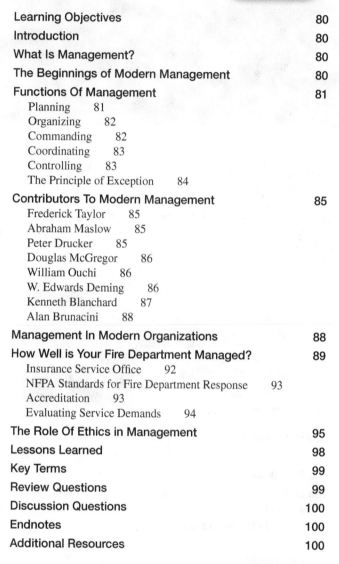

4

The Company Officer's Role in Management

5

The Company Officer's Role in Managing Resources

6

The Company Officer's Role: Principles of Leadership

7

The Company Officer's Role in Leading Others

8

The Company Officer's Role in Personnel Safety

9

The Company Officer's Role in Fire Prevention

10

The Company Officer's Role in Understanding Building Construction and Fire Behavior

11

The Company Officer's Role in Fire Investigation

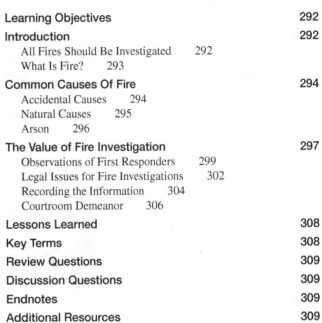

12

The Company Officer's Role in Planning and Readiness

13

The Company Officer's Role in Incident Management

THIRD EDITION

What a tremendous honor it is for me to be asked to prepare the foreword for the third edition of Delmar's textbook *Company Officer*. Following in the footsteps of both Chief Brunacini and Chief Coleman (the writers of the previous forewords) is one of the highest honors that I would ever hope to attain. Both Chiefs are truly among the most accomplished in the fire and rescue industry, and I am humbled by the opportunity to be the next person to prepare the foreword for such an important textbook. So, let me publically thank Mr. Clinton H. Smoke and Delmar, Cengage Learning for this great opportunity. The notion of preparing a foreword is to stimulate the reader's interest and urge the person to use the text in question to improve their performance or life. Because of the quality of work and the detailed content that went into this book, this is an easy assignment. Let me open by saying that this textbook is a must read and reference source for the modern fire-rescue officer. It is my opinion that the transition from firefighter to fire officer is the most difficult and important step that one can take in the American Fire–Rescue Service. I will expand this idea a little more, later in the Foreword.

Perhaps the best way to start the introduction of this outstanding textbook is to provide some insight about the author. Several decades ago, I had the chance to meet Clinton Smoke and work with him on several major training projects and national fire service committees. Clint's professionalism and leadership ability were instantly apparent. As I would learn, he was a retired senior U.S. Coast Guard officer who was now working as the program head at the Northern Virginia Community College Fire Science Program in Fairfax, Virginia. Having graduated from this college program in the early 1980s, I felt an instant connection with Clint and found him to be a great instructor and mentor. Smoke's military bearing and educational background provided the framework that has allowed him to write a textbook that is valuable and worthy of the time investment. I suggest that every fire officer or those who are aspiring to become fire officers obtain and use the information that Clint has published.

Now, a few thoughts about the value and importance of the fire–rescue officers have to our organizations. Arguably, the company officer is the most important level of supervision within our agencies. Typically, the first departmental representative who the customer meets is the company officer. The company officer is the connection between our labor force and management, making sure that the organizational goals and objectives are met, even when no one is watching. The impact that the company officer has on his or her crew and the overall health of the department cannot be overstated. If the company officer is well prepared and brings positive, effective leadership to the job, the impact that he or she will have becomes a tremendous force multiplier. The company officer must serve as the in-service trainer, personnel officer, safety advocate, coach, career advisor, and a host of other important titles that keep the company focused on what is important—"preventing harm in the community." Often, success or failure of a specific company is based on the quality of the supervision that is provided to that unit, by, you guessed it, the company officer.

The sad truth is that most departments do not have a structured way to prepare and develop their company officers. The third edition textbook *Company Officer* will go a long way to close the training and education gap as well as provide the needed information to help a member to be effective in this mission critical role. I recommend that this book be added to college and technical school curricula that teach any level of fire officer. Further, this is a textbook that everyone should add to their personal library if they are a fire officer or aspire to be an officer at any level. It would be great to think that every department can provide the needed training to be successful at the job at hand. However, in some cases, that is not possible, so the member must and should take on the responsible to prepare themselves. Clinton Smoke's newest textbook is a great place to start!

Chief Dennis L. Rubin
District of Columbia Fire and EMS Department,
Washington, DC

SECOND EDITION

It has been many, many years since I have served as a company officer, but I will never forget two things. The first is the influence my first company officer, Captain Larry May, had on my career. The second thing I will never forget is the day I was appointed as a fire captain and given the command of a fire company. Both events have had a profound influence over my current perspective on both creating fire officers and being one.

In the case of the former, Captain May was not just the guy in the right front seat. He was my mentor, my teacher, and my conscience as it were. I will never forget his speech to me one morning at Fire Station Two. I was an engineer. He asked me if I knew my job as an apparatus operator. I stated yes. He asked me if the unit was ready to roll. Again, affirmative. He then asked me if I knew the job of being a captain. I knew better than to bluff, so I said no! He answered, "Good, because I want to be the one to teach you how to be a fire officer." From that day forward, Captain May prepared me to be ready when my time came.

And when it did come time for me to accept the badge and the obligation, I was ready. I was prepared by the best to be able to deal with the worst. I was grateful to Captain May then for having the interest in helping me to be ready. I am grateful now to Clinton Smoke for creating a textbook that captures what is the core of being a good company officer. What Smoke has prepared in this text is the journalistic version of a thousand Captain Mays who are providing both wisdom and perspective in the written word.

Yet I am reasonably sure both Captain May and Clinton Smoke would agree that the act of being mentored or reading a textbook does not result in the finished product. What does create this final product is the person who absorbs this information and does everything possible to continue to become more and more proficient in the job.

The role of the company officer is mentioned many times in this text, ranging from the challenges and the opportunities of the position to the company officer's part in incident management. It is a role that has continued to evolve as the fire service has taken on more diversified responsibilities in society. It is not just a case of a single performance at a singular event, but it is also a need to be equal to the next greatest challenge.

Between the pages of this textbook is a lifetime of experience that will become part of your background and knowledge as you gain stature in taking on the responsibility of a fire officer. No matter how large or small your department is, the company officer is almost always the first one on the scene. Be they major emergencies, a problem in the firehouse, or something you have not expected to ever occur, it is your job as the company officer to do the right thing, the right way.

Good luck in your pursuit of that role in the future.

Ron Coleman
Retired State Fire Marshal and Business
Development Manager for Emergency Services
Consulting Incorporated (ESCI), Wilsonville, OR

FIRST EDITION

I am pleased that my friend Clinton Smoke asked me to write this foreword for his new textbook. Clinton has been teaching and writing about fire department operations for a long time. He brings that practical experience to his new book along with his excellent communications skills.

He has been associated with many important and progressive developments during his career. I have enjoyed our friendship and mutual involvement in many of these projects.

How we select, prepare, and support company officers is a challenge that is critical to the day-to-day delivery of fire department services. Throughout my entire career I have heard fire department managers (me included) articulate that challenge by saying "if I could 'have' any group within my department, make that group our company officers." To "have" a group means to somehow capture the consistent effectiveness of that group. Another indication of how important the company officer is to our overall effectiveness is hearing an experienced fire chief say that the quickest and most accurate way to evaluate the effectiveness of a fire company is simply to look at the profile of the company officer in charge of that unit.

A major reason the company officer is so important is that he or she is the only boss who has continuous access to our workforce. Company officers directly supervise the workers who deliver services to our customers. The company officer is the up-close and personal working boss who is always with and part of the crew of Engine 1 throughout the entire shift. There is no other middle person between the workers and the work. He rides in the right front seat, he sits at the head of the dinner table, he controls the TV remote (if he has enough seniority), he holds the IV bag and the flashlight, and he backs up the nozzle. When the workers do good, the boss smiles (big deal); when there is a problem, the company officer helps fix it. He can make a bad day better by whispering the answer to a worker who gets disconnected from the answer. He is the on-line safety officer for the crew and has the most direct control over how (and if) the crew goes home at the end of the shift with all their body parts and faculties intact.

Another major reason that the company officer is so critical to us is that he manages the business of our business, right at the point where the customer's problem occurs. He is in Mrs. Smith's kitchen at

3:00 A.M. when she smells smoke, and at that point, he is the entire department to her. He is able to create a positive feeling and memory in Mrs. Smith based on how quickly his company responds, how well they solve her problem, and, most of all, how nice they are to her. Simply, between 3:00 A.M. and about 3:20 A.M., we get one chance to win or lose, based on how well Engine 1 performs at Mrs. Smith's house. The memory that Mrs. Smith has of the night her kitchen caught fire is the product of how Engine 1's crew made her feel; how Engine 1's crew feels is the product of how their company officer treats them. This is how the company officer gets to connect the inside (crew) with the outside (customer) and this creates our most important relationship.

I am involved both professionally and personally with company officers, so my comments and feelings are not those of a casual observer. I was a company officer (at Engine 1) in an earlier life. The Phoenix Fire Department now employees 255 fire captains (our company officer title). Nearly one third are on duty as I write this foreword, and today they will respond to 450-plus requests for service. How they perform will regulate how 327 on-duty firefighters and thousands of customers will live, or die.

On a personal basis, my three children are all Phoenix firefighters. My son Nick was a company officer for 10 years before he was recently promoted to battalion chief. My son John is a captain on a busy downtown ladder company. My firefighter daughter Candi is going through the company officer promotion preparation process. Throughout the years, I have followed the standard going-on-duty send-off with them by saying, "be careful today," and the regular post-shift question, "what'd you guys do yesterday?" This book is directed at how to effectively manage the activity that happens in between these two rituals—the duty shift. Pay attention as you go through this material. It is the most important part of what we do.

Alan V. Brunacini,
Fire Chief, Phoenix Fire Department

Preface

Welcome firefighters and fire officers, to *Company Officer*, Third Edition! Whether you are seeking certification at the Fire Officer I level or are aspiring for a promotion to the Fire Officer II level, this comprehensive book provides insight into the role of the officer as defined by the 2009 Edition of NFPA 1021, *Standard for Fire Officer Professional Qualifications*. The basic principles of planning and prevention, communicating, managing resources, leading others, and incident management are covered, including an emphasis on safety in responding to incidents.

This book is a great resource for both paid and volunteer firefighters and fire officers to further their careers within the fire service. By the same token, we have attempted to make the material in this book useful and interesting for a broader audience. Issues of management and supervision for administrators, rescue squads and EMS organizations that are independent of a fire department are essentially the same as those in the fire service. Where appropriate, this book is intended to help these managers as well as those of the fire service—we hope that you will find the material interesting, informative, and useful. As an officer, you will be called upon to lead others in your role as an officer, and you will make a difference in the lives of many others. Be prepared, develop your skills, and never stop learning.

DEVELOPMENT OF THIS BOOK

Becoming certified as a fire officer is important, but becoming a proficient company officer in the fire service is even more important. *Company Officer*, Third Edition was created not only to address the requirements of the 2009 Edition of NFPA 1021 but also to focus on developing the soft-skills necessary to hit-the-ground running with the new responsibilities of being an officer. Through the dedication of our author, contributors, as well as content and technical reviewers, the new edition of *Company Officer* remains up to date with the revisions to the Job Performance Requirements (JPRs) in the 1021 Standard, and presents the content in a realistic, straightforward, and challenging way. *Street Stories* and *case studies* accompany each chapter, highlighting the theme of the content through accounts of actual incidents experienced by officers from across the United States—the decisions made, and the lessons learned. Full-color photographs bring to life the everyday activities of the officer, while the text presents the information in a clear and concise manner. Featured text boxes integrated throughout the book provide important advice from experienced officers, essential facts, tips on remaining safe on the fireground, and ethical considerations. All content is presented in an easy-to-understand method to ensure that you are properly equipped with the knowledge to not only accomplish your responsibilities but also excel in your role as officer.

OUR REVIEW AND VALIDATION COMMITTEES

Through the dedication of our authors, content and technical reviewers, as well as our Fire Advisory Board members and validation committee, the third edition of *Company Officer* continues to remain up to date with the changing landscape of the fire service world. As part of the development process, each chapter is carefully reviewed by a select number of practicing individuals in the fire service who offer their expertise and insight as we revised the content. Additionally, technical reviewers thoroughly check the manuscript in detail for clarity and accuracy.

We are excited to announce that for the third edition of this book, we have created a validation committee to ensure that the content meets 100% of the NFPA Standard 1021. Learning objectives, which have been validated by this committee of subject-matter experts, are included at the beginning of each essentials of firefighting chapter. The learning objectives tie all components of the learning solution (textbook, curriculum, test banks, supplements) to the NFPA standard, providing instructors and students a pathway to meet the intent of the JPRs outlined in the Standard.

THE VALIDATION PROCESS

The National Fire Protection Association (NFPA) professional qualifications standards identify the minimum JPRs for fire service positions. JPRs state the behaviors required to perform specific skill(s) on the job. The JPR statements must be converted into instructional objectives with behaviors, conditions, and standards that can be measured within the teaching/learning environment.

Our process includes the development of the learning objectives from the JPR by a subject-matter expert. The learning objectives are then reviewed and validated by a committee. The committee reviews the learning objectives to ensure that the JPR was correctly interpreted and also makes recommendation for additional learning objectives within our materials. Our authors are provided with the validated learning objectives to develop the materials. This ensures that 100% of the standard is met within our materials.

The learning objectives are used throughout the entire development process. This includes the development of the book, curriculum, and the certification test question banks. This process ensures consistency among all materials and offers the best possible materials available on the market.

For a complete list of our review and validation committee members, as well as other contributors to this book, please refer to the Acknowledgments section of this Preface.

HOW TO USE THIS BOOK

This book is intended as a text to help you gain competency in the requirements of NFPA 1021. That Standard lists the requirements for Fire Officer I and II in six categories: administration, community and government relations, human resource management, inspection and investigation, emergency service delivery, and safety. This book is organized into chapters covering those major topics. The early chapters deal with communications, organizations, management, and leadership. In later chapters, these tools are then applied to the process of preparing for and managing emergency incidents.

- *Chapter 1* starts with a discussion of the vital role of the company officer and examines the challenges that face company officers and outlines reasons why preparedness is so key when moving into this important position.

- *Chapter 2* is about effective communications, a vital topic that touches every aspect of the role of an officer. It is a tool that you will use throughout the remainder of the course, and throughout your career.

- *Chapter 3* addresses organizations and the company officer's role in the organization. Understanding your organization helps define your place and purpose in that organization. This chapter sets the stage for the discussions of management and leadership that follow.

- *Chapter 4* briefly reviews the field of management science and examines the company officer's role as a manager. Understanding the concepts of modern management can have a great impact on how effective you are as a leader.

- *Chapter 5* provides guidelines for practical applications of modern management principles at the company level. The ability to effectively manage your resources contributes to the smooth operation of your department.

- *Chapter 6* briefly examines leadership principles and the company officer's role as a leader. Understanding how groups work and function can contribute to how successful you are in managing people.

- *Chapter 7* offers some effective tools for making leadership work at the company level. Applying the concepts from Chapter 6 can help you to develop the leadership skills that you need for the officer level.

- *Chapter 8* addresses the important role of the company officer in administering the department's occupational safety and health programs at the company level. You must be able to implement such programs in order to prevent injuries or death occurring from accidents on the job.

- *Chapter 9* acknowledges that the prevention of civilian injuries, fatalities, and property loss due to fire is a part of the job, too, and what company officers should be doing more to support their department's efforts to prevent these losses from occurring.

- *Chapter 10* reviews how fire destroys, and how understanding the way various building types are constructed can help prevent fire loss.

- *Chapter 11* provides a brief overview of fire-cause determination, so company officers can accurately determine the cause of fires, especially when arson is indicated.

- *Chapter 12* covers the company officer's responsibilities for keeping themselves and their company trained, fit, and ready to deal with the inevitable emergencies. A proactive approach to training and preparation will reduce the negative consequences of these emergencies.

- Finally, *Chapter 13* deals with management of the emergency scene. When fires or other emergency events do occur, the company officer will likely be the first on the scene. His or her assessment of the situation and initial actions will have a major impact on all of the events that follow.

NEW TO THIS EDITION

- *Fire Officer I and II Levels:* Revised to meet the intent of the 2009 Edition of NFPA 1021, including new learning objectives validated by a committee of experts to ensure 100% compliance with the Standard. All-new design provides color-coded objectives to easily differentiate Fire Officer I and II requirements.

- *Safety:* Emphasized throughout the chapters, including expanded information on personnel safety in order to keep you and your team safe on the fireground through proactive actions.

- *Fire Prevention:* To help keep you, your team, and the community safe, included is a fully revised chapter on fire prevention. This chapter outlines the most up-to-date fire codes and inspection policies. It also includes how to better educate the community on how to prevent fires.

- *Communication:* Communicate effectively in the firehouse and on the fireground with new coverage on emerging forms of communications, such as email, emergency radio contact, and public relations. From verbal communication to written, this section shows how to get your message across in a clear and professional manner.

- *Effective Leadership:* Lead your team with confidence, by studying a new section on employee and self-assessments with the 360 Degree Assessment. Also included is a new section on the Leadership Tetrahedron, which covers the promotional processes within the fire service.

- *Ongoing Emerging Issues:* A new section on Standard Operating Procedures (SOPs) outlines the many important functions these SOPs serve, including examples to help illustrate the various types that are typical to the fire department—in both emergency and nonemergency situations.

FEATURES OF THIS BOOK

The following is a list of features that serve to enhance learning of the material and help you gain competence and confidence in mastering fire officer essentials.

NFPA 1021 Correlation Grid

This grid provides a correlation between *Company Officer*, Third Edition and the requirements for Fire Officer I and II, as stipulated by NFPA 1021, *Standard for Fire Officer Professional Qualifications*, 2009 Edition. These sections from the Standard are correlated to the book chapters and are referenced by page numbers.

Street Stories

Each chapter opens with personal experiences contributed by fire officers from across the nation. These personal accounts bring to life the content of the chapters that follow and highlight important lessons learned.

As a company officer, I serve many functions in my organization. It is sometimes easy to get caught up in finding and maintaining your proper place in the department's organizational scheme. In my fire service career, I managed to learn from other company officers some very valid truths that I have carried with me as I progressed to a leadership role.

During a fire in a single-story commercial structure, I was crawling behind my lieutenant into the building through the dark smoke-filled office area, looking for the seat of the fire. Flames were coming out the floor registers, and the building was getting hotter and hotter.

We were crawling toward what appeared to be an open area, and the floor seemed to be slanting. It was hard to tell in the severe smoke condition, but I kept following him, still searching for the fire. All of the sudden in a muffled yell I heard, "Get back, get the hell back, back. . . ." Immediately, we retreated as fast as we could. Once outside, I heard the blasting of truck air horns, which was our department's structure evacuation signal. As it turns out, the floor inside the building really was slanting. In fact, in the room my lieutenant and I were headed for, the floor had fallen into the basement! We almost went with it.

My lieutenant came up to me after the fire was contained and apologized for "cussing" at me. I just looked down and shook my head and said, "You don't need to apologize. You just saved my life. Thank you."

From that day forward, I knew that whatever place I occupied on my department's organizational chart, or whatever my job descriptions spelled out as my duties, my first and foremost responsibility over all others was the safety of the people on my team. Whenever I get bogged down with the various roles I play as a company officer, I will never lose sight of this most important "role" that I play: to protect the safety of my team members.

—*Sally L. McCann-Mirise, Lieutenant, Liberty Twp. Fire Prevention, Powell, Ohio*

Learning Objectives

Each chapter includes a list of objectives; satisfy yourself that you have learned each objective so that you are assured of gaining the appropriate competency for certification.

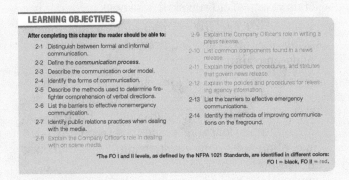

Key Terms

Throughout the text you will find key terms and notes of particular importance. These important items will provide you with the knowledge and terminology for effectively communicating with your superiors, subordinates, and your peers.

Officer Advice

These note boxes provide tips from experienced officers to guide you successfully through the challenges in your new position.

Fire departments would do well to publish a guide with some good examples of correspondence and report writing.

Fireground Fact

These note boxes outline important data, safety information, and considerations when responding to incidents.

Supervision is generally defined as the act of watching over the work or tasks of another. The NFPA defines a *supervisor* as an individual responsible for overseeing the performance or activity of other members.

Based on this definition, a company officer can be described as a first-line supervisor responsible for the performance and safety of assigned personnel in an emergency service organization.

Safety

These note boxes offer advice on how to react and, more important, be proactive in protecting the safety of yourself and your personnel.

Where the risk justifies entering the building, you should realize that just because you can safely enter the building does not mean that you can safely stay in the building.

Ethics

These note boxes emphasize the need to build respectful relationships with fellow firefighters within your department, with the public, and with administration in order to be an effective leader.

Two well-known authors, Dr. Kenneth Blanchard, the coauthor of *The One Minute Manager* and many other books, and Dr. Normal Vincent Peale, the author of *The Power of Positive Thinking* and many other books, collaborated on a book entitled *The Power of Ethical Management*.[2] In that book, they suggest a simple three-question test to help guide our actions. The three questions follow:

- Is it legal?
- Is it balanced?
- How will it make me feel about myself?

Case Studies

Each chapter ends with a report on an incident experienced by fire officers from across the nation. Each report is followed by a set of questions to develop critical thinking skills so that you are ready as an officer to respond effectively to incidents.

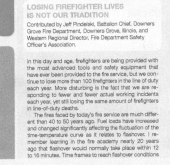

LOSING FIREFIGHTER LIVES IS NOT OUR TRADITION

Contributed by Jeff Pindelski, Battalion Chief, Downers Grove Fire Department, Downers Grove, Illinois, and Western Regional Director, Fire Department Safety Officer's Association.

In this day and age, firefighters are being provided with the most advanced tools and safety equipment that have ever been provided to the fire service, but we continue to lose more than 100 firefighters in the line of duty each year. More disturbing is the fact that we are responding to fewer and fewer actual working incidents each year, yet still losing the same amount of firefighters in line-of-duty deaths.

The fires faced by today's fire service are much different than 40 to 50 years ago. Fuel loads have increased and changed significantly affecting the fluctuation of the time-temperature curve as it relates to flashover. I remember learning in the fire academy nearly 20 years ago that flashover would normally take place within 12 to 16 minutes. Time frames to reach flashover conditions

losing more and more members to retirement who have served in a busier era of the fire service. Rapid promotion is also becoming part of the overall challenge. As we lose the experienced members, we are being left with a generation of firefighters that may be the most highly educated generation ever, due to their access to information through avenues such as the Internet, but may come up short in other areas that are necessary to successfully lead our firefighters. It is realistically possible, in today's fire service, that some departments may actually promote a firefighter to the company officer level who has never gained practical hands-on experience at an actual fire. These should not be looked upon as shortcomings but rather as challenges that we will need to overcome. Fault does not fall onto any particular generation—each one only has to deal with the economic, political, and social environments that influence the time period in which they live.

The fire service environment requires fire officers to make life-and-death decisions in a quick manner based on incomplete information. When we are exposed to a situation, our brain stores that experience in its own file, almost like a photographic slide within a tray. When another situation is experienced, our brain will try to

Review and Discussion Questions

There are questions at the end of each chapter. Some will prompt you to review the material, while others are intended to make you think and analyze what you have learned. When you complete the unit, you should be able to answer each question without referring back to the chapter.

Additional Resources

Each chapter includes a list of additional resources. Each of the suggested items offers additional information on the topic covered in that particular chapter. The goal is to give you as much information as possible to help you in your duties as a company officer.

This information will help you better perform your duties, not just to meet the requirements of the Standard, as important as they are. Whether you are reading this book to improve your skills or taking a structured program leading to certification, we hope that you will find this book informative, interesting, and useful.

A NOTE ON THE CERTIFICATION PROCESS

This book is the result of a need for a single up-to-date comprehensive text for teaching a certification course in any setting for persons seeking certification as fire officers. Your mastery of the material can be demonstrated in several ways. If you have seriously studied the material and satisfactorily answered all of the questions in the text and workbook had communicated effectively with your instructor, you should be ready for the challenges of certification and, ultimately, for the job.

CURRICULUM PACKAGE

This book was created not only as a stand-alone manual for firefighters and fire officers, but as a special package of materials for the full instructional experience. The supplement package provides a variety of tools for students and instructors to enhance the learning experience.

Instructor's Curriculum CD

The Instructor's Curriculum CD is designed to allow instructors to run programs according to the standards set by the authority having jurisdiction where the course is conducted. It contains the information necessary to conduct fire officer courses. It is divided into sections to facilitate its use for training:

- *Administration* provides the instructor with an overview of the various courses, student and instructor materials, and practical advice on how to set up courses and run skill sessions.

- *Equipment Checklist* offers a quick guide for ensuring the necessary equipment is available for hands-on training.

- *Modularized Lesson Plans* are ideal for instructors, whether they are teaching at fire departments, academies, or longer-format courses. Each lesson plan presents learning objectives, recommended time allotment, equipment, reading assignments, and an outline of key concepts presented in each chapter with coordinating PowerPoint presentation slides, Job Performance Requirements, and Skill Sheets.

- *Answers to Review Questions and Workbook Questions* are included for each chapter in the book and student workbook.

- *PowerPoint Presentations* outline key concepts from each chapter, and contain graphics and photos from the book, to bring the content to life.

- *Test Bank* containing hundreds of questions helps instructors prepare candidates to take the written portion of the certification exam for Fire Officer I or Fire Officer II.

- *Skill Sheets* outline important skills that each candidate must master to meet requirements for certification and provide the instructor with a handy checklist.

- *Progress Logs* provide a system to track the progress of individual candidates as they complete the required skills.

- *Quick Reference Guides* contain valuable information for instructors. The following grids are included:

  □ NFPA Standard 1021 Correlation Guide to cross-reference *Company Officer*, Third Edition with NFPA Standard 1021

  □ New Edition Correlation Guide to cross-reference the revisions between the second and third editions of *Company Officer*

- *Additional Resources* offer supplemental resources for important information on various topics presented in the textbook.

- An *Image Library* containing hundreds of graphics and photos from the book offers an additional resource for instructors to enhance their own classroom presentations or to modify the PowerPoint presentations provided on the CD.

Order #: 978-1-4354-2726-6

Study Guide

A classroom tool for assessing progress and preparation for the certification exam, the Study Guide consists of questions in multiple formats as well as homework assignments and additional information for each chapter to help guide students through lessons. It completely covers the 2009 Edition of NFPA 1021 requirements.

Order #: 978-1-4354-9311-7

Fire Officer I and II DVD

Adhering to NFPA 1021, *Standard for Fire Officer Professional Qualifications*, this four-part program focuses on the soft skills essential to Fire Officers at Levels I and II. Intended as visual reinforcement for *Company Officer* or other classroom learning materials, the *Fire Officer I and II DVD* offers challenges that an officer may face in relation to Human Resource Management, Community and Government, Inspection and Investigation, and Emergency Services—Delivery and Safety. Each program contains two to four scenarios depicting common situations through role-play and reenactments. The DVD encourages classroom discussion through a series of follow-up questions following each scenario, and presents the necessary information to effectively train the next generation of fire officers.

Key Features

- Adheres to NFPA Standard 1021, keeping up to date with the latest expectations in the industry. Actual lieutenants and captains reenact the segments, bringing their own experiences to the situations.

- Provides an Instructor's Guide in PDF format, including learning objectives, discussion questions for classroom distribution, and answers to discussion questions that present best practices to be used as a guideline.

- Lively scenarios bring common situations to life, providing candidates with a realistic view of the challenges that they will need to face as an officer.

- Provides a set of NFPA requirements with each program for a quick reference.

- Plays on DVD-equipped television or computer.

Order # 978-1-4018-8293-8

ABOUT THE AUTHORS

Lead Author

Clinton H. Smoke

Professor Clinton Smoke is the chair of the Fire Protection Department at Asheville-Buncombe Technical Community College in Asheville, North Carolina. Prior to this position, he had a similar position at Northern Virginia Community College in Annadale, Virginia. His fire service background includes over 20 years of volunteer and paid experience in several fire departments. He has also served over 20 years in active military duty in the U.S. Coast Guard, during which time he commanded four cutters. Other assignments included duties as the senior instructor at the Coast Guard's Officer Candidate School and as head of the service's Vessel Safety Branch. Professor Smoke has a B.A. in Fire Science Administration, an M.S. in Communications Management, and an MBA. He is a certified Fire Protection Specialist and a Fire Officer III.

Contributing Authors

Charles Keeton

Charlie Keeton served 35 years on the Wichita Fire Department, where he rose through the ranks to serve as the chief training/safety officer. Charlie coordinated the in-service training provided to the personnel of the Fire Operations Division, including management and oversight of all fire, hazardous materials, medical, and special operations training functions. Charlie also served as the department's safety officer, overseeing its accident and injury prevention programs. In addition, Charlie was the coordinator of the department's Heavy Rescue Team as well as a training coordinator for the University of Kansas Fire and Rescue Training Institute.

In 2008, Charlie was awarded the J. Faherty Casey Award for his activities in the National Fire Academy's Training Resource and Data Exchange (TRADE) program as well as the Insurance Women of Wichita "Firefighter of the Year" award in 2007. Charlie currently serves as the emergency preparedness coordinator for hospitals within a 19-county area in south central Kansas.

Charlie is a Wichita native and has been married to his wife, Dianna, for over 33 years. He has two children and four grandchildren.

Billy Jack Wenzel

Billy Jack Wenzel is a 29-year veteran of the Wichita Fire Department, currently serving as the division chief of safety and training. Chief Wenzel worked through the ranks as firefighter, lieutenant, and captain, earning multiple awards and letters of recognition. He worked as captain at one of Wichita's busiest houses and spent time as a member of Special Operations. Chief Wenzel holds certifications in many areas, including hazardous materials technician, high angle rescue, trench rescue, N.A.U.I. SCUBA, confined space rescue, and advanced crash rescue and holds a bachelors degree in business administration. He spent 19 years as a member of the Wichita Fire Department Honor Guard and served as the unit commander.

Chief Wenzel is the author of "Kansas Grain Dust Explosion" and "Training Tips" for *Fire Engineering* magazine. He has been published in several fire service trade publications. He is an adjunct instructor for the National Fire Academy, Department of Homeland Security, Emergency Management Institute, and National Fallen Firefighter Foundation and a past instructor at FDIC. He is a training manager for the University of Kansas Fire and Rescue Training Institute and is the co-developer of Wichita HoT, a regional fire school providing affordable hands-on-training to hundreds of firefighters every year. He is also a member of the Kansas Firefighters Association Training Committee, Kansas Historical Museum, and the South Central Kansas Incident Management Team.

Chief Wenzel has been married to his wife, Sheri, for 21 years and has three children: Anna, Jacob, and Jared.

Bradford Boyd

Bradford Boyd is a 16-year veteran of the Wichita Fire Department, currently serving as the medical officer in the Safety/Training Division. Captain Boyd worked through the ranks as firefighter, lieutenant, and captain, earning multiple awards and letters of recognition. He currently works as "C" shift medical officer and spends time in the classroom delivering medical and firefighter safety courses.

Captain Boyd is a national fire Instructor as well as a disaster medical adjunct instructor for Rescue Training Associates. He is also an NIMS 300-400 instructor, NFA emergency incident safety officer instructor, Kansas Board of Emergency Medical instructor coordinator, and National Registry paramedic and holds certifications in many other areas. Captain Boyd holds an associate's degree from Cowley County Community College and is currently working on a bachelor's degree.

ACKNOWLEDGMENTS

Many people made this work possible. I am pleased to share any credit that may be due with them; on the other hand, any errors or omissions are mine alone. Selecting a few to name here is challenging, but I'll try.

Jerry Laughlin helped me get started at NFPA. Working there opened many doors and set the stage for all that followed. My role in officer development started when Warren Isman, the late chief of the Fairfax County Fire and Rescue Department, asked me to help him start a Career Development Program for that organization. Battalion Chief Roy Nester and Captain Mike Ward, among others, helped me get that program started. Northern Virginia Community College cooperated and provided most of the college courses required in that program. Over the years I met Alan Brunacini; Ron Coleman; Dr. Denis Onieal and Edward Kaplin, both of the National Fire Academy; the late Jim Page; and many others who share my interest in officer development. I am grateful to all of them for their support, suggestions, and friendship.

I acknowledge that many of the great ideas here came from students and I am grateful to them for their contributions. I also thank the National Fire Protection Association, the National Fire Academy, the International Association of Fire Chiefs, and the International City Management Association for generously allowing me to use their material in this edition.

To Margaret Magnarelli, Andrea Walter, and Cheryl Lock: thanks for all your assistance in acquiring and interviewing candidates for the Street Stories and case studies.

To the fire officers, our gratitude goes to you for sharing your stories:

Dave Dodson, Lead Instructor, Response Solutions

Sally L. McCann-Mirise, Lieutenant, Liberty Twp. Fire Protection

James M. Grady III, Chief, Frankfort Fire Protection District

Greg Grayson, Fire Chief, Asheville Fire Department

Scott Windisch, Lieutenant, The Woodlands Fire Department

Linda F. Willing, Principal Trainer, Real World Training and Consulting

Dominic Depaolis, Captain, Metropolitan Washington Airports Authority

Craig Ibanez, Captain, Los Angeles County Fire Department

Steve Chikerotis, Battalion Chief/Assistant Director of Training, Chicago, Fire Department

David Short, Senior Training Analyst, MPRI, and Assistant Chief, Sterling Volunteer Fire Department

Steve Kropf, Captain, City of Eugene Fire Department

Melvin Byrne, Past Chief, Ashburn Volunteer Fire and Rescue

Bryan Tyner, Captain, Minneapolis Fire Department

Michael Washington, Fire Lieutenant, Cincinnati Fire Department

Jeff Lytle, Battalion Chief, Henderson Fire Department

Steve Mormino, Lieutenant, FDNY

Tony Tricarico, Captain, Squadron 252, Special Operations Command, FDNY

Stephan Solomon, Engine Captain, Village of Owego Volunteer Fire Department

Julie Miller, Captain, Inuvik Fire Department

Christopher Naum, Captain, Battalion 2, Moyers Corners Fire Department

George Lucia, Sr., Fire Marshall, City of Vista Fire Department;

Aaron Heller, Captain, Hamilton Township Fire District #9

Grant Mishoe, Captain, North Charleston Fire Department

The photographs that appear in this book are due in part to the services of Lieutenant Kevin Terry and the Fuller Road Fire Department in Albany, New York. Your technical expertise and friendship have been an invaluable gift and have contributed greatly to the education of the next generation of firefighters.

We wish also to thank the firefighters, models Kevin Craig West and Rodney Wiltshire, and photographer Mike Gallitelli for their participation in the photo shoot.

Finally, I thank my wife, Judith, for her encouragement, contributions, and support throughout both the first and second editions.

Our Reviewers and Contributors

Delmar, Cengage Learning and I wish to especially thank the many reviewers and contributors whose assistance and dedication to the project have made this book possible:

Michael Arnhart, High Ridge Fire Department, High Ridge, MO

James Barnes, Baltimore County Fire Department, Towson, MD

Tom Bentley, John Wood Community College, Quincy, IL

Jimmy Bryant, Shreveport Fire Training Academy, Shreveport, LA

Mike Bucy, Red Devil Training, Portage, IN

Dave Casey, Florida State Fire College, Ocala, FL

Dennis Childress, Orange County Fire Authority, Rancho Santiago Community College, Orange, CA

Mike Cox, Florida State Fire College, Ocala, FL

Mike Connors, Naperville Fire Department, Napervillle, IL

Brian Dodge, Kent Fire Department, Kent, WA

Pete Evers, Auburn Fire Department, Auburn, CA

Gary Fox, San Antonio College, San Antonio, TX

Christopher Gilbert, Alachua County Fire Rescue, Gainesville, FL

Robert Hancock, Punta Gorda Fire Department, Punta Gorda, FL

Tom Harmer, Titusville Fire and Emergency Services, Titusville, FL

Edward Hojnicki, Jr., Wilmington Fire Department, Wilmington, DE

Rudy Horist, Elgin Community College, Elgin Fire Department, Elgin, IL

Gail Ownby-Hughes, Chattanooga State Technical Community College, Chattanooga, TN

John Kingyens, Sarnia Fire Rescue Services, Sarnia, Ontario, Canada

Stephen Korpf, Eugene Fire and EMS, Eugene, OR

Robert Laford, Mt. Wachusett Community College, Petersham, MA

Bob Leigh, Aurora Fire Department, Aurora, CO

Murrey E. Loflin, Virginia Beach Fire Department, Virginia Beach, VA

Ron Marley, Shasta College, Reddig, CA

Jack K. McElfish, Chief, Richmond Fire Department, Richmond, VA

Mike Morrison, Consumnes River Community College, Elk Grove Community Services District Fire Department, Elk Grove, CA

Marion B. Oliver, Midland College, Midland, TX

Jeff Pindelski, Downers Grove Fire Department, Downers Grove, IL

Timothy Pitts, Guilford Technical Community College, Greensboro Fire Department, Greensboro, NC

Chris Reynolds, Hillsborough County Fire and Rescue, Hillsborough County Community College, Tampa, FL

Jake Rhoades, Jenks Fire Department, Jenks, OK

Marty Rutledge, Antarctica Fire Department, McMurdo Station, Antarctica

Alan Shelder, St. Petersburg College, Seminole, FL

W. G. Shelton, Jr., CEM, Virginia Department of Fire Programs, Richmond, VA

Jeff Simkins, Director Fire Service Ext., West Virginia University, Morgantown, WV

Shawn Smith, London Fire Department, London, Ontario, Canada

Captain Kevin P. Terry and members of the Fuller Road Fire Department, Albany, NY

Scott Vanderbrook, Oveido Fire Rescue Department, Oveido, FL

Cindy Willett, Baltimore County Fire Department, Towson, MD

Gary Wilson, Training Chief for the Olathe KS Fire Department

Tom Wutz, Deputy Chief Fire Services Bureau, Office of Fire Prevention and Control, Albany, NY

A special thank you goes to our Technical Editor, Matthew Thorpe from the King Fire Department in King, North Carolina, for all his work on the learning objectives and corresponding content.

We would also like to extend a special thank you to Dave Diamantes for his contributions to chapters 9 and 10 of the text

Our Advisory Board Members

To those experts who work behind the scenes and provide us with continuing guidance on this book as well as other learning materials on our emergency services list, we extend our gratitude:

Steve Chikerotis, Battalion Chief, Chicago Fire Department, Chicago, IL

Tom Labelle, Executive Director, New York State Association of Fire Chiefs, East Schodack, NY

Jerry Laughlin, Deputy Director of Education Services, Alabama Fire College, Tuscaloosa, AL

Jeff Pindelski, Battalion Chief, Downers Grove Fire Department, Downers Grove, IL

Peter Sells, District Chief of Operations Training, Toronto Fire Services, Toronto, Canada

Billy Shelton, Executive Director, Virginia Department of Fire Programs, Glen Allen, VA

Doug Fry, Fire Chief , City of San Carlos Fire Department, San Carlos, CA

Pat McCauliff, Director of Fire Science and EMS, Colin County Community College, McKinney, TX

Michael Petroff, Western Director for Fire Department Safety Officers Association, St. Louis, MO

Adam Piskura, Director of Training, Connecticut Fire Academy, Windsor Locks, CT

Theresa Staples, Program Manager, Firefighter and HazMat Certifications, Centennial, CO

Our Validation Committee Members

These committee members met and meticulously reviewed the 2009 Edition of the Standard 1021 along with the learning objectives of *Company Officer*, created by a JPR subject-matter expert for the third edition. The result of this meeting was a set of validated learning objectives that would tie the entire *Company Officer* package together with the new edition of the Standard. We whole-heartedly thank those who attended this meeting and helped to initiate a new learning solution for this series:

Brad Boyd, Captain, Medical Training Officer, City of Wichita Fire Department, Training/Safety Division, Wichita, KS

Donna Moucha Brackin, Ed.D., NBCT, Assistant Dean for Off-Campus Graduate Advising, The University of Mississippi, Southaven, MS

Mike Brackin, Ed.D., Fire Chief, Southaven Fire Department, Southaven, MS

Melvin Byrne, EFO–MS, Division Chief, Virginia Department of Fire Programs, Leesburg, VA

Derrick Clouston, Fire and Rescue Training Specialist, North Carolina Department of Insurance, Raleigh, NC

Patrick Draper, Fire Program Manager, Alabama Fire College, Tuscaloosa, AL

Doug Goodings, Executive Coordinator, Certification/Accreditation Program, Ministry of Community Safety and Correctional Services, North York, Ontario, Canada

Dave Hanneman, Deputy Chief–Fire Marshal, Boise Fire Department, Boise, ID

Eric Perry, Training Captain, Tucson Fire Department, Tucson, AZ

Tom Ruane, Fire and Life Safety Consultant, Peoria, AZ

NFPA 1021, 2009 Edition Correlation Guide

Fire Officer I

4.1.1
General Prerequisite Knowledge

Knowledge	Learning Objective(s)	Chapter(s)	Page(s)
Organizational structure of the department	3-1, 3-23, 3-19	3	48, 50, 51, 65, 66, 67, 70
Geographical configuration and characteristics of response districts	3-3, 3-17	3	51, 61, 63
Departmental operating procedures for administration, emergency operations, incident management system, and safety	4-2, 4-3, 8-8	4 and 8	90, 202
Departmental budget process	This component covered by LOs in JPR 4.4.3.		
Information management and record keeping	5-39, 5-40, 9-4	5 and 9	117, 118, 119, 240
The fire prevention and building codes and ordinances applicable to the jurisdiction	9-20	9	245
Current trends, technologies, and socioeconomic and political factors that impact the fire service	4-4, 5-43, 3-12	4 and 5	89, 92, 93, 94, 95, 122
Cultural diversity	6-16, 6-17	6	145, 148, 149, 150
Methods used by supervisors to obtain cooperation within a group of subordinates	6-1, 6-2, 6-6, 6-7, 6-8, 6-11, 6-12	6	135, 136, 139, 140, 141, 142, 143, 144
The rights of management and members	5-6, 5-7, 5-9	5	110, 111, 112
Agreements in force between the organization and members	5-8	5	111
Generally acceptable ethical practices, including a professional code of ethics	4-5, 4-6	4	95, 96, 97
Policies and procedures regarding the operation of the department as they involve supervisors and members	4-1	4	88

Reproduced with permission from NFPA 1021, *Fire Officer Professional Qualifications*, Copyright © 2009, National Fire Protection Association. This reprinted material is not the complete and official position of the NFPA on the referenced subject, which is represented only by the standard in its entirety.

Reproduced with permission from NFPA 1021, *Fire Officer Professional Qualifications*, Copyright © 2009, National Fire Protection Association. This reprinted material is not the complete and official position of the NFPA on the referenced subject, which is represented only by the standard in its entirety.

Reproduced with permission from NFPA 1021, *Fire Officer Professional Qualifications*, Copyright © 2009, National Fire Protection Association. This reprinted material is not the complete and official position of the NFPA on the referenced subject, which is represented only by the standard in its entirety.

4.3.1 Task Statement	Learning Objective(s)	Chapter(s)	Page(s)
Initiate action on a community need, given policies and procedures, so that the need is addressed.			
(A) Requisite Knowledge			
Additional LOs for the Task Statement above.	2-6, 9-2, 9-5, 9-14, 9-16	2 and 9	36, 37, 239, 240, 241, 242
Community demographics	9-15	9	241
Service organizations	9-7, 9-17	9	240, 242
Verbal and non-verbal communication	9-10	9	241
An understanding of the role and mission of the department	9-18	9	243, 244

4.3.2 Task Statement	Learning Objective(s)	Chapter(s)	Page(s)
Initiate action to a citizen's concern, given polices and procedures, so that the concern is answered or referred to the correct individual for action and all policies and procedures are complied with.			
(A) Requisite Knowledge			
Additional LOs for the Task Statement above.	9-9, 9-11	9	240, 241
Interpersonal relationships and verbal and non-verbal communications	Same as 4.2.2(A)(1) and 4.2.2(A)(2)		

4.3.3 Task Statement	Learning Objective(s)	Chapter(s)	Page(s)
Respond to a public inquiry, given policies and procedures, so that the inquiry is answered accurately, courteously, and in accordance with applicable policies and procedures.			
(A) Requisite Knowledge			
Additional LOs for the Task Statement above.	9-8, 9-12, 9-13	9	240, 241
Written communication techniques	Same as 4.1.2(a) and 4.1.2(b)		
Oral communication techniques	Same as 4.2.2(A)(1) and 4.2.2(A)(2)		

Reproduced with permission from NFPA 1021, *Fire Officer Professional Qualifications*, Copyright © 2009, National Fire Protection Association. This reprinted material is not the complete and official position of the NFPA on the referenced subject, which is represented only by the standard in its entirety.

4.4.1 Task Statement	Learning Objective(s)	Chapter(s)	Page(s)
Recommend changes to existing departmental policies and/or implement a new departmental policy at the unit level, given a new departmental policy, so that the policy is communicated to and understood by unit members.			
(A) Requisite Knowledge			
Additional LOs for the Task Statement above.	3-27, 3-28	3	73
Written communications	Same as 4.1.2(a) and 4.1.2(b)		
Oral communications	Same as 4.2.2(A)(1) and 4.2.2(A)(2)		

4.4.2 Task Statement	Learning Objective(s)	Chapter(s)	Page(s)
Execute routine unit-level administrative functions, given forms and record-management systems, so that the reports and logs are complete and files are maintained in accordance with policies and procedures.			
(A) Requisite Knowledge			
Additional LO for the Task Statement above.	3-24	3	70, 71, 72
Administrative polices and procedures	3-30, 9-31	3 and 9	73, 74, 75, 253, 254
Records management	9-32	9	253, 254

4.4.3 Task Statement	Learning Objective(s)	Chapter(s)	Page(s)
Prepare a budget request, given a need and budget forms, so that the request is in the proper format and is supported with data.			
(A) Requisite Knowledge			
Policies and procedures	5-3, 5-13, 5-29	5	107, 108, 113, 115
Revenue sources and budget process	5-18, 5-22	5	113, 114

4.4.4 Task Statement	Learning Objective(s)	Chapter(s)	Page(s)
Explain the purpose of each management component of the organization, given an organization chart, so that the explanation is current and accurate and clearly identifies the purpose and mission of the organization.			
(A) Requisite Knowledge			
Organizational structure of the department and functions of management	3-2, 3-20, 3-21, 3-22	3	50, 67, 68, 69

Reproduced with permission from NFPA 1021, *Fire Officer Professional Qualifications*, Copyright © 2009, National Fire Protection Association. This reprinted material is not the complete and official position of the NFPA on the referenced subject, which is represented only by the standard in its entirety.

4.4.5 Task Statement	Learning Objective(s)	Chapter(s)	Page(s)
Explain the needs and benefits of collecting incident response data, given the goals and mission of the organization, so that incident response reports are timely and accurate.			
(A) Requisite Knowledge			
Additional LO for the Task Statement above.	9-6, 9-33	9	240, 257
The agency's records management system	Same as 4.1.1(k), 4.1.1(l), and 4.1.1(m)		

4.5.1 Task Statement	Learning Objective(s)	Chapter(s)	Page(s)
Describe the procedures of the AHJ for conducting fire inspections, given any of the following occupancies, so that all hazards, including hazardous materials, are identified, approved forms are completed, and approved action is initiated:			
(1) Assembly			
(2) Educational			
(3) Health care			

4.5.1 Task Statement	Learning Objective(s)	Chapter(s)	Page(s)
(4) Detention and correctional			
(5) Residential			
(6) Mercantile			
(7) Business			
(8) Industrial			
(9) Storage			
(10) Unusual structures			
(11) Mixed occupancies			
(A) Requisite Knowledge			
Additional LOs for the Task Statement above.	9-29, 9-30	9	251, 252
Inspection procedures	9-19, 9-21, 9-22, 9-23, 9-24, 9-25	9	244, 247, 248, 249, 250
Fire detection, alarm, and protection systems	10-9, 10-12, 10-14, 10-15, 10-16, 10-17	10	269, 270, 272, 273, 274, 275, 276
Identification of fire and life safety hazards;	10-18	10	276
Marking and identification systems for hazardous materials	10-10, 10-11	10	272

Reproduced with permission from NFPA 1021, *Fire Officer Professional Qualifications*, Copyright © 2009, National Fire Protection Association. This reprinted material is not the complete and official position of the NFPA on the referenced subject, which is represented only by the standard in its entirety.

4.5.2 Task Statement	Learning Objective(s)	Chapter(s)	Page(s)
Identify construction, alarm, detection, and suppression features that contribute to or prevent the spread of fire, heat, and smoke throughout the building or from one building to another, given an occupancy, policies and forms of the AHJ so that a pre-incident plan for any of the following occupancies is developed:			
(1) Public Assembly			
(2) Educational			
(3) Institutional			
(4) Residential			
(5) Business			
(6) Industrial			
(7) Manufacturing			
(8) Storage			
(9) Mercantile			
(10) Special properties			
(A) Requisite Knowledge			
Additional LOs for the Task Statement above.	12-2, 12-3, 12-4, 12-11	12	315, 317, 320, 327
Fire Behavior	10-2, 10-5, 10-6	10	266, 267, 268, 269
Building construction	10-1, 10-3, 10-4, 10-19	10	266, 270, 271, 276, 277
Inspection and incident reports	12-1	12	314
Detection, alarm, and suppression systems	10-8, 12-5	10 and 12	269, 270, 322
Applicable codes, ordinances, and standards	12-7	12	324

4.5.3 Task Statement	Learning Objective(s)	Chapter(s)	Page(s)
Secure an incident scene, given rope or barrier tape, so that unauthorized persons can recognize the perimeters of the scene and are kept from restricted areas, and all evidence or potential evidence is protected from damage or destruction.			
(A) Requisite Knowledge			
Types of evidence	11-10	11	301
The importance of fire scene security	11-12	11	301
Evidence preservation	11-8, 11-9, 11-11	11	301

Reproduced with permission from NFPA 1021, *Fire Officer Professional Qualifications*, Copyright © 2009, National Fire Protection Association. This reprinted material is not the complete and official position of the NFPA on the referenced subject, which is represented only by the standard in its entirety.

4.6.1 Task Statement	Learning Objective(s)	Chapter(s)	Page(s)
Develop an initial action plan, given size-up information for an incident and assigned emergency response resources, so that resources are deployed to control the emergency.			
(A) Requisite Knowledge*			
Elements of a size-up	13-3	13	347, 348, 349, 350, 351, 352
Standard operating procedures for emergency operations	12-12, 13-10	12 and 13	330, 362
Fire behavior	Refer to 4.5.2(A)(1) and (A)(2)		

4.6.2 Task Statement*	Learning Objective(s)	Chapter(s)	Page(s)
Implement an action plan at an emergency operation, given assigned resources, type of incident, and a preliminary plan, so that resources are deployed to mitigate the situation.			
(A) Requisite Knowledge			
Standard operating procedures	Same as 4.6.2(A)(2)		
Resources available for the mitigation of fire and other emergency incidents,	13-6	13	356, 357
Incident management system	Same as 4.1.1(h) and 4.1.1(i)		
Scene safety	13-1	13	344, 345, 346
Personnel accountability system	13-2	13	350

4.6.3 Task Statement	Learning Objective(s)	Chapter(s)	Page(s)
Develop and conduct a post-incident analysis, given a single unit incident and post-incident analysis policies, procedures, and forms, so that all required critical elements are identified and communicated, and the approved forms are completed and processed in accordance with policies and procedures.			
(A) Requisite Knowledge*			
Elements of a post-incident analysis	13-17, 13-18	13	371, 372
Basic building construction	Same 4.6.1(A)(3)		
Basic fire protection systems, and features	Same as 4.6.1(A)(4)		
Basic water supply	12-6	12	322
Basic fuel loading	Same as 4.6.1(A)(7)		
Fire growth and development	Same as 4.5.2(A)(1)		
Departmental procedures relating to dispatch, response, tactics and operations, and customer service	12-13, 12-14, 12-15	12	330

Reproduced with permission from NFPA 1021, *Fire Officer Professional Qualifications*, Copyright © 2009, National Fire Protection Association. This reprinted material is not the complete and official position of the NFPA on the referenced subject, which is represented only by the standard in its entirety.

4.7.1 Task Statement	Learning Objective(s)	Chapter(s)	Page(s)
Apply safety regulations at the unit level, given safety policies and procedures, so that required reports are completed, in-service training is conducted, and member responsibilities are conveyed.			
(A) Requisite Knowledge			
The most common causes of personal injury and accident to members	8-2, 8-7	8	197, 201
Safety policies and procedures	8-3	8	198, 207
Basic workplace safety	8-9	8	202
The components of an infectious disease control program.	8-18	8	212

4.7.2 Task Statement	Learning Objective(s)	Chapter(s)	Page(s)
Conduct an initial accident investigation, given an incident and investigation forms, so that the incident is documented and reports are processed in accordance with policies and procedures of the AHJ.			
(A) Requisite Knowledge			
Additional LOs for the Task Statement above.	8-13	8	208
Procedures for conducting an accident investigation	8-14	8	208, 209
Safety policies and procedures	8-12	8	206, 207

4.7.3 Task Statement	Learning Objective(s)	Chapter(s)	Page(s)
Explain the benefits of being physically and medically capable of performing assigned duties and effectively functioning during peak physical demand activities, given current fire service trends and agency policies, so that the need to participate in wellness and fitness programs is explained to members.			
(A) Requisite Knowledge			
National death and injury statistics	8-1	8	196, 197
Fire service safety and wellness initiatives	8-6	8	200
Agency policies	8-5	8	200, 207

Reproduced with permission from NFPA 1021, *Fire Officer Professional Qualifications*, Copyright © 2009, National Fire Protection Association. This reprinted material is not the complete and official position of the NFPA on the referenced subject, which is represented only by the standard in its entirety.

5.1.1 General Prerequisite Knowledge	Learning Objective(s)	Chapter(s)	Page(s)
The organization of local government	3-7, 3-8, 3-9	3	53, 54, 56, 57
Enabling and regulatory legislation and the law-making process at the local, state/provincial, and federal levels	3-4, 3-5, 3-6, 3-11	3	53, 58
The functions of other bureaus, divisions, agencies, and organizations and their roles and responsibilities that relate to the fire service	3-10	3	57, 58

5.2.1 Task Statement	Learning Objective(s)	Chapter(s)	Page(s)
Initiate actions to maximize member performance and/or to correct unacceptable performance, given human resource policies and procedures, so that member and/or unit performance improves or the issue is referred to the next level of supervision.			

(A) Requisite Knowledge

	Learning Objective(s)	Chapter(s)	Page(s)
Human resource policies and procedures	7-27, 7-28	7	184
Problem identification	7-19, 7-20	7	181, 184
Organizational behavior	7-21, 7-29, 7-31	7	183, 184, 185 186, 187
Group dynamics	7-17, 7-18	7	181
Leadership styles	6-4, 7-2	6 and 7	138, 139, 162, 163
Types of power	6-5, 6-9, 6-10, 6-13, 6-15	6	139, 142, 143, 144, 145, 146, 147
Interpersonal dynamics	6-3, 6-14	6	136, 137, 138, 144, 145

5.2.2 Task Statement	Learning Objective(s)	Chapter(s)	Page(s)
Evaluate the job performance of assigned members, given personnel records and evaluation forms, so each member's performance is evaluated accurately and reported according to human resource policies and procedures.			

(A) Requisite Knowledge

	Learning Objective(s)	Chapter(s)	Page(s)
Additional LOs for the Task Statement above.	7-13, 7-16	7	179, 182, 183
Human resource policies and procedures	7-7, 7-9, 7-12	7	168, 169, 171, 172, 173, 175, 176
Job descriptions	7-1, 7-14	7	161, 162, 176
Objectives of a member evaluation program	7-8	7	169, 170, 171
Common errors in evaluating	7-10, 7-11	7	174

Reproduced with permission from NFPA 1021, *Fire Officer Professional Qualifications*, Copyright © 2009, National Fire Protection Association. This reprinted material is not the complete and official position of the NFPA on the referenced subject, which is represented only by the standard in its entirety.

5.2.3 Task Statement	Learning Objective(s)	Chapter(s)	Page(s)
Create a professional development plan for a member of the organization, given the requirements for promotion, so that the individual receives the necessary knowledge, skills, and abilities to be eligible for the examination for the position.			
(A) Requisite Knowledge			
Development of a career development guide	1-5, 1-6, 1-7, 7-15	1 and 7	8, 9, 10, 11, 12, 13, 14, 178, 179
Job shadowing	1-3	1	7, 14, 15

5.3.1 Task Statement	Learning Objective(s)	Chapter(s)	Page(s)
Explain the benefits to the organization for cooperating with allied organizations, given a specific problem or issue in the community, so that the purpose for establishing external agency relationships is clearly explained.			
(A) Requisite Knowledge			
Agency mission and goals	3-14, 3-18	3	60, 63
Types and functions of external agencies in the community	3-13, 3-15, 12-8, 12-9, 12-10	3 and 12	58, 59, 60, 61, 325, 326

5.4.2 Task Statement	Learning Objective(s)	Chapter(s)	Page(s)
Develop a project or divisional budget, given schedules and guidelines concerning its preparation, so that capital, operating, and personnel costs are determined and justified.			
(A) Requisite Knowledge			
The supplies and equipment necessary for ongoing or new projects	5-41	5	120, 121
Repairs to existing facilities			
New equipment, apparatus maintenance, and personnel costs	5-11, 5-16, 5-31, 5-34, 5-35, 5-36	5	112, 113, 115, 116,
Appropriate budgeting system	5-10, 5-12, 5-14, 5-15, 5-17, 5-30	5	112, 113, 115

5.4.3 Task Statement	Learning Objective(s)	Chapter(s)	Page(s)
Describe the process of purchasing, including soliciting and awarding bids, given established specifications, in order to ensure competitive bidding.			
(A) Requisite Knowledge			
Additional LO for the Task Statement above.	5-33	5	116
Purchasing laws	5-20, 5-21, 5-23, 5-25, 5-27	5	113, 114
Policies and procedures	5-24, 5-26, 5-28, 5-37	5	114, 115, 116

Reproduced with permission from NFPA 1021, *Fire Officer Professional Qualifications*, Copyright © 2009, National Fire Protection Association. This reprinted material is not the complete and official position of the NFPA on the referenced subject, which is represented only by the standard in its entirety.

5.4.4 Task Statement	Learning Objective(s)	Chapter(s)	Page(s)
Prepare a news release, given an event or topic, so that the information is accurate and formatted correctly.			
(A) Requisite Knowledge			
Policies and procedures	2-8, 2-9, 2-11, 2-12	2	37, 38, 39
The format used for news releases	2-10	2	37-38

5.4.5 Task Statement	Learning Objective(s)	Chapter(s)	Page(s)
Prepare a concise report for transmittal to a supervisor, given fire department record(s) and a specific request for details such as trends, variances, or other related topics.			
(A) Requisite Knowledge			
The data processing system	9-3	9	240, 244

5.4.6 Task Statement	Learning Objective(s)	Chapter(s)	Page(s)
Develop a plan to accomplish change in the organization, given an agency's change of policy or procedures, so that effective change is implemented in a positive manner.			
(A) Requisite Knowledge			
Planning and implementing change	3-25, 3-26, 3-29	3	72, 73

5.5.1 Task Statement	Learning Objective(s)	Chapter(s)	Page(s)
Determine the point of origin and preliminary cause of a fire, given a fire scene, photographs, diagrams, pertinent data and/or sketches, to determine if arson is suspected			
(A) Requisite Knowledge			
Additional LO for the "Task Statement" above.	11-3	11	298 299
Methods used by arsonists	11-1	11	297

5.5.1 Task Statement	Learning Objective(s)	Chapter(s)	Page(s)
Common causes of fire	9-1	9	239, 241
Basic cause and origin determination	11-2, 11-4, 11-5, 11-7, 11-13, 11-14	11	297, 300, 301, 302
Fire growth and development	11-6	11	299, 300, 301
Documentation of preliminary fire investigative procedures	11-15	11	305

Reproduced with permission from NFPA 1021, *Fire Officer Professional Qualifications*, Copyright © 2009, National Fire Protection Association. This reprinted material is not the complete and official position of the NFPA on the referenced subject, which is represented only by the standard in its entirety.

5.6.1 Task Statement	Learning Objective(s)	Chapter(s)	Page(s)
Produce operational plans, given an emergency incident requiring multi-unit operations, the current edition of NFPA 1600, and AHJ-approved safety procedures, so that required resources and their assignments are obtained and plans are carried out in compliance with NFPA 1600 and approved safety procedures resulting in the mitigation of the incident.			
(A) Requisite Knowledge			
Additional LO for the "Task Statement" above.	13-13	13	367
Standard operating procedures	13-9	13	361
National, state/provincial, and local information resources available for the mitigation of emergency incidents	13-20	13	375
Incident management system	13-14, 13-21	13	368, 369, 370, 374, 375, 376, 377, 378, 379
Personnel accountability system	13-4	13	355

5.6.2 Task Statement	Learning Objective(s)	Chapter(s)	Page(s)
Develop and conduct a post-incident analysis, given multi-unit incident and post-incident analysis policies, procedures, and forms, so that all required critical elements are identified and communicated and the approved forms are completed and processed.			
(A) Requisite Knowledge			
Elements of a post-incident analysis	13-15, 13-16, 13-19	13	371, 372
Basic building construction	10-21	10	282
Basic fire protection systems and features	10-13	10	273
Basic water supply	10-22	10	283, 284, 285
Basic fuel loading	10-7	10	269
Fire growth and development	10-20	10	278
Departmental procedures relating to dispatch response, strategy, tactics, and operations, and customer service.	13-12	13	362, 363, 364, 365, 366

5.6.3 Task Statement	Learning Objective(s)	Chapter(s)	Page(s)
Prepare a written report given incident reporting data from the jurisdiction, so that the major causes for service demands are identified for various planning areas within the service area of the organization.			
(A) Requisite Knowledge			
Analyzing data	9-26, 9-27, 9-28	9	250

Reproduced with permission from NFPA 1021, *Fire Officer Professional Qualifications*, Copyright © 2009, National Fire Protection Association. This reprinted material is not the complete and official position of the NFPA on the referenced subject, which is represented only by the standard in its entirety.

5.7.1 Task Statement	Learning Objective(s)	Chapter(s)	Page(s)
Analyze a member's accident, injury, or health exposure history, given a case study, so that a report including action taken and recommendations made is prepared for a supervisor.			
(A) Requisite Knowledge			
Additional LO for the "Task Statement" above.	8-10, 8-15, 8-16	8	202, 203, 204, 205, 209, 212
The causes of unsafe acts	8-4	8	198, 199
Health exposures	8-17	8	212
Conditions that result in accidents, injuries or occupational illnesses or deaths	8-11	8	205, 206

Reproduced with permission from NFPA 1021, *Fire Officer Professional Qualifications*, Copyright © 2009, National Fire Protection Association. This reprinted material is not the complete and official position of the NFPA on the referenced subject, which is represented only by the standard in its entirety.

1

The Company Officer's Role—Challenges and Opportunities

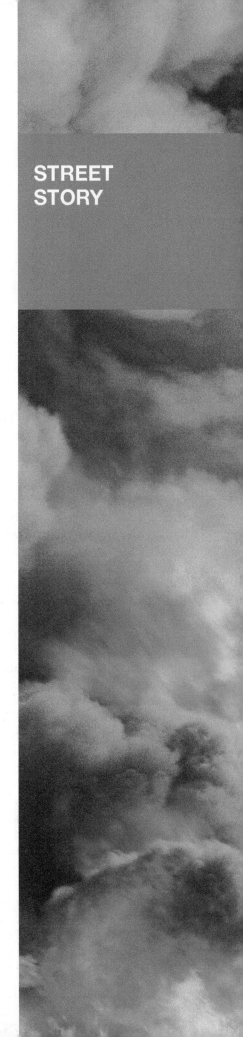

Too many people hire a yes-man in the type of position I'm in because they want to hear that they themselves are doing a great job. I don't want to hear that I'm doing a great job; I want to be challenged every day. I want to be able to change things for the better consistently.

I will have 20 years with the Portage Fire Department, in Portage, Indiana, this year, and I was fire chief at a volunteer department and worked for the Delaware County EMS prior to that. My role as the assistant chief is, first and foremost, to back up and support the fire chief. Our overall goal is to protect the citizens. On a daily basis, I work on fire operations, apparatus maintenance, and personnel policies. We have 61 full-time firefighters, and I interact with all of the department heads as well.

When I think about challenges that face new company officers, I always think back to my first days , as a new member of the team. That time can be difficult, because 90% of the departments out there are small, and you end up knowing everyone. On the other hand, if you're with a larger city, you end up working with people you do not know. Either one of those scenarios is a challenge. One of the main obstacles at the smaller department, where you might know everybody, is working with friends. That could be a benefit, but sometimes friends find it hard to see you as a boss. A lot of people want to walk into a situation like that and lay down the law right away to differentiate between being a boss and a friend. I think you can keep both of those titles and maintain respect by doing just that. Working with friends presents an opportunity also, because, as friends, people are more likely to work with you and work for you. That's key, because when you work with labor management and people aren't willing to work for or with the boss, then problems arise.

When I first took over at Portage, there were a lot of instances where people knew me ahead of time, and what my attitude was, and they didn't expect that to change. They were able to keep me grounded in that way. I told people that if I ever got to a point with them as their boss where I would have complained if I were in their position, they needed to let me know; there were a few times they did.

When it comes to separating friendship and work, it's important to remember that there are times when people will need their space if you discipline them. Some friends who have worked for me have needed some low-level discipline, and I've been able to give them that, and as soon as that's over we go back to our friendship because they understand when I'm talking about discipline that that's part of our work process, and then it's behind me. I don't hold that against them. It's not a perfect routine, though, and some things get under your skin more than others. Sometimes things are said and feelings are hurt, and you have to give the hurt person some space when that occurs.

In my job, and in life, I try to be as forthcoming and honest as possible, and sometimes that comes across the wrong way. In the end, I always just keep in mind that my ultimate job is to protect the citizens. It's a challenge for me sometimes because being brutally honest with people can put them on the defensive. When that happens, I make sure to just give them their space. Nothing gets accomplished when heated feelings are involved.

—*Mike Bucy, Assistant Chief,*
Portage Fire Department, Portgage, Indiana

LEARNING OBJECTIVES

After completing this chapter the reader should be able to:

1-1 Define the term *supervision*.

1-2 List the responsibilities of a supervisor.

1-3 List the responsibilities the company officer is required to fulfill.

1-4 Describe the role of a fire department supervisor.

1-5 Identify professional development opportunities for company officers.

1-6 List the components of a development plan for unit members.

1-7 Explain the rationale for assigning the various levels of professional qualifications.

*The FO I and II levels, as defined by the NFPA 1021 Standards, are identified in different colors: FO I = black, FO II = red.

INTRODUCTION

It is a dark and stormy night. The big fire truck moves slowly along the narrow mountain road, red lights flashing, spotlights drilling holes into the darkness along either side. The firefighters are looking for a car that was reported to have run off the road in this remote area.

Everyone in the truck is a little uneasy. It is raining hard. Trees and wires are down, and there is always a chance of flooding when it rains this hard. While the inside of the truck's cab is relatively dry, comfortable, and safe, there is an uncomfortable feeling of tension. Riding in the cab are two firefighters, a driver, and you, the company officer. In difficult conditions like these, you are expected to be calm and to reassure the others. But you have some anxieties, too. Not only are you sharing the same concerns about the weather as your young firefighters, but also you are responsible for managing them and the event. Suddenly one of your firefighters yells, "Stop!" Your driver puts on the brakes and brings the big vehicle to a safe stop. As you look out into the darkness, the spotlight illuminates a wrecked vehicle against a tree. The car door is open, and it looks like someone is inside. Now, not only are you responsible for those who came with you, but also you are expected to assess and manage this entire situation, assist those who might be trapped in the car or lying injured on the ground, and keep everyone, including your own personnel, from further injury. You think about the situation for a moment. You pick up the radio microphone; you are not yet sure of what to say. As you open the door, the cab lights come on, revealing the anxiety of your colleagues as they look to you for direction. A lot of responsibility is in your hands right now.

For the company officer, it is just another day at work.

NOTE

This book is for all fire officers. It is written with both volunteer and paid departments in mind. Throughout this text, we will refer to professional behavior, professional competencies, and similar terms that may suggest we are talking only about paid personnel. That is not the case: All references to professionalism address the calling of both volunteer and paid officers. NFPA 1021, the *Standard for Fire Officer Professional Qualifications*, applies to all fire officers.

Many of the concepts presented in this book are applicable to leaders in other emergency service organizations. This text has been used to help new officers in emergency medical services (EMS) and rescue organizations, as well as in the fire service, learn the communications, leadership, and management skills they needed to become effective officers.

THE ROLE OF THE COMPANY OFFICER

What is the role of the company officer? Company officers are first-line supervisors responsible for the performance and safety of assigned personnel in an emergency service organization. Company officers should be capable of leading their personnel during both emergency and nonemergency activities (as shown in **Figure 1-1**).

FIGURE 1-1 The company officer. One of the most exciting jobs in the fire service. The company officer manages resources during both emergency and nonemergency activities.

FIREGROUND FACT

Supervision is generally defined as the act of watching over the work or tasks of another. The NFPA defines a *supervisor* as an individual responsible for overseeing the performance or activity of other members.

Based on this definition, a company officer can be described as a first-line supervisor responsible for the performance and safety of assigned personnel in an emergency service organization.

As a first-line supervisor, the company officer has many responsibilities. The company officer's job is as varied as all of the activities of the organization. Company officers are often the senior representatives of the fire department at the scene of countless emergencies ranging from automobile accidents to fires and from treating sick and injured persons to mitigating hazardous materials spills. In nearly every case, the company officer is one of the first persons on the scene and is likely to be the last to leave.

But there is more to being a company officer than managing emergency situations. When the company is not involved in an emergency, the company officer essentially runs a small part of the fire department, rescue squad, or emergency medical services (EMS) organization. The company officer is the most influential member of the department and key to daily operations. At the company level, the company officer manages the resources: people, equipment, and time.

In most departments, companies spend less than 10 percent of their time dealing with emergencies. That means they spend 90 percent of their time in other activities, including providing fire safety public education programs for citizens, preplanning for fires and other emergencies, training, and maintaining good physical fitness. Regardless of the activity, the company officer is expected to plan, manage, and lead the company.

FIREGROUND FACT

In most departments, companies spend less than 10 percent of their time dealing with emergencies.

These duties may not sound as important as those in which the company is on the scene of an emergency, but they are. The company officer is also expected to be a full-time leader and instructor for the personnel assigned to the company, so human relations skills are very important for company officers. Look at the following list: How many of these activities deal with human relations?

THE MANY ROLES AND RESPONSIBILITIES OF THE COMPANY OFFICER

Coach	Manager
Communicator	Mediator
Counselor	Motivator
Decision maker	Planner
Evaluator	Public relations representative
Firefighter	Referee
Friend	Role model
Innovator	Safety officer
Instructor	Student
Leader	Supervisor
Listener	Writer

Company officers must learn how to perform these roles while retaining their previous skills and knowledge. Company officers do not have the luxury of standing around at the scene of emergencies with a radio in one hand, directing the firefighters' activities. They frequently find that they must help fight fires or help diagnose and treat patients, along with the rest of the crew. In addition, they are expected to manage and supervise their subordinates. The company officer has one of the most demanding jobs in the organization. Leading others and ensuring everyone goes home is the company officer's principal job (see **Figure 1-2**). The capabilities, efficiency, and morale

FIGURE 1-2 Leading others is the company officer's principal job.

FIGURE 1-3 When an officer takes a position as a firefighter on the attack hose line one has to ask "Who is in charge?" Who is looking out for the firefighter's accountability and safety?

of the company are direct reflections of the company officer's leadership ability. The personnel are the company officer's responsibility. If the company officer does not take care of them, no one else will.

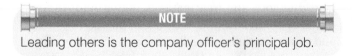

NOTE

Leading others is the company officer's principal job.

In the fire service we use the term "company" to describe the work teams. Clearly there is a need for teamwork at the company level. With the reduced staffing that many fire departments face, the company officer is not only the leader of the team but one of the players, too. Company officers are expected to lend a hand when needed, whether it be advancing hose, forcing doors, or performing similar activities that do not compromise their role as leader. There is nothing wrong with helping out, but be careful: When an officer takes a position as a firefighter on the attack hose line, one has to ask, "Who is leading? Who is in charge of the firefighters' accountability and safety?" (see **Figure 1-3**).

Company officers are expected to lend a hand when needed, whether it is advancing hose, forcing doors, or performing similar activities that do not compromise their role as leader.

Although company officers are required to fulfill many responsibilities, along with often helping to resolve the problem at hand, the officer's primary job is to lead. Many new officers find the most difficult part of becoming an officer is accepting the fact that they are no longer firefighters. The best firefighters are promoted and usually become good officers. But these same individuals sometimes find that the transition from doer to leader is difficult. They want to be involved with the work. In most cases they are very good at performing firefighting tasks and find a great sense of enjoyment in doing them. They find it difficult to give up the physical tasks and to think about supervision and management. For many, the physical tasks are easier and more enjoyable than the mental effort and the paperwork that accompany the officer's job.

The opposite situation occurs from time to time when an officer will not have anything to do with the firefighters or their work. These company officers seem to forget where they came from and what life was like as a firefighter. There are various reasons for this situation, but sometimes firefighters see these officers as placing their own careers ahead of those of the firefighters. Officers should remember that firefighters are the backbone of the service delivery system. Firefighters are valuable assets to the community who have the knowledge, skills, and abilities to serve the citizens of their communities. The officer has to take care of them, nurture them, and help them develop their full potential.

Firefighters are the backbone of the service delivery system. The officer must take care of them, nurture them, and help them develop their full potential.

THE COMPANY OFFICER'S ROLE AS A SUPERVISOR IN THE DEPARTMENT

In later chapters, we discuss organization and management principles. For now we use a simple graphic device to represent a typical fire service organization (**Figure 1-4**). Notice that the company officer plays an important role in the organizational structure. For firefighters, the company officer is not only their supervisor but also their next link in the chain of command. That link connects them with the rest

FIGURE 1-4 The company officer occupies a vital link in the fire department's structure.

FIGURE 1-5 The company officer is expected to manage resources, people, time, and money, at the company level.

of the department. The company officer may also be considered their link to the public, for it is the company officer who likely speaks for the company and the department both at the scene of an emergency and during nonemergency situations.

To the fire chief and other senior officers, the company officer represents the company. If the chain of command is being followed, the company officer is both the voice of the chief to the firefighters and the voice of the firefighters to the chief.

Two other factors affect the company officer's position in the organization. First, we often take relatively inexperienced personnel (at least in a leadership role) and place them in one of the most challenging positions in the fire service (**Figure 1-5**). Second, we often place these new company officers at a worksite that is remote from their supervisor for all of their normal nonemergency activities and for many minor emergencies, too. Although this practice is not unique to the fire service, we should consider the benefits to the individual officer and to the organization of being closer to the supervisor.

Where a company is co-located with fire department headquarters, the company officer has greater access to his or her supervisor and should always keep the lines of communication open. There are certainly pros and cons of being a company officer in such an environment, but many companies and many company officers in such environments tend to be above average and these officers are often selected for advancement ahead of their peers. Could it be that these officers are better because they have greater access to their supervisor? Finally, let us look at the role of the company itself. In fire departments or other emergency response organizations, the majority of the employees work at the company level, making the organization's mission statement work at the company level, and hence making the organization's mission statement come alive for the citizens. The department is there to provide a service to the community. Service delivery is the sole reason these organizations exist. To put it another way, most of the organization's service delivery or, to use a more current term, most of the organization's *customer service* is delivered by personnel at the company level.

PROMOTION TO COMPANY OFFICER

Departments use a variety of methods to select their company officers. Some use a selection process with tests and other assessment tools to determine the best qualified individual for promotion. Other departments base the promotion process on seniority, implying that the person with the longest service is the best qualified for advancement. (A few departments carry this process all the way to the fire chief's position.) Some departments are not so subtle: Relatives and friends are promoted ahead of others.

Regardless of the promotion system in place, many of these officers were not fully prepared for the new duties they faced. We should not criticize these individuals; rather, we should admire those who were successful despite these conditions. We should understand that those who have struggled may have been placed into positions without the experience or training needed for the job. Imagine the consequences of letting other professionals, such as civil engineers or physicians, practice under similar conditions.

Regardless of whether you are serving in a paid or volunteer capacity, you would like to be considered a professional. The public is entitled to have qualified, competent, and professional personnel in emergency service organizations. Fire departments should be sure that their personnel are professionals. This can be accomplished through certification, experience, and education.

For purposes of this text, **certification** means that an individual has been tested by an accredited examining body on clearly identified material and found to meet a minimum standard. Certification provides a yardstick by which to measure competency in every type of department from the largest to the smallest and from all-paid to all-volunteer. There are other reasons to certify:

- Protection from liability—due diligence
- Recognition of demonstrated proficiency
- Recognition of professionalism
- Budget and salary justification
- Standardized process

We have already suggested that it would be difficult to provide a concise yet complete definition of the company officer's job, but we should try. One way to do this might be to ask a committee of respected fire service professionals to define the company officer's job and to define the competency requirements that address the knowledge, skills, and abilities needed to perform that job. Then we can develop a training program to help prepare individuals to move effectively into the supervisory ranks.

A fire department or other emergency response organization is a complex business involving science, good business practices, and human relations. For the most part, fire protection is a local problem. However, many national organizations and federal agencies have a role in fire protection as well. The same can be said for EMS, hazardous materials response, and urban rescue operations.

You should be aware of the existence and role of these organizations, for they have had a significant impact on changing the character of the fire service over the past 35 years. Ten of these organizations gathered together for a meeting in 1970. The leaders of these organizations became known as the Joint Council of National Fire Service Organizations.[1] Their meeting was significant because it was the first time they had organized to present a united effort to address the nation's fire problems. One of the council's first tasks was to develop nationally accepted standards for firefighters, fire officers, and others. The importance of this role is clearly seen in the first paragraph of the council's goals statement.

NOTE

Officers should be aware of the existence and role of these national and federal organizations, for they have had a significant impact on changing the character of the fire service over the past 35 years.

THE NATIONAL COMPETENCY STANDARD FOR FIRE OFFICERS

A significant step in the certification process occurred when the Joint Council established the National Professional Qualifications System in 1972. That system provided a means of representing the professional opportunities in the fire service and specified a particular body of knowledge required for each level and for each specialty area in the fire service professional ladder. The standards apply to volunteer and paid personnel alike. Volunteers, just as paid personnel, need to know what to do and how to do it at the scene of any emergency. The lives of their colleagues as well as the lives and property of the citizens of their community are often at stake.

These standards started and remain a product of a peer group process and are not the result of any regulatory action. The present standards provide nationally recognized competency criteria for firefighters, fire service instructors, fire officers, inspectors, investigators, public fire educators, and others.

The present certification standards are part of the National Fire Protection Association's (NFPA) **codes** and standards-making process. There are approximately 300 NFPA codes and **standards** dealing with every aspect of fire protection. These documents have a profound impact on improving safety where we live and work. They are widely adopted for legislation and regulation at the federal, state, and local levels. Many federal agencies, including the Occupational

PROFESSIONAL COMPETENCIES AND COLLEGE PROGRAMS PROVIDE SIMILAR OUTCOMES

The traditional way to prepare to enter a profession is to attend college or a technical school. Community college–based fire science programs have been around for more than 30 years. They were created to help individuals enter and move up the fire service organizational ladder. A 1969 publication entitled *Guideline for Fire Service Education Programs in Community and Junior Colleges* identified basic courses that should be included in all fire science programs. These included communications skills, human relations, government, fire prevention, firefighting strategy and tactics, and building construction, among others. (Note the date of publication—1969. These issues have been around for 35 years!)

These topics are still included in fire science programs today, and they are also part of the national competency standards addressed in the next section. Interestingly, that report listed nearly 150 degree programs in existence. Today, there are nearly twice as many. (Refer to Appendix A for a special note regarding the future of training in the fire service, by Denis Onieal, Superintendent of the National Fire Academy.)

GOALS OF THE JOINT COUNCIL OF NATIONAL FIRE SERVICE ORGANIZATIONS

- To develop nationally recognized standards for competency and achievement of skills development, technical proficiency, and academic knowledge appropriate to every level of the fire service career ladder
- To make the public aware of the significant contributions made by the fire service of this nation in protecting life by providing increased financial and moral support to aid the fire service in carrying out its mission
- To make public officials at every level of government more aware of their responsibilities in providing increased financial and moral support to aid the fire service in carrying out its mission
- To reassess public fire protection in light of contemporary demands, ensuring appropriate fire protection for all communities at a reasonable cost
- To establish realistic standards of educational achievement and to provide to every member of the fire service equal educational opportunities commensurate with professional requirements
- To identify and establish nationwide information systems that will enable improved analysis of the fire problem with particular emphasis on the life and safety factors for the public and its firefighters
- To encourage and undertake the research and development necessary for the prompt and successful implementation of these goals

Safety and Health Administration (OSHA), National Institutes of Health (NIH), and Department of Energy (DOE), reference NFPA's codes and standards in their own regulations. Many insurance companies use NFPA documents for guidelines in assessing risks and setting premiums. NFPA's codes and standards are the result of more than 200 committees consisting of more than 5,000 individuals who serve voluntarily. Built into the codes and standards-making process is an opportunity for the public to comment on documents at every stage of their development. Only after the public comment phase is complete is the document brought before the membership of NFPA for formal adoption. Once this phase is completed, the standards are published and available for voluntary adoption by local jurisdictions.

Ten of the NFPA documents deal with the qualifications of those who serve in the fire service. As public organizations, fire service organizations are open to public scrutiny and are held accountable for their actions. There is considerable value in being able to demonstrate that the personnel of these agencies are certified as meeting the competency standards of an entity that has itself been evaluated by an independent, thorough, objective, and public process and approved or accredited as meeting the requirements of the process.[2]

NFPA published the first national standard for fire service officers in 1976. It defined the knowledge, skills, and abilities needed by the company officer. Over the years the document has been revised several times, with each revision improving the list of requirements for those seeking certification as fire officers (see **Figure 1-6**).

The present version of NFPA 1021, *Standard for Fire Officer Professional Qualifications*, outlines the requirements for fire officers at four levels of competency. The first level, Fire Officer I, focuses on the needs of the first-line supervisor, clearly including company officers. For Fire Officer II, the requirements focus on the management aspects of the company officer's job and help prepare individuals to move into staff assignments. The requirements for Officer III deal mostly with administration and management, preparing officers to move into positions of increasing responsibility at the mid-management level within their organizations. At the top of the officer certification ladder, the requirements for Officer IV satisfy the needs of senior staff and chief officers (see **Figure 1-7**).

NOTE

The present version of NFPA 1021, Standard for Fire Officer Professional Qualifications, outlines the requirements for fire officers at four levels of competency.

Comparing NFPA 1021 and College Programs

The requirements in NFPA 1021 are the result of a task analysis that determined the work done by typical fire officers at four levels within their organization.

For example, for Fire Officer I, that standard requires the following:

- Effective oral and written communications skills
- Leadership skills so that one can make assignments, supervise work, and evaluate performance

FIGURE 1-6 Good communication skills are important for the company officer as indicated by the NFPA Standard 1021.

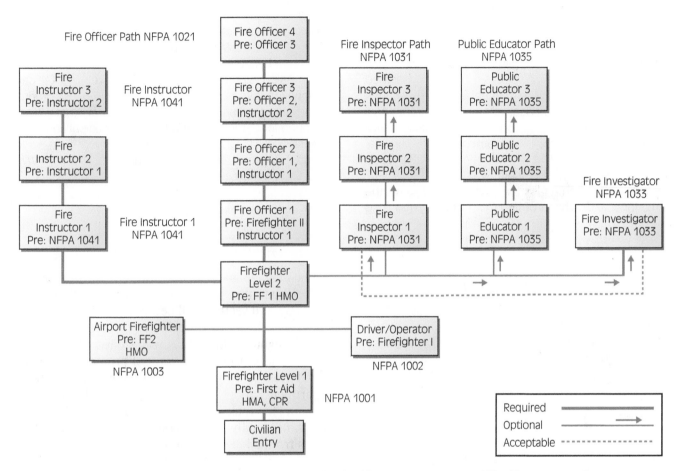

FIGURE 1-7 An example of a certification ladder utilized by the Virginia Department of Fire Programs.

- Skills to deal with member-related problems
- Skills to respond to a citizen's concern and deal with public inquiries
- Skills to solve problems and recommend changes in department policy
- Skills to manage safety by demonstrating knowledge about the most common causes of injury accidents and the spread of infectious disease
- Skills to conduct an initial accident investigation and report the results in writing
- Skills to develop a pre-emergency plan, understand the community's emergency action plan, and explain fire growth and development, building constructions, relevant fire and building code requirements, and the value of fire protection features
- When emergencies do occur, the skills needed to implement a plan so that resources are safely and effectively deployed to mitigate the incident

If you will look at the preceding list of requirements, you will see that all of these topics are addressed in courses offered in most college fire science programs. Indeed, the courses in college fire science programs and the requirements of the fire officer certification process are similar, and both represent the needs of individuals moving into and ascending the supervisory ranks of the fire service.

In addition to pragmatic courses (for example, building construction and hydraulics) taught in most college-level fire science courses, most community college programs provide the following learning outcomes for all students:

- The ability to effectively communicate through speaking, writing, reading, and listening
- The ability to locate, evaluate, and use information to analyze problems and make decisions
- The ability to apply math skills to organize, analyze, and use information appropriately
- The ability to apply critical thinking skills to analyze and solve problems ranging from physical to emotional

CHIEF FIRE OFFICER DESIGNATION

In addition to meeting the requirements for certification set forth in NFPA 1021, one can become designated as a chief fire officer. This is a voluntary program designed to recognize individuals who can show their excellence in seven areas: experience, education, professional development, professional contributions, association membership, community involvement, and technical competencies. Several fire departments use a variation of this model to promote officers at all levels.

Since 1993, when a task force was formed within the International Association of Fire Chiefs to begin looking at recognition for chief fire officers, many dedicated individuals have labored to create a program that recognizes individual excellence throughout the fire and emergency services. A commission composed of fire service, municipal, and academic leaders administers the program. The mission of the Commission for Chief Fire Officer Designation is to assist in the professional development of the fire and emergency service personnel by providing guidance for career planning through participation in the Professional Designation Program.

For further information, go to http://www.publicsafetyexcellence.org/.

NOTE

It is envisioned that in addition to the requirements of NFPA 1021, the authority having jurisdiction may require additional credentials. These can include fire degree programs and general education in business, management, science, and associated degree curricula.

PROFESSIONAL DEVELOPMENT: AN OPPORTUNITY FOR COMPANY OFFICERS

The public's need to resolve any problems encountered places a high level of responsibility on every member, including the newest firefighter. For these reasons, demanding entry-level requirements and rigorous recruit training are vitally important. To maintain these skills and learn new ones, a continuing program of training and education should be provided to all members. This training helps members maintain proficiency at their present levels, meet certification requirements, learn new procedures, and keep up with emerging technology.

Training to maintain skills should be provided by the department. Preparing for advancement should involve some individual initiative on the part of the member. In other words, if you want to get promoted, you have to put a little extra effort into the process. Although advancement brings greater prestige, it also brings added responsibilities. You should have an opportunity for promotion, but at the same time, you should take some personal initiative.

The company officer's job is vast and varied, and it covers many diverse topics. Company officers are expected to walk into the new job, know all the answers, and be ready to go to work. However, new promotions are not quite that way. Officers learn something new every day, and yet they will never know all the answers. Company officers can prepare to make the transition from firefighter to fire officer with study and effort.

There have been many ways of promoting officers over the last century, but we are seeing more and more departments adopt an organized career development program for their members (see **Figure 1-8**). Such programs usually require courses from various disciplines from a local community college. The program

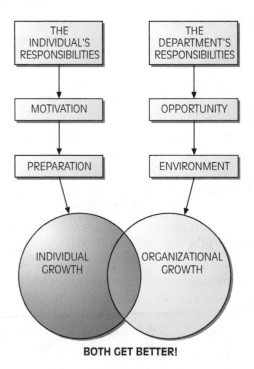

BOTH GET BETTER!

FIGURE 1-8 Professional development is a shared responsibility. Both you and your community have an obligation, and both will benefit from your growth. (*Courtesy of the Phoenix, Arizona, Fire Department*)

content should be carefully considered so that once it is in place, it remains constant. Thus, members can know exactly what is required for advancement. To help members prepare for advancement and, more important, prepare to capably serve in the new position after advancement, the new skills, knowledge, and abilities should be mastered before the member becomes eligible for promotion to the next rank.

The value of a structured professional development program is further reinforced in a recent change to NFPA 1021. Not only should a department have such a program, with officers enrolled in such a program, but officers are expected to help others. The standard states that for certification at the Officer II level, officers should "create a professional development plan for a member of the organization, given the requirements for promotion, so that the individual acquires the necessary knowledge, skills, and abilities to be eligible for the examination for the position".[3]

OFFICER ADVICE

To help members prepare for advancement and, more important, to prepare to capably serve in the new position after advancement, the new skills, knowledge, and abilities should be mastered before the member becomes eligible for promotion to the next rank.

Clearly, and appropriately, this places a burden on those who seek promotion. The department can ease this burden by encouraging participation through tuition assistance and arranging with a local community college for courses to be scheduled at convenient times and locations. Where members work alternating shifts, this means asking the college to offer courses to accommodate the members' work schedules. Thus, members can be assured of both the convenience and the opportunity to take courses without having to take time off from duty.

Some personnel may view these training and education requirements as barriers to their advancement; others will see them as opportunities. Taking the courses helps them develop new skills, knowledge, and abilities so that they can perform duties in additional areas, perform current duties in a more effective and safe manner, and learn new procedures as they are introduced. Much of the training and education required of a professional development program can also help in one's personal life. The information obtained while learning computer skills, improving communications skills, or becoming more effective in interpersonal relations does not have to be left in the station—it can be used off

FIGURE 1-9 Today's fire officer must be competent in using the computer. Computers are needed for e-mail, access to the Internet, writing reports, and countless other applications.

duty as well as throughout the rest of one's life (see **Figure 1-9**).

In the final analysis, professional development provides members with an established program of advancement opportunity. For the individual, advancement means increased responsibilities. As advancement occurs, the department and ultimately the citizens of the community benefit from having knowledgeable, productive, and effective members. A well-planned professional development program is a good deal for everyone: It benefits the member, the department, and the community.

However, these benefits imply some responsibilities. The department should encourage participation, provide an environment that offers opportunities for preparing for advancement, and establish a valid system to select candidates for advancement. The member must be motivated to prepare for and accept the additional responsibilities and duties that come with promotion.

High standards should be set for promotion to the first supervisory ranks. These standards reflect the significant demands placed on supervisory personnel while performing their duties. The primary purpose of the requirements is to ensure that personnel are fully prepared to face the complex challenges of the new positions they are assuming. Most of the requirements focus on administration, management, and supervisory issues.

OFFICER ADVICE

Be aware that a program leading to fire officer certification is only the first step; additional training and education are strongly recommended.

NOTE

NFPA 1021 does not mention fire officer ranks as such, allowing individual departments to align the four levels of qualification with their organization as they see fit. However, for purpose of illustration, we will use the traditional company officer titles of lieutenant and captain. Regardless of a department's rank structure, an officer in charge of a company should meet the requirements for Fire Officer I, and one could make a strong case that they should also meet the requirements for Fire Officer II.

In the NFPA standard, the focus of the requirements for Fire Officer I is on supervision. The requirements for Fire Officer II continue that focus but add managerial requirements as well. Many departments are now expecting company officers to manage their company (people), including the physical (station and equipment) and sometimes even the financial resources (money) needed to accomplish that task. The requirements for Fire Officer II address the skills, knowledge, and abilities needed to accomplish these tasks.

Lieutenants

- Fire Officer I certified (NFPA 1021)
- Instructor I certified (NFPA 1041)
- A semester (or similar length course) of the following:
  - □ College English
  - □ Building Construction for the Fire Service
  - □ Fire Suppression (sometimes called Strategy and Tactics)

Captains

- Fire Officer II certified (NFPA 1021)
- Instructor II certified (NFPA 1041)
- A semester (or similar length course) of the following:
  - □ A second English course, preferably in report writing
  - □ College mathematics
  - □ Administration or management
  - □ Leadership or supervision

Many of these courses are part of a college-level degree program. Completion of these courses and the other requirements listed here will place you well along the path toward a college degree. Transfer credit for prior college efforts and relevant courses taken at the National Fire Academy and state training programs should always be accepted.

NOTE
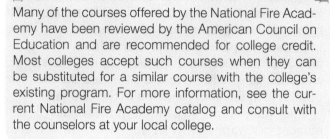

Many of the courses offered by the National Fire Academy have been reviewed by the American Council on Education and are recommended for college credit. Most colleges accept such courses when they can be substituted for a similar course with the college's existing program. For more information, see the current National Fire Academy catalog and consult with the counselors at your local college.

To make the system work best, these courses should be completed before you take the test for advancement. To validate the process, the test or assessment process should carefully consider the content of all of these courses in their selection criteria. Because of the upfront effort on the part of the candidates, fewer candidates may appear for the test, but it is more likely that all will be qualified. On the test day, all that is required is to select the best qualified and rank them. Once promoted, these individuals should be ready to assume the duties of their new position.

Once promoted, you should continue your professional development by reading relevant books and magazines and by attending professional conferences. These activities will help keep you current in the ever-changing world and let you see how others are solving problems that are similar to yours. Reading for pleasure and professional growth is much different than reading a textbook for a course in which you are enrolled, but both are important.

STRATEGIES FOR SUCCESS AS A COMPANY OFFICER

We have talked about the qualities of a good officer and how you can be prepared to meet the challenges of this very responsible position. Before we close this chapter, let us leave you with a short list of qualities that will enhance your success. These qualities are foundational and the key to success as a company officer.

- *Be a professional.* Being a professional means that you are dedicated and committed to the job, learning all that you can and giving as much as you can. Work to make yourself, your company, and your department the very best possible, whether you are paid or a volunteer.

- *Set personal goals.* Every time you go to the station, make some contribution to the improvement of the place and your organization.

- *Continuously work on your own training and education and encourage the same in others.* Training is a part of the job, so obtain all of the training and education that you can. Set personal training and education goals, and use your time and energy to attain them.

■ *Be loyal to your colleagues and your department.* Speak well of your job, your department, and your associates. Although the U.S. Constitution guarantees freedom of speech, good judgment suggests that this freedom must be tempered with reason and sometimes with constraint.

■ *Be a role model.* You will likely be able to recall the first officer or first instructor you had as a firefighter, EMT, or rescue squad member. These individuals leave a lasting impression with us—hopefully positive. For those who are in your charge, the same holds true; be a role model for them. Help them to develop a professional attitude about the job, and help them to be all that they can be. Being a good role model does not mean that you must be perfect. It means you try hard and work to improve so that you can do better tomorrow. Being a role model means being a professional. Professionalism encompasses attitude, behavior, communication style, demeanor, and ethical beliefs. It is as simple as A, B, C, D, and E.

  □ *Attitude is at the core of your performance.* A good positive attitude suggests that you are working to be a role model.

  □ *Behavior is how you act.* You are watched by others, both while you are on duty and when you are off duty. Your actions reflect upon yourself, your department, and your profession (see **Figure 1-10**).

  □ *Communication* is how you get your ideas across to others. Communication can be oral, written, or nonverbal. We are in a people business; we work with people, and we serve people. Today's emergency service organizations spend a great deal of time working to improve the human relations side of us so that we work together better and serve our citizens better. Communication also involves effective listening and seeking to understand before being understood.[4]

  □ *Demeanor embraces all three of the previously listed items.* Focus your energies on the mission of

FIGURE 1-10 The successful company officer is a role model for others.

your organization and use positive attitude, good personal behavior, and effective communications to accomplish the goals of your organization.

  □ *Ethics deals with conforming to the highest professional standards of your organization.* It is more than just words and it is more than just acting out words. It is doing the right thing, every time, every day.

ACCEPTING THE COMPANY OFFICER CHALLENGE

Jake Rhoades, Deputy Chief for Rogers Fire Department, Arkansas (effective July 2009)

The company officer is the most important position within the fire department.

This statement is profound and will garner many opinions within the fire service, but I firmly believe it to be true. Balancing the objectives of chief officers—whether you agree with them or not—with the needs of firefighters that help company officers achieve ultimate harmony and performance on a daily basis is a critical talent that is learned over time. The company officer is the link between these two disparate groups and requires a well-developed skill set. Where do these skills come from, and how are they achieved?

When I was promoted to lieutenant, I was like many other first-time officers, receiving little to no training and making an amazingly fast transition from riding backward to supervising in the span of a few days based on my success on a promotional exam. Like many other departments across the nation, the department I worked for at the time did not have an officer development program to teach me a new skill set, and the strategies and tactics had to be learned on the job. But now I was a company officer and was assigned to an unfamiliar station and tasked with the supervision of several seasoned veterans who were set in their ways as firefighters. My thought process changed quickly as I soon realized the responsibilities placed on company officers: administrative duties, documentation, time management, and, most important, the safety of my personnel. The actions of firefighters reflect their company officer, and their shortcomings and training were my responsibility. The truth is, company officers have a direct influence on the performance, professionalism, and morale of the department, and this is reflected throughout the organization and, ultimately, in the community in the quality of service that is delivered. These realizations set in, and I realized that I had a lot of work to do in order to become an effective company officer, and that work would never stop if I wanted to become a successful company officer.

Over the next several years, I became a student of the fire service, reading everything that I could, attending training events and conferences, obtaining additional formal education, and, more important, learning my trade on the job through the daily experiences that it provided. I became active in many organizations such as the Everyone Goes Home program, Firefighter Near Miss, and the Fraternal Order of Leatherheads. These organizations provided tools that I could use on a daily basis, as well as a group of like-minded individuals with whom I could confer about my thoughts and problems. These individuals reconfirmed what we all know—firefighters are the same all over the world and the fire service is evolving—and to be effective, company officers must continue to learn and be willing to embrace these changes.

The evolution of the fire service has caused me to continue to grow and pursue the knowledge and skills necessary to be an effective officer. The lessons in leadership that I gained from my first day as a lieutenant still follow me today, and serve as the foundation for my career as an officer. I am now an assistant chief with the City of Jenks, and, even considering my new position, I continue to profess that the key to our organizational success is the company officer.

The fire service will continue to change, and it is the responsibility of every company officer to continue to welcome these changes and to do whatever it takes to become an effective company officer. You should strive to guarantee that everyone for whom you are responsible has the knowledge and the skills to perform his or her job as a firefighter—it is your responsibility to ensure that everyone goes home.

Questions

1. What challenges did Jake Rhoades face in his first promotion to a company officer position?

2. What did Lieutenant Rhoades do to overcome the challenges he faced as a new company officer?

3. How does Assistant Fire Chief Rhoades describe the role of a company officer?

4. Do you agree with Assistant Chief Fire Rhoades that the company officer is the most vital position in a fire department? Why or why not?

5. What advice would you give a newly promoted company officer to help in overcoming the challenges he or she might face?

LESSONS LEARNED

Being an officer in the fire service is one of the greatest jobs on earth. At times you are your own boss. You also supervise others. Often you are empowered to do many of the things you came into the fire service to do. There is no higher calling. Being a company officer is not easy, but it has its rewards. One of those rewards is that you get to work with others, often as a leader, to do the very things that motivated you to enter this profession in the first place—to save lives and property.

There are significant rewards associated with the position of company officer, but there are also significant responsibilities. Our purpose here is to help you prepare to meet these responsibilities.

The topics of officer development and increasing the professionalism in the fire service are getting a lot of attention these days. You can turn to any of the leading trade journals or look on the Internet and find plenty of evidence to substantiate this statement.

Look at the National Fire Academy's Web site, where you will find information about professionalism and higher education in the fire service. National Fire Academy Superintendent Denis Onieal speaks about this topic at many conferences and wrote about the subject in 2003. A condensed version of Dr. Onieal's essay appears in Appendix B of this textbook. You can read the article in its entirety in the following magazines: *Firehouse, FireRescue,* and *The Voice* (the member magazine of the International Society of Fire Service Instructors).

At the end of this chapter, you will find a list of additional resources. Note how much material has been recently published on the relevant topics of higher education and professional development for the fire service. Several of the nationally circulated trade journals have an article on these topics every month.

KEY TERMS

certification   document that attests that a person has demonstrated the knowledge and skills necessary to function in a particular craft or trade

codes   systematic arrangement of a body of rules

standard   rule for measuring or model to be followed

REVIEW QUESTIONS

1. List five roles of the company officer.
2. Where does the company officer fit into the fire department organization?
3. List some of the rewards of being a company officer.
4. List some of the challenges that face company officers.
5. What is the name of the document that identifies competency standards for fire officers?

6. Why should fire officers be certified?
7. List the five strategies for success.
8. What is the company officer's primary job?
9. What is the company officer's role in leading the company?
10. What is the company officer's role in the department?

DISCUSSION QUESTIONS

1. What knowledge do you expect a good company officer to possess?
2. What skills do you expect a good company officer to possess?
3. What abilities do you expect a good company officer to possess?

4. What behaviors do you expect a good company officer to demonstrate?
5. Are there any other qualities you expect to see in a good company officer?
6. Other than those listed in the text, what are the roles of a good company officer?

7. If you could design a course that addresses the items listed in questions 1–6, which would you include?

8. What are the qualities you personally need to develop to be a good company officer?

9. What are your personal (as opposed to the professional) goals? What are you doing to meet these goals?

10. What are your motives in becoming a company officer?

ENDNOTES

1. Among the organizations represented were the International Association of Fire Chiefs, the International Association of Firefighters, the International Fire Service Training Association, the International Society of Fire Service Instructors, the National Fire Protection Association, and the National Volunteer Fire Council. The council, having accomplished its initial goals, disbanded in 1992.

2. Reproduced with permission from NFPA 1000, *Fire Service Professional Qualifications Accreditation and Certification Systems*, Copyright © 2006, National Fire Protection Association. This reprinted material is not the complete and official position of the NFPA on the referenced subject, which is represented only by the standard in its entirety.

3. NFPA 1021, *Standard for Fire Officer Professional Qualifications*, 2009, Section 5.2.3.

4. Steven Covey, *7 Habits of Highly Effective People* (New York: Fireside, 1989).

ADDITIONAL RESOURCES

Bennett, Jack A. "Effectiveness at the Company Officer Level." *American Fire Journal*, August 1996.

Carter, Harry. "What Higher Education Means for My Fire Service Career." *Firehouse*, August 2007.

Carter, Harry. "A Personnel Growth Plan for Future Development." *Firehouse*, February 2009.

Carter, Harry R. "Developing a Plan for Self-Evaluation." *Firehouse*, November 2008.

Clark, Burton. "Who Needs a Ph.D.?" *Fire Chief*, February 2004.

Clark, Burton A. "Higher Education and Fire Service Professionalism." *Fire Chief*, September 1993.

Curmode, Rick, and John Leslie. "Promotional Progress." *Fire Chief*, March 2007.

DeMint, Tom. "So You Want to Be Promoted." *Fire Engineering*, July 2006.

Favreau, Donald F. *Guidelines for Fire Service Education Programs in Community and Junior Colleges*. Washington, DC: American Association of Junior Colleges, 1969.

Ford, Travis. "What Is FESHE?" *Firehouse*, February 2007.

Gayk, Ray. Company Officer Development." *Fire Rescue*, April 2008.

Goodwin, Scott. "What They Did and Didn't Give You with the Badge," *Fire Engineering* July 2004.

"Higher Education," www.usfa.fema.gov/fireservice/nfa/highereducation.

Leath, Eric. "Higher Education = Better Skills, Better Future." *Fire Engineering*, April 2007.

Mourchid, Younes. "Earning a Fire Science Degree FESHE Style." *Firehouse*, August 2008.

Mourchid, Younes. "Where Public Safety Policy Meets Higher Education." *Firehouse*, November 2008.

NFPA 1000. *Fire Service Professional Qualifications Accreditation and Certification Systems*. Quincy, MA: National Fire Protection Association.

Onieal, Denis. "Professional Status: The Future of Fire Service Training and Education," *Firehouse*, August 2003.

Prziborowski, Steve. "What Every 'Route' Company Officer Needs to Know." *Firehouse*, February 2009.

Prziborowski, Steve. "Prepare for Your Promotion Now!" *Fire Engineering*, July 2007.

Prziborowski, Steve. "What's Keeping You from Getting Promoted?" *Firehouse*, August 2008.

Rubin, Dennis, and Edward Kaplan. "Could Higher Education Be in Your Future?" *Firehouse*, February 2008.

Schaefer, James J. "University and Community-Based Interoperability." *Firehouse*, November 2008.

Snodgrass, Paul. "Building an Educational Foundation to a Fire Service Career." *Firehouse*, February 2009.

Snodgrass, Paul. "Rich Media: Enhancing Content in Online Courses. *Firehouse*, August 2008.

Snodgrass, Paul. "Characteristics of Successful E-Learners." *Firehouse*, October 2007.

Stocker, Alan. "Help Make Your Employees Successful." *Fire Engineering*, July 2008

Terpak, Michael. "Promotional Assessment Centers: Understanding the Process." *Fire Engineering*, July 2006.

Thompson, Scott, and Eddie Bachanan. "It's All about Attitude." *Fire Engineering*, July 2004.

Wakeham, Ron, and Bill Lowe. "Getting Educated about Being College Educated." *Firehouse*, August 2004.

2

The Company Officer's Role in Effective Communications

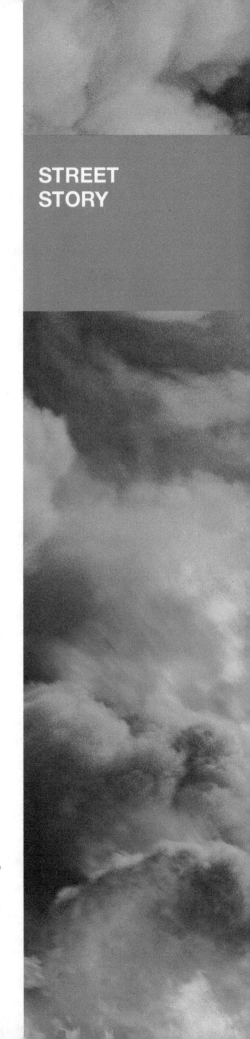

One of my early weaknesses was trying to control the adrenaline buzz I would get when a structure fire was dispatched. I would get excited, and this was reflected in my communications—whether on the radio, with my crew, or with victims. More often than not, my style did not help matters. In an effort to make up for this lack of experience, I spent off-duty hours riding with engine and truck companies in busy fire departments. These professional visitations taught me much about company leadership and communications. One communication trait I noticed right away was the calm, confident nature that experienced company officers used when dealing with station issues, incidents, and conflict. One particular captain became a mentor of sorts for me—he was always quiet, firm, and clear. I was amazed how easily he switched his communications style to fit specific situations, yet he always remained calm and confident. When responding to incidents, he would communicate any known information to his crew and would ask if they had knowledge of the building or area to contribute. When the apparatus turned the corner and the incident came into view, he would always tell his driver to "slow down." After hearing this a few times, I asked him why he always told the driver to slow. He simply replied, "I'm stalling!" Then he went on to explain that it was his way to control his excitement and to channel that energy into evaluating the environment. In doing this, he could pick up on the presence of electrical lines, hydrant locations, bystander reactions, and other hazards—not to mention analyzing the building construction and features and reading smoke conditions.

Of course, I tried to emulate that in my own role as a lieutenant and started using the "slow down" technique. After several incidents, my driver became very frustrated. It was not long before he cornered me and defensively asked if I felt his driving was bad! I realized right there that my communication was not clear and the message was interpreted differently than I had intended. After a bit of discussion, we found an understanding of my needs and the intent of the message. A few months later, my company was dispatched to a structure fire in a residential area. As my driver turned the corner, we could see a large crowd gathering near a working fire that was blowing through the front windows of a house. As if on cue, my driver said, "Don't worry, lieutenant, I'm slowing down!"

Being mindful of the message, audience, and tone is a learned communication skill. As a company officer, I have learned that the WAY you communicate is more important than the message itself. Over time, I realized that calm is usually best—even when things are getting hairy. I have learned that "telling" is not "teaching"—creating a discovery environment is worth the extra effort. I have learned that a brief second used to compose oral communications is a wise investment. I have learned that critical written communications should be read by someone outside the intended reader BEFORE sending it. Most important, I have learned that "hearing" and "listening" are very different.

—Dave Dodson, Lead Instructor, Response Solutions, Colorado

LEARNING OBJECTIVES

After completing this chapter the reader should be able to:

2-1 Distinguish between formal and informal communication.

2-2 Define the *communication process.*

2-3 Describe the communication order model.

2-4 Identify the forms of communication.

2-5 Describe the methods used to determine firefighter comprehension of verbal directions.

2-6 List the barriers to effective nonemergency communication.

2-7 Identify public relations practices when dealing with the media.

2-8 Explain the Company Officer's role in dealing with on scene media.

2-9 Explain the Company Officer's role in writing a press release.

2-10 List common components found in a news release.

2-11 Explain the policies, procedures, and statutes that govern news release.

2-12 Explain the policies and procedures for releasing agency information.

2-13 List the barriers to effective emergency communications.

2-14 Identify the methods of improving communications on the fireground.

*The FO I and II levels, as defined by the NFPA 1021 Standards, are identified in different colors: FO I = black, FO II = red.

INTRODUCTION

Effective communications are a vital part of your professional life. Consider these generally accepted truths:

■ Those who can communicate effectively get more out of life.

■ Those who cannot communicate effectively will not reach their full potential.

Given these statements, you should realize that good communications skills are essential in both your work and your personal life. Many fire officers will gladly sign up for a weekend of training on a technical aspect of firefighting. Many would also benefit from making a similar commitment to improve their communications skills. Effective communications skills will also add value to your personal life.

NOTE

Good communications skills are essential in both your work and your personal life.

This chapter appears at the front of the text for several good reasons. First, it is one of the most important topics in the book. Many fire chiefs report that the one weakness prevalent among their officers is their inability to communicate effectively. A lack of good communications skills will hurt even the best of individuals. You may have the best ideas, but to implement them you must communicate them to others. You have to communicate to make recommendations to your superiors and to motivate your subordinates. You have to communicate to a variety of audiences, including the public you serve. When you represent yourself, your communications shortcomings will only hurt yourself. However, when you represent your department, community, or profession, your inabilities to communicate your ideas and knowledge have a much greater impact.

NOTE

A lack of good communications skills will hurt even the best of individuals.

The second reason for placing this material in the front of the text is that communications skills are a tool needed to complete the requirements for certification. According to NFPA 1021, *Standard for Fire Officer Professional Qualifications,* good oral and written communications skills are a prerequisite for most of the certification requirements for Fire Officer I.[1]

NOTE

Communications can be defined as a process by which information is exchanged between two individuals.

THE IMPORTANCE OF COMMUNICATIONS IN OUR WORK AND LIFE

Being able to communicate effectively is important for both you and your fire department. Think about how much of your time is spent communicating; it is an important part of your life.

Formal versus Informal Communications

To begin our discussion of personal communications, we start by distinguishing between formal and informal communications. Because we are focusing on the needs of the job as an officer in the fire service, we are focusing on the formal aspects of both oral and written communications.

Formal communications are conducted according to established standards. They tend to follow the customs, rules, and practices of the industry or workplace. The fire service has its established practices. The medical community has its practices. The legal profession has quite a different set of practices. Formal communications transmit official information. Formal orders and directives, standard operating procedures, and official correspondence are examples of formal communications. Such formal communications usually have legal standing within the organization.

Informal communications are simpler and more spontaneous. We write with less formality when sending memos, short notes, and e-mail to others. Likewise, we tend to be less formal during social events such as meals and other situations where we might be together with colleagues.

Written communications can be either formal or informal. We have already cited some examples of formal communications (for example, formal orders), but many organizations use informal written communications as well. We usually communicate up and down the organization using formal communications. When we need to communicate across the organization, we usually use a more informal communications style.

For example, you might ask a peer in another section for information or an opinion on a new idea. Because this communication is based more on friendship than on any formal relationship, informal communications are more comfortable.

Informal communications works well in established organizations where there are stable relationships among the work units and among the individual members. When work units are emerging, when there

ELEMENTS OF THE COMMUNICATIONS PROCESS

Regardless of whether you are speaking or writing, the communications process is generally thought to include five elements, called the *communication model.*

Communication Model

- *Sender:* The process starts with the sender. The sender is the person who has the information and who wishes to send it to another person. Selecting the method of communications, forming thoughts, and expressing these thoughts in words form an important part of this process. All of this activity is accomplished by the sender.

- *Message:* The sender selects the message. The message is the information, which is usually communicated using words, although certain symbols, such as the signs used in mathematics, are also considered part of the message. The proper selection of the message is vital to the communications process.

- *Medium:* The sender also selects the method of transmission, sometimes referred to as the medium. The medium is the means by which the message is transmitted. Media include oral and written communications, signs, and graphic representations. A stop sign is a medium. When a police officer is standing in the center of an intersection, the officer's upraised hand is a medium. The sound of a siren might also be a medium. All send the same message. Other media of communication include books, the spoken word, and images on the screen in class.

- *Receiver:* The receiver is the recipient of the message. Although the receiver is thought to be a passive role, we shall soon see that the receiver plays an active role in the communications process.

- *Feedback:* The final step in the communications process is called **feedback.** Feedback tells the sender that the **message** was received and understood. Feedback comes in many ways (see **Figure 2-1** and **Figure 2-2**). Compliance with a request is an obvious example of feedback. If you ask someone to turn on the light, and the person does, that is feedback. The act of turning on the light indicates that the person received the message, understood the message, and complied with the request.

In this situation, the feedback was obvious. Sometimes getting feedback is much more difficult. The lack of feedback leaves the sender wondering if the message was received, if it was understood, and if anyone is going to respond. We discuss feedback later in this chapter.

FIGURE 2-1 Effective communications skills are an important tool for officers.

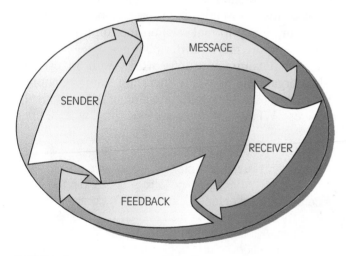

FIGURE 2-2 The communications model represents a continuous process.

is conflict, or when there is a lack of trust, we tend to see greater use of formal communications.

Forms of Personal Communications

The spoken word is the most commonly used method of personal communications. It is fast, practical, and usually effective. Oral communications can be one-way or two-way; for example, a recorded message on an answering machine is considered one-way communication, while a conversation on the telephone is a form of two-way communications. Oral communication can be accomplished with one **receiver** or a large audience in a conference or classroom. Or it can be done in daylight or total darkness. However, oral communications are limited—they may be inhibited by barriers such as language difficulties or background noise, and they may not be documented.

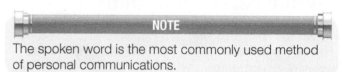

NOTE

The spoken word is the most commonly used method of personal communications.

The second form of personal communications is written communications. Written communications provide a record that can be used for future reference, showing that the message was sent, and in some cases there is evidence to show that the message was received.

The third form of communications is communicating without words, sometimes referred to as body language. A smile or a frown can express more than words, or an extended hand can make a clear statement. When appropriate, a pat on the back can send an obvious message of support and confidence.

There are other forms of communications, such as symbols or signs, that can be used to relay a message or emotion (see **Figure 2-3**).

ORAL COMMUNICATIONS

Of all communication forms listed, oral communication is easiest, most commonly used, and effective. It is effective because there is immediate feedback from the listener; as a result, most of us prefer face-to-face contact when speaking with others. Face-to-face contact allows the feedback process to work at its best. Good eye contact allows us to look directly at the other individual and watch that person's facial expressions and gestures. A nod of understanding encourages the speaker to continue. On the other hand, a frown should cause the speaker to pause and cover the material in another way. Even the most effective speakers should realize that there will be times when their message is not understood.

FIGURE 2-3 Signs are a form of communications. Some are more personal than others.

Although most of us like to communicate face-to-face, it is not always possible. When a telephone or radio is used to communicate with words, there is no face-to-face contact (see **Figure 2-4**). Facial expressions and gestures cannot be seen; therefore, communications is deprived of some feedback that may be essential to the communications process. When using such forms of oral exchanges, it is important the **sender** make sure the receiver has received and has understood the correct message that was communicated.

In any case, the speaker should stop from time to time and ask the listener a question. This practice allows the sender and the receiver to trade places for a moment and makes the communications process easier for both parties. This works, regardless of whether you are speaking to one or one hundred.

When speaking and making a habit of pausing occasionally and asking a question, several things happen. As we have already indicated, this practice allows the sender and the receiver to momentarily trade places. The change is usually welcomed on both sides. Also, you will get questions, and with those questions, you get additional feedback that helps to confirm understanding.

NOTE

The speaker should stop from time to time and ask the listener a question.

Another advantage of this process is that the receiver will anticipate that a pause is coming and will hold any questions until the appropriate opportunity is presented. So this practice allows the sender and the receiver to have a short break from their respective activity, it allows for feedback, and it allows the listener a chance to ask questions, if needed.

It is best if the listener only has to concentrate on listening, and at times that can be difficult. This means removing the barriers that get in the way of good communications and also means that the receiver should avoid anticipating what the sender will say or interrupting the speaker with questions. Often, when a person attempts to multitask, such as talking on the telephone and working on the computer at the same time, the listener becomes distracted and the messages are lost.

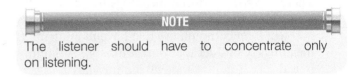

NOTE

The listener should have to concentrate only on listening.

Barriers to Effective Communications

We have discussed some of the characteristics of effective communication. Now let us consider some of the barriers that make communications difficult. Any obstacle in the communications process is called a **barrier.** Consider a barrier to be like a filter. One or more filters, or barriers, reduces the information flow between the sender and the receiver. There are several types of barriers (see **Figure 2-5**).

- *Physical barriers:* Physical barriers are environmental factors that prevent or reduce the sending and receiving of communications. Walls, distance, and background noise are examples of physical barriers. These are usually obvious barriers.

- *Personal barriers:* Personal barriers are less obvious. They arise from the judgments, emotions, and

FIGURE 2-4 Communicating by phone or radio can be more difficult than communicating face-to-face.

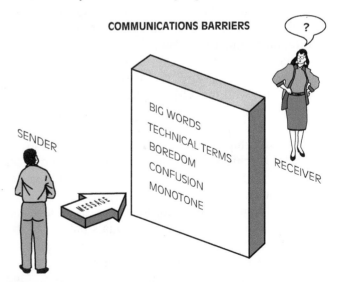

FIGURE 2-5 There are many barriers to effective communications.

social values we place on people. These factors cause a psychological barrier, which can be just as real as a physical barrier. Personal barriers act as filters in nearly all of our interactions with others. We see and hear what we want to see and hear, and we remain selectively "tuned out" to that which we do not wish to see or hear.

- *Semantic barriers:* Semantic barriers arise from language problems. Our language is filled with words that have multiple meanings and words that have vague meanings. Consider the word "round." There are several dozen meanings to this word in a standard dictionary. In fact, there are more than 500 words in common use that have more than 20 meanings. Consider the word "dummy." A dummy is a rescue training tool. A dummy has a different meaning in the publishing industry. The word "dummy" is also used in the game of bridge. None of these terms is offensive. But just call someone a "dummy" and see what happens. No wonder we sometimes have trouble communicating!

You should be aware of the impact of barriers and change the conditions that impose the barriers when the communications process is adversely affected. Some things can be easily controlled. You can control the choice of words; the use of technical terms, acronyms, or trade jargon; and the speed of delivery. You can sometimes control the time and place of the communication. You can enhance understanding and retention by repeating vital information and showing the same information in some graphic form. All of these techniques will have considerable impact on the listener.

So it is with a new firefighter who may be working for you. New firefighters may be bright and highly motivated, but when you are teaching them something new and you talk over their heads or speak too fast for their comprehension, you will not be effectively communicating. Place yourself in the role of that new firefighter. Provide the information in small bits and ask questions to be sure that the communications process is still working. Other things to consider include the age, educational levels, possible language differences, the background and experience, and the working relationship. Overcoming communications barriers can be achieved by being adaptive to the audience, by having a specific purpose, and by staying focused, brief, and clear.

When you are speaking to groups of people and when you are speaking to individuals outside of the fire department, the process is even more challenging. For example, consider the delivery of the department's public life safety education message. One day you might be asked to speak to a group of third graders. That evening you might be asked to speak to a group of senior citizens. The next day, you may go into a community where English is the second language. Although the purpose of your visit is similar on each occasion, the way you communicate your safety message should be quite different.

Part of the Communications Process Is Listening

We have indicated the importance of good, effective communications and that the sender has quite a responsibility to select the medium and the message, to check for feedback, and to be aware of the potential barriers. Communicating also involves listening. In fact, listening may be the most important part for the company officer. Understanding others requires an active role on the part of the listener. This role is called **active listening**. As a listener, you can show the sender that you are actively listening by focusing all of your attention on the speaker and by showing genuine interest in the speaker's message. This usually means looking at the speaker, doing nothing but listening, and allowing your facial expressions to indicate interest and understanding (see **Figure 2-6**). Alert facial expression and good posture indicate a good listener. This is what the world renowned authority on leadership, Dr. Steven Covey refers to in his book *7 Habits of Highly Effective People*. The fifth principle is "Seek first to understand before being understood." Mr. Covey writes:

> If you're like most people, you probably seek first to be understood; you want to get your point across. And in doing so, you may ignore the other person completely, pretend that you're listening, selectively hear only certain parts of the conversation or attentively focus on only

FIGURE 2-6   Good listening skills are part of communicating.

the words being said, but miss the meaning entirely. So why does this happen? Because most people listen with the intent to reply, not to understand. You listen to yourself as you prepare in your mind what you are going to say, the questions you are going to ask, etc. You filter everything you hear through your life experiences, your frame of reference. You check what you hear against your autobiography and see how it measures up. And consequently, you decide prematurely what the other person means before he/she finishes communicating.[2]

NOTE

Understanding others requires an active role on the part of the listener.

When the message is unclear or not understood, the listener should ask questions when appropriate. Questions and comments usually indicate interest and encourage the speaker to expand on an area that was not clearly presented.

While questions are signs of good listening, interrupting with too many questions or offering solutions to the speaker's problems before the speaker has finished is not good practice. These activities show that the listener is impatient to bring the conversation to a close.

Certainly there are times when the process must be hurried and even interrupted. That is part of life in the fire service. When interruptions occur, both parties should make an effort to resume the conversation as soon as circumstances permit. The real mark of a good leader is a comment that one sometime hears from one of the firefighters: "I like the captain. She listens to me when I have a problem."

NOTE

Alert facial expression and good posture indicate a good listener.

Consider the following situation: You walk into your supervisor's office with a question. If your boss stops work, looks at you, and gives you 100 percent of his/her attention, it is likely that he/she is actively listening. On the other hand, if one or more of these conditions is not met, it is likely that he/she is not actively listening. Which situation would you rather encounter?

NOTE

When you are actively listening, three important things happen:
- First, you actually hear what is being said.
- Second, you are more likely to remember what was said.
- Third, you show respect for the sender.

Listening to others is an important part of your job and probably takes up a good part of your day as company officer. Good listening skills are important. Research has shown that we take in a very small part of what we hear and that we remember only a small part of what we take in. Listening is an active process that requires considerable effort.

OFFICER ADVICE

At several points in this text, we depart from preparing you to become an officer in the fire service and offer some friendly advice. This is one of those occasions. Consider the preceding information about active listening in your personal life. How well do you communicate in an informal setting? How well do you listen to your friends, your spouse, or your children? They also deserve 100 percent of your listening talent. Try some active listening at home or in a social setting.

Good Communications Is a Part of Leadership

As a supervisor, the way you communicate with your people has a lot to do with your leadership style. Suppose a subordinate stops at your office with a question. If you take time to focus on that person, show respect for and interest in him/her as a person, and actively listen to his/her idea or question, you are doing a lot to enhance both the communications process and the personal relationship. Not only are you listening but also you are showing a genuine interest in the visitor and his/her thoughts. Whether you do this or not, the time and energy spent in the conversation are about the same. But look at the difference in the outcomes. With active listening, you hear better and you understand better, and you say, "I care about you."

Regardless of whether you are communicating through the spoken word or through your writing, there is a need to consider the human relations aspect in every communications activity. Human relations are discussed in greater detail in Chapters 6 and 7, but we should note here that consideration for the other person is important to the communications process.

When we communicate, we have an opportunity to show respect for the person or persons with whom we are communicating. That respect, or the lack of it, will be clearly apparent in the speaker's tone as well as the words that are used. When the sender uses words and phrases that show respect and consideration for the receiver's interests and point of view, it is far more likely that the message will be accepted, even when it brings bad news.

NOTE

When we communicate, we have an opportunity to show respect for the person or persons with whom we are communicating.

Although you should show a concern for the human relations aspect of communications, there must come a time in the conversation when you have to get to the point of the meeting. Getting to the point does not mean that there is any lack of respect, nor does it mean that the sender has to be blunt or abrupt. This problem is especially true when dealing with performance issues in the workplace. We are often afraid of hurting the other person's feelings.

It is important to remember the human relations aspect of communications, but you have to communicate a message. That requires effective communications and the ability to focus on specifics. After a brief warm-up, get to the heart of the message and tell it like it is. Avoid generalities and exaggerations. "I've told you a million times, don't exaggerate!" is a perfect example. In the workplace, a supervisor's comment such as "You are late again! It seems that you are late every day" is likely to start a confrontation rather than a productive session in which solutions can be achieved.

As a supervisor, you (sender) should be able to talk about performance issues without talking about the personality or personal traits of the subordinate (receiver). If the subordinate's work is not up to standard, you should talk about the standard and the fact that the subordinate is not meeting the standard. You might show how this impacts others and what this does to the organization's overall performance. These are facts. What does this do to the conversation? In this case, the member's performance becomes the issue, not the member's personality, attitude, or ability.

As the sender, you should be able to express your own feelings and reactions, when appropriate, especially when dealing with performance problems. For example, you might say, "I am disappointed with this work."

NOTE

Remember to make your communications a two-way process.

Finally, remember to make your communications a two-way process (see **Figure 2-7**). As the supervisor, you can usually ask a question that will bring more information to light or help the other person understand the issue. Likewise, when writing, always close with an offer to continue the discussion or provide additional information.

FIGURE 2-7   When discussing a members' performance remember to let the member tell his/her side as well.

Making Practical Use of Effective Communications

Consider the following example: "Bill, I have noticed that on our last three calls you were the last one getting to the truck. On our most recent call, the driver and I had to wait 30 seconds for you to arrive and get properly seated. Those delays may seem minor, but as you well know, there are times when every second counts. Although no one else has noticed it, this concerns me. Is there a problem? Can I do anything that might help you get there a little faster?"

Notice that the officer has not directly accused Bill of anything wrong, nor has the officer threatened Bill in any way. The officer has only stated professional observations and expressed a personal concern about the situation. The officer asked Bill two open-ended questions. Hopefully, these will give Bill an opportunity to recognize that there is a problem and to realize that the boss is both concerned and willing to help. In this case, the officer has shown concern for the firefighter and has given Bill a chance to identify and solve his own problems.

Contrast that action with the following: "Bill, you are too slow on all of our alarms! I guess you just don't think that others are very important. That is, if you can think . . . ."

In summary, show respect, get to the point, focus on specifics, express your own personal feelings, and make the discussion a two-way process.

OFFICER ADVICE

Show respect, get to the point, focus on specifics, express your own personal feelings, and make the discussion a two-way process.

At this point you may likely say, "Well, that is okay in the fire station, but when we get out on the fireground, things change." Surely the tempo picks up, and the communications tend to be very directive in nature. "Do this! Do that!" Some of that is justified by the urgency of the situation. But even under these conditions, your leadership style is heard in the tone of your voice and in what you say. If you remain polite and reasonable in your communications, you will get just as much, if not more, from the troops.

OFFICER ADVICE

Be careful what you say. Speak as if the radio or a tape recorder were capturing your every word. Even though these things rarely happen by accident, it's always a good practice to be careful what you say. This is good advice for everyone. Officers should set a good example, withholding comments that are inappropriate or unprofessional.

One of the company officer's many jobs is to facilitate the communications process. Firefighters tend to be interested in matters that are going on elsewhere within their department, and they should also be interested in what is going on elsewhere in the fire service. You should facilitate this process. By virtue of your rank and experience, you probably have more access to what is happening at headquarters and elsewhere. You should be reading the directives from headquarters and the trade magazines. In both cases, you should pass on to the troops appropriate information as soon as practical.

If firefighters have a question, you should take the initiative to help them find the answer (see **Figure 2-8**). That answer may be to simply (and politely) tell them where to look for help. Some questions will require a call to headquarters. In many cases, it certainly would be far more appropriate for you to make the call. Besides, you are more likely to know whom to call and what questions to ask.

Other Ways to Improve Your Speech Communications

Most adults speak well. However, many adults would benefit from some further training on this most commonly used form of communications. One way to improve your communications skills is to watch yourself on videotape. Make a speech and record it. Then sit down by yourself and critically watch what you did. Be forewarned: It can be a painful experience.

Effective oral communications require some effort. Your speech should be clear and distinct. The pace should be such that the listener has time to hear and understand what you are saying. Common problems in oral communications include not looking

FIGURE 2-8 When you do not know the answer, admit it. Then, make an effort to find the answer.

at the person to whom you are speaking and using distracting activities. Distracting activities can range from using a word or phrase to the point that it becomes annoying or repeating some physical activity such as jingling keys or coins.

As a part of your officer training, you should be looking for ways to improve your oral communications skills. Taking a training course leading to becoming a fire instructor is a good start. Other options include courses by Dale Carnegie and Toastmasters International. Both programs offer opportunities for individuals to come together to become friends and gain experience in communicating orally with others in a friendly environment. College courses are also available.

IMPROVING WRITING SKILLS

Effective fire officers must be able to write. You must be able to express your ideas to your supervisors. Most of the significant recommendations and requests you submit are written. It is also important that you be able to express your ideas to your subordinates. This is how most policies, procedures, and directives are published. Finally, you have to complete a variety of reports (see **Figure 2-9**). All of this activity requires writing skill.

FIGURE 2-9 Writing reports is another important job for the company officer.

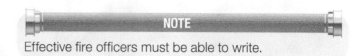

NOTE

Effective fire officers must be able to write.

Principles of Effective Writing

Use the following principles whenever you employ writing in your communications.

- *Consider the reader:* First and foremost, consider who the target reader is. Use plain language. Use terms and abbreviations that readers will clearly understand. If in doubt, spell out the term at its first use followed by the abbreviation or acronym. Technical terms, jargon, and abbreviations are like hurdles on the track: They slow the reader rather than helping.

- *Emphasis:* Memos, letters, and directives should usually be limited to one topic. This practice allows the writer and the reader to focus on just one issue and discuss it as needed until the issue is understood and resolved. This rule also helps satisfy the next item.

- *Brevity:* Brevity is desired. Consider the material you get in the mail every day. The letters that are brief and to the point usually get read right away. Brevity is often misunderstood. Brevity does not mean too short; it just means not too long. It means that it is long enough to do the job.

- *Simplicity:* Use everyday words when possible. Use the same words you would use while speaking. When writing, we frequently resort to bigger words, thinking that it will impress the reader. Our efforts should be to impress the reader with the information we are sharing, not with the writing style. In many cases, the use of larger words actually slows the communications process. Here is an example: "Long, rambling, and elaborate dissertations will usually languish in the receiver's mailbox until they die of old age."

  Here is another: "It is the policy of the fire chief that all activities of this fire department shall be carried out in a manner that will ensure the protection and enhancement of the environment through control and abatement of environmental pollution." It might have been easier to say, "Don't pollute."

  On the other hand, use specific terms when appropriate. A report that contains ". . . it was a large fire" leaves too much to the imagination of the reader. The message here is use language that is appropriate for the occasion and the reader.

USING EASIER WORDS AND PHRASES

Instead of	Try
commence	start
facilitate	help
enhanced method	better
finalize	end
improved costs	less expensive
in order to	to
in the event of	if
it is requested	please
it is the recommendation of this office	we recommend
it is necessary that you	you should
optimum	best
prioritize	arrange
prorogate	issue
utilize	use

OFFICER ADVICE

Use language that is appropriate for the occasion and the reader.

■ *Objectivity:* Most of your writing requires an impersonal viewpoint. In most situations where facts are being reported, there is little place for personal bias. Some reporting formats invite a personal opinion, and at that point the writer should clearly establish that these statements are her own opinion. For example, a standard reporting format for an accident investigation includes a provision for the writer's personal opinion. After determining and outlining the facts, the writer has more knowledge about the accident than anyone else on earth. The writer's opinion is valuable at this point and should be accepted in light of her expertise. However, under normal conditions, we are simply part of the communication process: Our own bias should not be apparent in our writings.

ETHICS

Our own bias should not be apparent in our writings.

■ *Mechanical accuracy:* This term describes the various rules of good written communication. Most of us have been taught the rules, but some of us need to review them from time to time. We will cover the most common problems here. At the end of the chapter is a list of additional references that may be useful.

Underlining or italics are generally used to identify the titles of books, plays, works of art, magazines, and newspapers. An article within one of these publications is usually set off in quotation marks. For example, we might have a sentence that reads: Those interested in firefighter safety issues should read the report entitled "U.S. Firefighter Injuries," published annually in the *NFPA Journal*.

There are generally accepted rules for using numbers in text. For the numbers one through nine, you spell out the word. Starting with 10, you use the digits.

Many writers make mistakes using capital letters; most of the mistakes result from using capitals when they are not needed, such as when the word is not associated with any specific person. Capitals are usually used when referring to a specific individual, place, or thing, whereas lowercase letters generally refer to the generic term, for example, the training officer, the lieutenant, and the fire chief, but Captain Wright and Chief Coffman. Consider the following examples:

a captain	Captain Joe Friday
a chief	Chief Jim Flynn
a lake	Lake Michigan
a street	Main Street
a textbook	*The Company Officer*

CAPITALS ARE USED AT THE START OF A

Sentence
Proper noun
Line of verse
Name of a human race
Name of a government body
Noun used to refer to a deity
Direct quote when within a sentence
Name of a planet, constellation, or star
Holiday, day of the week, or month of the year

Another communication problem is the use of passive voice. Consider the following conversation:

Doctor: When did you first notice your use of verbs in the passive voice?

Patient: The use was first noticed by me shortly after the fire service was entered. The fire service was also entered by my brother. The same condition has been noticed by him.

Doctor: Do you know that most of the verbs we read and speak with are active voice?

Patient: Well, it is believed by me that most verbs are made passive by fire service writers. In the letters and reports that have been prepared by this speaker, passive voice has been used extensively.

Are problems caused? Yes, problems are caused. It would be better to say that writing in the passive voice causes problems. Using passive voice usually results in wordy, roundabout, and confusing writing. Avoid these problems by putting the subject of the sentence (the doer) ahead of the verb. For example: The station was inspected by the fire chief. Who did the action? The fire chief. This sentence would be better if it read as follows: The chief inspected the station. There are times when the passive voice may be the only choice. For example, if the person who performed the act is unknown, the passive voice may be appropriate. For example: The fire was started by unknown causes. Also the passive may be more effective when the receiver of the action is more important than the doer. For example: "Any person who violates this rule will be fired." Your writing will be more interesting and more effective when you use the active voice.

We have limited this discussion to the most common problems. There are others, of course. Fire departments would do well to publish a guide with some good examples of correspondence and report writing. The guide should also include information regarding the details of style that may be unique to the organization. Regardless of whether such information is available from your agency, it would be smart to obtain one of the several generally accepted writers' guides. These can be found in bookstores everywhere. Several are listed at the end of this chapter.

> **OFFICER ADVICE**
>
> Fire departments would do well to publish a guide with some good examples of correspondence and report writing.

In addition to these standard references, most dictionaries contain several pages of good information on grammar. In addition, a good modern dictionary is invaluable for checking the spelling and the exact meaning of words. Do not be ashamed to look at these references when you have a question. The fact that so many references are available suggests that many writers need to look things up from time to time.

Putting It All Together

The following are some suggestions for effective writing.

- *Organize your thoughts:* Organization may be the most important part of writing. Unfortunately, for many writers it is also the most difficult part of writing. Like building a house, organizing does not have to be done all at once. Start with what you know, and add the other facts as they come to mind. Outlines help some writers. Word processing helps, too. The ability to easily arrange and rearrange sentences and paragraphs has been one of the significant improvements brought by the word processor. Rewrite! Paper is cheap. Time and good impressions are priceless.

- *Start with facts, explain as necessary, then stop!* When you write, think about the one sentence you would keep if you were allowed to keep only one. That sentence should be in the first paragraph and may even be the first sentence of the first paragraph. With this approach, you put requests before justifications, answers before explanations, conclusions before discussions, and summaries before details.

  You might be saying, "Wait a minute! That's not what we were taught in school." In today's fast-paced business world, managers want answers and information in the shortest possible form. If your answer is reasonable, they do not need a lot of detail. Those who can provide information in this style are appreciated.

> **OFFICER ADVICE**
>
> When you write, think about the one sentence you would keep if you were allowed to keep only one.

- *Consider busy readers:* If the answer can be provided in one or two sentences, put those at the top of the page. If that information satisfies the readers' needs, they may stop as soon as they have what they need. If they need more explanation or justification or want to read more about the way you obtained your information, they can read on. While it is important to provide the "big bang" up front, it is also important to provide the other information. If facts are presented, they should be documented. If conclusions are drawn, they should be supported with logic.

- *Use short paragraphs:* Long paragraphs often swamp ideas and overwhelm the reader. Cover one topic before starting another, and let the topic take several paragraphs if necessary. The white space between the paragraphs helps to separate your thoughts, makes the reading easier, and encourages your reader to continue. Listings are a useful technique to provide information in a quick and easy-to-read format. This is often done with bullets. This technique is used throughout this book.

■ *Use a natural writing style:* Write as if you were speaking. Imagine that the reader is sitting next to you and that you are speaking aloud the same words you are writing. Use personal pronouns, everyday words, and short sentences. Try to avoid "I" except when speaking about yourself or expressing an opinion that is clearly your own. When representing your organization, the use of "we," "our," and "us" is more appropriate.

■ *Be direct:* Try to avoid roundabout sentences. Again, think about the situation as if you were speaking, and imagine the reader sitting next to you. Consider the following two sentences as examples:

☐ It is necessary that your reply be received by the first of next week.

☐ We want your reply by the first of the week.

Which statement are you more likely to say in a conversation? Which statement is easier to understand?

Review Your Work

When it comes to writing, several habits will help. First, avoid waiting until the last minute to do your work. The pressure of the deadline will force you to make compromises that will not be in your best interests. By doing the work in advance, you also have the opportunity to review and edit your work before submitting it.

With computers in widespread use, the ability to use word processing software is available to nearly everyone (see **Figure 2-10**). Use this technology to enhance your writing. Write your letter or memo. Review it on the screen and run the spell-check. Then print a draft copy. Read it and mark up the copy with corrections or areas where you think it needs improvement. Wait about 24 hours (at least until the next day) and repeat the process. In nearly every case you will make additional corrections. When you reread the document, you will want to add a word or two or rearrange several sentences to enhance the reader's understanding. Make the corrections and print a second draft. If time permits, repeat the process again.

Although your second reading is always important, for real benefit try to get someone else to read the document. That person does not have to be an editor or a grammar teacher but should be someone who has reasonably good communications skills and who has the patience to read very carefully. Good proofreading is not like reading the sports pages of the paper; it requires a slow, deliberate reading. Some writers even read aloud to themselves when proofreading. In so doing, they will hear mistakes more readily than if they were just reading in the usual manner. In any case, you (or someone else) should read the document carefully and make corrections where necessary. These steps will significantly enhance your writing. Good writers perform both all of the time.

FIGURE 2-10 Computers make writing a lot easier. But you still need to understand the rules to be an effective writer.

Take care with the final form and appearance of your product. In many cases, the paper you submit is like a sales effort; it represents you and your organization. You want it to make a good impression. Be sure that the final product is clean and well groomed. Make sure that the printer is working properly. When you submit good work, you can be proud of your accomplishments.

OFFICER ADVICE

Take care with the final form and appearance of your product.

Other Ways to Improve Your Writing Skills

In Chapter 1, we suggested that you might benefit from taking a writing course. Let us mention that again:

A good course, workshop, or other formal learning activity in which you are asked to write will always be beneficial. Many high schools, libraries, community centers, and colleges offer short courses as well as the traditional semester-long classes in writing skills for adults. Most colleges offer a variety of communications courses, and many provide a testing service to help you to identify your own personal needs. Take advantage of these opportunities.

In addition, bookstores and libraries have a wealth of good information on this topic. Several useful books of particular value for fire service writers are listed at the end of this chapter. Take advantage of as much of this information as you can and work to improve your communications skills. You will be more effective at work and in other activities where communications are important. You may even want to try writing an article for one of the trade journals (they are all looking for fresh authors) or something even more ambitious.

COMPARATIVE WRITING EXAMPLES

Example 1
Unedited: This program promotes life safety in homes and businesses throughout the country by offering a simple way to make purchasing decisions by providing additional information about life safety products without sacrificing quality or product performance.

Edited: This program helps homeowners and businesses buy life safety products without sacrificing quality or performance.

Example 2
Unedited: Right of use means any authorization issued under this part that allows use of Department related equipment or property.

Edited: "Right of use" means any authorization under this part to use department property.

Example 3
Unedited: Although the aforementioned language was not included in the FY2007-2008 budget authorization, we believe that the continuation of this special initiative is important.

Edited: Although this language is not in the FY2007-2008 budget authorization, we believe it is important to continue this special initiative.

Example 4
Unedited: All issues and questions were discussed and explained very clearly. Following the completion of every task I received a full feedback from my supervisor which gave me an opportunity to improve and reflect on my performance continuously. I was given support in terms of addressing my personal objectives such as improvement of interviewing skills and building technical knowledge.

Edited: After completing every task, my supervisor critiqued all the issues and fully answered my questions. His comments helped me reflect on and improve my performance. He constantly encouraged me to address my objectives, such as improving my interviewing skills and building my technical knowledge.

Example 5
Unedited: Procedures for electronic processing of multiple department forms are being explored and may be developed in the future.

Edited: We are exploring and may develop procedures for electronically processing of department forms.

Example 6
Unedited: When a filing is prescribed to be filed with more than one of the foregoing, the filing shall be deemed filed as of the day the last one actually receives the same.

Edited: A document is considered "filed" only when all those who are supposed to receive it actually receive it.

USING E-MAIL

Electronic communication, or e-mail, is fundamentally different from paper-based communication because of its speed and broadcasting ability. E-mail is cheaper and faster than a letter, less intrusive than a telephone call, and less hassle than a fax. With e-mail, differences in location and time zone are less of an obstacle to communication. E-mail tends to be a more open process, encouraging the sharing of thoughts and information. Because of these advantages, e-mail is common in both business and personal communications.

This is not about the mechanics of sending e-mail. Those details are different for every e-mail software package and are better handled at the local level. We will focus here on the content of an e-mail message: how to say what you need to say. While these guidelines may show you how to be a nice person while communicating, the real purpose is to show you how to be more efficient, clear, and effective.

In a paper document, it is absolutely essential to make everything completely clear because your audience may not have a chance to ask for clarification. With e-mail documents, your recipient can ask questions immediately. E-mail thus tends, like conversational speech, to be sloppier than communications on paper. This is not always bad. It makes little sense to slave over a message for hours, making sure that your spelling is faultless, your words eloquent, and your grammar beyond reproach, if the point of the message is to tell your coworker that you are ready to go to lunch.

You should be aware, however, of when your e-mail can be casual and when it has to be meticulous. For the latter, your facility with language is crucial so that your correspondents understand your message. Because e-mail messages lack face-to-face cues such as facial expressions, body language, tone of voice, and a shared environment, you must rely on language itself to communicate clearly. So you need to choose words carefully and spell them correctly. You need to avoid anything—for example, humor or sarcasm—that could diminish clarity. Finally, you need to proofread your message.

Another difference between e-mail and older media is that what the sender sees when composing a message might not look like what the reader sees. Your vocal cords make sound waves that are perceived basically the same by both your ears and those of your audience. The paper that you write your love note on is the same paper that the object of your affection sees. But with e-mail, the software and hardware that you use for composing, sending, storing, downloading, and reading may be completely different from what your correspondent uses. Your message's visual qualities may be quite different by the time it gets to someone else's screen. Thus, your e-mail compositions should be different from both your paper compositions and your speech. Even when writing business communications, make it personal. Write for an audience of one. Use common conversation, and avoid formal speech.

Here are some suggestions for writing effective e-mails:

- Use useful subject lines. Many people are swamped with e-mail; useful subject lines will help them sort their mail and will help your message get noticed. Your subject line determines whether your e-mail gets read or not.

- Give complete information, but try to keep your message short. That will help your readers respond to you most effectively.

- If you are responding to previous e-mail, quote the text that you are referring to. Instead of sending an e-mail that says "yes," indicate the question you are replying to with your response. When troubleshooting, we often include all of the details so no significant detail is missed.

- Replace pronouns with detail. Carefully examine the pronouns in the first few sentences of your e-mail. If they do not refer to objects specifically stated in your e-mail, replace them with words that are more descriptive.

- Use shorter paragraphs. They are easier to read than long ones in most e-mail programs. It is a good idea to pare your paragraphs down to a couple of sentences, using a line of space between them.

- Use capital letters sparingly. In e-mail, ALL CAPITAL LETTERS READ LIKE SHOUTING; please do not shout unless you feel STRONGLY about something.

- If there is something you want the reader to do, include a call to action. Tell people exactly what you want them to do. Do not leave them wondering what to do next. Point them to your "most desired action." Otherwise, they may just close your message and forget it.

- Never send an e-mail when you are angry or upset. E-mails can become permanent admissible records that can be attained by anyone through the open records act. So be careful what you send to whom. If you are unsure, let it set overnight and reread it the next day when your mood may be different.

For more information, check any of the many sources on the Web. Using your favorite search engine, enter "using e-mail."

PUBLIC RELATIONS
Dealing with Individuals

So far our discussion on communications has focused on internal communications—communications with others within our organization. Let us turn now to how well we deal with those outside the organization.

Most likely, we do fairly well in one-on-one communications during an emergency situation. Almost everyone in public service realizes that we have to be civilized with others, sometimes even when they are uncivilized. Unfortunately, we often fall short when dealing with citizens who seek our assistance for nonemergency needs and when speaking to members of the press. Let us consider the citizens' needs first.

A common situation occurs when citizens call or visit a fire station asking for information. They may have a concern or a complaint or just want information. Company officers should be prepared to handle these situations as well. In many cases, the citizens will visit the fire station because they identify you as a very visible and accessible part of local government. Maybe they just want directions. We should be able to do that—we should know our city well. Maybe they want to get a burn permit or to ask a question about the fire code. These are clearly issues for the fire department. On the other hand, they may want to complain about their water bill. It would be easy to dismiss them with "That's not our job." How would you feel in a similar situation?

Good, effective, active listening skills are needed here. Invite them in, offer them a cup of coffee, and take them to your office. Take time to listen to them and help them get a handle on their problem. Show understanding and support. If an apology is warranted, make it on the behalf of your government. It does not cost anything, and in many cases, the citizens will feel a lot better, maybe even satisfied. Sometimes simply telling them where the appropriate government office is located can help them.

In many cases, the issue can be resolved with a telephone call. If appropriate, make the call for them. You probably have a better knowledge of local government than they do, and your listing of government telephone numbers is probably better than the one in the local telephone book. In any case, take a personal interest in seeing that their concerns, complaints, or inquiries are resolved satisfactorily. It is easy to get defensive when people are complaining, and it is easy to tell them that it is not your responsibility. Make it your responsibility to satisfy their needs. Provide understanding, offer assistance, and try to help make the situation better.

Dealing with the Media

Now let us look at the needs of the media. The media (newspapers, radio, and television) provide a link to the public. Instead of speaking to one individual, you might be speaking to 100,000. Sounds a bit intimidating; and you might say, well, we have a public relations officer who takes care of the media. Most departments do have a public relations person, but that individual may not be present when the press wants its story. So it falls back on the company officer.

Certainly, you want to see your organization favorably presented on the late evening news or in the morning newspaper. Fire departments, like it or not, are in the news-making business. And we can safely add that fire departments engage in public relations activities every day. Some of these public relations opportunities are the result of an incident that is covered by the press. (Remember, members of the press are listening to the dispatch frequencies for fire, EMS, and police; they know what is going on.) Sometimes we seek the media's attention to our activities: fire prevention activities, the graduation of a new class of rookie firefighters, a new fire station, and similar activities we want to share with the citizens we serve (see **Figure 2-11**).

But many departments, or at least some individuals within these departments, shun the press. Officers (and firefighters) are told that only the chief talks to the media. If the chief and the media do not talk, they do not get a story, and we are disappointed that our late-evening efforts to save someone's house are not on the front page of the morning newspaper.

FIGURE 2-11 Television crews will show up at the scene of a large fire and want an interview. An officer should be designated, briefed, and prepared to tell the reporter what happened. While the fire itself is a tragic event, a well-presented story provides an opportunity to present a very positive image of the fire department to a very large audience.

There are two issues here—first, we have to understand that the press is looking for a story, and members of the press will talk to anyone we put in front of the camera provided that the person is articulate and reasonably well informed. Second, we have to realize that inaccuracies or incompleteness in a story may be our fault. If we give the press good information, it will usually report good information.

As company officers, especially the first company officer to arrive at the scene of any emergency, you are most likely the best-informed individuals available. Given a few moments to compose your thoughts, you can provide an excellent perspective on what happened. At large-scale events, with a fully staffed Incident Management System, the Public Information Sector will help with this activity, but for the average event, you may be not only the best informed but also the most available person present.

Our space here limits the amount of detail we can cover. The topic is worthy of at least a chapter, if not more. In fact, several good books are available on this topic. (Please see the listing at the end of the chapter.)

Let us conclude this section by offering a few pointers to help company officers:

- First, build a positive relationship with members of the media. You likely know who they are; you see their faces or read their articles every day. Let them get to know you. Invite them into your fire station. Share a training activity or a meal. Some departments routinely offer ride-along opportunities. (Make sure you clear this sort of thing in advance.)

- Remember that it is the media's job to report the news. If the fire department makes a mistake, you might expect the media to report it, just as, you hope, the media reports your successes. If you have done a good job with the items listed previously in terms of building a good relationship with members of the media, they might be more willing to hold off their report until you have had a chance to tell them your side of the story.

- Learn how to give members of the media the kind of information they want. They do not want technical stuff, and they do not want gruesome stuff. But there is much you can say. If upon entering the house, the smoke was so thick that you could not see your hand a foot in front of your face, tell them that in just so many words. If the heat made you feel as if you were in a furnace, use those words. The reporters and, more important, their audience can relate to the conditions you experienced.

- The most important take-away point when dealing with the media is to follow your department's guidelines. Attempts to overdramatize the incident

FIGURE 2-12 Representatives of the press frequently interview fire investigators and other officers at the scene of fires and other emergencies. Try to include a Safety Message in the interview. (*Photo courtesy of Asheville Fire and Rescue Department*)

should be avoided, and company officers should follow simple department media guidelines—for example, describe what crews found upon arrival and explain how crews mitigated the situation. Company officers should try to relay to the citizens a safety message when giving an interview as well (see **Figure 2-12**).

- Stick to the facts. If you are not certain about how a fire started or why it may have spread so fast, say you don't know or that it is under investigation. But do not editorialize or offer up an unfounded opinion. Bad information is worse than no information!

- Remember, anyone can capture pictures, video, and audio. You and your crew are always "on stage." Conduct yourself accordingly!

The News Release

The news release is a standard form of communication used by organizations to alert the media to news pertaining to the organization's activities. News releases are often used to announce a new product or service. Public organizations, such as fire departments, often use a news release to announce public events.

The news release is usually printed on a form that makes it conspicuous (see **Figure 2-13**). The text includes a description of the event, date, time, and location of the first session and any pertinent background information. To be successful, the news release should be no more than one page and should be sent so that the information can be processed and published in time to benefit the public. Copies are normally mailed to the media (television, radio, and newspapers that serve your community).

USFA and IAFF Release Voice Radio Communications Guide for the Fire Service

Contact: USFA Press Office: (301) 447-1853

Emmitsburg, MD. – The United States Fire Administration (USFA) has completed a project with the International Association of Fire Fighters (IAFF) to study what important areas of safety and technology discussed in the USFA manual *Fire Department Communications Manual - A Basic Guide to System Concepts and Equipment (FA-160)* needed to be updated or revised, as well as what topics and technology related to fire department communications not discussed in the manual needed to be added since its development in 1996. This joint USFA and IAFF study was conducted with support from the Department of Homeland Security's (DHS) SAFECOM Program Office.

"The need for an understanding of today's modern communications concepts and technology for firefighter and citizen life safety and operational effectiveness remains as valid, if not more so today, than it did when this USFA document was first produced in 1996," said U.S. Fire Administrator Greg Cade. "USFA was pleased to work with DHS SAFECOM and the IAFF on this study to provide critical information to the fire service."

The new manual, *Voice Radio Communications Guide for the Fire Service*, provides updated information on communications technology and discusses critical homeland security issues and concepts, such as SAFECOM, that did not exist when the original manual was first published. It also provides a wide fire service audience with a minimum level of familiarity with basic communications issues such as hardware, policy and procedures, and human interface.

"The safety of both firefighters and citizens depends on reliable, functional communication tools that work in the harshest and most hostile of environments," said IAFF General President Harold Schaitberger. "The IAFF was pleased to work with USFA on this important project."

Further information about this partnership effort may be found under the Research section of the USFA Web site.

FIGURE 2-13　Sample news release. (*Courtesy of U.S. Fire Administration*)

A good news release will:

- Be on recognizable letterhead
- Contain the name, telephone number, and e-mail address of the contact person
- Be simple and lack technical jargon
- Be well-written, free of factual or grammatical error
- Contain all the information needed to write a story about your event

The first paragraph should be a summary of what the news release is all about. In three or four lines, you should try to answer all the questions and sell your story. Try to address the what, when, where, and who. For example:

The Plain City Volunteer Fire Department invites the public to its annual fundraising barbeque at the County Fairgrounds on Saturday, October 8, from 4 until 8 P.M.

In one sentence we have answered all of those questions.

The next several paragraphs should follow a logical order. Use simple sentences, and short paragraphs. Expand upon the basics as needed—the what, where, when, why, who, and, if appropriate, the how. Cover all the facts, but don't bury the reader in details. You also should ensure the news release has value—show that a community need or good is being met, or that your people or organization are providing a valuable service.

As with all forms of written communication, you should always proofread your work before making copies and mailing to the media. Make sure that the release meets all of the department's policies and procedures that govern the release of information. In some instances, it may be necessary to have the department's legal counselor approve the coverage before it is released to the media. There have been some recent cases where confidential information was released to the media without approval. This is why it is essential that departments have policies in place for the release of information to the media.

ETHICS

There are some things you should not talk to the media about. Internal issues, especially personnel matters, are properly referred to the public information officer or to the fire chief. Other examples would include statements regarding a change in the department's policies, budget information, and staffing or unit deployment changes. Clearly these are not "breaking news" types of events or stories. They should be handled during business hours by the staff at fire department headquarters.

Whether you are speaking to one citizen at your fire station or representing your department in front of a television camera, your job is to address the concern or inquiry accurately, courteously, and in accordance with your organization's policy.

EMERGENCY COMMUNICATIONS

The technology of communications has changed over the years. In the early twentieth century, fire alarms were transmitted by fire alarm boxes, and fire companies were dispatched by telegraphing the box number to the fire stations. At the fireground, officers used a megaphone to convey commands. Things have changed a bit.

Today, reports of fire are transmitted by automatic detection/reporting equipment and by citizens using a telephone. The information is processed by an emergency communications center with elaborate computer support and instantly forwarded to fire stations. Voice radios and data terminals connect all responding units, and portable radios are used extensively at the scene. It would seem as if this technology has solved all of our communications problems. But we still read reports about communications problems during emergency operations.

As fire officers, you should understand the basics of the communications system in which you operate. Starting with the emergency communications center, you should understand how calls are received and processed and how companies are assigned and dispatched. Regardless of the source of the call, the information about the location and nature of the emergency must be quickly and accurately gathered and rapidly followed by a timely dispatch or appropriate resources. The time allocated for processing the call is quite precious. During this time, a life might be fading or a fire might be growing (see **Figure 2-14**).

Computer-aided dispatch can help the dispatcher get accurate information and assign appropriate resources to the call. Computer-aided dispatch can provide occupancy information to responding units, assisting in their size-up activities, track the

FIGURE 2-14 The dispatcher is a vital link in the emergency communications process. They provide the first notification of the event to responding units, provide information while the units are en route, and provide on-scene units with additional information and resources.

IMPROVING FIREFIGHTER COMMUNICATIONS

Several recent incidents involving firefighter fatalities demonstrate that, despite technological advances in two-way radio communications, important information is not always adequately communicated on the fireground or emergency incident scene. (Several of these incidents are discussed later in this text.) Inadequate communication has a definite negative impact on the safety of emergency personnel and may contribute to injuries or deaths of firefighters, rescue workers, and civilians. Several key issues follow:

■ *Unsuitable equipment:* The primary communication problem reported by firefighters and company officers is the difficulty in communicating when using full personal protective equipment, including SCBA. The majority of portable radios currently used by fire departments are ill suited for the task.

■ *Portable radios needed for all firefighters:* Despite some technical limitations, portable radios are a proven lifesaver during fires and emergency incidents; they should be considered a critical item of personal protective equipment akin to SCBA. Ideally, every firefighter working in a hostile environment should have a portable radio with an emergency distress feature.

■ *Little attention paid to human factors:* There is a dearth of available literature pertaining to the impact of human factors on effective fireground communication.

■ Furthermore, while fire departments devote substantial time to manipulative skill training, relatively little training is provided to help firefighters develop stress-tempered communication skills.

■ *Importance of active listening:* All firefighters on an emergency incident should actively monitor their radios for important information at all times, not just when specifically queried. Communications should be emphasized as an essential part of firefighters' functioning as a tactical team, not just operating as individual entities.

■ *Standard message formats and language:* Fire departments can enhance fireground communication by creating standard message formats and keywords used consistently. Plain English is usually preferred over codes, especially when transmitting a complex message.

■ *Tiered message priority:* Keywords to prompt immediate action can be tiered based on their priority; for example, "Mayday" signals a life-or-death situation, whereas "urgent" may be used to signify a potentially serious problem. Such message headers prompt each crew's listening priorities and radio discipline.

■ *Attention to cultural factors:* If necessary, firefighters are not usually reluctant to circumvent the chain of command to report critical safety issues. There may be greater hesitation to communicate problems in completing an assigned task; however, this is usually due to a lack of situational awareness and not to a fear of reprisal from other members. Studies on firefighter communications show that sometimes the culture of bravery in the fire service is reflected in a reluctance to communicate quickly enough when help is needed.[3]

responding units, and provide dispatchers information regarding the status of the remaining units.

At the company level, firefighters must get information quickly and accurately to respond promptly. In Chapter 12, we will explore some of the things that can be done to safely shorten the response time.

Once the response is under way, radio communications help the process immeasurably. Responders will hear other companies responding, and the first-arriving company will mark on the scene, providing a benchmark for all that follows. One of the very first actions needed is a good size-up report, an accurate description of what is going on. Here clear communications are absolutely essential. The first arriving officer must describe in a sentence or two what the scene is, what the company is doing, and what help, if any, is needed. We will talk more about this in Chapter 13, but the point here is that, once again, a company officer must have good communications skills.

At the scene, radio communications help connect people and equipment that may be spread over long distances. Gone are the days of shouting commands (with or without a megaphone). In fact, a well-run emergency scene will be amazingly quiet. Ideally, all firefighters at the scene will have a radio. They can receive directions and information, call for help if needed, or transmit vital information that otherwise would be lost. But the fact that everyone has a radio leads to two problems. Sometimes there is too much radio traffic. This can be corrected by good training for company officers and effective leadership (and apt radio discipline) at all levels. The other problem is one of human nature. When firefighters miss a message, they may be inclined to let it go, rather than to risk the embarrassment of going on the radio to ask for a repeat of information they should have received. The best way to manage this is to make sure that all messages are transmitted in a clear, calm voice and

FIGURE 2-15 Good fireground communications are vital. Radio transmissions should be made in a clear, calm voice and at a deliberate pace which conveys to all who are listening that someone is in charge, and that they know what they are doing.

with a deliberate pace. Again, good training and well-managed communications are keys to the solution (see **Figure 2-15**).

Even with the best technology, we still have communications problems. Unless someone is using one of the new self-contained breathing apparatus (SCBA) units that has a built-in microphone, communicating effectively by radio while wearing a facemask is challenging. Radio communications in large buildings is also challenging, and frequently the available radio channels are overwhelmed with radio traffic. In many cases, officers have to understand the limits of the equipment and must work with what they have.

Good communications are a critical part of effective emergency activities. As company officer, you play a vital role in this process, considering that you probably send and receive a majority of the reports transmitted during emergency activities.

THE VALUE OF GOOD COMMUNICATIONS

Melvin Byrne, Past Chief, Ashburn Volunteer Fire and Rescue, Loudon County, Virginia

A few years ago, the volunteer fire department of which Melvin Byrne was chief started experiencing rapid growth. The years preceding had seen limited change in personnel, and since most people had been in their roles for some time, they knew their job. But as the community expanded, the fire department needed to do so, too; that meant heavy recruitment.

The only problem was that the new people would show up and would not know what to do. The chief found he would have to walk each individual from one duty to the next, which used to work fine when there would be one new person every few months or so. But having so many new people to coordinate at the same time was a struggle, and it took 6 to 8 months before all were fully organized in their responsibilities. The old system—word of mouth—just was not working on this scale.

Then Byrne came up with a solution: putting in place written expectations of each position. The operational officers got together to decide what the expectations were. That meant getting in sync with the county's prerequisites and paperwork and letting new officers know even the basics of when and how to get their CPR certification.

The chief acknowledged that he should have started the process sooner and that he should have asked for more input from his officers. Good two-way communications are the key to effective management in any organization. And communications are a two-way process: Someone may have a great idea, but that person has to be able to communicate that idea. And someone else has to be willing and able to listen to the idea.

Questions

1. What are the communications issues here?

2. What are the advantages of providing written policies?

3. Chief Byrne acknowledged that he should have gotten more input. How could he have invited that input?

4. What else could he have done to get his officers involved?

5. How could the officers let their concerns be known to the chief?

LESSONS LEARNED

Good communications skills have a positive impact on every aspect of your work. Being able to communicate effectively enhances your leadership ability, helps you gain respect from your supervisors and peers, and makes you more effective in talking to the public, the media, and others. In all cases, your ability to communicate creates an impression about you, your organization, and your profession. Work to be an effective communicator. And remember—practice what you preach!

KEY TERMS

active listening   deliberate and apparent process by which one focuses attention on the communications of another

barrier   obstacle; in communications, a barrier prevents the message from being understood by the receiver

feedback   reaction to a process that may alter or reinforce that process

message   in communications; the message is the information being sent to another

receiver   in communications, the receiver is the intended recipient of the message

sender   part of the communications process, the sender transmits a thought or message to the receiver

REVIEW QUESTIONS

1. What is the communication process?
2. What is the role of the company officer in communications?
3. What are the five parts of the communications model?
4. What are the common barriers to communication?
5. How can these barriers be minimized?
6. What is meant by the term "active listening"?
7. How does your communication style indicate your leadership style?
8. Name several of the principles of effective writing.
9. What steps can you take to improve your writing?
10. What is the company officer's role in the communication process?

DISCUSSION QUESTIONS

1. What are the benefits of effective communication?
2. When writing, what steps can you take to enhance the finished product?
3. How do improved communication skills help you in your professional life?
4. How do improved communication skills help you in your personal life?
5. List any barriers to effective communication found in your organization or workplace.
6. How has the communication process changed the way emergency response organizations are notified of an emergency?
7. How has the communication process changed for emergency response organizations in the way they are dispatched to the emergency?
8. How has the communication process changed for emergency response organizations are notified of an emergency the way they communicate at the scene?
9. While communication technology has improved in recent years, many issues are still apparent. What are the issues in emergency communication in your community?
10. What do you think the future will bring in the way of changes in communication technology?

ENDNOTES

1. There are several ways to assess proficiency in communications. One alternative would be to ask you to successfully complete a college-level English class. This idea has a lot of merit and many departments use this approach.

2. Steven Covey, *7 Habits of Highly Effective People* (New York: Fireside, 1989).

3. Adam Thiel, *Improving Firefighter Communications* (Emmitsburg, MD: U.S. Fire Administration, 1999).

ADDITIONAL RESOURCES

"A Beginner's Guide to Effective E-Mail." http://webfoot.com/advice/email.top.php

Bigrigg, Brad, and Cynthia Tustin. "From Note Taking to Note Making." *Fire Engineering,* July 2006.

The Chicago Manual of Style. 15th ed. Chicago: The University of Chicago Press, 2003.

Curtis, Timothy. "Personality Puzzle." *Fire Chief,* August 2007.

Dils, Jan. "Writing Fire and Non-fire Report Narratives," 1990. Published by the author, P.O. Box 50544, Pasadena, CA 91115.

Dunne, Thomas. "In the Spotlight: Preparing for the Media." *Fire Engineering,* March 2008.

Fowler, H. Ramsey, and Jane E. Aaron. *Little, Brown Handbook.* 10th ed. New York: Addison Wesley Education Publisher, Inc., 1998.

Hunschuh, Rita, Bob Hunschuh, and David Hunschuh. *National Media Guide for Emergency and Disaster Incidents.* Durham, NC: National Press Photographers Association, 1995.

Jerz, Dennis. "Writing Effective E-Mail: Top 10 Tips." jerz.setonhill.edu/writing/e-text/e-mail.

Lasky, Rick. "Starting the Shift Off Right." *Fire Engineering,* March 2007.

Markley, Rick. "The Great Communicator." *Fire Chief,* October 2008.

"Media Primer for the Fire Service." www.nfpa.org/PressRoom/Primer.

Sovick, Mary. "Importance of Workplace Writing Skills in Fire and EMS Services." *The Instructor,* April 2004.

Thiel, Adam. *Improving Firefighter Communications.* Emmitsburg, MD: U.S. Fire Administration, 1999.

Turabian, Kate L. *A Manual for Writers of Term Papers, Theses and Dissertations.* 5th ed. Chicago: The University of Chicago Press, 1996.

Werner, Charles. "Communications: The 'Avalanche Effect,'" *Firehouse,* March 2004.

Writing Effective E-Mail. www.customnet.on.ca.

3

The Company Officer's Role in the Organization

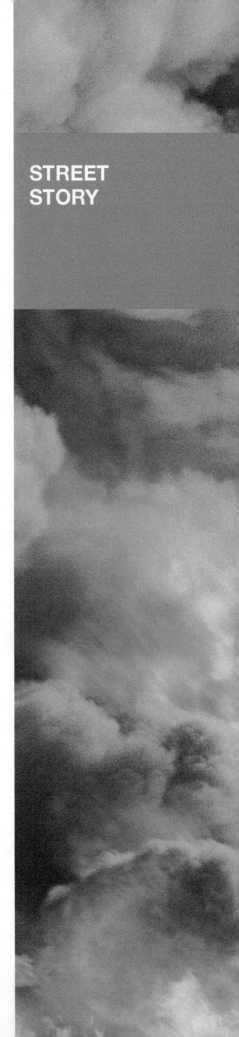

As a company officer, I serve many functions in my organization. It is sometimes easy to get caught up in finding and maintaining your proper place in the department's organizational scheme. In my fire service career, I managed to learn from other company officers some very valid truths that I have carried with me as I progressed to a leadership role.

During a fire in a single-story commercial structure, I was crawling behind my lieutenant into the building through the dark smoke-filled office area, looking for the seat of the fire. Flames were coming out the floor registers, and the building was getting hotter and hotter.

We were crawling toward what appeared to be an open area, and the floor seemed to be slanting. It was hard to tell in the severe smoke condition, but I kept following him, still searching for the fire. All of the sudden in a muffled yell I heard, "Get back, get the hell back, back. . . ." Immediately, we retreated as fast as we could. Once outside, I heard the blasting of truck air horns, which was our department's structure evacuation signal. As it turns out, the floor inside the building really was slanting. In fact, in the room my lieutenant and I were headed for, the floor had fallen into the basement! We almost went with it.

My lieutenant came up to me after the fire was contained and apologized for "cussing" at me. I just looked down and shook my head and said, "You don't need to apologize. You just saved my life. Thank you."

From that day forward, I knew that whatever place I occupied on my department's organizational chart, or whatever my job descriptions spelled out as my duties, my first and foremost responsibility over all others was the safety of the people on my team. Whenever I get bogged down with the various roles I play as a company officer, I will never lose sight of this most important "role" that I play: to protect the safety of my team members.

—Sally L. McCann-Mirise, Lieutenant, Liberty Twp.
Fire Prevention, Powell, Ohio

LEARNING OBJECTIVES

After completing this chapter the reader should be able to:

3-1 Describe the organizational structure of the department.

3-2 Describe the purpose of an organizational chart.

3-3 Identify the jurisdictional boundaries of the agency.

3-4 Define the terms *laws, statutes, regulations,* and *ordinances.*

3-5 Describe the law-making process at the local level.

3-6 Explain how the state/provincial and/or federal laws may affect local laws and ordinances.

3-7 Describe the different types of local government.

3-8 Identify the chain of command within the local government structure.

3-9 Describe the major departments and divisions that comprise local governmental structure.

3-10 List the functions of local, state, and federal bureaus, divisions, and agencies that affect the fire service.

3-11 Define *enabling* and *regulatory legislation.*

3-12 Explain the socioeconomic and political influences that impact the fire service.

3-13 List the allied agencies that the fire department may need to cooperate with to address community issues.

3-14 Correlate the mission and goals of the agency to those of allied agencies.

3-15 Categorize by function, the allied agencies within the community that assist the fire department.

3-16 Identify the geographic configuration and composition of the department's response district.

3-17 Classify the demographic composition of the department's response district.

3-18 Describe the strategic plan for the agency.

3-19 Identify each level of supervision and management in the department.

3-20 Explain the role of each position on the chart.

3-21 Distinguish between line and staff functions.

3-22 Differentiate the roles of operation (line) managers and administration managers.

3-23 Describe the chain of command of the department.

3-24 Identify those routine administrative functions pertinent to unit-level supervision.

3-25 Explain the company officer's role in planning and implementing change.

3-26 Plan the steps required for implementing a change of policy and/or procedure.

3-27 Determine how the new policy affects existing policy.

3-28 Identify the differences between the new and existing policies.

3-29 Describe the process for evaluating change.

3-30 Identify the policies and procedures related to unit-level administrative functions.

*The FO I and II levels, as defined by the NFPA 1021 Standards, are identified in different colors: FO I = black, FO II = red.

INTRODUCTION

There are more than 30,000 fire departments in this country. Fire departments range in size from small departments with a dozen members and one vehicle to departments with thousands of members and hundreds of vehicles, but they are all organizations. Despite the vast variation, many characteristics are common to all fire departments. Understanding the characteristics of effective organizations will enhance your role in your organization.

Looking at **Table 3-1,** we see a list of the largest fire departments in the United States and Canada. While these departments represent less than 1 percent of the fire departments, they serve over 15 percent of the population.

Although large cities are protected by large and often well-known fire departments, most of America is protected by small fire departments, many of them comprised entirely of volunteer members. Most of the departments are staffed by volunteer members, have

TABLE 3-1   Largest Fire Departments in United States and Canada

Jurisdiction	Fire Deptartment Members	Population Served	Total Calls
United States			
New York, NY	15,600	8,000,000	1,854,660
Chicago, IL	5,101	2,800,000	667,375
Los Angeles County, CA	4,780	4,040,392	284,744
Houston, Tx	4,025	2,016,582	278,345
Los Angeles City, CA	3,694	4,018,000	363,022
Miami-Dade, FL	2,541	1,715,574	231,581
Washington, DC	2,071	588,000	153,778
Phoenix, AZ	2,060	1,551,009	145,279
Prince Georges Cty, MD	2,037	828,770	131,007
Memphis, TN	1,892	646,356	115,714
Baltimore City, MD	1,679	675,000	235,000
Dallas, TX	1,670	1,210,390	113,895
San Francisco, CA	1,658	800,000	98,959
San Antinio, TX	1,622	1,322,900	159,965
Columbus, OH	1,549	769,934	131,007
Boston, MA	1,491	650,000	71,630
Fairfax County, VA	1,392	1,077,000	92,087
Jacksonville, FL	1,289	817,210	113,710
San Diego, CA	1,153	1,300,000	104,191
Nashville, TN	1,152	580,450	95,910
Seattle, WA	1,100	580,000	79,827
Charlotte, NC	1,083	658,848	87,796
Baltimore County, MD	1,082	809,300	113,800
Honolulu, HA	1,054	371,657	38,669
Atlanta, GA	1,038	420,000	61.956
Canada			
Toranto, ON	3,144	2,700,000	142,515
Montreal, QC	2,671	1,854,442	53,786
Winnipeg, MB	1,165	706,800	79,817
Halifax Region, NS	1,129	382,200	11,540
Edmonton, AB	944	730,372	33,569

Courtesy of Firehouse Magazine.

PROFILE OF THE U.S. FIRE SERVICE

NFPA estimates that there were approximately 1,148,800 firefighters in the U.S. in 2007.

Of the total number of firefighters 323,350 or 28% were career firefighters and 825,450 (72%) were volunteer firefighters. Most of the career firefighters (74%) are in communities that protect 25,000 or more people. Most of the volunteer firefighters (95%) are in departments that protect fewer than 2,500 people.

There are an estimated 30,185 fire departments in the U.S. Of these, 2,263 departments or 8% are all career, 1,765 (6%) mostly career, 4,989 (17%) are mostly volunteer and 21,168 (70%) are all volunteer. In the U.S., 13,221 or 44% of departments provide EMS service, 4,472 departments or 15% provide EMS service and advance life support, while 12,492 departments or 41% provide no EMS support.

Source: U.S. Fire Department Profile through 2007. Quincy, MA: National Fire Protection Association, 2008. Reprinted with permission.

one or two pieces of equipment, and serve communities of fewer than 10,000 people.

Even in this environment, good leadership and management skills are important (see **Figure 3-1A-D**).

GROUPS AND ORGANIZATIONS

Formal Organizations

Groups exist whenever two or more people share a common goal. Organizations are groups of people. Typically, these people share a common goal, have formal rules, and have designated leaders. Obviously this is a rather simplistic description of an organization, but organizations are really nothing more than just large groups of people. But organizations can easily become quite complicated. Just think about all the one-on-one relationships that might be possible in an organization of five or ten people. As organizations grow, the number of relationships multiplies quickly.[1]

We would like to think that each of these relationships is positive and that the outcome of any interaction between two or more members of the organization has positive results, but we know from experience that this is not always the case. Good communications plays an important role in organizations. Understanding the relationships within the organization is also important to the success of the individuals, their relationships, and the success of the entire organization.

Understanding organizations is more than just understanding our fellow members. We look at the human aspect of organizational relationships in Chapters 6 and 7, but here let us look at organizations from a theoretical point of view and examine some of the principles of organizations. In due time, we will see how people fit into and behave within organizational structures.

When studying any organization, we should start by looking at the organizational chart. Organization charts can be a lot more than a bunch of boxes connected with lines. If the chart is prepared properly, the actual structure of the formal organization is reflected in the chart. The organizational chart indicates the formal lines of authority and responsibility, lines of communications, and so forth. We could take all persons in the organization to a large field, arrange them neatly according to the way they fit into the organization, and take their picture, but this action is usually not practical. An organizational chart provides a similar graphic representation on paper.

NOTE

The organizational chart indicates the formal lines of authority and responsibility.

An organizational chart is only a graphic representation of what the organization should look like. In larger organizations, there are always changes, as new activities are added and people are transferred from one assignment to another. And there are always a few vacancies due to leave, promotions, and retirements. Despite these fluctuations, it is always a good idea to have an organization chart and to have it posted where members can see it. Keeping it up-to-date and putting the members' names in the blocks bring the chart to life and help everyone associate the organization with the real people who occupy its ranks. From the chart we should be able to see who works together, as well as to see the lines of communications and authority within the organization.

Understanding organizational structure and one's place in the organization is important.

Informal Organizations

So far we have talked only about the formal organization. In any workplace, there is also an informal organization. Informal organizations are not bad. Informal organizations exist because certain people like certain other people. They chat over a cup of coffee or have lunch together. Sometimes members even take time to get together after work for softball or bowling. Trying to define this type of organization

(A)

(B)

(C)

(D)

FIGURE 3-1 In many communities, the fire station is a prominent structure. Some are quite fancy and others rather simple. All provide a shelter to the fire department that provides service to the community.

and represent it with a chart would probably be impossible. Although we cannot represent it on paper, informal organizations really do exist concurrently with most formal organizations.

AUTHORITY FOR THE LOCAL FIRE DEPARTMENT

In the past, fire officers and, sometimes, even fire chiefs did not realize the importance of understanding how government works. Fire chiefs went to city hall only when the mayor called them, and they generally took a low profile in anything pertaining to the

government process. Fortunately, that has changed in most jurisdictions.

The fire department is a part of local government. We could think of our organization as being at the center of the universe, but closer study will reveal that we are just a small part of something that is much larger (see **Figure 3-2**). Let us review some of the things we learned in school about government.

Government can be defined as a political organization comprised of individuals and institutions authorized to formulate public policy to improve the public welfare. To that end, governments are empowered to establish and regulate the action of the people within their jurisdiction.

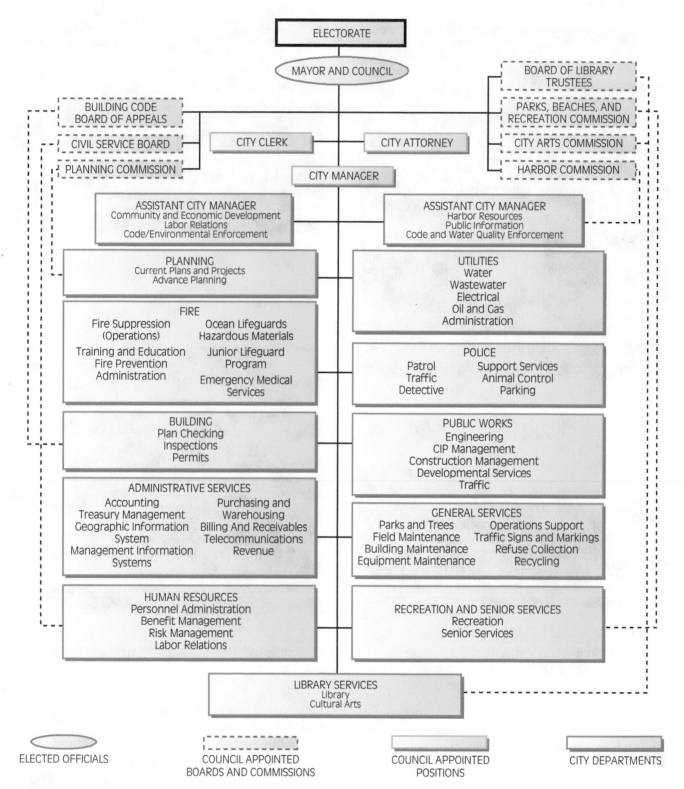

ELECTORATE

MAYOR AND COUNCIL

BOARD OF LIBRARY TRUSTEES

BUILDING CODE BOARD OF APPEALS

PARKS, BEACHES, AND RECREATION COMMISSION

CIVIL SERVICE BOARD

CITY CLERK

CITY ATTORNEY

CITY ARTS COMMISSION

PLANNING COMMISSION

HARBOR COMMISSION

CITY MANAGER

ASSISTANT CITY MANAGER
Community and Economic Development
Labor Relations
Code/Environmental Enforcement

ASSISTANT CITY MANAGER
Harbor Resources
Public Information
Code and Water Quality Enforcement

PLANNING
Current Plans and Projects
Advance Planning

UTILITIES
Water
Wastewater
Electrical
Oil and Gas
Administration

FIRE
Fire Suppression (Operations)
Training and Education
Fire Prevention
Administration

Ocean Lifeguards
Hazardous Materials
Junior Lifeguard Program
Emergency Medical Services

POLICE
Patrol
Traffic
Detective

Support Services
Animal Control
Parking

BUILDING
Plan Checking
Inspections
Permits

PUBLIC WORKS
Engineering
CIP Management
Construction Management
Developmental Services
Traffic

ADMINISTRATIVE SERVICES
Accounting
Treasury Management
Geographic Information System
Management Information Systems

Purchasing and Warehousing
Billing And Receivables
Telecommunications
Revenue

GENERAL SERVICES
Parks and Trees
Field Maintenance
Building Maintenance
Equipment Maintenance

Operations Support
Traffic Signs and Markings
Refuse Collection
Recycling

HUMAN RESOURCES
Personnel Administration
Benefit Management
Risk Management
Labor Relations

RECREATION AND SENIOR SERVICES
Recreation
Senior Services

LIBRARY SERVICES
Library
Cultural Arts

ELECTED OFFICIALS

COUNCIL APPOINTED BOARDS AND COMMISSIONS

COUNCIL APPOINTED POSITIONS

CITY DEPARTMENTS

FIGURE 3-2   This is a representative organizational chart for a city. The citizens are at the top of the chart; they elect the council to represent them. The council will select the city manager who usually hires the assistant managers and department heads. (*Graphic courtesy of ICMA Washington, D.C.*)

Our System of Government

Our governments are built around the U.S. Constitution. That remarkable document establishes the form of our federal government. It also limits the role of the federal government. The Constitution authorizes the three branches of the federal government—the executive, the legislative, and the judicial. The federal government provides a variety of services to promote the general welfare of the citizens of the United States. To this end, it conducts foreign relations, controls foreign and interstate commerce, provides currency, provides for the national defense, and collects taxes to pay for these activities.

The executive branch, comprised of the president and the various departments, provides most of these services. One of the newest departments in the federal government is the Department of Homeland Security. That department includes the Federal Emergency Management Agency, the U.S. Fire Administration, and many other organizations that provide security at our borders and within our homeland.

The legislative branch of the federal government is the U.S. Congress, made up of the House of Representatives and the Senate. Congress enacts legislation. Legislation that passes both houses of Congress is sent to the president for approval. If approved, it becomes a law.

The executive branch creates laws, too. The various departments within the executive branch create rules and regulations that affect all of us. The Internal Revenue Service collects taxes to pay for government services. Under the authority granted to it by Congress, it publishes and enforces rules pertaining to the tax code. Likewise, the Equal Employment Opportunity and the Occupational Health and Safety Administration, both a part of the Department of Labor, have rules and regulations that affect nearly every workplace. These rules are generally referred to as **administrative laws.** As we will see later in this book, administrative laws are very important to us as citizens and to us as a tool in fire prevention.

The U.S. Constitution also provides that all powers not specifically established for the federal government are reserved to the states and to the people. Thus, a great part of the burden for providing services to citizens is left to state government.

Most states are governed by a government that is similar in structure to the federal government. They have an executive, called the governor, and the various departments that report to the governor. State governments also have a legislative body, usually made up of two bodies, and a judicial system. The state also influences our lives by enacting laws and regulations that affect our daily lives.

Like the federal government, the state governments turn many of the details of governance and providing services to the citizens over to local governments. Cities, towns, and counties represent local governments. Local governments also have an executive and a legislative branch and may have a judicial branch. The executive branch is usually made up of a city or county manager and various departments. Local governments can enact laws in the form of local ordinances. A local ordinance is a law usually found in a municipal code. In the United States, these laws are enforced locally in addition to state law and federal law. The fire department is usually one of the departments of local government.

THE LAW—DEFINED

There are a variety of definitions of law, but for our purposes we will consider a *law* to be that which must be obeyed subject to sanction or consequences by the government. There are a wide variety of things that meet such a definition. For example, there are laws that authorize firefighters to order buildings and areas evacuated.

The United States Congress is responsible for passing laws. The laws that Congress passes are called *statues.* Federal statutes are organized into a large body of laws called the United States Code. State legislatures enact laws which are also called statutes or *general laws.* State statutes must comply with the United States Constitution as well as the state constitution. On the local level, the legislative body of a municipality, such as city or town council, has the authority to pass laws called *ordinances.* Ordinances must comply with both the state constitution and the United States Constitution. In addition, local ordinances can not violate any state statues.

Source: Definitions from *Legal Considerations for Fire and Emergency Services* by J. Curtis Varone (Clifton Park, NY: Delmar, Cengage Learning, © 2007).

Types of Local Government

At the local level, there are five forms of government: council-manager, mayor-council, commission, town meeting, and representative town meeting.

Council-Manager

In the council-manager form of government, the council is the governing body of the city elected by the public, and the manager is hired by the council to carry out its policies. The council usually consists of five to nine members including a mayor (or council president) who is either selected by the council or elected by the people as defined in the city charter. Typically, the mayor is recognized as the political head of the municipality; however, he or she is also a member of the legislative body (council) and does not have the power of veto. The size of the council is generally smaller than that of a mayor-council municipality, and council elections are usually nonpartisan.

The council provides legislative direction, whereas the manager is responsible for day-to-day administrative operation of the city based on the council's recommendations. The mayor and council as a collegial body are responsible for setting policy, approving the budget, and determining the tax rate. The manager serves as the council's chief advisor. Managers also serve at the pleasure of the council and are responsible for preparing the budget, directing day-to-day operations, and hiring and firing personnel.

This form of government provides professional leadership at the highest level and is increasing in popularity. The article by former ICMA Executive Director William H. Hansell Jr. outlines some of the benefits of this form of government (see feature box, "Evolution and Change Characterize Council-Manager Government").

Mayor-Council

The mayor-council (council-elected executive counties) form of government is the form that most closely parallels the U.S. federal government, with an elected legislature and a separately elected executive. Mayors or elected executives are designated as the heads of city or county governments; however, the extent of their authority can range from purely ceremonial to full-scale responsibility for day-to-day operations. The mayor's (or elected executive's) duties and powers generally include hiring and firing department heads, preparation and administration of the budget, and, in some cases, veto power (which may be overridden) over acts of the legislature. In some communities, the mayor or executive may assume a larger policy-making role, and responsibility for day-to-day operations is delegated to an administrator who is appointed by and is responsible to the chief executive.

The council adopts the budget, passes resolutions with legislation, audits the performance of the government, and adopts general policy positions.

Commission

The commission form of government, characterized by an elected governing board that holds both legislative and executive powers, is the oldest form of government in America. A descendent of the old English shire-moot, or county governing board, the board is usually comprised of three to five members, although the number varies.

The board of commissioners is the county governing board, and it serves as the head of the government. It has responsibility for adopting the budget, passing resolutions, and enacting ordinances and regulations. A number of other officials are also popularly elected and serve as heads of some of the major county departments (although the number of these independently elected officials varies considerably). Some of the most common are the sheriff, treasurer, and clerk.

Town Meeting

In many smaller towns, especially in New England, all of the qualified voters of the town gather on a given day (usually once a year, but more often if necessary) to elect a board of officers and to make major policy decisions. The board, in turn, has the responsibility for carrying out the policies set by the citizens. In some towns, a manager or administrator is appointed to carry out the administrative operations of the town.

Representative Town Meeting

The representative town meeting form of government is structured in much the same way as the town meeting form, with the exception that a large number of citizens are chosen by the general electorate to represent them in voting. All citizens can attend the meetings and participate in debates, but only those chosen as representatives have a direct vote.

Although all of the branches of all levels of government affect us as citizens, probably the lawmaking process is the one area that is understood the least. The writers of the Constitution very carefully created the powers held by the three branches of the federal government. Their intent was to provide a system in which the powers of each of the three branches would be kept in check by the power of the other two. Thus, it takes the action of both houses of Congress and the president for legislation to become law. These laws are statutory laws—laws that become a part of the federal statutes. As we have already seen, the various departments within the executive branch publish rules and regulations called administrative laws.

EVOLUTION AND CHANGE CHARACTERIZE COUNCIL-MANAGER GOVERNMENT

William H. Hansell Jr., former Executive Director
International City/County Management Association

The 20th century could well be called the century of local government in American democracy. The American system of governance is based on a federal system, which divides the powers of sovereign governance between, on the one hand, a central or national government and, on the other, 50 state entities. The powers of the national government are limited to those specifically delineated in the United States Constitution. All other powers are reserved for the people or given to each of the 50 states.

This preference for decentralized government has been advanced by the 50 states as they empower their local governments, resulting in the strongest and most decentralized local governments in history. Throughout the 50 states, for example, there are 27 large consolidated cities/counties, 3,043 independent counties, 19,279 municipalities, 16,656 towns and townships (in 20 of the states), 14,422 independent school districts educating American children, and 31,555 special districts.

This is a significant amount of local government for 295 million people. But as the French observer Alexis de Tocqueville said in the early 1830s, "Local institutions are to liberty what primary schools are to science: they put it within the people's reach; they teach people to appreciate its peaceful enjoyment and accustom them to make use of it. Without local institutions, a nation may give itself a free government, but it has not got the spirit of liberty."

Stability and Change

By the end of the 20th century, local governments in the United States had become the most trusted level of American government. Year after year, citizens ranked local government higher than state or national government as the level from which they felt they got the greatest value for their money.

This has not always been the case. During the 1900s, local governments evolved from holding the dubious distinction of the most troubled component of the American system of governance to becoming the most trusted. This evolution has been accomplished through a series of radical changes that began in the early part of the century, continued throughout it, and is generally known as the progressive reform movement. The reforms have included:

- Instituting civil-service personnel systems.
- Promoting personnel on a merit basis.
- Mandating competitive tendering or purchasing of goods and supplies.
- Requiring performance budgeting and measurement.
- Ensuring open, transparent government that is fully accessible to citizens and the media.
- Implementing alternative service-delivery mechanisms, including contracting, volunteerism, franchising, and privatization.
- Developing the council-manager system of organizing and operating local government.

Of the reforms, the development of the council-manager system may have been the most significant step in improving the performance and credibility of local government.

The Flexible System

The council-manager form continues to prove its adaptability. The system's developers originally conceived elected councils as consisting of only five to seven members, elected at large on a nonpartisan basis. The council would select one member to serve as mayor. Initially, elected council members received little or no compensation for what was viewed as volunteer service to the community.

Data collected by International City/County Management Association (ICMA) in 1996 show that council-manager communities in the United States elect an average of six council members; however, councils consisting of 12 to 13 members have become increasingly common among larger jurisdictions. Often, some or all council members are elected to represent a particular section of the community. Seventeen percent of council-manager communities nominated and elected council members by ward or district; an additional 18.5 percent used a combination of at-large and by-district elections. Seventeen percent of responding council-manager communities also held partisan elections.

To fulfill development needs in public works and capital infrastructure, local governments initially recruited managers with engineering backgrounds or undergraduate degrees. In the book *The Rise of the City Manager,* author Richard Stillman reported that in 1934, 77 percent of responding managers with college degrees had majored in engineering and that 51 percent listed a bachelor's degree as their highest level of educational attainment.

According to information in Stillman's book and in ICMA's 1996 *Municipal Year Book,* appointed managers often are trained in the general management field of public administration, and the proportion of managers with a graduate or professional degree has risen steadily, from 27 percent in 1971 to nearly 73 percent in 1995. A manager's course of study may include such diverse topics as public finance, resource allocation, economic development, technology, intergovernmental relations, planning, public policy, and environmental and human resource management.

Roles and Responsibilities

An appointed manager constantly must be aware that the powers of the local government belong to the council. Any authority or responsibility assigned to the manager by the council or by the citizens through a local charter can be removed at any time, for any reason. The illusion of power that may appear to rest with a manager must be accepted as a temporary acquisition based on the manager's knowledge and expertise.

In fact, the manager can be removed at any time. Many managers serve without a guaranteed term or tenure, however, the number of managers with employment agreements which set the terms and conditions of employment and separation and give clear guidelines for performance evaluation continues to grow. The results of a 1998 ICMA survey showed that nearly 80 percent of responding local government managers and chief administrative officers reported having a contract or letter of agreement with their councils, compared with 14 percent in 1974.

Managers who serve a jurisdiction for an extended period come to understand this community's values, traditions, and goals. Yet, the manager must remember that the council represents the community, and they must defer to the elected officials as reflecting the policy wishes of the citizenry.

Elected Leadership

In most communities, the citizens perceive their mayor as their most visible leader, and the elected leaders of council-manager communities are no exception. In this way, mayors are analogous to corporate board chair persons.

Mayors in council-manager communities fulfill two critical leadership functions. The first is that of consensus building: the mayor unites the community's disparate constituencies so they can work together successfully. The second role held by the mayor in council-manager communities is to guide the development and implementation of policies that improve community service delivery.

Over the past 86 years, the position of mayor within council-manager communities has been strengthened in a variety of ways. Under the original understanding of the form, the mayor was selected annually by the council from among its members. Councilors frequently rotated into the mayor's position, over time leaving no clearly identifiable political leader. But ICMA's 1996 form-of-government survey showed that 62 percent of all council-manager communities now elect their mayors directly by a vote of the citizens, for a two or four year term. The people elect their political leader, and the council must accept that selection.

By 1918, there were 98 council-manager localities; in 1930, ICMA recognized 418 U.S. cities and seven counties as operating under the council-manager form. And by 1985, the number of council-manager communities had grown to 548 cities and 86 counties in the United States. Currently, 3,302 U.S. cities with populations greater than 2,500 and 371 U.S. counties operate under this system of local government.

- More than 75.5 million individuals live in communities operating under council-manager governments.
- During the past 16 years, an average of 63 U.S. communities per year have adopted the council-manager form.
- Sixty-three percent of U.S. cities with populations of 25,000 or more have adopted the council-manager form.

In recent years, some communities have chosen to strengthen the mayor's authority, providing the mayor with one or more of these powers:

- Veto actions of the council that require an extraordinary majority to overturn the veto.
- Power (1) to organize the council by assigning councilors to chair or serve on committees and (2) to assign matters to these committees.
- Power to appoint citizens to serve on advisory or quasi-judicial authorities, boards, or commissions.
- Power to receive the annual budget prepared by the manager and to present that budget, with comments and suggestions, to the council for their consideration.
- Power to make an annual report to the council and the citizens on the state of the community.
- Power to initiate the hiring and/or involuntary termination of the person serving as manager.

The manager remains the chief executive officer with clear authority to:

- Hire and terminate all senior officers and staff.
- Purchase all goods, supplies, equipment and services required by the government.
- Prepare an annual budget and financial plan for consideration by the council.
- Administer and enforce all contracts involving the local government.
- Enforce all laws adopted by the council.

Communities with a delegation of authority to an appointed manager and a strongly empowered mayor might be seen as working under a mayor-council-manager form. Under this scenario, the elected leadership comes "from the people" via the election process. The citizens oversee the operations of the government through an open and transparent system that guarantees that it will continue to be "by the people."

Day-to-day management of the government is directed by an experienced professional manager who is selected by the elected officials and who facilitates the government by operating "for the people" who live in the community and who will live there in the future. Here are the values these managers add to the system of governance for the communities they serve:

- Establishment of policy and service delivery strategies on a basis of need rather than of demand.
- An emphasis on the long-term interests of the community as a whole.
- Promotion of equity and fairness.
- Recognition of the interconnection among policies.
- Advancement of broad and inclusive citizens' participation.

For nearly 90 years, the council-manager form has successfully adapted to American community needs. Cities and counties are not static, and the changes taking place in them involve the core of our values. Professional managers and the council-manager form, however, continue to evolve so that today, as in the early 20th century, this system offers government of the people, by the people, and for the people. In short, council-manager government is a system of reform that will continue to serve communities well in the 21st century.

THE LEGISLATIVE PROCESS

Mr. Carl Peterson

Every state has a constitution that is the basic plan for its government, along with numerous supporting laws. The constitution of each state describes the process for making, deleting, or amending the laws. Laws are generally adopted because of a perceived need. The need for making a law is generally transmitted to the legislative body by a group or individual activity, publicity generated by interest groups or powerful individuals, a crisis generating action, or by personal contact.

A sponsor introduces proposed legislation, usually referred to as a bill. The more powerful a sponsor of the bill is in terms of legislative assignments, affiliations, knowledge, and seniority, the better chance of a piece of legislation becoming law. The sponsors are usually members of one of the houses of the legislature.* The sponsor of the proposed legislation presents it to the chair of their legislative body who presents it to the legislative body for its first reading. The chair assigns the bill to a committee. The chair of the committee has considerable discretion on what bills will be considered, the order of consideration, and the amount of time to debate the merits of the bill. The committee is usually a bipartisan committee, but the chair is a member of the party that controls that house of the legislative body.

Many bills die in committee and no further action will be taken on that proposed legislation during that session of the legislature; however, some legislatures have the ability to call the bill out of committee with a "super" majority. If the bill makes it out of committee, it goes to the floor of the house for a second reading. At this time the bill is debated and is subject to amendment. The presiding officer then schedules a third reading, which may or may not have significant priority depending on the attitude of the presiding officer concerning the legislation. After the third reading, the bill is put to a vote. If it is adopted, the bill, now referred to as an "act," is submitted to the other body of the legislature where the same process occurs.

If there is a modification of the bill by the other body of the legislature, a conference committee of both houses is formed to attempt to reconcile the differences. If the differences cannot be reconciled in the conference committee, the bill can die in committee and is generally dead until a later legislative session.

If the members of both houses approve the act it is presented to the governor. Every state has different rules concerning the effect of actions or inactions of the governor concerning the act at this juncture. Generally, if the legislature is still in session the governor must either sign the bill or veto it. In many states the act becomes law if the governor fails to take any action after a period of days. Vetoed bills are returned to the house and senate for reconsideration. If a "super" majority (the number differs in different states) of the legislature overrides the veto the act will become law in spite of the veto of the governor.

Fire service leaders can be a valuable resource for articulate and well-reasoned testimony in committee hearings. A registered lobbyist can be an effective tool in influencing the voting behavior of a legislator. While it is appropriate as an individual to discuss pending legislation with a legislator, in most states it is illegal to make the same request on behalf of a group or organization unless that person is a registered lobbyist. Enlisting the aid of other interested groups that may be sympathetic to the proposed legislation is an effective tool in the persuasive process.

Many firefighters are members of local fire service organizations, state organizations, and national organizations. Fire chiefs are often members of their state and regional fire chiefs' associations. Regional associations are often affiliates of national associations. These organizations are generally politically active and can also be a valuable resource for providing information and technical support. Depending on the issues, trade organizations or other professional organizations should be recruited to assist in mutually using their influence to enhance the possibility of the adoption and passage of the proposed legislation. The process of influencing the passage of desired legislation is not an exact science, but the tools that are available must be utilized to provide any hope that the desired legislation will become law.

—Mr. Peterson is a retired judge. He teaches Fire Service Law at Asheville-Buncombe Technical Community College in Asheville, NC.

Most states have a bicameral legislature, meaning that their legislative branch is comprised of two groups of elected officials, like the U.S. Congress. However, the names of these two bodies vary from state to state.

THE LAW—DEFINED

Enabling acts [or *enabling legislation*]: Statutes enacted by Congress or state legislatures that authorize the creation of administrative agencies and provide the legal authority for the agency to operate.

Regulations [or *regulatory legislation*]: Laws that are created by administrative agencies pursuant to authority delegated by the legislature. They are rules that carry the weight of law and are designed to carry out the quasi-legislative authority of the agency.

Source: Definitions from *Legal Considerations for Fire and Emergency Services* by J. Curtis Varone (Clifton park, NY: Delmar, Cengage Learning, © 2007).

The Role of the Courts

So where do the Supreme Court and the federal courts fit into all of this? Their job is to make sure that the provisions of the Constitution are followed. Generally, the federal appeals courts are concerned with matters of a constitutional nature. When these courts hear cases in which the law is deemed to be unconstitutional, Congress may react by amending the law to prevent similar occurrences in the future. An example of this was the Do-Not-Call legislation enacted in 2003. Several federal courts found the law exceeded the authority of the executive branch. Congress and the president responded by amending the law. Of greater significance were the actions of the Supreme Court with regard to the Civil Rights Act. Once again, Congress and the president responded by amending the law.

It is important that we understand the safeguards provided by the Constitution. They ensure an effective government balanced with democracy, individual freedom, and equality for all. These protections are fundamental to life in our great country.

THE FIRE SERVICE IN AMERICA TODAY

As a result of the decentralization of power, most of the matters pertaining to fire protection are delegated to the states and to local government. As a result, there are approximately 30,000 fire departments in the United States individually organized to meet the specific needs of the communities they serve. (Although public fire protection is usually a function of local government, provincial, state, and federal property may also have organized fire departments for their protection.)

Public Fire Departments

A **public fire department** is part of local government. The head of the fire department (fire chief) reports to the chief executive officer (city manager) or to the senior elected officer (mayor). As such, the chief is responsible for all things related to the fire department's operations. The most common type of public fire protection in larger cities is the public fire department. A less common form of municipal fire protection is provided by a fire bureau, a part of the department of public safety. A variation of this is provided when employees of the public safety department provide both police and fire protection. In some jurisdictions, these personnel are called public safety officers.

County Fire Departments and Fire Districts

County fire departments are a little more complicated. County fire departments are becoming more common as suburban communities grow and the demands for emergency service response outpace the capabilities of the volunteer fire companies. A model that works well is to have a county fire chief or administrator responsible for countywide fire protection, with a responsibility for coordinating the efforts of the independent paid and volunteer fire companies. As the need arises for hiring paid staff for the volunteers' fire stations, the paid fire chief is involved with selecting, training, and assigning these people, and supervising them once they are hired. (This model has proven to work better than hiring the firefighters first and then looking for a fire chief to lead them.) As the organization grows, the county fire chief will likely add staff assistants to deal with training and communications. As the county continues to grow, the demand for service increases proportionally. Gradually, more functions, including emergency response activities, are shifted from the volunteers to the paid staff as the volume of calls and the demand for additional service increase.

The Volunteer Fire Service

We sometimes forget that America's fire service was once all-volunteer. With the growth of large cities and the accompanying development of paid departments in the late nineteenth century, America had both volunteer and paid firefighters. That statement is still true, but when one looks at fire departments as organizations, the picture becomes a bit more complicated. There are paid departments with some volunteers, and there are volunteer departments with some paid personnel. And there is an infinite array of combinations in between.

Today, volunteer firefighters make up about 74 percent of all firefighters. Just like their paid colleagues, volunteer firefighters are the first line of defense against fires, medical emergencies, hazardous materials incidents, all forms of specialized rescue, and responses to natural disasters and terrorist attacks.

Volunteer firefighters face several major challenges:

■ During the past 25 years, many volunteer departments, like many other organizations, have faced the problem of retaining veteran members and attracting new members. During the same time, the number of emergency calls has increased as well as the requirements for training. Although the problem is national in scope and applicable to many departments, we should note that some volunteer emergency response organizations have had no problems in this area. The U.S. Fire Administration is also well aware of these issues. They recently published *Retention and Recruitment for the Volunteer Emergency Services* which discusses the concerns and offers solutions.

■ Another critical issue is the lack of communication between some volunteer fire departments and the communities they serve. In many areas, community leaders have little understanding of the significant issues facing volunteer emergency responders. Unlike most other agencies, volunteer departments are independent and must survive on their own. They often raise the bulk of their funding with little support from local government.

FIREGROUND FACT

Nearly three-quarters of the nation's 1 million firefighters are volunteers. Almost half of these volunteers serve in communities with a population of less than 2,500. They are found in cities, towns, fire districts, and counties. There are all-volunteer departments in cites with a population over 100,000, and there are all paid departments in communities of fewer than 2,500 people.

■ A third critical issue is firefighter health and safety. Although the overall number of firefighter deaths and injuries has declined over the years, a majority of the deaths and injuries in recent years are in the volunteer fire service. Most of the deaths are from heart attacks and vehicle accidents. The volunteer fire services, represented by the National Volunteer Fire Council and the U.S. Fire Administration, are working together on both of these issues and have issued several reports that highlight the problems and offer solutions.[3]

In many states, the counties are divided into fire districts as provided by state law. These fire districts become essentially separate, independent units of government, administered by some form of commission. The citizens of a fire district can elect the type of fire protection they want, and, in some cases, they can elect the level of protection afforded. Usually, these citizens pay a special tax for the level of service they desire.

Another form of fire protection is established with the creation of a fire protection district. A fire protection district is a legally established tax-supported unit that contracts for fire protection with a neighboring fire department.

Volunteer organizations may be supported by taxes, subscriptions, or fund-raising activities.

Regardless of the basis of authority and how the department is organized and funded, it is important to know the laws that pertain to its existence. When a department is not legally organized or when it does not operate in full accord with state laws, local ordinances, and its own regulations, it operates outside the scope of the local government and misrepresents itself to the citizens of that community. As such, it exposes itself to considerable legal risk.

Federal Fire Departments

Although most firefighters are a part of local government or a volunteer fire department, there are federal firefighters who are employees of the federal government. They work at government installations, including military bases, around the world. Most federal fire departments are relatively small, but, collectively, federal firefighters comprise the largest fire department in the United States. In addition to their traditional duties of fire suppression and emergency medical response, most federal firefighters are very actively involved with fire prevention activities. As a result of these and other efforts, fire loss on government property is usually lower than average.

Both the U.S. Forest Service and the U.S. Park Service have fire personnel, including dispatchers, firefighters, fire managers, safety specialists, researchers, and others, who work together for the common goal of fire management, fire use, fire prevention, and fire suppression. Many states have similar opportunities.

Several branches of the military also have firefighter specialties among their enlisted personnel. The U.S. Army, Air Force, and Marine Corps all have firefighters. They are primarily involved with providing fire and rescue support for aircraft operations, both at home and while deployed abroad. The U.S. Navy and Coast Guard have a specialty rating that designates personnel who maintain the firefighting equipment on ships and smaller shore units, but actual firefighting is generally an all-hands event, especially on ships at sea.

Most departments have some sort of automatic/mutual aid agreement set up with allied agencies. For smaller organizations, it becomes essential to have these types of agreements so that the required protection can be given to their citizens. For these agreements to work, however, it becomes necessary for allied agencies to develop goals and objectives that mirror one another. For example, departments that respond together should have an accountability system that emulates one another so that all members can be adequately tracked. Without mission specific goals and objectives, automatic and mutual aid departments will often lack the ability to effectively serve.

Allied agencies do not necessarily have to be emergency-related departments. Just about any type of business or community-based organization can be an ally to the fire department. Take, for instance, an incident that involves a trench rescue. Where would you find the necessary lumber and tools needed to properly shore a trench? Who would you call if your department was in immediate need of a large shelter for abandoned citizens? These are questions that need to be asked ahead of time and then put together in an Emergency Services Plan for the organization. This plan will list all of the allied agencies within the jurisdiction and list what supplies and services they can offer.

The following is a small list of those allied agencies that assist with fire department functions and emergency response:

- *Police department:* The police department and the fire department work hand-in-hand on numerous types of incidents. However, not all of their interaction is on an emergency scene. Many organizations have banded together to teach the community about public safety. By working together to educate their citizens, these organizations have seen a drastic decrease in fire or police related emergencies.

- *Local industry:* Building relationships with local industry can have many advantages. These types of relationships can lead to something as simple as assistance with a local fire prevention campaign to providing materials for catastrophic events.

- *Private non-profit agencies:* These types of agencies can range from a local group that cleans the streets to a volunteer fire department. The American Red Cross is a good example of such an organization. This organization can be essential during response to a natural disaster, since this agency is equipped to provide relief and resources to victims requiring aide. The fire department is recommended to affiliate themselves with as many of these organizations as possible. By assisting them they will in-turn assist the fire department in a time of need.

- *EMS/hospitals/clinics:* Emergency situations happen every day in every community across the nation. But what happens when the emergency involves a large number of people? It is essential to work with not only the local EMS agencies, but also the local hospitals and clinics to devise a response plan for these types of situations.

- *Local/state/federal emergency management agencies:* These agencies are critical during response to a disaster. First response to a disaster is responsibility of local and state government emergency services, but upon request from the governor of a state as a result of a catastrophic event, federal assistance can also be called in to work with the local municipalities. These agencies have the resources required for

FIREGROUND FACT

Until World War I, paid members of fire departments worked a continuous-duty system with limited time off. By World War II, most departments had adopted a two-platoon system that allowed additional time off and reduced the amount of time worked per week. Since then, the number of hours has continued to decline; however, paid firefighters still work more hours per week than do other workers in municipal employment or private industry. Today, the most common arrangement is to have three platoons each working 24-hour shifts in some sort of rotation. This works out to an average of approximately 56 hours per week or 10 days on per month.

FIGURE 3-3 Volunteer firefighters make up a large majority of all firefighters.

FIREGROUND FACT

Historically, America was protected entirely by volunteer fire departments. During the second half of the nineteenth century, the arrival of steam-powered equipment and the growth of large cities gradually set the stage for the arrival of paid firefighters. As cities grew, so did the need for fire protection, and the demands—in terms of call volume and work hours—were, in many cases, beyond the reasonable capabilities of the volunteer fire departments.

both response and recovery to a disaster and are an essential mutual aid partner.

We have learned that there are about 30,000 fire departments and about 1 million firefighters in the nation. These departments, and their firefighters, respond to nearly 20 million calls for assistance annually. Fire calls make up less than 10 percent of the total call volume. (See **Figure 3-3.**)

Across the country, calls for emergency medical service continue to increase. In communities where the fire department can provide such service, calls for hazardous materials incidents and technical rescue are also increasing.

- Most U.S. fire departments are volunteer fire departments; however, most of the United States is protected by paid firefighters.

- Three of every four U.S. fire departments are all-volunteer fire departments; such departments protect one-fourth of U.S. residents.

- One in seventeen departments is all-paid; these departments protect 40 percent of the nation's population.

Generally speaking, volunteers are concentrated in rural communities, whereas paid firefighters are found disproportionally in large communities. Rural communities, defined by the U.S. Bureau of Census as a community with a population of less than 2,500, are 99 percent protected by all-volunteer or mostly volunteer departments. Collectively, these communities account for about half of the all-volunteer or mostly volunteer departments in the United States.

The size and type of fire department in a given community are usually related to community size. As we have seen, the mix of paid and volunteer personnel is generally a function of community size. And one might tend to make conclusions about the level of fire protection and the degree of risk from the consequences of fire based on community size. However, one should be careful in making such a conclusion, for even a small community may have a large factory complex, a sports stadium, or a highrise building with all the technical complexities and potential for high concentrations of people or valued property that such property entails. Even a large city can have a wildland/urban interface region and exposures to the unique fire dangers attendant to such an area. It is likely that every fire department will need to have some familiarity with every type of fire and every type of emergency, if not in protecting its own community, then at least in its role as a source of mutual aid or as a component in a regional or national response to a major incident.

In the final analysis, fire burns the same way in the open or in enclosed spaces in all communities. Fire harms people and their property in the same ways in all communities (see **Table 3-2**). And the resources and best practices required to safely

TABLE 3-2   Fires, Deaths, Injuries, and Dollar Loss in the United States 1997–2007

| Year | National | | | |
	Fires	Deaths	Injuries	Direct Dollar Loss In Billions
1997	1,795,000	4,050	23,750	$8.5
1998	1,755,000	4,035	23,100	$8.6
1999	1,823,000	3,570	21,875	$10.0
2000	1,708,000	4,045	22,350	$11.2
2001[1]	1,734,500	3,745	20,300	$10.6
2001[2]	-	2,451	800	$33.4
2002	1,687,500	3,380	18,425	$10.3
2003	1,584,500	3,925	18,125	$12.3
2004	1,550,500	3,900	17,875	$9.8
2005	1,602,000	3,675	17,925	$10.7
2006	1,642,500	3,245	16,400	$11.3
2007	1,557,500	3,430	17,675	$14.6 billion

[1] Excludes the events of September 11, 2001.
[2] These estimates reflect the number of deaths, injuries and dollar loss directly related to the events of September 11, 2001.

Note: The decrease in direct dollar loss in 2004 reflects the Southern California wildfires with an estimated loss of $2,040,000,000 that occurred in 2003.
The 2007 dollar loss includes the California Fire Storm 2007 with an estimated damage of $1.8 billion.

Source: Michael J. Karter, Jr., Fire Loss in the United States 2007, NFPA, August 2008 and previous reports in series.

address the fire problem—or any other emergency—tend to be the same everywhere. What may differ are the defined scope and the quality and quantity of resources available to the department to perform those responsibilities.

THE EVOLVING ROLE OF FIRE DEPARTMENTS

Many fire departments are doing more than just fighting fires. Total responses may be increasing, but, in most jurisdictions, the number of fire responses is down significantly (see **Table 3-3**).

In many communities, fire departments are responding to an ever-increasing variety of situations, including natural and human-made disasters. In some cases, the fire department reacts to the situation and does the best it can to serve the needs of the community. Fire departments are among the first to respond to natural disasters—floods, earthquakes, tornadoes, and similar events—that often tax the resources of

the organization. Fire departments are also among the first on the scene at man-made disasters—terrorist events.

Some departments are trying to be proactive rather than just reactive and are developing plans and procedures to deal with these threats. For many progressive departments, planning the community's, including the fire department's response activities for such events is now at the forefront. Much of the effort at the National Emergency Training Center in Emmitsburg, Maryland, where the National Fire Academy and the Emergency Management Institute are located, is focused on planning for responses to chemical and biological agents and weapons of mass destruction. Clearly, the fire department has taken on roles that would not have been considered just a generation ago.

NOTE

Like other organizations, a fire department consists of people working together in a coordinated effort to achieve a common goal.

Table 3-3   Fire Department Calls 1986–2007

	Total	Fires	Medical aid	False alarms	Mutual aid	Hazardous materials	Hazardous condition	Other
1986	11,890,000	2,271,500	6,437,500	992,500	441,000	171,500	318,000	1,258,000
1987	12,237,500	2,330,000	6,405,000	1,238,500	428,000	193,000	315,000	1,328,000
1988	13,308,000	2,436,500	7,169,500	1,404,500	490,500	204,000	333,000	1,270,000
1989	13,409,500	2,115,000	7,337,000	1,467,000	500,000	207,000	381,500	1,402,000
1990	13,707,500	2,019,000	7,650,000	1,476,000	486,500	210,000	423,000	1,443,000
1991	14,556,500	2,041,500	8,176,000	1,578,500	494,000	221,000	428,500	1,617,000
1992	14,684,500	1,964,500	8,263,000	1,598,000	514,000	220,500	400,000	1,724,500
1993	15,318,500	1,952,500	8,743,500	1,646,500	542,000	245,000	432,500	1,756,500
1994	16,127,000	2,054,500	9,189,000	1,666,000	586,500	250,000	432,500	1,948,500
1995	16,391,500	1,965,500	9,381,000	1,672,500	615,500	254,500	469,500	2,033,000
1996	17,503,000	1,975,000	9,841,500	1,816,500	688,000	285,000	536,500	2,360,500
1997	17,957,500	1,795,000	10,483,000	1,814,500	705,500	271,500	498,500	2,389,500
1998	18,753,000	1,755,500	10,936,000	1,956,000	707,500	301,000	559,000	2,538,000
1999	19,667,000	1,823,000	11,484,000	2,039,000	824,000	297,500	560,000	2,639,500
2000	20,520,000	1,708,000	12,251,000	2,126,500	864,000	319,000	543,500	2,708,000
2001	20,965,500	1,734,500	12,331,000	2,157,500	838,500	381,500	605,000	2,917,500
2002	21,303,500	1,687,500	12,903,000	2,116,000	888,500	361,000	603,500	2,744,000
2003	22,406,000	1,584,500	13,631,500	2,189,500	987,000	349,500	660,500	3,003,500
2004	22,616,500	1,550,500	14,100,000	2,106,000	984,000	354,000	671,000	2,851,000
2005	23,251,500	1,602,000	14,373,500	2,134,000	1,091,000	375,000	667,000	3,009,000
2006	24,470,000	1,642,500	15,062,500	2,119,500	1,159,500	388,500	659,000	3,438,500
2007	25,334,500	1,557,500	15,784,000	2,208,500	1,109,500	395,500	686,500	3,593,000

Source: Quincy, MA: National Fire Protection Association, www.nfpa.org. Used with permission.

ORGANIZATIONAL STRUCTURE

Like other organizations, a fire department consists of people working together in a coordinated effort to achieve a common goal (see **Figure 3-4**). To function effectively, a fire department must have an organizational plan that shows the relationship between the various divisions and activities. Work should be divided among the divisions and the individuals within these units according to a plan. The plan should be based on functional activities such as prevention, emergency operations, and support activities. This plan is often referred to as the strategic plan. The strategic plan defines the strategy of the organization and gives direction by allocating resources to pursue the strategy. While the strategic plan is often defined as a 3- to 5-year, or sometimes 20-year, budgetary plan, most see it as a way to direct the entire organization into a new era.

Coordination of these activities is important. In small departments, coordination can usually be easily accomplished because of the relatively few number of people involved. However, as the size of the department increases, the number of relationships increases, and the effort required to coordinate their efforts grows accordingly. There should be a job description for every position in the organization including chief officers. The job description should identify the skills required, the principal work expected, and the working relationships required of the position. In addition to the job description, organizations usually have rules and regulations that define the policies of the organization, the conduct that is expected, and standard procedures for handling frequently occurring situations. Many people find all of this organizational structure rather distasteful. The alternative may be less pleasant and certainly less effective. Many organizations have lost much time and money

FIGURE 3-4 In many departments, having a well-prepared dinner is a well-established tradition and helps foster relationships. Here we see paid members of two fire departments and the crew of a county EMS unit eating together.

because they lacked the essential tools for good organizational management.

> **NOTE**
>
> There should be a job description for every position in the organization.

To aid our discussion, let us invent a fire department, the Plain City Fire Department (see the feature box, "The Plain City Fire Department (PCFD)". The Plain City Fire Department could be a paid or a volunteer organization; at this point, it does not really make any difference. Firefighting personnel are comprised of paid, volunteer, and paid-on-call firefighters. When one or more of these groups is combined into one department, we have what is referred to as a combination department. Many factors determine the type of arrangement that might work best in a given community. Community size and financial resources

THE PLAIN CITY FIRE DEPARTMENT (PCFD)

What does it take to organize a fire department?

Suppose that your community needs fire protection. Someone says, "Let's organize a fire department." The local government has been planning for this event and has some money set aside to help get started.

- How do you start a fire department?
- How do you establish organized fire protection?

Let us look at what is represented in an organization and what represents organized fire protection. Based on your previous experiences, you have been selected to help start the Plain City Fire Department. Knowing that

it takes several firefighters to staff one company, you get a couple of friends to join you in this venture. And because you want to be able to provide both fire and EMS services, you realize that you will actually need to divide the company according to function. Because a major highway is being built through the city, it will soon be necessary to have separate stations on each side of the highway. Even though the department is small, you can see some of the issues of organizations starting to arise.

- What are the legal requirements for starting a department?
- What services will we provide?
- Where do we get the people?
- Should we hire any paid personnel?
- Who will be in charge?
- Where do we get a station?

must always be considered. It is also important to consider the services that are expected, the risks that can be expected, the frequency of incidents, and the availability of the volunteer firefighters. All of these factors must be considered by a community to determine what works best.

Increasing demands for service and limited funding have prompted fire chiefs to attempt new approaches to service delivery. For example, fire chiefs have tried various combinations of mixing firefighters and EMS personnel—using volunteers to staff stations when paid personnel are on vacation, and using paid personnel to staff stations when volunteers are away from their community at work—to provide effective and efficient emergency services. These initiatives have occasionally met with resistance. Tradition, turf battles, and political issues arise when new ideas are being considered and when change is introduced. Strong leadership and open communications are vital to the success of any organization.

> **NOTE**
>
> Strong leadership and open communications are vital to the success of any organization.

To effectively manage the Plain City Fire Department, it is necessary to create some sort of organizational hierarchy. Most organizations have a pyramid structure, with one person in charge and an increasing number of subordinates at each level as you move downward in the organization. There are good reasons for this traditional form of structure. Here we see our familiar pyramid representing the organizational structure of the Plain City Fire Department (see **Figure 3-5**).

> **NOTE**
>
> Most organizations have a pyramid structure, with one person in charge and an increasing number of subordinates at each level as you move downward in the organization.

The Fire Chief

Let us examine the pyramid from the top. In the fire service, the top of the organization is usually referred to as the fire chief. Some organizations are using other terms, such as director or chief executive officer (CEO), to designate the top official in the fire department, but we use the traditional term. The fire chief is the head of the agency. The fire

FIGURE 3-5 Organizational structure of the Plain City Fire Department.

chief can be selected by any one of several methods, but, once installed, the fire chief is expected to perform the duties as the leader of the fire department.

Fire chiefs are usually tasked in their job description with providing all the services required, both for the department itself and for the public they serve. Few fire chiefs can perform all of these functions all of the time, so they delegate some of the work. When done properly, the fire chief has an organization to undertake these activities in a structured manner. To make things work well, the chief **delegates** some of the responsibility to the subordinates in the organization. The fire chief is usually supported by staff that includes other officers and an administrative support staff to make the entire operation function effectively. The size and function of the senior staff will vary based on the size of the department and on something we will call "corporate attitude" (see **Figure 3-6**).

In these days of increased concern over the cost of government, many fire departments are taking a hard look at the size of their senior staff, especially in paid fire departments. In many older departments, fire chiefs have become dependent upon large staffs to help them run the department, yet the cost of maintaining that staff is significant. One senior officer may be paid as much as several firefighters. When forced to reduce budgets, chiefs are forced to make hard decisions between eliminating a senior staff position or eliminating line firefighters. Ideally, departments with more than 10 percent of their employees in staff assignments should look at the way they are doing business.

FIGURE 3-6 The fire chief should meet regularly with senior staff members. Such activities keep all of the senior staff informed of the activities and plans of the organization.

THE ROLE OF THE CHIEF

Fire departments tend to be very conscious of rank within their own organization. Clearly the fire chief is the boss, and all others are subordinate. But when they are out of their own town, all fire chiefs seem to be considered equal. You can see this at a fire chiefs' conference. The chief of a single-station fire department might wear a uniform that is just like the uniform worn by the chief in a city of 250,000. An assistant chief (or other rank) may be a far more conspicuous player in a state or national organization than the chief of the same department.

When these personnel work together in committees and other activities, they have an equal voice and are respected based on their contribution, not on their organizational rank or on the number of fire stations back home.

Assistant Chiefs and Deputy Chiefs

The senior staff represents middle management. This is where the vice presidents are located in the corporate world. In the fire service, we have assistant, deputy, and battalion chiefs. These officers fill many of the middle-management positions within the department. They serve as division officers and heads of major components of the headquarters staff. In many departments, captains and lieutenants are also assigned to the headquarters staff as section heads.

We realize that this is not a text for fire chiefs, although we hope that you are interested in being a fire chief some day. However, there is an important lesson here for everybody in the organization. Regardless of where you are in the organization, the treatment of your second in command is important. Keeping that person in the information flow, delegating the fun-to-do as well as the not-so-fun jobs, and preparing that person for the position you hold are significant parts of what organizational relationships are all about and should involve every aspect of training, management, and leadership, for yourself and for your unit. Delegating some of this activity not only prepares another to take your place but likely increases overall understanding and support of the work at hand. It also provides you with a very well-qualified assistant.

Battalion Chiefs

Battalion chiefs usually represent the middle-management layer of a fire department. In larger organizations, they may have responsibility for part of the department and supervise three to six companies. In smaller communities, they may be responsible for all on-duty emergency response personnel. In either case, they will probably become the incident commander at an emergency response of any consequence.

> **OFFICER ADVICE**
>
> Regardless of where you are in the organization, the treatment of your second in command is important.

Company Officers

Moving down the pyramid to the next level, we have the company officers. Most company officers are supervisors of firefighters and, as a result, are referred to as first-line or front-line supervisors. As we have already noted, company officers wear many hats. Company officers are expected to be proficient in their jobs as firefighter. They are also expected to have the managerial ability to understand the organization and the supervisory skills needed to effectively lead their subordinates.

Firefighters

Throughout the fire department organization are firefighters. Most firefighters are assigned to companies and work at fire stations. However, firefighters can also be found throughout the headquarters staff in many fire departments.

THE ORGANIZATIONAL CHART

The pyramid diagram shown in Figure 3-5 was adequate for the discussion in Chapter 1, but you can quickly see the limitations of this representation, even for our small fire department. It might be more appropriate to represent our department as shown in **Figure 3-7**. This chart more accurately represents the organizational relationships of the department. For this small department, it represents the line authority. An officer with **line authority** manages one or more of the functions that are essential to the department's mission. When we see an organizational chart, we usually think of the authority that officers have over the activities that fall under them. In addition

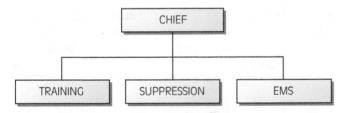

FIGURE 3-7 A more traditional organizational chart representing the Plain City Fire Department.

to authority, officers also have **accountability** and **responsibility.**

We have taken care of the essential activities of the department. However, as the new department takes shape, the fire chief recognizes the need for prevention and training, so a prevention officer and a training officer are established (see **Figure 3-8**). We could debate the placement of these activities but consider that the prevention and training positions are staff assignments. We discuss the details of line and staff authority in the next section.

Let us go back to the Plain City Fire Department. Suppose that there is no real organization, that everyone is the same rank and has the same responsibilities, and that the department is called to a structural fire. What do you think would happen?

As a department grows, we see a continuation of the evolution that has already started. **Figure 3-9** is an example of an organizational chart for a medium-size fire department.

As our hypothetical community grows, so does the need for fire protection. The department shown in Figure 3-9 can no longer keep up with the demand for services. The fire chief and the senior staff are overwhelmed with just keeping up with the growth, and they find that they cannot really efficiently supervise those personnel assigned in the field. In addition, delays in the arrival of senior officers responding to large-scale emergencies are affecting the way the department deals with these emergencies.

As the department grows, the number of relationships also continues to grow. In addition, as the organization grows, there are more options for further growth (see **Figure 3-10**). We see that the department has added a senior position, a person who is called the shift commander. This position takes some of the burden off the senior staff of actually running the department 24 hours a day. The shift commander has command authority over all except the largest events. Even more important, the shift commander has authority to deal with most of the administrative issues that arise within the department that were formerly approved by officers at headquarters.

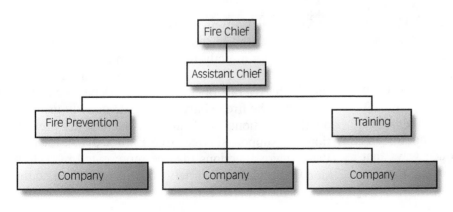

FIGURE 3-8 Chart of the Plain City Fire Department showing fire prevention and training as staff functions.

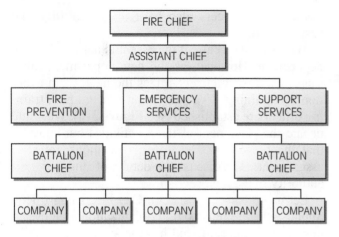

FIGURE 3-9 Chart of the Plain City Fire Department showing the addition of three shift commanders in the emergency services division.

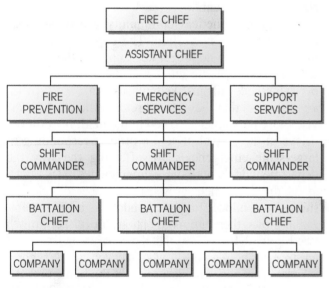

FIGURE 3-10 As more companies are added in the Plain City Fire Department, another layer of managers is needed.

Another alternative is to divide the community into three geographical divisions and place a senior officer in charge of each division (see **Figure 3-11**). The division commander has complete responsibility for all fire department functions within a certain part of the city.

Both of these approaches are widely used. Each has its advantages and each has its disadvantages. A third alternative would be to establish a separate EMS support organization. Our purpose here is not to reorganize your department but to introduce ways that departments are organized and some of the reasons that these organizations are essential.

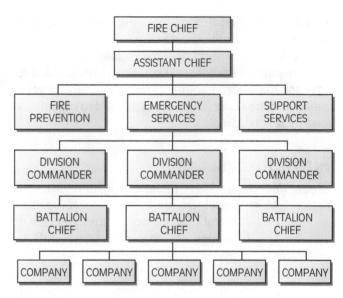

FIGURE 3-11 As an alternative to shift commanders, the fire department's resources in Plain City might be geographically divided among three division commanders. The division commanders would work a normal work week. Each division commander would have three battalion chiefs, possibly one for each shift.

Staff versus Line Organization

In the field of industrial management, **line functions** refer to the activities that accomplish the manufacturing process. Many fire service textbooks draw a parallel here, suggesting that the fire service has generally considered the line function to be synonymous with providing emergency services. This service delivery activity is called by various names, including suppression, operations, and emergency services. This definition of the line organization suggests that everything else is a supporting role, performed by staff personnel.

NOTE

The support staff is essential in the overall operation of the department, and without it the fire department would quickly grind to a halt.

Staff functions are those activities that support the department's basic purpose. Staff functions do not normally get directly involved with delivering emergency services. However, the support staff is essential in the overall operation of the department, and without it the fire department would quickly grind to a halt. Traditional staff functions include fire prevention and support services (see **Figure 3-12**). Each of these major functions may be represented by a division that performs a variety of activities, based on the size of the department.

FIGURE 3-12 Staff functions are important to the life of any organization.

The terms "line" and "staff" are traditionally used to describe the activities of many organizations, including fire departments. Many fire service textbooks follow this pattern and put fire prevention on this staff side of the organization.

We tend to disagree with this approach. As shown elsewhere in this textbook, for most fire departments, fire prevention activities are on the increase and firefighting is decreasing.

In most cases, the decrease in the number of fires is the result of fire prevention efforts. In some departments, fire prevention plays a greater role than fire suppression, both in terms of the daily lives of the firefighters and in the way the department serves the community.

In a large fire department, the prevention division may have separate sections dealing with inspections, investigations, life safety education, plans review, and systems testing. Likewise, in a large fire department, support services may include communications, payroll, personnel, research, recruiting, resource management, safety and health, and training.

On the other hand, in small communities, many of the support activities common to all government agencies are gathered together under a central agency. Purchasing and personnel activities are good examples. As the fire department grows, it eventually becomes beneficial to perform these activities within the fire department's organization.

Before we conclude this discussion, it is important to note that titles vary from department to department. In some departments the second level officers are deputies, and in other departments they are referred to as assistant chiefs. Where there is only one level of middle management, they could be called by either title. In smaller departments, the second-in-command officer may be a battalion chief. Titles are important but sometimes confusing. Checking the organizational chart usually clears up these issues.

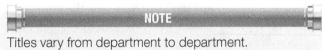

NOTE

Titles vary from department to department.

THE COMPANY OFFICER

The company officer is usually the first supervisory level in a fire service organization. In many ways, company officers have an opportunity to exercise more influence than that of any other rank when one considers the role of the company for all of the internal activities of the department as well as the activities associated with the department's delivery of emergency services. The company officer provides the tone of the company in terms of morale, preparedness, and service delivery.

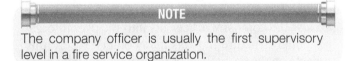

NOTE

The company officer is usually the first supervisory level in a fire service organization.

The Fire Chief's Perspective of the Company Officer

Most fire chief officers see their company officers as a part of the organization's management team. Many fire chiefs allow and even encourage their company officers to get involved in the decisions that have strategic impact on the department. Such fire chiefs value the input from their company officers and support their ideas. This implies trust in one another. Effective team building is encouraged in this organizational environment. This environment also places a value on education. Education enhances your ability to think through complex issues and to effectively express your ideas to others.

The Firefighter's Perspective of the Company Officer

Regardless of the relationship of the company officer to the fire chief and the other members of the department's senior staff, firefighters usually see their company officer as part of the department's management team. We have already indicated that the company officer has a greater impact on company members than does any other officer in the department. After all, it is the company officer who enforces department policies, grants the firefighters' requests, and evaluates their performance. Indeed, the company officers hold a lot of the strings that control the firefighters' work and attitude.

Firefighters informally evaluate their company officers based on how well they (the officers) perform their duties. The firefighters evaluate how well the company officers can perform their job, how well they apply policy, and how fair they are in their actions among the members of the company. They can assess whether the officer can teach and coach the members. They can evaluate if the company officer is up-to-date on new ideas and techniques and is composed under pressure. They can evaluate communications skills. The list goes on and on. Unfortunately, firefighters seldom get to express their opinions on these matters.

Most company officers would like to be respected by both their bosses and their subordinates. Some company officers are focused on pleasing the boss and looking good. Other company officers focus their energies on the welfare of their subordinates, sometimes at the expense of their relationship with their own supervisors. Being successful in both areas can be quite a challenge and essential.

How the company officer fits into the overall organizational structure impacts many others. It certainly impacts on the daily lives of the firefighters who work for that officer. Eventually, it will impact on the company's readiness to deal with anything other than the simplest emergencies. Good companies become standouts in the organization. Companies that perform poorly tend to pull from the talents of other companies and drag down the average of all the others. Poor performing companies eventually affect the entire department and possibly even the community they serve.

NOTE

How the company officer fits into the overall organizational structure impacts many others.

PRINCIPLES OF GOOD ORGANIZATIONS

There are several principles of organizations. Successful organizations seem to follow most of these "rules" quite well. Sometimes organizations are successful, even though they do not follow all of these

rules. However, most organizations run into problems when these principles are forgotten.

Officers in the fire service should be concerned about the organization in which they function and the job they perform within that organization. Most positions are essential to the department's mission: Few fire departments have positions that exist just to fill a block on an organization chart. People in some of those positions perform duties directly in support of the mission, whereas others perform duties in support of those who are out on the street. Both are essential!

NOTE

Some positions perform duties directly in support of the mission, whereas others perform duties in support of those who are out on the street. All positions and duties should support the mission of the department.

Let us look at the organization we call our fire department. Even though it is likely that you did not invent it, and even though it is almost as likely you will not change it right away, you should understand the organization and how your particular position fits into the overall scheme of things.

Scalar Organization Structure

The **scalar principle** relates to the continuous chain of command in the organization. Like playing a scale on a musical instrument where every note is sounded, the scalar principle suggests that every level in the organization is considered in the flow of communications, authority, and responsibility. Most fire departments are established along the traditional lines of a scalar organization (see **Figure 3-13**).

Unity of Command

This concept is frequently confused with chain of command. **Unity of command** refers to the concept of having one boss. In examining the organizational structures in this chapter, you will note that each subordinate reports to only one boss. This concept is known as unity of command. Unity of command is an essential organizational concept.

Division of Labor

Organizations break the accountability process into small units that can be managed. A fire company, such as an engine or a truck company, is usually assigned a specific task and is usually assigned to cover a specific part of the community. Every company has a job to do at the scene of a fire or other emergency. Every

Assignment of Specific Responsibilities

Another good principle of organizations is that no specific responsibility should be assigned to more than one person. Now, this does not mean that we no longer work in teams, nor does it mean that we no longer have backup personnel. We are saying that the primary responsibility is vested in one person. What happens when that person is out for the day? Clearly the assignment is moved to another member, but our standard procedures normally take care of this. Typically, every riding position on a truck or engine company has a specific duty upon arrival at a fire or other emergency event. It is important to have the position covered, and it is important that the person in that position is aware of the overall responsibilities before fulfilling that assignment. Likewise, those in staff assignments should have an alternate trained and prepared to fulfill their duties when they are absent.

NOTE

No specific responsibility should be assigned to more than one person. The primary responsibility is vested in one person.

Control at the Proper Level

Supervisors should not get lost in a maze of details. Violations of this rule are a common complaint. The fire chief should not be dealing with the details that are rightly the job of the company officer. Likewise, the company officer should avoid overmanaging firefighter activities at the station level. This practice is often referred to as micromanaging.

Span of Control

The principle of span of control is simple: Do not have too many subordinates. The number of people you can effectively supervise is usually referred to as the **span of control**. The number of people you can effectively supervise varies, of course, based on many factors, but generally that number is between three and seven with five being the optimal span.

Delegation

Along with the other principles, delegation plays into the workings of an organization. As the organization expands, we assign certain duties to subordinates. This activity is called **delegation**. It is important that the delegation be mutually understood and that authority be provided along with the responsibilities.

FIGURE 3-13 The badges shown represent all of the steps from firefighter to chief, as does the principle of scalar organizations. (*Photo courtesy of the Fairfax County Fire and Rescue Department*)

company has a place where it is expected to do that job. These are essential components of the fire department's organization and are essential to fire department management. When companies are divided into geographic areas and assigned to specific tasks, they illustrate the first principle of organization—division of labor.

As the department becomes larger, the administrative activities must be organized into manageable sections. This activity also illustrates the division of labor. Division of labor allows an organization to divide large jobs into specific smaller tasks. Most fire departments provide EMS. Where this occurs, a fairly senior officer probably coordinates the department's overall EMS policy. Likewise, at the successive levels in the organization, there are officers who manage the EMS program and supervise some or all of the EMS responders' activities. For departments of any size, this management/supervisory activity of field personnel would be impossible for one person to manage. However, by logically dividing the activity among several people, the task is quite manageable.

Delegation allows dividing or sharing the work and allows subordinates to take on part of the supervisor's work. However, the supervisor is still ultimately responsible for the assignment. We discuss the art of delegating in Chapter 5. For our purposes here, it is an important part of the overall concept of organizational structure.

Good supervisors take the heat from their supervisors without passing it down to subordinates. The supervisor acts as a buffer in the communications process, knowing that the troops will get the job done without using pressure from the fire chief's position to make it happen. Consider the following sentences. Which sentence will do more to strengthen the officer's position?

"The chief said that we need to do more preplanning."

"I think that we should be doing more preplanning."

A final thought about delegation. You should feel free to delegate to your subordinates. Delegation is an important empowerment tool and generally makes subordinates feel better about their jobs.

OFFICER ADVICE

Delegation is an important empowerment tool and generally makes subordinates feel better about their jobs. However, the supervisor is still ultimately responsible for the assignment.

THE ORGANIZATION IS TRANSPARENT TO THE AVERAGE CUSTOMER

Most citizens are not concerned about your organization's structure or rules. Nor do average citizens know or care whether the department has an assistant chief in support services or whether it uses a fully integrated incident management system. They do know and care about the service you offer, and they expect the service to be fast and dependable. Understanding your organization's structure may not be important to citizens, but it should be important to its members.

Organizations Are Ever Changing

The number one initiative listed in the National Fallen Firefighters Life Safety Initiatives declares that the fire service needs a cultural change. However, when you mention the word "change," people tend to become defensive. They become defensive because people fear the uncertainties of change. Nonetheless, it is the company officers duty to lead subordinates through the ever revolving door of change. To accomplish this, the company officer must do the following:

- *Evaluate yourself:* For subordinates to accept change they must first see their superior accept it. Negative comments made by the company officer will only discourage subordinates and turn them away from any possibility of acceptance. To help with this, the company officer needs to understand that they themselves are also fearful of unknowing the outcomes. They must talk with other officers within the organization and make sure that they understand all aspects of the proposed adjustment.

- *Celebrate the past:* It needs to be understood that just because there is a change forthcoming, it does not necessarily mean that the old way was the wrong way of doing it. However, we all need to understand that as the fire service progresses, we must also. New technologies and equipment are making the job of firefighting safer.

- *Involve them in the process:* It is important to ask subordinates their opinions on change. Depending on the situation, you will find that the majority of them will be tolerable of change if they understand why change is being introduced. Feed them as much information as possible. Show them that this is a positive change and that it will improve the safety of the organization.

The process of introducing change starts with an evaluation of why a change is considered necessary. Change that appears to have occurred for no apparent reason will often be resented by all. This is why an evaluation period must take place. Take, for instance, an outdated policy on SCBA safety inspections. To move forward with a recommended change, the company officer should first review the current standards and trends need to conduct a thorough inspection. Next the company officer should evaluate how the change will affect the organization as a whole. Will the change improve safety? Will the change have any costs associated? Will the change require additional training of personnel? These are just a few of the questions that will need to be answered before a recommendation for change can take place. Once all of the questions have been answered, the company officer should place the change recommendation in writing and pass it up the chain of command.

Another item to look at that relates to changing policies and/or procedures would be how the new policy affects existing policies. It may prove evident that because of a new or updated policy, other policies may need to change as well. Take, for instance, a change in a promotional policy. Changing or adding to a policy of this nature would entail numerous other changes throughout the organization's policies and procedures. It becomes essential to monitor all changes to policies so that the differences between the new and existing policies are found. This is often done with a focus group that evaluates each new addition or change to a policy.

Once all of these steps have been completed, it is important to evaluate the change. Evaluating change can prove to be difficult. However, using the above example, a department might find that the SCBAs are now in full working order on 98 percent of all incidents. In other instances, change can be hard to evaluate without some indicators. It might be necessary to conduct some interviews or surveys with all members of the department to see if the outcome has been attained. In other instances, it might be as obvious as a reduction in injuries. Whatever the case, a documented review of the change should take place.

Introducing change is often not an easy process. However, even if we ignore it, change is still occurring.

Flat and Tall Organizations

Linked to the span of control principle is the concept of flat and tall organizations. When the span of control is smaller, that is when supervisors have fewer people to supervise, we tend to see organizations with a narrower or "taller" structure. This works well, but with the concern over the cost of doing business these days, organizations, especially government organizations, are taking a hard look at the span of control of their managers and the size of the middle-management structure. Many organizations are eliminating positions at the middle-management level. This results in an organization that appears flatter. Reduced costs are one benefit of flatter organizations; another benefit is that communications are usually improved because there are fewer layers in the organization, fewer levels of management between the fire chief and the firefighters. In the early 1980s, Thomas Peters and Robert Waterman Jr. published *In Search of Excellence,* a book that became a best seller. It is still in bookstores and still worth reading. The authors looked at some of America's best-run companies and found eight practices that seemed to be characteristic of these companies. Some of the items on their list may fit better into our upcoming chapters on management and leadership, but several are directly tied to the subject of organizational style.

One of the eight items they listed was called "simple form, lean staff." The idea is to keep the company organization simple. This involves reducing the layers in the organization, reducing middle-management positions, and minimizing the number of staff positions.

OFFICER ADVICE

Keep the company organization simple.

Along with the smaller staff is the idea of keeping things as simple as possible. This means that you do not have a staff specialist for every conceivable idea and problem that might come along. You have a few good people, and you give them authority over a broad range of management issues.

Autonomy and Entrepreneurship

A couple of big words describe a very simple but often overlooked principle. Again, the idea comes from *In Search of Excellence.* The authors use the words "autonomy" and "entrepreneurship" to describe a management style. This concept follows the characteristics discussed in the preceding paragraph. With a well-developed organization that is lean and simple, the next step is to develop an understanding of who is responsible for the various functions. If all decisions have to flow up to the fire chief, why have a staff? The manager (fire chief) should assign specific responsibilities and give the person who gets the assignment a feeling that the activity is his or hers to manage, a feeling of entrepreneurship, if you will. Once this is done, problems should be addressed and resolved by the responsible officer.

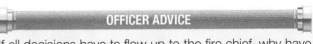
OFFICER ADVICE

If all decisions have to flow up to the fire chief, why have a staff? The manager (fire chief) should assign specific responsibilities and give the person who gets the assignment a feeling that the activity is his or hers to manage.

Discipline

Before we conclude our discussion on the principles of management, we need to mention one more word: **discipline.** Discipline is usually considered to be synonymous with punishment. We look at discipline/punishment as a leadership tool in Chapter 7. In this discussion, we are looking at discipline as the organizational and individual responsibility to do the assigned job.

The Plain City Fire Department is now operating and has a mission statement. Officers are assigned specific responsibility in support of that statement. Companies are expected to respond to emergencies and take action in accordance with the department's policies and procedures. You expect this to happen. It happens because of discipline, not because you will be punished if you fail to perform, but rather, because you accept the duty. You also expect others to accept their share of the duty.

Suppose that you, as a company officer, find that all of this stuff about training, pre-emergency planning, building surveys, physical fitness, and so forth are not really important. You might get away with that for awhile, but one day an alarm is transmitted for a working fire in your first-due area. The first-arriving companies are expected on the scene within 4 or 5 minutes, ready to quickly mount an aggressive fire attack. But because of your approach to company readiness, your people are not ready, your equipment is not ready, and you are not ready. By the time you get things sorted out, second-due companies are already at the scene, and so is the battalion chief. The fire puts the department to quite a test. The fire is in an occupied apartment building. Because of delays in the fire attack, the fire extends into several adjacent exposures, and what should have been a simple job becomes a major event. Several people are injured during the rescue operations and there is significant property loss. Organizational discipline helps keep this sort of thing from happening.

Every company in a fire department has a job and a place to perform that job, but it is also important that the company be competent in the performance of that job. We all know that some companies are better than others in certain tasks. Companies in larger cities often get more experience and are probably more proficient in those tasks they routinely encounter. But even within a city, some companies are better than others. Regardless of the size of the city, company preplanning and training activities are important. Well-disciplined companies do both and usually perform better during emergency activities.

ORGANIZING YOUR OWN COMPANY

Even within the company, there are ways to organize the few people that work there. Suppose that you are the company officer. You have four firefighters working for you. One is an seasoned and experienced driver. Another is a respected firefighter who recently became a paramedic. One is a firefighter with 5 years on the job. The fourth is a firefighter with 5 weeks

on the job. Using the traditional approach, you might organize as shown in **Figure 3-14**.

This organization certainly follows some of the rules about flat and lean organizations and allows for good communications among you, the officer, and all of the firefighters. It reinforces the rules about unity of command and span of control.

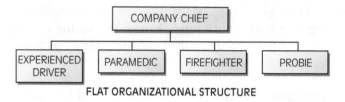

FLAT ORGANIZATIONAL STRUCTURE

FIGURE 3-14 A traditional approach to organizing a company.

SCALAR COMPANY STRUCTURE

FIGURE 3-15 This organizational approach reinforces the scalar principle.

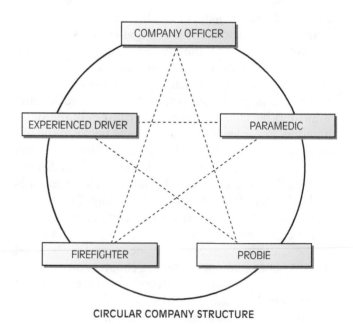

CIRCULAR COMPANY STRUCTURE

FIGURE 3-16 A circular organizational structure may be appropriate occasionally.

You could also organize along the lines shown in **Figure 3-15.** This diagram illustrates the scalar organizational structure. Here we have a tall vertical structure. You can sort the four personnel into the slots as you feel appropriate. This organization reinforces the chain of command.

Another alternative might be to organize along the lines of a circular organization (see **Figure 3-16**). In this environment, everyone is at an equal level in the organization, and everyone has the opportunity to communicate with all of the others. As King Arthur explained to Lancelot as the latter was shown the round table at Camelot for the first time, "We are all equal here."

Each of these organizational structures has appropriate uses within the fire service. As an officer, you should understand the advantages and the limitations of each and use them appropriately.

BECOMING A SUPERVISOR

Bryan Tyner, Captain,
Minneapolis Fire Department, Minneapolis,
Minnesota

Bryan Tyner recognized that things were going to change when he became a company officer—he would now have to distance himself from his firefighters in order to supervise them and be respected. What he did not know was that part of that distance meant aligning himself more with his own supervisors. It means that he now was the voice of the administration in his firehouse.

Tyner has found that this comes into play especially when a new rule or policy comes down from headquarters. As captain of his station, it is his role to pass that information down. He must organize a meeting that everyone is required to attend, then read the department communication of the rule, and require all members to sign and acknowledge that they know the policy. The rule is then posted in the station as a constant reminder. His superiors, including the fire chief, will expect him to make this communication clear to his department.

Tyner does not always agree with the policies, but his role requires him to hold his tongue. That happened recently when a new sick leave rule came out—now when people take a day off, they must see their doctor and bring in a note confirming their illness. Tyner knew this rule would be unpopular, especially considering that the department's new insurance requires members to pay the whole medical bill. And, indeed, it did meet with resistance when he announced the policy. But he could not let on that he thought this was unfair, because his staff looks to him for guidance. If he were to indicate that he would not follow the policy, neither would they.

Although not supporting his firefighters may make Tyner uncomfortable, Tyner also knows it would reflect badly on him to his superiors if his department disregarded the rules. He will be held accountable for his department's infractions. As a further emphasis on the importance of the policy, he followed through by imposing the appropriate punishment on anyone who ignored the rule.

Questions

1. When a firefighter becomes a supervisor, what changes does management expect?

2. When a firefighter becomes a supervisor, what changes might other firefighters expect?

3. What are the officer's responsibilities in a situation like this?

4. Where should the officer's loyalties lie in a situation like this?

5. How should officers communicate a policy they may not personally agree with?

LESSONS LEARNED

The company officer has two roles in the organization. First, as a company officer, you represent the fire department to the average customer. You are often the first on the scene and often the last to leave and, for many events, are the only person the customer talks to.

At larger events, other units of the department may be represented. Other companies may be present, and other parts of the organization are supporting the operation, whether or not they are at the scene of the emergency. As company officers, you must understand how the entire organization works, how you (as a company) fit into the total organization, what resources are available and appropriate for any type of incident, and, if appropriate, how to help the customer connect to the appropriate part to get the service needed.

Second, as a company officer, you are a vital part of the organization that you represent. You should understand your role, how you represent your company and its needs to the department's management team, and how you represent the rest of the organization to the firefighters within your company.

Understanding organizations is important. In the past, people worshiped organizational structure, paid homage to the chain of command, and measured organizations by the number of rules they had. Those are good things, and we should understand their value, both in today's organizations and for those who grew up in this culture. But today we are learning that there may be better ways to run an organization. We are seeing more organizations move away from rules and structure. They are focusing more on values, especially where the members are concerned.

Modern management concepts are discussed in Chapter 4. We do not want to leave you with an impression that all of this information about organizational principles is the end of the story. Far from it—it is only the beginning!

KEY TERMS

accountability   being responsible for one's personal activities; in the organizational context, accountability includes being responsible for the actions of one's subordinates

administrative law   body of law created by administrative agencies in the form of rules, regulations, orders, and court decisions

delegate   to grant to another a part of one's authority or power

delegation   act of assigning duties to subordinates

discipline   system of rules and regulations

line authority   characteristic of organizational structures denoting the relationship between supervisors and subordinates

line functions   those activities that provide emergency services

local ordinance   law usually found in a municipal code. In the United States, these laws are enforced locally in addition to state law and federal law

public fire department   part of local government

responsibility   being accountable for actions and activities; having a moral and perhaps legal obligation to carry out certain activities

scalar principle   organizational concept that refers to the interrupted series of steps or layers within an organization

span of control   organizational principle that addresses the number of personnel a supervisor can effectively manage

staff functions   activities that support those providing emergency services

unity of command   organizational principle whereby there is only one boss

REVIEW QUESTIONS

1. Define *organization*.
2. Why do we need structure in organizations?
3. How do organizational charts show lines of authority?
4. How do organizational charts show lines of communication?
5. Define *division of labor*.
6. Define *span of control*.

7. Define *unity of command.*

8. What is meant by " delegation"?

9. Distinguish between "line" and "staff" in an organization.

10. What is the role of the company officer in the organization?

DISCUSSION QUESTIONS

1. What type of local government do you have where you live/work?

2. Describe the composition of the fire service in the United States.

3. What are the challenges facing the volunteer fire service today?

4. The statistics in Table 3-3 suggest that total call volume is up while the number of fires is down. What is making up the difference? Will this trend continue?

5. The above statistics represent the national picture. What is happening in your community?

6. What is the legal authority for fire departments where you live?

7. What are the roles of your department's senior officers?

8. Where are these roles defined?

9. What are the roles of the company officer within the department?

10. Why is the company officer important to the organization?

ENDNOTES

1. Although this is not a math text, remember that in math we learned a formula for determining the combinations possible from a certain number of objects taken two or more at a time. For example, with a group of 25 people, there are 300 separate two-way relationships possible. With a group of 100 people, nearly 5,000 such combinations are possible. As the group gets larger, the number of relationships increases dramatically.

2. Based on information published by the National Volunteer Fire Council and the U.S. Fire Administration and on a report from the National Volunteer Fire Summit held at the National Fire Academy, December 2–3, 2000.

3. *A Needs Assessment of the U.S. Fire Service.* Emmitsburg, MD: U.S. Fire Administration, 2002.

ADDITIONAL RESOURCES

A Needs Assessment of the U.S. Fire Service. Emmitsburg, MD: U.S. Fire Administration, 2002. In cooperation with NFPA.

Bruegman, Randy. *Fire Administration.* Upper Saddle River, NJ: Prentice Hall, 2009.

Four Years Later—A Second Needs Assessment of the U.S. Fire Service. Emmitsburg, MD: U.S. Fire Administration, 2006, in cooperation with NFPA.

Klinoff, Robert. *Introduction to Fire Protection,* 2nd edition. Albany, NY: Delmar Publishers, 2003.

National Fire Protection Association, Standard 1201. *Developing Fire Protection Services for the Public.* latest ed. Quincy, MA: National Fire Protection Association.

Paulsgrove, Robin. "Fire Department Administration and Operations," *Fire Protection Handbook.* 19th ed. Quincy, MA: National Fire Protection Association, 2003.

Peters, Thomas J., and Robert H. Waterman, Jr. *In Search of Excellence.* New York: Harper and Row, 1982.

4

The Company Officer's Role in Management

When I was promoted to captain, I was in charge of managing Fire Station 6 on the West side of town in Wilmington, which meant I was in charge of 18 firefighters and all of the responsibilities in the station. A few years ago I was promoted to battalion chief. I am now assigned to the South district, in charge of three stations, five companies, three engine companies, the ladder and heavy rescue unit, and I'm in charge of the personnel of those shifts.

When I was thrown into management, I did not feel as though I had been properly trained—one day you can be a firefighter and the next day you can be promoted and be in charge of supervising three firefighters. As a front line supervisor, you're given this promotion, but there is also a definite increase in responsibility and accountability.

As a chief officer you have to wear two hats. You are management, and you're in charge of disciplinary action, but you also have to work and live with these guys, so you want to be their friend as well.

There are three things I suggest when you are thrown into management. One, you have to seek advice from a senior person, or someone who has been in the position as an officer who can give you advice and guidance for certain situations.

The second thing is to attend classes, even if you have to go on your own time. There are a lot of classes available on management skills for dealing with people, which can be especially helpful.

The third, and most important to me, is to keep your core ethics strong. My father was a captain, and he always said you have to be firm but fair. I live by that. The rulebook will always be there, but you have to keep in mind that certain circumstances lend themselves to flexibility, and every situation is unique.

We get all the training that we need for fires, EMS, all kinds of emergency response situations. What we sometimes lack is the personnel skills.

One other thing to keep in mind when managing is that it's always best when everyone stays busy. The old cliché that busy hands are happy hands is true. If your department ends up going for six months at a time with no fires, that is a lot of down time, and that's the time when I get my issues. When it comes to dealing with fires, my guys are great, but when there is nothing going on, that's when we have trouble. You need to have a backup plan for that.

Sometimes the job isn't all glory. Firefighters are now delivering babies *and* putting out fires—we're multifaceted individuals now, and that comes with a lot of other issues that can really trip up a company if there is a lack of training. If you try to get some advice ahead of time, take a class or two and keep your core ethics strong, then you'll be on the right track for management.

—Ed Hojnicki, Battalion Chief, District 2, A Platoon,
Wilmington Fire Department, Wilmington, Delaware

LEARNING OBJECTIVES

Upon completion of this chapter, you should be able to do the following:

4-1   Identify the standard operating procedures (SOPs) pertaining to supervisors and members regarding the operation of the fire department.

4-2   Identify within departmental operating procedures those SOPs governing emergency operations, incident management systems, and safety.

4-3   Identify within departmental operating procedures those that pertain to administration.

4-4   Recognize the current trends that impact the fire service.

4-5   Identify the various ethical dilemmas that may confront a supervisor serving in a public or private agency.

4-6   Explain the ethical obligations of a supervisor serving in a public or private agency.

*The FO I and II levels, as defined by the NFPA 1021 Standards, are identified in different colors: FO I = black, FO II = red.

INTRODUCTION

Many words describe you as the company officer. You are a supervisor, a leader, and a manager. Certainly you are expected to effectively manage the department's resources at the station level. To be effective as a manager, you should understand what management is and the basic management concepts used in modern organizations.

WHAT IS MANAGEMENT?

The traditional definition of management is the activity of getting things done through people. Updating this definition a little, we can say that **management** is the act of guiding the human and physical resources of an organization to attain the organization's objectives. Management includes determining what needs to be done and the accomplishment of the task itself. From this definition it is clear that managers plan and make decisions about how others will use the organization's resources. Managers should also be committed to accomplishing these activities efficiently and effectively.

NOTE

Management includes determining what needs to be done and the accomplishment of the task itself.

THE BEGINNINGS OF MODERN MANAGEMENT

There is evidence throughout recorded history of the use of management concepts. During the past 100 years, much attention has been focused on the science of management. One of the first contributors was Henry Fayol (1841–1925), a very successful manager of a French coal mine. In addition to his ongoing management activities, Fayol was interested in understanding the process by which work was accomplished. Fayol's writings on this topic first appeared in 1900.

By 1916, Fayol had enough material for a book. Though originally published in French, the work was soon translated into English under the title *General and Industrial Management*. Fayol proposed some 14 principles of management. Some of his principles related to organizational structure, and some related to the work process itself. Even at this early date, Fayol was concerned about the importance of the human resources in the organization. (We see an evolution of his ideas in several other writers who follow.)

As you look at Fayol's list, you will recognize that a few of these items were addressed in Chapter 3. Some writers are critical of Fayol where his ideas cross between organizational structure and the management process. But let us be kind here and remember that Fayol's ideas were developed a century ago. To say that he was ahead of his time would be an understatement: Fayol is called the "Father of Professional Management." Many of Fayol's ideas are still valid and quite relevant to our discussion here on management in the fire service.

Fayol also noted that management activities increase as one moves up in rank in an organization (see **Figure 4-1**). That was certainly the case in his time, and it remains true in many organizations today. However, where modern management is practiced, there is a clear pattern to decentralize management activities as much as possible. Where these modern

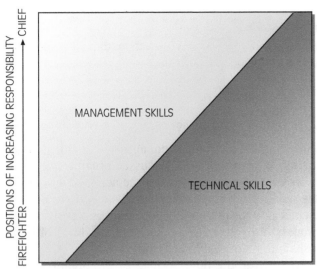

THE NEED FOR MANAGEMENT SKILLS INCREASES AS ONE MOVES UP THE ORGANIZATIONAL LADDER

FIGURE 4-1 Management activity increases as you move up the organizational ladder.

management ideas are accepted, organizations are putting increasing responsibilities upon a greater number of managers, especially on first-line supervisors. In the fire service, the first-line supervisors are the company officers.

NOTE

Fayol is called the "Father of Professional Management."

FUNCTIONS OF MANAGEMENT

Many years ago, Fayol and the other pioneers of management science recognized that management activities could be broken down into several discrete components: planning, organizing, commanding, coordinating, and controlling. Over the years, various authors have renamed these activities, and some have repackaged them a bit. However, Fayol's concept of five basic activities or functions fit our needs here. Let us look at Fayol's five basic functions.

NOTE

Management activities can be divided into five discrete components: planning, organizing, commanding, coordinating, and controlling.

Planning

Any activity should start with **planning.** You may be among the many who find planning to be a difficult and frustrating experience. Planning is largely a

FAYOL'S PRINCIPLES

1. *Division of labor.* Divide the work into manageable portions and specialize the activities to improve efficiency.

2. *Authority and responsibility.* These terms go hand in hand. Where authority arises, responsibility must follow. Managers must have the authority to make decisions and give directions.

3. *Discipline,* as a characteristic of authoritarian organizations, includes good leadership and clear agreement between management and labor on the role of each.

4. *Unity of command* suggests that one should receive orders from only one supervisor.

5. Unity of direction suggests a commitment from all to focus on the same objectives.

6. *Subordination of individual interests* to general interests in an organization suggests that the general interest must take precedence over the individual interests of the members.

7. *Proper remuneration,* a fair wage for fair work.

8. *Centralization of authority,* like the division of work, brings order to the organization. It provides accountability and responsibility where necessary. However, decentralization (we might think of it more as delegating) usually increases the worker's importance and value.

9. *Scalar (continuous) chain of ranks* or layers in an organization, and the notion that authority flows from the highest to the lowest ranks in the organization's chain of command.

10. *Order,* or a place for everything and everything in its place. Fayol was interested in both material order and social order. Today, we would say that order refers to having the right personnel and the right materials ready to do the job so that employees are effective and efficient.

11. *Equity and fairness* are important in the treatment of the employees of any organization.

12. *Initiative* is important to the organization and its members. The ability to think out a plan and see it through to completion is one of man's greatest satisfactions.

13. *Stability of our personnel,* recognizing that there is value in a long-term relationship with our employees.

14. *Esprit de corps,* an organizational spirit, is essential to the survival of any organization.

mental process and often requires the ability to envision things that have not yet happened. When planning is done properly, you challenge these visions with "what if" situations that make you think through the consequences of your proposals. This envisioning can be difficult. You can be frustrated when things do not turn out exactly as you planned or your ideas are not approved. As a result of such experiences, you may see planning as a waste of time and shy away from good planning activity.

Planning covers everything from the next hour to the next decade. We usually divide planning into three areas (see **Figure 4-2**):

- *Short-range planning* looks at the rest of today, tomorrow, and the rest of the year. Many planning activities are driven by annual events. As a result, budgets, training, certification, and such are usually scheduled on a 1-year cycle. For our purposes, we arbitrarily limit short-range planning to 1 year.

- *Mid-range planning* reaches out from 1 year to 5 years. During this time, we can usually convert goals into definitive plans. Most budget, procurement, recruiting, and training activities fall into the realm of mid-range planning. Most departments have plans in the mid-range category, although some do this planning better than others. All departments should be looking well ahead at the needs for personnel, equipment, and facilities.

- *Long-range planning* looks at the needs of the organization beyond 5 years. Many departments do not get involved with long-range planning, but those that do see the benefits of keeping ahead of the community's needs. When communities are changing, long-range planning is essential if the department is to change and provide service as the community changes. As planners, you must look at present growth and anticipate future growth. You have to increase (or relocate) resources to serve the expanding community. As managers, some of the things that you should consider are fire station location or relocation, access to as well as the impact of major highways, and the increasing concern for the environment.

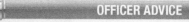
OFFICER ADVICE

When planning precedes the action, the action is easier and is often accomplished faster.

Organizing

The second function of management is called organizing. We have already discussed organizations, but in this case we are looking at the activity of organizing itself. The organization breaks down the department's activities into manageable tasks. Likewise, the process of **organizing** is the breaking down of large tasks into manageable activities.

In a sense, the development of the organizational structure of the department is the result of organizing. We indicated the benefits of this process when we talked about organizations in Chapter 3. When the Plain City Fire Department is established, the development of a structure and the assignment of tasks form the start of the organizing process.

Organizing keeps the department's functions in mind and, when done properly, focuses energy on the department's goals and objectives. Organizing includes managing the resources available to the department. At the company level, these resources typically include human resources, physical resources, time, and in some cases, money.

Commanding

Commanding is the third management function. Because others become involved in this area, the job of commanding can be difficult. Here we see the application of leadership skills and other human relations activities. We discuss the details of leadership in Chapters 6 and 7.

Commanding does not mean that you stand in the hall or in the street and shout orders. **Commanding** means that you exercise your delegated authority to get things accomplished, and because you are also

SHORT-RANGE—GOALS
1 HOUR TO 1
YEAR DISTANT

MEDIUM-RANGE—GOALS
1 TO 5
YEARS DISTANT

LONG-RANGE—GOALS
5, 10, 20 YEARS DISTANT

FIGURE 4-2 Planning activities are generally divided into short, medium, and long range.

responsible for controlling the resources of an organization, it means that you are using the talents of others to attain these goals. The essence if command is leading, motivating, and building teamwork among your team members.

Coordinating

The fourth management function involves the coordination of the available resources. These resources can include money, equipment, and personnel (see **Figure 4-3**). You should always consider time as a resource as well. To some extent, these resources can be substituted for each other. For example, if time and money are available, and personnel or equipment are scarce, you might consider contracting for a particular service you need.

As managers, you should be constantly looking for the proper mix of resources to best meet the needs of the organization. Effective managers are constantly finding ways to do more for less. In the private sector, this is important if the company is to remain competitive. In the public sector, it is important because the citizens and their elected representatives are expecting good management practices from everyone in government.

Although most fire departments are nonprofit organizations, they operate in a competitive environment. Public fire departments are constantly competing with other government organizations, such as the police, libraries, and schools, for funds and support.

Coordinating is putting the various functions together into a smooth and well-operating organization. For the manager who is effective in planning, organizing, and directing, it would appear that coordinating would be easy. Indeed, if these three activities have been well accomplished, coordination will be a lot easier.

Any one of several factors may interfere with good coordination. Poor communications, poor timing, resistance to change, lack of clear-cut objectives or at least the lack of understanding or acceptance of these objectives, and ineffective policies, practices, and procedures all can hinder your efforts to coordinate activities. The group's acceptance of you as a leader plays an important part, too. If the team members accept you as the leader, you will have a greater opportunity to coordinate their efforts.

THINGS THAT CAN BE COORDINATED

The resources of a station

The resources of a department

The resources of a community

Controlling

Controlling is the fifth and final item on Fayol's list. **Controlling** is monitoring our progress. If we say that planning was looking ahead to see what we are going

FIGURE 4-3 For company officers, the greatest resource is your teammates. Effective coordination of their efforts is essential for company excellence.

FIGURE 4-4 Managers should always be watching performance, measuring effectiveness, and looking for ways to improve.

to do, then we can also say that controlling is looking back to see how we have done. Controlling is similar to the feedback we talked about in Chapter 2; it tells us how we are doing (see **Figure 4-4**).

Controlling allows you to measure the effectiveness of your efforts, to help you maintain your goals, to seek new ways to improve, to increase production, and to help plan for future undertakings. It should also give you an opportunity to identify those who should be recognized for their accomplishments.

CONTROLLING HELPS US GET TO THE RIGHT PLACE AT THE RIGHT TIME

When aircraft fly to a destination of any distance, they file a flight plan. Ships file a similar document when they leave port. There are several reasons for this information, some associated with safety, but for the crew members involved in navigating the trip, it gives them a plan to follow during that trip. If they find that they are off schedule or off course, they can take appropriate corrective action early in the trip so that they arrive in the right place at the right time. Controlling is similar for you as managers; you want the organization's management actions to arrive at the right place and at the right time.

The Principle of Exception

Some textbooks mention another management principle—the principle of exception. The principle of exception suggests that management control is greatly

increased and corrective action greatly expedited when you concentrate on the exceptions to expected results. In other words, most of your controlling effort should be directed to the exceptions or to those situations that do not conform to the standards. This makes sense for two reasons: First, the exceptions are relatively few in number, and, second, the exceptions are the cases that require remedial action.

Why spend time reviewing the results that are consistent with expectations and require no corrective action? Instead, you should concentrate on the few that are out of line and will require corrective action.

MANAGEMENT BY EXCEPTION

Management by exception (MBE) suggests that the manager gets involved only when something unusual occurs. This approach allows you to focus on complex and urgent matters. The idea is to define the normal parameters. When something gets outside these boundaries, attention is called to the problem. Budget tracking is a good example. If expenses are tracking according to plans, there is little need to spend time going over the numbers. A lot of things are like this. For example, most cars no longer have an engine temperature gauge on the dashboard. Most of us really do not need to know how hot the engine is, as long as it is not too hot! A red light will warn of this condition. We only need to take action when the situation is outside the normal.

Most managers understand this concept and many practice it without realizing that they have applied this principle.

> **NOTE**
>
> Management control is greatly expedited and increases when managers concentrate on the exceptions to expected results.

CONTRIBUTORS TO MODERN MANAGEMENT

During the twentieth century, industry sought ways to increase production. The assembly line was established, and machines were invented to increase the individual worker's productivity. As the pace increased, workers became subordinate to the machines they operated. Whereas Fayol and other management scientists had set out to create more pleasant work conditions, management had created situations that reduced the work to simple repetitive tasks. Workers became detached from both the product and from the work itself. Worker satisfaction decreased, and with it, the quality of work decreased.

Many studies of management took place during the first half of the twentieth century. They make interesting reading to the serious student of management, and you will find a good discussion of these events in any good text on management. We briefly cover several of the significant contributors to this evolution, for they help us understand where organizations are headed today.

Frederick Taylor

Frederick W. Taylor (1856–1915) was an American engineer working as a superintendent in a steel mill. Like Fayol, Taylor was concerned with increasing the productivity of both humans and machines. However, Taylor's focus was on the first-line manager and the efficiency of the worker. He reasoned that if efficiency could be improved, the workers' productively could be improved, and with increased productivity, workers would have a good, steady job and would be well paid for their efforts. Taylor is called the "Father of Scientific Management."

Dr. Taylor studied the resources—the machines, tools, equipment, materials, methods, and workers' skills. He analyzed the sequence of events needed to perform a job, varied the inputs, and collected data. In a sense, Taylor was applying one of the characteristics of the Industrial Revolution to human behavior:

He saw machines as having standardized interchangeable parts. He saw humans as parts of a "human organization" with standardized tasks.

Taylor made a science of looking at each element of work and looking for ways to improve the worker's productivity. Taylor's studies included the interaction of humans in the workplace—the social environment, if you will, as well as the physical environment. The results were profound: With the application of his recommendations, productivity went up dramatically. Many of the core elements of his efforts remain popular today.

> **NOTE**
>
> Dr. Taylor reasoned that one of the major problems in the workplace was the lack of expression by managers of what they expected in terms of a day's productivity, and of the employees in knowing and understanding what that expectancy was.

Application of Taylor's work is evident in the fire service. We have division of labor, of course, but for many departments, the tasks have been narrowly defined so that little flexibility is allowed. Taylor would be pleased to know that many departments design their apparatus with standardized hose-bed and compartment arrangements "to increase productivity" and have SOPs that provide standardized solutions to routine situations.

Abraham Maslow

Abraham H. Maslow (1908–1970) concerned himself with the full range of human relationships. In 1954, he published his *Hierarchy of Needs*. We discuss the application of Maslow's theory in Chapter 6, but we introduce it here because of the role he played in the evolution of management theory. Maslow recognized that most of us have five levels of needs, and that the very first of these are the basic physiological needs—the need for air, food, and water, and for clothing and shelter to protect us from the elements. Maslow said that these basic needs must be satisfied before any worker can be productive.

Peter Drucker

Peter Drucker (1909–2005) is credited with inventing a concept known as management by objectives (MBO). Drucker was a professor who had real-world experience as a consultant in a variety of settings in the United States and elsewhere. He wrote several books including *The Practice of Management*, published in 1954.

The basis for MBO lies in managers defining areas of responsibility for their subordinates. Measurable goals are set for these areas of responsibility, and these goals are used as a standard to evaluate the results obtained. MBO requires planning and thus reduces some of the surprises that face managers. It encourages innovation and improves performance because people are working together to meet the unit's goal. MBO has been challenged because it is labor intensive for managers; it places a heavy burden on them to plan, set goals, and provide a frank appraisal of the results. Many managers find that it is not worth the effort.

Douglas McGregor

Douglas McGregor (1906–1964) was also a professor. In 1960, McGregor wrote a book entitled *Human Side of Enterprise*. In that work, McGregor introduced the now well-known concept of Theory X and Theory Y. McGregor said that many managers feel that they need to have tight controls over their organization because workers are lazy and resist the demands of their managers. McGregor called this traditional style of management **Theory X** and said that it was not a good way to manage.

McGregor proposed an alternative style, something he called **Theory Y.** He felt that people would accept work and responsibility if given a chance, and that management should provide suitable working conditions so that people could achieve personal goals while satisfying the organization's goals. We will return to McGregor in Chapter 6.

William Ouchi

In the early 1980s yet another professor, William G. Ouchi, developed a concept he called **Theory Z.** Professor Ouchi focused on workers' needs. Building on Maslow's and McGregor's ideas, Ouchi said that workers really want more than just the satisfaction of working productively; workers want to be involved in their own destiny. He also found that workers wanted to be trusted and to be entrusted. He found that many Japanese companies using this approach were realizing higher productivity and often getting better products.

With Theory Z, the focus is on the employees rather than on the work itself. The company encourages employee involvement at every level of decision making and accepts these ideas. Teams, rather than individuals, assemble products, rather than parts. As a result, workers share in the satisfaction that comes in seeing the results of their work and share in contributing to make the product even better.

In Theory Z, the relationship goes beyond the worker—family needs are recognized, too. The workers reciprocate by being totally involved with the company over a long period of time.

COMPARING THEORIES X, Y, AND Z

Theory X
- People dislike work.
- People work only for money.

Theory Y
- People want to feel important.
- People want to be recognized.

Theory Z
- People want to contribute and be part of the process.
- People represent unlimited potential.

W. Edwards Deming

We would be remiss if we failed to mention the work of W. Edwards Deming (1900–1995). Deming wrote a book entitled *Quality, Productivity, and Competitive Position.* Deming's ideas were used by the Japanese in their manufacturing process (similar to Theory Z), and he eventually brought his ideas to the American workplace. The acceptance of his ideas is still ongoing. In some circles, Deming's ideas are referred to as **total quality management,** or TQM.

NOTE

Total quality management was a fad that passed through many organizations in the latter twentieth century. Although the term may be passé, many of the ideas embraced by TQM remain very much a part of modern management. We will use the term TQM as a shorthand way to refer collectively to these ideas and concepts.

Deming's Fourteen Points (altered a bit for the fire service)

1. Understand and accept the organization's mission, goals, and objectives.
2. Seek ways to improve the process and train the people to do it better. All improvements, big and small, are important.
3. Use inspections as an improvement tool, not a threat.
4. Think beyond the bottom line. Consider quality and personal value along with the cost in every decision.

5. Demand quality, not better numbers, in everything you do.

6. Accept the idea of quality. Do every job right, the first time, every time.

7. Use training time effectively; some organizations train just to meet quotas.

8. Promote pride in the job. As supervisors, be sure that your team members have the equipment and materials they need, and recognize their good work.

9. Promote leadership. Work on giving supervisors the best tools possible to deal with the organization's most valuable resource—its people.

10. Create an atmosphere of trust and open communication.

11. Get rid of the organizational barriers that inhibit trust and effective communications, and force people to cooperate rather than compete.

12. Use analytical tools to help better understand and monitor the job.

13. Work as a team.

14. Use TQM all the time.

NOTE

Consistency of these purposes must be defined and maintained.

To be successful, TQM must put the focus on the customer's needs. Long-term TQM commitments include establishing a vision, defining a mission, and providing guiding principles for the organization. Consistency of these purposes must be defined and maintained. The TQM process involves everyone in the organization, but, because of traditional values (we expect the leaders to lead), it is usually necessary for top management to sell TQM to the other members of the organization.

NOTE

TQM is based on employee participation and the concept that all must work together to achieve quality goals while also focusing on a continuing process of improvement.

But it takes more than just words of inspiration from the leaders. TQM is based on teammate participation and the concept that all must work together to achieve quality goals. Teamwork is essential. Effective communication, both vertical and horizontal, is critical to the process. TQM uses analytical techniques to provide the data necessary for timely and accurate deci-sions. Data can be used to define the process, identify problems, indicate solutions, and monitor improvements. TQM is not driven by analytical data; it is merely supported by such information.

Total quality management focuses on a continuing process of improvement. This approach combines all that we have learned about human resource management and adds the ability to use modern quantitative analytical techniques to provide timely data for making accurate decisions.

Kenneth Blanchard

The One Minute Manager, written by Kenneth Blanchard and Spencer Johnson in 1981, caused quite a stir when published. One of the key items in the book was the suggestion that we should try to find more ways to make people feel good about themselves. The authors suggest that people who feel good about themselves do good work. The lesson is that as managers, you should focus on what is going well. The key action is this: Find someone doing something right and give this person recognition.

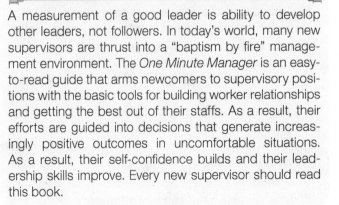

SOME COMPARISONS

Traditional values	Total quality management
Looks for quick answers	Takes new management approach
Maintains old ways	Uses small innovations to improve
Focuses on short term	Focuses on long term
Looks for errors	Prevents errors
Decides using opinions	Decides using facts
Motivated by cost factors	Motivated by customer satisfaction

OFFICER ADVICE

A measurement of a good leader is ability to develop other leaders, not followers. In today's world, many new supervisors are thrust into a "baptism by fire" management environment. The *One Minute Manager* is an easy-to-read guide that arms newcomers to supervisory positions with the basic tools for building worker relationships and getting the best out of their staffs. As a result, their efforts are guided into decisions that generate increasingly positive outcomes in uncomfortable situations. As a result, their self-confidence builds and their leadership skills improve. Every new supervisor should read this book.

Alan Brunacini

Management evolution has led managers in many organizations to focus on a concept called **customer service.** Many companies are focusing on serving the customer better and starting that process with attention to the members of the organization, often called the **internal customers.** The logic here is that if you make the internal customers (the members) happy, they will make the paying customers (the citizens) happy. The goal is to create an environment where every member wants to delight the customer. For businesses, it means they want the customer to come back. The employee handbook of a major restaurant chain illustrates the point beautifully. The job description for waiters is simply this: "Do whatever it takes to convince each of our guests to come back here for their next meal."

Alan Brunacini, a former chief of the Phoenix (Arizona) Fire Department, played a key role in introducing the concept of customer service to the fire service. He suggested that you look closely at what you are doing and think about how you treat our customers, both those inside the organization as well as the public you serve. While you are not looking for repeat business, you do want your customers to be delighted with your service. After all, they are paying the bill, including your salary.

His department's mission statement read something like this: "Prevent harm, survive, be nice." Clearly, turning this concept into a reality is not easy nor is it quickly accomplished; it takes time and energy to change attitudes. However, being nice and taking care of the customers should be an essential part of every organization, especially for a public-service organization. We explore this topic further in Chapter 5.

MANAGEMENT IN MODERN ORGANIZATIONS

Management is the active part of making an organization run and includes planning, organizing, commanding, coordinating, and controlling. We usually think of management as something done by the senior staff of an organization, and as a result, we usually relate management to the administrative or nonoperational activities essential to the organization. For many organizations, this includes the support activities, such as human resources management, training, purchasing, and communications. But management activities are found at all levels and functions of modern organizations. Most fire chiefs

FIGURE 4-5 Fire chiefs do most of their managing from their desks, not from the middle of the street.

spend more time working at their desks than commanding emergency activities (see **Figure 4-5**). Most senior officers spend more time managing the organization than managing emergency incidents. Most company officers spend more time managing than fighting fires.

Management can be enhanced through good policy, procedures, and even by personal observation. **Policy** is usually developed by fire department administration. Such policy prescribes what should be done in response to various situations. Policies are broad in nature and should be clearly understood. Personnel need to know where they stand with regard to the department's directions. The department's statement regarding harassment in the workplace is an example of a policy.

Procedure may be also be developed by the administration, but a good part of procedure is initiated at the company level. Established procedures, like established policy, simplify the daily routine for firefighters and supervisory personnel alike. Instead of each problem having to be resolved each day it comes up, established procedures make these activities routine.

Procedure has a lot to do with organization and management. Procedure is how tasks are assigned. Procedure deals with questions regarding responsibilities and the order in which tasks are accomplished. Established procedure provides standardized solutions to similar problems and usually simplifies life at work. People like having solutions; members tend to be more productive when they have standard solutions to recurring events. Some procedures are essential to the mission of the organization and to the safety of those who work there. The procedure for raising an extension ladder must always be the same. It takes practice and teamwork to do it well. Practice and teamwork do not bring rewards every day but they increase the likelihood that the ladder will be raised quickly, correctly, and safely when it is needed most.

Procedures written for administrative personnel are often developed by the city or town administration. These procedures often spell out what is required of the individual, what duties they are required to perform, and their place within the city or town's organizational structure.

Some procedures may not be so firmly fixed. For example, items dealing with the fire station maintenance program and meal preparation may vary, depending on your management style as the company officer. Some company officers may treat these procedures the same as that dealing with ladder raising, and never allow for variations. Others will allow their members to try alternative approaches to routine tasks as long as the job gets done. It is important that the members be able to separate those essential tasks that have very fixed procedures from those where some discretionary latitude is appropriate.

Another management technique is observation of the work and the workers. When company officers sit in their office doing necessary paperwork, they may think that they are managing. When this situation occurs to excess, these company officers are overlooking the most important resource they manage—the human resources under their command. Company officers who dedicate their energies to their personnel will find that the overall productivity of the company is improved.

OFFICER ADVICE

Company officers who dedicate their energies to their personnel will find that the overall productivity of the company improves.

Much has been written recently about the art of managing by walking around and observing the work and the workers firsthand. Where practiced, it is important to remember the respective role of everyone in the organization and not violate the rules we have covered. Personal observation allows you to see what is going on, engage team members in polite conversation, and make inquiries about the job as well as personnel issues. But be careful about saying things that might be inferred as direction or as responding to an organizational or personal need. You may be bypassing that chain of command.

Comparing Organizational Values

Traditional	Modern
Focused on function	Focused on process
Centralized decision making	Decentralized decision making
Tall structure, many layers	Flat structure, few layers
Distinction between line and staff	Both are joined as teams
Narrow job definitions	Wide-ranging job descriptions
Heavy reliance on rules	Heavy reliance on innovation
Emphasis on vertical communications	Emphasis on horizontal communications
Emphasis on individuals	Emphasis on teams
Focus on stability	Focus on change

HOW WELL IS YOUR FIRE DEPARTMENT MANAGED?

We introduced organizations in Chapter 3. Fire departments are organizations, and their ability to perform as such is a reflection of how well they are managed. Some departments provide excellent service to the citizens they serve, whereas others just exist. In the past, it was difficult to accurately assess the worth of a fire department: We could count personnel and apparatus and get a general feel for how well the department was prepared to deal with emergency situations. Today, there are several good analytical tools to help determine the quality of fire protection provided in a community.

Because the safety of people, property, and the environment is so vital, questions that focus on the organization and deployment of firefighting resources are especially important to communities. Fire prevention activity is also important. Although fire prevention and life-and-fire safety education efforts are less costly than fire suppression, many do not see the value in these preventive activities. As a result, we tend to focus on fire suppression resources to deal with emergencies that do occur. Both are important.

Many communities are faced with significant budget issues: public safety, comprising both fire and police protection, is a relatively expensive undertaking. Consequently, the fire department is in competition for the community's limited financial resources that also fund schools, parks, and better streets. As a result, the department's budget is usually less than it should be and in turn, limits the resources that might be available for prevention and suppression activity.

THE ROLE OF STANDARD OPERATING PROCEDURES (SOPs)

Standard Operating Procedures (SOPs) have many important functions in fire service organizations. In particular, a complete set of SOPs is the best method for departments to "operationalize" other organizational documents—by-laws, plans, policies, operational strategies, mutual aid agreements, etc. In simple terms, SOPs "boil down" the important concepts, techniques, and requirements contained in these documents into a format that can be readily used by personnel in their daily jobs. As such, SOPs help to integrate departmental operations, linking the work of managers and planners with the activities of other workers. SOPs are also essential for addressing the diverse legislative and regulatory requirements that affect fire service operations. Many laws and regulations state that fire departments must follow "safe work practices." Courts have generally agreed that this includes the development and maintenance of SOPs. In addition, many communities have ordinances that mandate compliance with standards developed by NFPA or other professional organizations. A very effective way to implement these legal requirements and national standards is to incorporate them into SOPs. By documenting the department's effort and intention to be compliant, SOPs may also help to limit liability when incidents go awry. SOPs have many other applications and benefits for fire service organizations, including:

- *Explanation of performance expectations:* SOPs describe and document what is expected of personnel in the performance of their official duties. As such, they provide a benchmark for personnel, an objective mechanism for evaluating operational performance, and a tool for promoting a positive organizational culture.
- *Standardization of activities:* SOPs identify planned and agreed-upon roles and actions. This information helps standardize activities and promote coordination and communications among personnel. SOPs also simplify decision-making requirements under potentially stressful conditions.
- *Training and reference document:* Written SOPs can provide the framework for training programs, member briefings, drills, and exercises. These activities, in turn, improve the understanding of work requirements and help identify potential problems. A comprehensive SOP manual also serves as a self-study and reference document for personnel.
- *Systems analysis and feedback:* The process of researching and developing SOPs provides opportunities for managers to compare current work practices with the state-of-the art in the fire service. Feedback from outside groups, technical experts, and staff can help to identify potential problems and innovative solutions.
- *External communications:* SOPs clarify the department's operational philosophy and recommended practices. As such, they may prove useful in communicating organizational intentions and requirements to outside groups, or enhancing the public's understanding of the fire service.

The SOP Development Process

The SOP development process can be viewed as essentially eight sequential steps that address the most important organizational and management considerations for department personnel. The methodology used for these steps can vary, depending on the scope of the project, local needs and resources, and other variables. The steps are listed below, with additional detail on each presented in the following sections:

1. Build the development team
2. Provide organizational support
3. Establish team procedures
4. Gather information and identify alternatives
5. Analyze and select alternatives
6. Write the SOP
7. Review and test the SOP
8. Ratify and approve the SOP

Build the Development Team

The constantly changing environment in which fire and EMS agencies operate requires that departments regularly analyze and update their SOPs to meet current needs. To manage this ongoing requirement, managers may find it helpful to appoint a standing group or committee to oversee SOP-related work. This team is given authority and responsibility for gathering information from department members and other sources; identifying, analyzing, and selecting from among alternative operating procedures; writing SOPs that address the department's needs; distributing draft materials for peer review and testing; and submitting final SOPs to the fire chief or agency head for approval.

SOPs are usually most effective when members of the organization are included in every step of the development process. As a general rule, departments should get input from all groups potentially affected by the SOPs. This strategy borrows on concepts described in *Total Quality Management,* a business philosophy that encourages managers to get feedback from those using a service, as well as from workers who provide it. Member participation helps boost employee morale, increase "buy-in," and promote a better understanding of the final product.

Whenever possible, all affected viewpoints should be represented on the SOP development team. Regular members of the team might include:

- Operational or line emergency responders, who carry out SOPs and have the greatest stake in their success
- Safety program representatives, such as the health and safety officer, incident safety officer, operational medical director, department physician, and safety and health committee members
- Managers, who have a broad organizational perspective that includes budgeting, public relations, political considerations, and other factors
- Representatives of labor unions and employee associations, who understand contractual issues, articulate the viewpoints of their members, and provide insight on alternative procedures
- Mutual-aid and regional response agency representatives, to help integrate and coordinate multijurisdictional responses
- Representatives from other local government agencies that may interface with department operations, such as a third-service EMS agency or municipal health agency
- Legal counsel or other personnel familiar with the content and implications of related laws, regulations, and standards
- People with technical expertise in policy or systems analysis, operations research, evaluation theory, survey design, statistical analysis, technical writing, etc.

Implementation Planning

The development or modification of SOPs must be accompanied by a plan to implement the new procedure within the department. The implementation plan provides an opportunity to think through related tasks, assignments, schedules, and resource needs. The planning process may be formal or informal, depending on the requirement.

The first step is to decide on the purpose and scope of the task to be accomplished. Several important questions need to be considered:

- How many SOPs are being implemented? A whole new set of SOPs? A major portion of the SOPs? One or two SOPs?
- How significant are the changes to existing SOPs? What are the potential consequences if the SOPs are not implemented quickly or effectively?
- Who needs to know about the changes to the SOPs? Do different groups need different types of information?
- What are effective ways of disseminating information within the department? What methods have worked before and what methods have been unsuccessful?
- Is training necessary to ensure competence in the new SOPs? How will performance be monitored and enhanced?
- Would the SOPs be most effectively implemented by using a well publicized changeover date or a defined phase-in period?

The answers to these questions determine the strategy and methods in the implementation plan. The approach might vary from simple notification during member meetings to on-the-job training for selected members to formal classroom sessions for the entire department. For example, a small volunteer fire department creating its first set of SOPs will take a drastically different approach than a larger career department doing a minor update of one or two SOPs affecting only the hazardous materials team.

NOTE: SOPs can prove to be difficult to write without a guideline. The information listed here will assist in the writing and implementation process. In addition, Appendix B provides examples using the standard SOP format for both emergency and non-emergency operations.

Source: Excerpts from "Developing Effective Standard Operating Procedures for Fire and EMS Departments," FA-197, December 1999, Federal Emergency Management Agency, 500 C Street, SW Washington, DC 20472.

In evaluating the effectiveness of fire protection, there are several issues: First is the already mentioned high cost. Fire stations, fire apparatus, and on-duty personnel are indeed expensive. A second problem is determining the level of protection required. Issues include determining the number of on-duty personnel, the maximum allowable time allowed for response, and the capabilities of these responders once they arrive on the scene.

In the United States, fire protection has always been a function of local government, but sometimes, local governments and the citizens they serve lack an understanding of the complex nature of fire protection and may not be qualified to determine what is best. This raises a significant question: Who can best judge? Should local communities determine their own needs, or should some sort of national standard be applied?

These are challenging questions. Determining the appropriate level of protection, assessing the cost of such protection, and determining the means by which these decisions are made are daunting tasks. Citizens are often presented information about the cost of fire protection, but many have no real interest, and those who are interested often lack the expertise to answer these questions. The benefits of fire prevention, public safety education, and code enforcement are not well publicized, and, therefore, there is often little pressure exerted on their behalf. Thus, relatively few resources are allocated to these activities. On the other hand, most fire protection experts know that investments in these very activities will pay off and eventually reduce the need for more expensive fire suppression resources.

Several organizations have developed methods to assist communities in evaluating the effectiveness of their local fire department. We will introduce three of them here.

Insurance Service Office

Fire insurance is unique in that the cost of the product (insurance payout) is unknown at the time of the sale when payment is made in the form of a premium. To help the insurance industry better determine the probability of future losses, several fire insurance rating organizations joined together in 1971 to form the Insurance Service Office (ISO). Today, ISO publishes the Fire Suppression Rating Schedule, a successor to the previously mentioned Grading Schedule. The rating schedule evaluates three areas of a fire department: receiving and handling of fire alarms, the fire department resources, and the community's water supply.

Each of these areas is important in the timely and effective intervention of the fire department at a typical structural fire. ISO representatives conduct field surveys to determine appropriate scores in each of these categories. Their survey involves an on-site review of the fire department's records, an evaluation of the fire department's personnel and equipment, and an assessment of the community's water supply system. Normally, this evaluation is conducted every 5 years.

The results of these surveys provide a rating. A rating of 1 through 8 indicates that there is some sort of recognized fire department, that the nearest engine company is fewer than 5 miles travel distance from any protected property, and that there is either an adequate fire hydrant within 1,000 feet or other water supply resource. (The "other water supply" is a new concept in the rating schedule; in the past, ISO gave credit only for hydrants. The new schedule recognizes the availability of surface water such as lakes and ponds. In such cases, ISO allows fire departments to demonstrate their ability to deliver an adequate water supply using tankers.) A class 9 rating is similar to a rating of 1 through 8 but it lacks an adequate water supply. A class 10 rating indicates that there is essentially no fire protection. Such a situation would typically exist where the protected property is beyond the 5-mile travel distance of the nearest fire station, and there is no water supply.

In each category, the field evaluators are guided by a detailed list of performance indicators upon which to base their evaluations.

- *Receiving and handling of fire alarms* (10 percent of the total score) considers telephone service, including the number of incoming lines, equipment, and listings in the local telephone directory, the number of personnel on duty, and the number and types of dispatch circuits.

- *Fire department resources* (50 percent of the total score) considers engine companies, reserve pumpers, pump capacity, ladder companies, reserve ladders, the distribution of these resources, the staffing of these companies, and the training of assigned personnel.

- *Community's water supply* (40 percent of the total score) includes water main capacity and hydrant distribution, types and method of installation of hydrants, and hydrant inspection and maintenance activity.

These scores are converted to ratings as shown in **Table 4-1**.

ISO evaluations provide a method of determining fire insurance classifications for the purposes of determining fire insurance premium calculations. Some think, incorrectly, that ISO mandates certain elements of fire protection. In fact, ISO sets no mandates, nor does it provide recommendations regarding loss prevention and life safety.

How does all of this improve the fire department and the fire protection provided within a community? The fire protection grading process not only determines insurance rates for individual properties but also points out deficiencies in the community's fire defenses. A good fire department will see that certain actions will improve its rating. Recognizing these needs, the department can systematically go about the process of making incremental steps to improve its next evaluation. It is important to note that this rating is only applicable to a department's fire suppression activities and does not provide an evaluation of how well a department may be able to provide other services such as medical.

FIREGROUND FACT

There is a clear correlation: Better fire protection leads to lower insurance costs. Local jurisdictions can take steps to improve their capabilities and lower the cost of fire insurance for their citizens. Few, if any, other municipal services have this opportunity.

TABLE 4-1	How ISO Ratings Can Affect Fire Insurance Premiums		

Score as evaluated by ISO Surveyors	ISO Classification	Insurance Premium $150,000 residence	1,000,000 office building
90 to 100	1	$650	$2,950
80 to 89	2	$650	$2,980
70 to 79	3	$650	$3,020
60 to 69	4	$650	$3,040
50 to 59	5	$650	$3,060
40 to 49	6	$650	$3,120
30 to 39	7	$670	$3,230
20 to 29	8	$777	$3,330
10 to 19	9	$972	$3,440
0 to 9	10	$1,072	$3,710

While these figures are reasonably representative of the difference an ISO rating can have in the midwest portion of the United States, the actual premium and credit for the various ISO classifications will vary based on the insurance company utilized and the geographic region in which the structure is located.

NFPA Standards for Fire Department Response

The ISO rating system provides a valuable service; however, because of ISO's focus on fire suppression and because of the relatively infrequent on-site evaluations, some feel that it is too limited to help communities face future issues. In response to this need, NFPA developed two documents.

NFPA 1710

The first, NFPA 1710, *Standard for the Organization and Deployment of Fire Suppression Operations, Emergency Medical Operations, and Special Operations to the Public by Career Fire Departments*, defines response capabilities for fire and medical incidents for the first-due engine company, for the entire first-alarm assignment, and for first-level, basic, and advanced emergency medical responders. The number of firefighters necessary for initial fire attack and other roles are specified in the standard, as are other aspects of the organization, suppression, operations, and emergency medical and other specialized services.

NFPA 1720

NFPA 1720, *Standard for the Organization and Deployment of Fire Suppression Operations, Emer-*

gency Medical Operations, and Special Operations to the Public by Volunteer Fire Departments, deals with similar issues for organizations that are staffed predominately by volunteers.

While these standards provide benchmarks, it is up to the local jurisdictions to determine their level of performance and to determine what changes, if any, are needed to improve services.

Accreditation

Accreditation is a process by which an agency or institution evaluates and recognizes a course of study, such as Fire Science Program, or an institution, such as a college or university, as meeting certain predetermined standards. Accreditation is awarded only to programs or institutions, not to individuals. Thus, a program may be accredited by an agency, such as the International Fire Service Accreditation Congress (IFSAC) or the National Professional Qualifications Board (NPQB); a college may be accredited by one of the several regional accrediting agencies. For colleges and universities, accreditation is essential to their ability to award recognizable degrees.

An accreditation process is now available for fire departments through Certified Fire Accreditation, International (CFAI). CFAI is affiliated with the International Association of Fire Chiefs. In this case, accreditation is purely voluntary.

The objectives of accreditation follow:

- To create organizational motivation and self-improvement
- To provide a voluntary activity focused on self-evaluation and education as a viable means to improve service delivery
- To provide a means to recognize quality performance
- To protect the interests of the general public[1]

Generally, accrediting agencies set standards by which agencies, programs, and institutions are measured, and they define the evaluation process. Most accreditation programs involve a self-assessment by the agency, program, or institution itself, a process that can take several years. This is followed by a brief (2 to 3 days) on-site visit by a team representing an independent accrediting body to validate the self-assessment. The team will issue a report of its findings, commenting on the strengths and weaknesses of the organization and will make recommendations for further improvement. The report is usually first presented to the applicant for validation and then to the accrediting agency.

The intent of the CFAI program is to provide an accreditation process that is credible, realistic, and achievable. It can be used by fire departments, city and county managers, and elected officials to evaluate community fire risks and to improve the delivery of emergency services. The thrust of the program is to improve these services and to reduce the risks associated with fire and related hazards. The self-study process provides a list of questions to be answered by the department seeking accreditation. Certain core-value questions must be answered affirmatively.

ISO evaluations, meeting NFPA standards, and seeking accreditation through CFAI are all good tools for managers. All three use accepted benchmarks to objectively measure a department's abilities. Using these accepted tools, a department can determine its own strengths and weaknesses and can undertake action to make improvements. With these improvements, the department will better serve the needs of the citizens of its community.

Evaluating Service Demands

Fire management should continually monitor the response characteristics of its department using the best available data. Not only should the response of the first-in units be measured, but the average time for the second-in units as well as the time to get a complete response on the scene (according to the appropriate NFPA standard presented earlier) should be analyzed. In addition, the workload of individual units should also be measured to see if there are any excess workloads. Excess workloads can lead to low response reliability for a company, meaning they may already be out on a call within their immediate response area when another call comes in. Consequently, the next closest unit needs to be dispatched, resulting in longer response times. If an overload exists, the first strategy considered should be to share workloads among existing co-located units, e.g., have another unit from the same station pick up some of the call overloaded. This observes Fayol's first principal, which is to "Divide the work into manageable portions and specialize the activities to improve efficiency." This may go against the grain of some departments where certain companies have function-specific assignments and may not normally respond to call of a particular nature.

If a particular area of the city develops unexpectedly heavier demand than can be handled by the units in the stations, then adding another unit to the overloaded station or neighboring station should be considered to relieve the load. The next option would be to consider opening a new station.

Fire risk levels should also be evaluated as part of the response demand. Some areas may not be equipped with fire hydrants, which create additional challenges related to response and firefighting tactics.

Other factors that should be included into evaluating a department's response capability is the level of training that personnel are trained to. In other words, does your department respond to medical calls as first responder or advanced life support? The level that your department's personnel are trained to will dictate if or how well they can provide a service.

Finally, does your department have mutual aid or first response agreements with other departments within fringe areas of your protection district? First-response agreements can have both a positive and negative impact on your department's ability to provide service in its first in coverage area.

Data available to evaluate and manage these issues can be collected from a number of sources. One of the most valuable services is a Geographic Information System, commonly referred to as GIS. This system uses statistical response data to create maps that can graphically illustrate: where the most responses occur, average time for certain response areas, call types, call demands and many other services. The advantage of the graphic displays is that they are quick and easily to read. So it is easy to make an analysis about an issue or even a number of very complex issues without having to be a statistician. Data that are normally evaluated include number of responses, call type, time of call, time to be en route to a call, response time, and time of arrival of additional units (if required). Most of this information is available from dispatch. However, valuable information can also be garnered from the incident reports. For example, the type of call that a unit actually responded to can be collected from the report and compared to the type of call that the unit was dispatched to. This can be used to evaluate the effectiveness of the call taker.

THE ROLE OF ETHICS IN MANAGEMENT

This discussion on management would not be complete without some mention of ethics.

The fire service has a long and rich history of providing service to the citizens of our communities and generally enjoys an excellent reputation, a reputation that has been well earned and that we do not want to lose. The ethical standards of a fire department are influenced by what society expects and what the local community believes is the function of the department. Recent polls indicate that more people have trust in their fire department than almost any other public institution.

Anyone in public office faces a challenge of having a greater ethical obligation than that of the average citizen. Everyone in the fire service is serving in a public office. For officers, there is an increased obligation, because, on or off duty, you are expected to lead by personal, positive example and take appropriate corrective action when things are wrong.

What are ethics? **Ethics** can be described as a system of conduct, principles of honor and morality, or guidelines for human actions. For the most part, these definitions are rather vague. Even where a professional organization sets standards for its individual members, there is always some degree of variance in how these "rules" are to be interpreted, and there are always a few who will try to beat the system.

FAYOL'S BRIDGE

If members precisely followed the formal rules for organizational behavior, the only people they could talk to would be their own supervisor and their own subordinates. All other communications should flow through the chain of command.

Suppose the officer of Ladder Company 1 wants to use the training academy's facilities during the coming weekend. Following all the formal rules, you, as the company officer, would have to ask your boss who in turn would ask another boss up to a point where there is a person who has supervisory authority over both the department's field activities (the ladder company) and the academy. In many departments, that person might be the fire chief. Then the request flows down the organization until it gets to the academy. The answer would have to flow back the same way.

This arrangement is obviously impractical. While we usually think about delegation and decentralization as relatively modern concepts, we can look back to Henry Fayol for a solution to this basic organizational problem. Fayol realized a century ago that rigidly following the chain of command was an impractical arrangement. He said that for organizations to be effective, the manager in one section should be able to communicate with the manager in another section, without having to go all the way to the top of the organizational structure. The shorter path is called **Fayol's bridge** (see **Figure 4-6**).

If the fire chief authorizes the officer of Ladder Company 1 to call the training officer of the academy to use the academy's facilities, and if the fire chief authorizes the training officer to respond directly to such requests, the chief has created an environment where members do not have to refer routine issues up the organizational chain of command for resolution. This speeds the process and leaves the senior officers free to deal with more important issues.

For those who work in organizations where this takes place daily, you may be wondering why we have taken half a page of text to describe the process. For those of you who work in organizations where this direct liaison never occurs, you may be wondering, "How can this work?" Our purpose here is to point out that this practice is common in well-run organizations, yet there are a lot of places where using Fayol's bridge would be unthinkable. How about your department?

Modern management suggests that we should do away with as many of the obstacles to getting our business done as we can. Certainly we need to have organizations and understand organizational structure. Certainly we need to have some rules of conduct for those who work in these organizations. But at the same time, we need to provide a workplace where business can be done effectively.

There will always be exceptional situations that are outside the parameters of the normal day-to-day business that should be referred to upper management. But resolving the routine issue at the lowest possible level adds to the quality of life of the organization and for all who work within it.

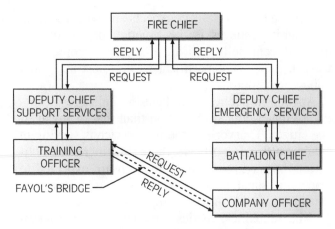

FIGURE 4-6 Information relationships and effective delegation within organizations facilitate crossing the organizational lines to get things done. This concept is known as Fayol's bridge.

Ethics are not new. We usually have a feeling of guilt when we do something wrong. This feeling is manifested in many ways, including physical reactions on our skin and in our stomach. The history of ethics goes back thousands of years. You can find evidence of ethics in the earliest recorded history. Many civic, fraternal, professional, and even religious organizations have a set of ethical values and principles. These organizations provide guidelines for their members and encourage members to follow these practices at all times. These rules cover our actions with others: honesty, courage, justice, tolerance, and the full use of our talents. Doing so improves the relationships within the group, and the group's image also enhances our relations with others.

Two well-known authors, Dr. Kenneth Blanchard, the coauthor of *The One Minute Manager* and many other books, and Dr. Normal Vincent Peale, the author of *The Power of Positive Thinking* and many other books, collaborated on a book entitled *The Power of Ethical Management*.[2] In that book, they suggest a simple three-question test to help guide our actions. The three questions follow:

- Is it legal?
- Is it balanced?
- How will it make me feel about myself?

Ethics have a direct impact on the management of the fire service, from the chief to the company officer. At the company level, you, as the company officer, must often make decisions for which there are no clear-cut guidelines. At the same time, through personal example and positive leadership qualities, you are expected to set an example for others in the department.

There are laws and regulations that address ethical issues, but these have little influence on those who wish to be unethical. What is effective is a well-defined culture or a value system within the department that acts as a regulator of conduct. The laws and regulations that impact us as citizens change. Ethics often begin where laws leave off. In the area of team member relations, issues surrounding diversity and harassment are very much in the news these days. These are ethical issues. We read stories of allegations being brought against individuals regarding their personal conduct on these matters. Not too many years ago, such conduct was accepted, although it was just as improper then as it is now.

Ethics often begin where laws leave off.

With a change in public attitude, increased litigation, and the attention brought on by the press, what used to be acceptable no longer is. These changes in attitude impact our delivery of services as well as the way we operate within our own organization.

Laws and regulations and even an organizational code of ethics can only serve as guidelines. Our personal system of values is what really determines our behavior. This system is influenced by many factors, including family values and culture, community

NFPA 1021 INCLUDES A FEW WORDS ON ETHICS

Fire officers are expected to be ethical in their conduct. Ethical conduct includes being honest, doing "what's right," and performing to the best of one's ability. For public safety personnel, ethical responsibility extends beyond one's individual performance. In serving the citizens, public safety personnel are charged with the responsibility of ensuring the provision of the best possible safety and service.

Ethical conduct requires honesty on the part of all public safety personnel. Choices must be made on the basis of maximum benefit to the citizens and the community. The process of making these decisions must also be open to the public. The means of providing service, as well as the quality of the service provided, must be above question and must maximize the principles of fairness and equity as well as those of efficiency and effectiveness.[3]

attitudes, the influence of laws, our own set of values, and life's experiences. We all have ethical values. Where we get into trouble is when our values do not coincide with those of our organization or society.

The decisions faced by officers in the fire service do not fit neatly into a tidy framework of values. Often operating without supervision, you, as the company officer, are responsible not only for the conduct of the company members but also for their productivity. It is up to you to see that the company remains ready to deal with emergencies and that the many activities needed to maintain readiness are accomplished.

THE INTERNATIONAL ASSOCIATION OF FIRE CHIEFS' CODE OF ETHICS

- Maintain the highest standards of personal integrity, be honest and straightforward in dealings with others, and avoid conflicts of interest.
- Place the public's safety and welfare and the safety of firefighters above all other concerns; be supportive of training and education that promote safer living in occupational conduct and habits.
- Ensure that the lifesaving services offered under the members' direction are provided fairly and equitably to all without regard to other considerations.
- Be mindful of the needs of peers and subordinates and assist them freely in developing their skills, abilities, and talents to the fullest extent; offer encouragement to those trying to better themselves and the fire service.
- Foster creativity and be open to consider innovations that may better enable the performance of our duties and responsibilities.[4]

If company personnel are expected to have daily physical activity, conduct fire safety inspections, and prepare preplans for their first-due area, the zeal and effectiveness with which these activities are conducted will probably go unnoticed by the public and to a large extent by the department. It becomes a question of doing what is right. If you fail to properly accomplish these tasks, who will step forward to see that they are done?

The standards for ethics change. Today, there seems to be increased attention on ethics, especially for those in public office. Nearly every day's news contains at least one story of alleged unethical conduct on the part of someone in public office. Although our Constitution guarantees that all citizens are presumed innocent until proved guilty, public opinion can quickly condemn an individual or group without the benefit of a trial. Once an individual or organization is tainted by allegations of unethical conduct, restitution is difficult. A lifetime of ethical behavior can be wiped out by a single breach of the ethical principles.

NOTE
The standards for ethics change.

As officers, you have an obligation to act in an ethical manner at all times, to demonstrate integrity, honesty, courage, and faithful productivity. Some will say, "That's not the way we have been doing things here." We acknowledge that, and we note that few of us can make major changes in history. However, all of us certainly can control the present and maybe even shape the future in some small positive way. Ethics is an issue today and must be faced by all organizations. As managers, you must accept the fact that you are responsible for your actions as well as those of your team members. You must define your ethical values and follow these values through personal example.

ETHICS
Ethics are an important management issue.

CODE OF ETHICS

- Be loyal to the highest moral principles and place allegiance to country above loyalty to any person, party, or government department.
- Uphold the Constitution, laws, and regulations of the United States and all of the governments therein.
- Give a full day's labor for a full day's pay, giving earnest effort and best thought to the performance of duties.
- Seek to find and employ more efficient and economical ways to accomplish tasks.
- Never discriminate by dispensing special favors or privileges to anyone, and never accept favors or benefits under circumstances that might be construed by reasonable people as influencing the performance of duties.
- Make no private promises of any kind that are binding on the duties of your office because your private word cannot be considered as binding on your public duty.
- Engage in no business that is inconsistent with the conscientious performance of duties.
- Never use any information gained confidentially in the performance of duties as a means to make a private profit.
- Maintain high standards of personal integrity, honesty, and straightforwardness.
- Provide fair and equal lifesaving service to all.
- Expose corruption wherever you discover it.
- Uphold these principles, ever conscious that a public position is a public trust.[5]

INTRODUCING CHANGE

Michael Washington, Fire Lieutenant, Cincinnati Fire Department, Cincinnati, Ohio

When Michael Washington was promoted to lieutenant in 1999, he was moved to a new station where he knew none of the people who would be working with him. He knew this would be a challenge, earning their respect and also getting them to think of him as the head of their team, all aligned for the same purpose: to get the job done.

As a new officer he thought that it might be best for him to run things at status quo—the way the department was used to operating—for several months before instituting any new rules or programs. It kept the team comfortable, but it also gave him a chance to get to know the company—its people, its assets, and its flaws.

He made it a point to meet with each of the firefighters, letting them know his expectations of their positions, setting a bar for them to aim for. The meeting immediately established that Washington, though young, was not to be taken lightly; the meeting also let him get to know what each person liked best about his or her job. Combined with watching each person on calls, he was able to determine what each individual's strengths were: One person was particularly good at technical operations like rope-rescue extrication, another was the go-to for

emergency medical, another had an unmatched brute strength. Understanding this was an asset because he could lead the company more smoothly, putting in the right person for the job. He has followed up regularly during the review process, asking all persons what grade they feel they deserve.

When he eventually started implementing new procedures, a few people did leave, but Washington expected that; he knew he could not make everyone happy. Their departures gave him the opportunity to get people who truly wanted to be working for him.

And in the end, Washington ended up with a smoothly run company. In fact, when he had to take a month of leave last year, the company actually ran itself.

Questions

1. What do you think about Lieutenant Washington's approach—go with the "status quo" for a while?

2. What do you think about his approach for sizing up each member of the team?

3. Why did people leave after Lieutenant Washington started to implement changes?

4. When Lieutenant Washington took leave, he found that the company "ran itself." Do you think this is desirable?

5. What had he done to make this happen?

LESSONS LEARNED

As a company officer, you are part of the department's management team. In keeping with contemporary management trends, many departments are passing much of the management process down to you, and beyond. When permitted to do so, you can have a great impact on the effectiveness and efficiency of the entire organization.

The information in this chapter can be used at the company level. As company officers, you should be aware of and should be using all the functions of management. You should be aware of the models of management style that have evolved through the science of management and use them as needed.

As with other tools you use on the job, no one uses all of the tools all of the time. However, it nice to know which tool is the right tool for the job at hand and to know how to use it effectively.

Ethics are important for you too. The best way to have good, fair, and honest people in the station environment is for you, the company officer, to set a good personal example and to expect the same level of performance and behavior from all others. We further explore practical applications of management in Chapter 5.

KEY TERMS

commanding third step in the management process, commanding, involves using the talents of others, giving them directions and setting them to work

controlling fifth step in the management process, controlling, involves monitoring the process to ensure that the work is accomplishing the intended goals and objectives, and taking corrective action when it is not

coordinating fourth step in the management process, coordinating, involves the manager's controlling the efforts of others

customer service service to people in the community

ethics system of values; a standard of conduct

Fayol's bridge organizational principle that recognizes the practical necessity for horizontal as well as vertical communications within an organization

internal customers members of the organization

management accomplishment of the organization's goals by utilizing the resources available

management by exception (MBE) management approach whereby attention is focused only on the exceptional situations when performance expectations are not being met

organizing second step in the management process, organizing, involves bringing together and arranging the essential resources to get a job done

planning first step in the management process, planning, involves looking into the future and determining objectives

policy formal statement that defines a course or method of action

procedure defined course of action

Theory X management style in which the manager believes that people dislike work and cannot be trusted

Theory Y management style in which the manager believes that people like work and can be trusted

Theory Z management style in which the manager believes that people not only like to work and can be trusted, but that they want to be collectively involved in the management process and recognized when successful

total quality management focus of the organization on continuous improvement geared to customer satisfaction

REVIEW QUESTIONS

1. Define *management.*
2. What are the functions of management?
3. Among the functions of management, planning seems to get a lot of attention. Why is planning so important?
4. Are company officers managers?
5. What resources are available for management by company officers?
6. Describe the characteristics associated with Theories X, Y, and Z.
7. Who are the customers of a fire department?
8. What resources are available to assess the quality of a fire department?
9. What is meant by the term *ethics?*
10. How do ethics tie in with the company officer's position?

DISCUSSION QUESTIONS

1. How does management theory apply to the fire service?

2. Discuss Fayol's principles in the context of a modern fire department.

3. Which of Fayol's functions would be represented in each of the following:

   - Selecting the site for a new fire station
   - Arranging resources at the scene of an emergency event
   - Examining the most efficient use of existing apparatus
   - Determining the cause of a gradual increase in response time
   - Assigning tasks to accomplish the daily needs of a fire station

4. What lessons can be learned from a study of management theory?

5. How has management practices changed over the past 50 years?

6. How do you think management practices will change over the next 10 years?

7. The text discussed several ways to assess a fire department's capabilities. What value is there in using such external benchmarks in measuring such an organization?

8. How do you feel about ethics in public service organizations?

9. Why are ethics important in such organizations?

10. Explain the management roles of a company officer.

ENDNOTES

1. Commission on Fire Accreditation International. *Fire and Emergency Service Self-Assessment Manual.* Fairfax, VA: The Commission, 2003.

2. Kenneth Blanchard and Norman Vincent Peale. *The Power of Ethical Management.* New York: Fawcett Crest, 1988.

3. Reproduced with permission from NFPA 1021, *Fire Officer Professional Qualifications,* Copyright©2009, National Fire Protection Association. This reprinted material is not the complete and official position of the NFPA on the referenced subject, which is represented only by the standard in its entirety.

4. Code of Ethics of the International Association of Fire Chiefs. Used with permission.

5. Robert D. Carnahan, "A Code of Honor." *FireRescue Magazine,* January 1998, p. 78. Reprinted with permission.

ADDITIONAL RESOURCES

Blanchard, Kenneth, and Spencer Johnson. *The One Minute Manager.* New York: William Morrow, 1981.

Blanchard, Kenneth, and Norman Vincent Peale. *The Power of Ethical Management.* New York: Fawcett Crest, 1988.

Bryan, Peter, and Pamela Pane. "Evaluating Fire Service Delivery." *Fire Engineering,* April 2008.

Compton, Dennis, and John Granito, eds. *Managing Fire and Rescue Services.* Washington, DC: International City Management Association, 2002.

Favreau, Donald. *Fire Service Management.* New York: Dun-Donnelley, 1973.

Hickey, Harry. *Fire Suppression Rating Handbook.* Louisville, KY: Chicago Spectrum Press, 2002.

McGregor, Douglas. *The Human Side of Enterprise.* New York: McGraw-Hill, 1960.

McLaughlin, Patrick. "Growing Pains: Strained Resources Challenge Fire Departments and Insurers Alike." *Fire Chief,* January 2004.

Newman, Jerry. *My Secret Life on the McJob.* New York: McGraw-Hill, 2007.

Peter, Lawrence J., and Raymond Hull. *The Peter Principle.* New York: William Morrow, 1969.

Peters, Tom. *Thriving on Chaos.* New York: Alfred A. Knopf, 1987.

5

The Company Officer's Role in Managing Resources

While I am the captain at the fire department and deal with fires, I also specialize in EMS and other related incidents, most commonly Hazardous Materials Response. I have had several incidents where I have commanded the Haz-Mat response portion of the situation, and what I have found is that while each incident is similar, all are unique and come with their own set of management issues.

The first thing I have always found to be important is to know ahead of time what resources you will need in any job to accomplish the goals you set for yourself. The goals and tasks may keep growing as the incident moves forward and develops, but if you have the right resources to start you moving in the correct direction, you're off to a good start. I have also learned throughout the years that the resources you have may be the only ones you will get, so you need to make the most of them. Many times our Haz-Mat team travels to other fire department districts to assist them, and what we bring with us is truly all we will have to accomplish the task.

Another resource at any team leader's disposal, and perhaps the most important one, is his or her personnel. One thing that an officer must do when moving to get a job done is take stock of all the resources, including personnel, equipment, and time. Personnel will most likely be the toughest one to get and deal with, but choosing the right person for each job can make all of the other resources more effective. In larger incidents, I have found that you certainly cannot manage it all by yourself. I realize that even as the ranking officer at the scene, that does not automatically make you the smartest person in the room. When you share the goals that need to be met and ask for input from your crew, that's making the most of all of your resources. It also usually turns out that your span of control is more limited in relation to how difficult the task is. When I am at a Haz-Mat incident, I am always limited in the number of qualified Haz-Mat responders whom I have on the scene. I have to be able to look at the people I have available and know which person to set to which tasks.

As an officer, you have to know your crew members and their capabilities, and you have to know that some jobs will run more smoothly with some people leading compared with others. Managing those kinds of skills and resources will make for a more effective operation.

—*Tom Bentley, Captain, Quincy Fire Department, Quincy, Illinois*

LEARNING OBJECTIVES

After completing this chapter the reader should be able to:

5-1   Describe the importance of organizational vision and mission statements.

5-2   Understand how goals and objectives support the department's mission.

5-3   Describe the business plan.

5-4   Identify principles of human resource management.

5-5   Define the principles of project management.

5-6   Compare management and member rights.

5-7   Describe the rights of members within the scope of their duties.

5-8   Describe the types of labor agreements that are commonly utilized between the organization and its members.

5-9   Describe management's rights within the scope of their duties.

5-10   Describe the purpose of a budget.

5-11   Determine the resources needed to establish a budget.

5-12   Explain the parts of a budget.

5-13   Identify the various data fields found on budget forms.

5-14   Identify the various types of budgets.

5-15   Describe how a budget affects the overall development of the agency business plan.

5-16   List material needs for annual departmental programs.

5-17   Describe the approval process for the budget.

5-18   Describe the timeframes for budget development.

5-19   Describe the purchasing process guided by the laws governing purchasing.

5-20   Differentiate between a vendor and a contractor.

5-21   Apply the purchasing laws including soliciting and awarding bids for an item or a service to be purchased.

5-22   Describe the parts of a budget request.

5-23   Explain how to legally expedite purchases.

5-24   Explain the guidelines needed to submit a request for purchasing equipment.

5-25   Define Request for Purchase (RFP).

5-26   Explain the approval process for a purchase request.

5-27   Define Request for Quote (RFQ).

5-28   List the criteria for bidding purchases.

5-29   Describe the departmental policies as they relate to requesting a budget item.

5-30   Describe how to assemble the information for budget inclusion.

5-31   Identify agency policies required for funding a budget.

5-32   Define how to effectively manage human resources, time, and the department's financial resources.

5-33   Identify procedures for the development of item specification requirements.

5-34   Identify potential sources of needed resources.

5-35   Estimate the cost of proposed needed resources.

5-36   Construct a timeline for ordering materials.

5-37   Identify agency policies required for acquiring resources.

5-38   Identify time management skills that will help make efficient use of human resources.

5-39   Identify the types of records that are retained by the department.

5-40   Understand the necessity of maintaining accurate, up-to-date records.

5-41   Explain the process for implementing project management principles.

5-42   Describe the steps for solving problems.

5-43   Identify the emerging technologies within the fire service.

* The FO I and II levels, as defined by the NFPA 1021, the *Standard for Fire Officer Professional Qualifications*, are identified in different colors: FO I = black, FO II = red.

INTRODUCTION

This chapter brings together everything we have covered so far. We look at your role as the company officer in the organization, how the organization works, how management helps the organization do its assigned mission, and finally how we communicate effectively within the organization. Some of this material may seem like a repeat of the previous chapter, but the focus here is on your practical application of the material presented in previous chapters and on how you apply these to being a supervisor and function as the officer in charge of a fire company or operation. As the company officer, you have direct regular contact with the employees you supervise. In fact, you probably have more contact and influence on the station firefighters than any other person within the organization. Often, company officers do not have subordinate supervisors or other supervisors under them. As a result, they are not only required to assist in performing the work functions but also provide leadership to get it accomplished. It is a demanding but very rewarding position.

VISION AND MISSION STATEMENTS

Let us start with the big picture, a **vision** of what you are and where you are going. With a clear understanding of your organization's vision, you can set goals and plan your own destiny. Remember that personal life and professional life are equal and should be interchangeable. In this chapter, we look at how we do this for your work position, and to some extent, for your personal life.

Organizations should have a vision, too. As members, we should be part of that organizational vision and realize that if we want to be comfortable and effective in our activities at work, we need to share at least some of the same vision that has been adopted in our workplace.

Mission statements declare the vision for the organization. A mission statement is like putting up a sign that says what you are and what you do. Mission statements tell your members and your customers the services you provide. Many mission statements define the priorities and values of the organization.

In the past, a fire department may have had it a little easier in writing its mission statement. In those days, a completely adequate mission statement may have read something like this:

> Our job is to prevent fires, but when fire occurs, to protect life and property and confine and extinguish the fire.

In recent years, the job of the fire service has become more complicated (see **Figure 5-1**). Many departments have worked hard to define their mission in the face of these changes. Here are two examples:

> Our mission is to provide emergency and nonemergency services for the protection of the lives and property of the citizens entrusted to our care.

> Our mission is to save lives, preserve property and the environment, and ensure the health and safety of ourselves and the community.

Expanding the basic list of services for the Plain City Fire Department, we get the following:

> The mission of The Plain City Fire Department is to provide a range of programs to protect the lives and property of the inhabitants of Plain City from the adverse effects of fire, sudden medical emergencies, or the exposure to dangerous conditions created by humans or nature.

Some mission statements go beyond defining the scope of services that are provided to the community to say something about the organization's philosophy

FIGURE 5-1 As fire departments take on new missions, their personnel are expected to take on new skills. *(Photo courtesy of Mark D. Baker)*

or to suggest the way that these services will be provided. The following is an example:

> This department will provide emergency services to the citizens and businesses of the city at fires and other disasters. We will also attempt to prevent such emergencies through safety education programs and code enforcement. These services will be provided in a prompt, courteous, and professional manner.

A mission statement may also suggest that there is a continuing effort to improve on what the organization does as well as a comment about management style. The mission should also suggest everyone is committed to the mission. This commitment must be a voluntary all-hands everyday action. Here is an example:

> We serve with honor and integrity. We treat everyone with fairness and respect. We provide friendly and efficient service. We value the public's confidence and trust. We are responsible for our actions. We pursue excellence. We work as a team. We make our community even better.

Another mission statement follows:

> Our mission is to protect the lives, property, and environment of all the people within our community by preventing the occurrence of and minimizing the adverse effects of fires, accidents, and other emergencies. This will be accomplished through community education, preparedness and professionalism, and a focus on quality customer service.

Mission statements should be more than just nice words and a sign: They should accurately reflect the focus and the philosophy of the organization and should be the result of a collaborative effort to gather ideas from representatives of the entire department, working together in a committee process. The mission statement should be accepted and supported from the top down, not just with words but with actions that clearly indicate that these words are really what we are all about. Most importantly, the words in the statement should be understood and practiced by everyone in the organization. Does your organization have a mission statement? Do you know it? Review it to see if you think it fits the concept as described above. If not, why?

We will see more about management style, employee commitment, and a focus on improvement in the following sections.

NOTE

An effective mission statement needs to be more than just nice words and a sign. It needs to be something that you can put into action.

GOALS AND OBJECTIVES

Following the mission statement, goals and objectives must be established to bring the vision of the mission statement to life. **Goals** are the larger, strategic statements that may take several years to accomplish. Goals help you identify where you are going.

Objectives are usually an element within a goal and represent the efforts of a shorter period. Goals and objectives should be consistent with the department's mission, and in fact, they should support it. Goals and objectives help us identify where we are going and help us measure progress as we make the trip.

To illustrate, let us suppose that you are undertaking a significant personal trip, one that will require several days of driving. Because the purpose of the trip is to attend an event that is occurring at your destination, let us say your grandmother's surprise birthday party next Saturday, you must be there at a certain time, or the whole effort is a waste of time. As you plan the trip, your goal is to arrive at your destination on Friday afternoon. Your objective may be to arrive at specific intermediate points by suppertime of each day while en route.

You might question the value of all this goal-setting activity and say that it appears to be a waste of time and energy. Most successful organizations have mission statements and set goals. They train their members on the importance of these visions and provide training to prepare them to do their duties consistent with these goals.

NOTE

Goals and objectives help us identify where we are going and help us measure progress as we make the trip.

You may think that all of this goal-setting activity is done by the department's top management, but it is better if everyone has an opportunity to contribute to the setting of goals and objectives for the organization. The more participation in the process, the better the results. Participation also enhances acceptance by those upon whom these goals and objective will impact.

SETTING PERSONAL GOALS

- GOAL: Do something each day to make this place better.
- OBJECTIVE: Make each day productive.

- QUESTIONS you should ask yourself: What can I do at the company level
  - ☐ To define work that needs to be done?
  - ☐ To provide the training, if needed?
  - ☐ To provide the necessary materials?
  - ☐ To plan and schedule the activity?
  - ☐ To check the progress of our efforts?
  - ☐ To recognize the accomplishments of individuals?

NOTE

The more participation in the process, the better the results. Participation also enhances acceptance by those upon whom these goals and objectives will impact.

USING THE FUNCTIONS OF MANAGEMENT

In Chapter 4, we listed the five functions of management as planning, organizing, commanding, coordinating, and controlling. Many modern managers also reference the *Project Management Body of Knowledge,* or the *PMBOK® Guide.* The *PMBOK® Guide* is a recognized standard that identifies the fundamentals of project management across a wide range of responsibilities and projects, including the automotive, construction, and engineering industries. The guide describes processes by which to accomplish the work and is consistent with other management standards such as *ISO 9000®. ISO 9000* is a family of standards for quality management systems. *ISO 9000* is maintained by **ISO,** the International Organization for Standardization, and is administered by accreditation and certification bodies. It is a good guide by which to manage most regularly scheduled projects. The five processes the guide identifies are Initiating, Planning, Executing, Controlling and Monitoring, and Closing.

Let us see how planning and organizing relate to goal setting at the company level.

Planning

Planning is a systematic process and should be ongoing. Planning should always start with a clear understanding of the goals and objectives. A good plan has clearly defined goals, a logical set of objectives, a list of resources needed, and a method to measure progress.

NOTE

Planning should always start with a clear understanding of the goals and objectives.

Most organizations and successful company officers develop and operate off of a plan. A **business plan** is a written formal set of goals that convey the mission of the organization, the company, or the individual; the reasons why these goals should be attained; and how and when the organization, company, or individual intends to reach those goals. Do you want to attain an associate's degree this year because you want to eventually attain a bachelor's degree 3 years from now? Most business plans also contain background information about where the organization, company, or individual has been.

The most important part of any business plan is that it puts down on paper where you want to be 1, 3, or 5 years from now. Do you know for sure? If you do not know where you intend to go, how are you going to get there? It is not possible to plan without knowing where you are going. Sometimes it takes a little creative thinking or dreaming, but in the end it is worth it. A good way to get started is think about where you were a year ago and what you have accomplished and how it came about. Then build on that and figure out where you want to be a year from now and how you are going to get there.

A business plan can focus on a number of different issues, including financial goals, organizational goals, or project goals. Do you want to get a degree or a certification? What steps are you going to take to get it? Maybe you do not think it is accomplishable in a year. That is when your 1-year plan becomes a multiyear plan! Your individual plan is based on how determined you are to succeed, how willing you are to invest your own time and possibly money, and what you intend to do if it does not work out as planned.

CHARACTERISTICS OF A GOOD PLAN

A good plan has
- A clearly defined goal
- A logical set of objectives
- A list of resources needed
- A method to measure progress

Proper planning also involves research and data analysis, looking for emerging trends and variances. There are many sources available from which to gather this data. It is important when accessing data that you use trusted and reliable sources. It is critical that data from these systems be validated for accuracy. Only data that are accurate, complete, and validated should be used. Otherwise, what you are using is just arbitrary information and figures. Two of the most notable national resources are the National Fire Academy and the National Fire Protection Association. Both agencies publish an inordinate amount of data related to departmental activities. The internal and state reporting systems are also a valuable source of information that can be accessed for relative data. Most of the data are available in electronic format, making it easy to insert into tables or convert to graphs for further analysis.

There are many reasons that the data that officers input into such systems must be as accurate and complete as possible. For example, if you respond to a grass fire that extended to a shed but you only complete a report for a grass fire, because you did not want to complete the details of the report associated with a structure fire, the data is skewed for the entire department. How can the department recognize that there has been additional dollar loss? How can the owner verify the event for insurance purposes? In addition, it is extremely important that medical records be entered as accurately and completely as possible. This is the officer's responsibility. Even if the task is delegated to a subordinate, the company officer should review the report to verify the information within.

Quite often the data by themselves do not tell us what we need to know or deliver the big picture we may need for planning purposes. It is therefore necessary to be able to analyze the data to provide the answers we seek. It is important that the data not be interpreted with a bias or distorted for self-serving purposes. Not only is it immoral and does it potentially provide misdirection for the department, but if it is discovered that you have done this (intentionally or not), you will lose creditability that you may never be able to regain.

Organizing

Organizing involves the integration of human and physical resources, money, and time.

Managing Human Resources

As company officers, you are assigned personnel to staff your company. How effectively you use these personnel is a function of your leadership style. We discuss leadership in Chapters 6 and 7, but let us look here at how you manage these human resources.

In Chapter 4, we listed Fayol's Principles of Management. In larger organizations, these principles cover the way we manage in an organization where a management hierarchy is inevitable. As an organization gets larger, increasing number of managers are needed to supervise the work. The consequences are additional layers of managers. One way to slow this process is through delegation.

Delegation is a process of entrusting some of our work to others. Delegation happens in large organizations and in small ones. A mom-and-pop ice cream stand has some degree of delegation, especially if mom and pop hire their kids to work there during the summers. In the delegation process, two things happen:

1. The supervisor assigns duties to the subordinate.
2. The subordinate takes on an obligation or responsibility to perform these duties.

The supervisor should also assign **authority** to do the job. To be effective, delegation of the duty and acceptance of the obligation for the job should be coupled with the authority to do the job. Limits are usually set upon how much authority is granted, but the authority should be sufficient to allow the subordinate to be productive in the assigned task.

Delegation is often referred to as the granting of authority to another to carry out an assigned task. Delegating requires that the task be clearly identified and that the duties and limits of the delegation be spelled out. Effective delegation does not mean that the task is eliminated from the supervisor's responsibilities. Rather it implies that partnership has been established in which the actual accomplishment of the task may be done by another. However, the supervisor ultimately remains responsible for the work being accomplished. As such, the supervisor is expected to continue to manage the activity and provide appropriate guidelines (see **Figure 5-2**).

With the exception of a few specifically identified tasks, most of your duties as the company officer can be delegated. How well this delegation occurs is a clear reflection of how well you understand yourself and your team members. In both cases, it is important to understand the strengths and weaknesses involved and the experience and motivation of all concerned for effective project management.

Once delegation has taken place, you must continue to manage. Hopefully, planning and organization have already taken place. But you should continue to coordinate and control the activity, monitoring the progress. Provide help and assistance where needed. Give encouragement and recognition when appropriate.

FIGURE 5-2 Effective management of any organization involves the ability to delegate certain tasks.

There are many benefits of delegating. Certainly it will take some of the work off of the supervisor. It will also make the supervisor look good—the company is a team, not the work of individuals. Individuals that are part of that team will be motivated to do more by their own accomplishments.

Delegation is difficult for some people, especially for new first-time managers. And that is quite understandable, for in many cases they have never had opportunities to delegate work before and are hesitant, either due to a lack of knowledge or experience in the process.

OFFICER ADVICE

When delegating to trained and qualified members, define the purpose of the work and leave the details to the member.

member into the picture to explain what you want. When delegating to trained and qualified members, define the purpose of the work and leave the details to the member.

BARRIERS TO DELEGATION

- ◾ I can do it better myself.
- ◾ It will take time to train someone.
- ◾ I am not sure that anyone else can do this.
- ◾ I really cannot take a chance on this project.
- ◾ I think my boss would want me to do this myself.

Delegation is like other management activity: It should follow a regular pattern. You start by planning, giving some thought to the delegation process. Maybe you are bringing a new member into the team or maybe you are reallocating some of the work among existing members. Some planning must precede the actual handoff of work. You should define the work and establish procedures for its accomplishment. At some point, you should actually bring the

In many cases, some training is needed to prepare the new person for the duties at hand. But eventually, it will become time for the new member to start the work. Usually, the new member works with someone else, then gradually takes on the task alone. Finally, some sort of feedback is necessary to make sure that our expectations as supervisors are being met. You might ask for a regular meeting time or some sort of reporting system so that you can effectively monitor the member's progress. You have probably been through this process many times, as the delegated member.

Managing Human Resources on a Larger Scale

Personnel within some departments are represented by a labor union. As a result, management is required to enter into formal relations with the union concerning such things as wages, hours, and conditions of employment as well as the interpretation and administration of the contract over the period of the contract.

The administration of the contract lies with both the personnel specialists of the organization and the organization's managers. Personnel specialists deal with the details of administering the union contract. Hopefully, these specialists provide some guidelines to the managers on how to make the contract work effectively.

NOTE

The attitude of top administrative and labor management filters down throughout the organization and permeates the relationship at every level.

As with most everything else in an organization, the attitude of top administrative and labor management filters down throughout the organization and permeates the relationship at every level. It is important for the fire chief and other top officials to look upon labor relations as a positive opportunity. Likewise, union leaders should look at their relationship with the department's management as a positive opportunity.

When you think about the benefits that unions have advocated for their members, you should think of several major issues: improved wages and benefits and formal rules for hiring, promotions, pay, discipline, and remuneration. Ideally, management and union leadership will have developed a collaborative and partnering relationship so they can effectively and efficiently deal with any issues that impact union members.

The questions of unionization and the selection of the union to represent the workers are subject to an election among the workers. Collective agreements are of fixed duration and provide a binding contract between the workers and management. Collective bargaining allows the workers to be represented by the union.

NOTE

A characteristic of unions is that they have exclusive representation—only one union is authorized in a given job territory. The questions of unionization and the selection of the union to represent the workers are subject to an election among the workers. Labor contracts are of fixed duration and provide a binding contract between the workers and management. Collective bargaining allows the workers to be represented by the union.

In the private sector, union membership has been declining, and, today, less than 20 percent of all non-farm workers are union members. However, some employment sectors are seeing growth in union membership. Whatever their status, unions remain a strong social, political, and organizational force. Where unions exist, management must work with the union rather than directly with the workers on many issues.

Union representation among public sector employees is a significant part of total union membership. These employees work for federal, state, and local governments. At the federal level, labor relations are regulated by the president. At the state and local level, labor relations are determined by state law. Most states have passed legislation pertaining to labor relations issues for state and local government employees. The laws differ in the states, and you would be prudent to understand the laws in your particular state. Most firefighters are represented by the International Association of Fire Fighters (IAFF), a union with 260,000 members.

FIREGROUND FACT

IAFF, the International Association of Fire Fighters, is the AFL-CIO affiliated labor union for the fire service. The union has more than 2,700 affiliates in the United States and Canada. While its main purpose is to negotiate for the rights, health, and safety of paid firefighters, the union provides free training for all firefighters and fire departments, with worthwhile programs addressing hazardous materials, special rescue response, terrorism response and health and fitness.

In addition to these training activities, the union is supporting a revolutionary program to help recruit and train qualified candidates for fire departments. A major component of this initiative is the Candidate Physical Aptitude Test (CPAT) that helps the fire service recruit candidates prepare for the rigors of the test, and provides a validated measure of their ability to perform job related skills.

For most paid fire and emergency medical services (EMS) personnel, the right to form, join, or assist a labor organization is governed by state law. These laws are generally modeled after the federal laws. The most conspicuous law in this area is the Labor Management Relations Act. Provisions of this act protect the interests of both sides. For example, the act protects the rights of individuals to form, join, or assist in the activities of the union. The act also protects the unions from interfering with an employee's rights as a citizen and as an employee.[2]

Most labor relations contracts spell out a grievance procedure. Grievances occur when there is a differing interpretation of the contract between representatives of labor and management; when a violation of the contract has occurred; when there is a perceived violation of a law, regulation, or procedure; or even when there is just a perception of unfair treatment. The employee grievance process is a systematic process for resolving these differences. Most union contracts define how the grievance process should work. The idea is to bring the problem to resolution as quickly and as fairly as possible. There are generally three steps in the grievance process. The first step involves filing a complaint with the supervisor. Most grievances are resolved at this level.

LABOR RELATIONS ARRANGEMENTS

A **closed shop** usually results from an agreement between their union and the employer that requires prospective employees to become members of the union prior to being hired. Such agreements are generally unlawful.

A **union shop** usually results from an agreement between the employer and the union that requires membership within a specified period of being hired. This practice is lawful.

In an **agency shop,** the agreement provides that union membership is not required, but payment of union dues is mandatory for all employees.

Finally, there is the **open shop** where employees are free to decide if they will join the union. If they do join the union, they must pay union dues. If they do not join, they do not have to pay dues. Usually, employees can terminate membership only during certain periods, such as when the contract is up for renewal. In those states where right-to-work laws have been adopted, only the open shop is authorized.

Other provisions of labor relations contracts usually address several major issues as follows:

- *Compensation and working conditions:* Many of these provisions cover personnel administration issues—hiring, pay, benefits. However, many also address items related to working conditions, items that are under the purview of the supervisor.

- *Employee security:* Issues addressed here include such things as protecting an employee's position during times of promotions and layoff.

- *Union security:* These provisions pertain to what the union can do to encourage employees to become members.

- *Management security:* In this area, the contract defines those issues that are the exclusive domain of management.

- *Contract duration:* The length of the contract. Contracts usually run from 1 to 5 years.

NOTE

Typical labor–management issues include the following: discharge, discipline, discrimination, grievance process, health and safety, hiring, insurance, leave, promotion, and shift duties.

The role of the federal government in union contracts is relatively passive. The National Labor Relations Act of 1935 provided employees with the right to engage in union activities. That act also established the National Labor Relations Board to supervise representation elections and to investigate allegations of unfair labor practices.

The power of strike has been reduced in recent times. Growing antiunion sentiment, court decisions, and a pool of workers ready to replace the strikers have generally reduced the unions' power through strike action. We frequently hear about strikes in the news, but in reality a very small percentage of employee time is lost through labor conflict.

Typical labor–management issues include discharge, discipline, discrimination, grievance process, health and safety, hiring, insurance, leave, promotion, and shift duties. When the union and management cannot agree on the issues, they may agree to a third party coming in to help them settle their differences. There are three levels of involvement: mediation, fact finding, and interest arbitration.

Mediation involves a third party acting as a facilitator. The mediator opens communications channels, persuades the two sides to meet, helps the parties readjust their positions on issues, and controls the process by scheduling the meeting time and agenda. The Federal Mediation and Conciliation Service may be invited to participate in the negotiations. There is no specific time in the process at which it enters, and in most cases, it does not enter at all. Most labor contracts are settled without any assistance.

The next level of involvement is **fact finding**. The third-party fact finder investigates the truthfulness of facts that are relevant to the impasse. The fact finder can also make a recommendation to the parties regarding the reasonable settlement of the issue.

Arbitration is the final step. Arbitration allows the third party to settle the issue. The arbitrator hears both sides and determines the results. Arbitration is a last resort: It provides a solution, but neither party may like the results. In some states, the findings of an arbitrator are nonbinding, meaning the employer can choose to ignore the findings.

To achieve long-term success, labor and management must learn to collaborate with each other rather than to confront one another. By so doing, labor and management can learn to increase productivity and improve the quality of work life. Unions and management groups should work to achieve established goals and initiatives and avoid establishing an adversarial relationship.

NOTE

To achieve long-term success, labor and management must learn to collaborate with each other rather than to confront one another.

Research has shown that good labor relations occur when there is mutual trust and respect, honest and open communications, mutual problem solving, good (consistent) contract administration, and timely resolution of grievances.

Managing Financial Resources

Money is another resource you manage. The principal tool for managing money is called a **budget.** In the simplest of terms, budgets are a management tool that represents nearly all of the five functions of management: Budgets allow us to plan, organize, coordinate, and control our financial activities and to represent the overall plan of the department in terms of intended revenues and expenditures. Budgets tend to be unpopular topics and consume a lot of time and effort, yet they are a very critical part of running a department. If you do not understand the budget process and life cycle, it is unlikely that you will be successful in implementing new projects or acquiring new equipment; without proper planning, it is unlikely that the financial support will materialize, as budgets seldom contain "extra" money that can be used to fund unplanned projects or purchases. Ideally, funding for new projects or equipment acquisitions should be justified and built into the upcoming budget cycle. This is especially true in light of the rising cost for liability coverage, compensation packages, medical costs, and workers' compensation.

Budgets are usually broken down into two categories: operating and capital. The **operating budget** of a department usually encompasses a single year and is tied to a "fiscal year" or a "calendar year." During this timeframe, the budget usually goes through several phases in its life cycle: planning/preparation,

FIGURE 5-3 A new fire truck represents a significant long-term investment.

adoption, implementation, and analysis or evaluation. Most commonly, items or programs within the budget are tied to the department's goals and objectives and identify the resources needed to achieve these goals. If you want to provide basic or advanced EMS capabilities on every engine and truck company in your department, you have to think about the training and equipment needed to support this program. You should plan for the procurement of these resources, determine the costs involved, and calculate these costs into the budget for the coming years.

Capital budget items are large, expensive items that last longer than 1 year. Most capital budgets are set up on a 10-year span. For the fire service, classic examples of capital budget items are fire stations and apparatus (see **Figure 5-3**). Local jurisdictions usually have a definition for capital budget items in term of cost and expected life span. Because of their cost, capital budget items get a lot of attention and are usually clearly identified and carefully defended in the overall budget process. Smaller departments' and a lot of volunteer departments' capital budgets are set up on 5-year spans due to smaller tax bases. The operating budget includes nearly everything not covered in the capital budget. Most of the department's expenses are in the operating budget. For most paid fire departments, 90 percent or more of the operating budget is typically dedicated to salaries and related personnel costs. Within a budget, this is typically referred to as FTEs, or full-time equivalent positions. Other obvious operating expenses include fuel, utilities, supplies, and equipment. It is common for operating budgets to also reflect multiple years. Not only will it include the actual year for which the department is operating, but also it can reflect an adopted budget for the coming year, a proposed budget for the year after that, and perhaps even a revised budget showing changes that had to be made to the current operating budget due to unforeseen circumstances. Once approved, the budget becomes the plan or guide for the coming year's activities. The same fields are almost always defined with a budget. The beginning balance or operating budget is typically the overall dollar amount that you or the department has to operate within. Revenues show the sources for the incoming money. For most fire departments, this is in the form of special funds, taxes, or mill levies. Some departments are functions of a larger government organization and are simply allocated a percentage of an organization's overall income stream (General Fund). There is also a column for expenditures or expenses. This identifies how much the department is going to spend and on what. This can be broken down into great detail and in this fashion is often referred to a line-item budget. Today, most organizations use some form of **line-item budgeting.** With line-item budgeting and accounting, like items are collected for purposes of keeping track of the expenditures during the year. Suppose we consider the cost of electrical utility service as a line item. We can plan for a certain expenditure for electricity, but extreme weather conditions may cause this cost to increase. A comparison of weather data and the electric bills will show that the expenditures were unpredicted but appropriate. On the other hand, suppose the department planned on spending money on urgent replacements for turnout gear. Six months into the year, the money is still in the account. Managers should be asking the personnel responsible for this purchase some questions about their "urgent" procurement needs.

Like the management process, developing the budget or a budget proposal starts with planning. The planning should focus on what resources are needed to turn the organization's goals and objectives into reality. Depending on the size of the department and the complexity of local government, this may take anywhere from a month to a year to accomplish and involves a lot of estimating. However, because it is a direct reflection of you and your knowledge of what is required to accomplish the proposed project or acquisition, you should be as accurate as possible. Most approving authorities are reluctant to provide any additional monies to complete a project. As a result, if you start a project and come up short of funds, you may have to cut corners or go back to the approving authority and beg for more, while having to explain why you did not request enough in the first place. The worst-case scenario is that you may have to terminate the project.

This is why good planning is a necessity before submitting a finalized budget proposal. All departmental programs should be analyzed for material needs so that adequate funding is available. It may be necessary to perform a priority ranking of all budgetary categories. This ranking, while often used for capital items, can be used as a decision-making tool for all departmental programs. In time of economic hardships, fire department programs often have reduced

funding or, in many cases, have no funding at all. Therefore, the priority ranking of these programs becomes even more necessary.

Most budgets must undergo a two- or three-phase approval process. The three-phase system involves approval by the fire chief or department head, and then the budget must be approved by the city or town administration and, finally, by the city's or town's elected officials. Once approved, the budget will begin at the start of the fiscal year and must be managed throughout that year.

If you are dealing with multiyear projects, be sure to include inflationary costs into the budget. Rule of thumb is to add 5 percent of the overall cost for each project year. This is also true if it may be a year or longer before the funding becomes available (i.e., Assistance to Firefighters Grant). By the time you get the funding to start the project, the supplies, materials, and labor may have increased in cost and as a result you will start out underfunded if you did not account for inflation. Once funding is approved, the department's purchasing agent or the individual in charge of the project will begin the acquisition process. This typically means that specifications are developed for the purpose of soliciting a formal bid, a request for quote, or a request for proposal.

A **Formal Bid** or a **Request for Bid** is a formal bid for products and/or services that are typically over a certain dollar amount. Most organizations now accept bids online as well as on hardcopy. A formal bid process usually works the best when the contractor is soliciting competitive prices from vendors for a specific piece or pieces of equipment. Most commonly, specifications or specific brands and model numbers relating to the equipment is sent out at least 2 weeks in advance. Vendors then provide a written response identifying what they will sell and for what price. All bidders should have the opportunity to bid on the same items on equal terms and have bids judged by the same standards as set forth in the specifications. The purchasing agent for the department, otherwise known as the contractor, is responsible for reviewing the bids, not only looking at the costs but also ensuring that the vendor's proposal and/or product meets all of the specified criteria. Controversy related to bids most commonly centers around a department having to settle for less than what they intended to purchase because details regarding what they wanted were not identified in the specifications. That is why it is essential that bid specifications be written to define exactly what the department needs. If a specification is written too loosely, then a vendor may interpret it differently than intended and supply a product that does not meet the department's need. Readers are encouraged to review recent specifications used by their department and compare them against the identified need.

Once accepted, the bid becomes a formal contract between the contractor and the successful vendor. The contractor typically then develops and sends the vendor a purchase order. The vendor fulfills the purchase order. Unless other arrangements have been made, full payment is typically made once all items have been received and the order is complete.

Occasionally, there is a need to sole source the purchase of an item. Most commonly, this is due to the overarching need to buy a certain type of equipment for compatibility or safety reasons. Be aware, however, that most grant programs require that a department solicit competitive bids for any items purchased with grant funds. Failure to comply with this requirement could result in a competing vendor in filing a complaint, causing the department to reject all bids and send the item back out for bid, which can result in a lengthy delay in the acquisition process. At the very worst, you could lose funding altogether.

A Request for Purchase, also referred to as **Request for Proposal (RFP),** is a solicitation used when you need design, consultant, products, or services and you are looking for answers. An RFP is most commonly used when the specifications are broader in nature and the department is open to more than that one specific brand of equipment or vendor or the department is requesting a rather complex commodity rather than a single commodity or tangible goods. A prime example is the difference between purchasing computer hardware and selecting a software designer or consultant. One is rather straightforward and can be specified by brand or exacting performance specifications. The other is rather arbitrary and as a result may include things like past work references, performance measures, and time lines in which to get the job done. As you can tell, one is more complex than the other and therefore more difficult to write specifications for and evaluate. An RFP should describe all your requirements, terms, and conditions that will apply to the contract from prospective vendors or contractors. The advantage of the RFP over a formal bid process is that it allows the contractor to compare what each vendor has to offer relating to different components of the bid in comparison to the other bidders and make a selection based on the overall proposal, rather than price alone. For example, you can evaluate things such as warranties, timeframe in which the service is to be performed, etc.

Award of an RFP is not based on low bid but on what is in the best interest of and value for the department. Depending on the solicitation, you may be required to advertise. Typically, a screening and/or selection committee meets, reviews the proposals,

selects companies to interview (depending on the total amount of the proposed purchase or proposal), interviews companies, and makes a recommendation to the purchasing agent. The committee should use a standard evaluation process. The purchasing agent then negotiates and prepares the contract, and takes it through the appropriate approval process.

The approval process often differs based on the total dollar amount. If the total dollar amount is below a certain amount, an administrator can typically approve the process. However, if the total dollar amount exceeds that amount, then the approval by an auditor and governing body (commission or council) is often required.

Like an RFP, a **Request for Quote,** or RFQ, communicates requirements from vendors and contractors. However, it is typically just for informational purposes and not necessarily binding in nature. Basically, it is an informal bid for a smaller quantity of products and services. An RFQ may or may not need to be advertised based on your agency's purchasing guidelines. Most commonly, it is not, but rather solicitations are sent to selected vendors. The approval process for RFQ's is typically less stringent due to the lesser dollar amounts involved.

Two factors to assist in deciding whether to use an RFP, an RFQ, or a bid is to determine if you are dealing with vendors or contractors and the overall dollar amount. Typically, a vendor sells a commodity or commodities. It is usually a one-time sell and no service is required, so a RFQ is most appropriate. However, if a significant total dollar amount is involved, then you may want to consider doing a formal bid process. If you are negotiating for a contractor to provide a service over a given period of time you should perhaps consider doing a RFP. The following is an example of a bid request: "The Fire Department is soliciting Request for Proposals (RFPs) from qualified contractors to provide physical fitness assessment evaluations and screenings. Assessments are to meet the specifications described herein by the end of the performance period." Be sure to write performance requirements that the contactor is required to meet within the contract. It is also important to recognize the difference between a vendor and a contractor. A vendor is typically one who supplies goods or services. This can be a manufacturer, importer, or wholesale distributor. A contractor is usually hired or makes an agreement to perform a specific service or piece of work. Consultants, independent contractors, physicians, lawyers, and construction contractors are common professions that are contracted. Be aware that most states allow contractors and subcontractors to place liens on departments, cities, and counties until they are paid.

Occasionally, there is a need to expedite purchases or make emergency purchases, either as the result of an actual emergency or because a commodity or service is needed to keep your organization operational. In these times, there may be justification to step outside of normal purchasing guidelines, but it cannot be an habitual process, nor can it violate purchasing laws, as they are in effect even during an emergency. The best way to handle these situations is to have pre-established contracts or **Memorandum of Understanding (MOUs)** in place with these vendors prior to an emergent event. This is one of the reasons as soon as you recognize the need, it is critical you place qualified staff in the finance section of the command structure.

One final word is needed about soliciting bids and awarding contracts. Be sure to consult with your purchasing agent and follow the procedural framework and purchasing guidelines of your governing agency. Violation of city, county, state, or federal purchasing laws is a very serious offense that could ultimately result in termination or, as a worse case scenario, incarceration. Your legal department should always review final draft before soliciting and/or awarding bids or contracts.

As the weeks and months unfold, it is appropriate to check the level of activity in the budget. The budget can be set up in a variety of ways to facilitate this check. This not only allows you to make sure you are financially on target; it affords you the chance to assess the performance of the department relative to its stated mission and goals.

You might note that the controlling function of management is necessary as the budget year passes. Suppose that the extreme cold weather added significantly to the total cost of electricity and heating fuel. With the cold weather, it is likely that during this same season there was an increased amount of alarm activity, further adding to departmental costs. These added costs may not be apparent right away but will show up in time. These obligations may require some adjustments to the budget to keep the department operating as the end of the year approaches. Purchases may be deferred. Travel and training may be cancelled. Other discretionary costs may be delayed or eliminated to keep the department in business. It is necessary to have some sort of budget controls in place to make sure that the department adheres to its adopted plan and the priorities, policies, and programs are appropriately funded.

Many organizations use a **program budget.** A program budget gathers expenses that support a particular program. For example, if the department provides EMS, a program budget would identify all of the costs associated with the EMS program.

The Role of the Company Officer in Budgeting. Your level of participation as the company officer in budgeting varies. Some departments control budget planning and administration at the very top, whereas other departments maintain control at the top but keep everyone informed through a variety of timely reports. This approach allows everyone to at least see what is going on and to have some understanding of why things are or are not happening.

As the control of the budget is more decentralized, you, as the company officer, have increasing involvement with the procurement process. Many departments allow you to identify needs for the coming year. These are forwarded up the chain of command and then ranked in some order of priority. As money becomes available, procurement is authorized. The funding for these projects are often spelled out in policies set forth by the governing body. However, in some instances, as in private fire departments, budgetary policies are developed by the department. These policies must outline where and how received monies are to be allocated.

SAVING MONEY HELPS EVERYONE

Even when traditional budget controls are in place, there should be incentives for saving money. Turning out lights that are not being used, setting the thermostat at a reasonable level, easing off the "gas" pedal, and shutting down engines that may idle for long periods may seem like small gestures, but collectively, they can have a significant impact on the department's finances. These small savings can be converted into purchases that are really needed.

For routine supplies, many departments allow you to draw supplies against a specified account. Some departments take care of this supply activity by providing a stocking service where needed supplies are provided and stocked by a private contractor.

If the department makes it a practice to look at its spending activities regularly, it can determine if the spending is following the planned budget. Deviation reports, monthly position reports, and other methods are available to help managers keep track of the budget.

When the expenditures are racing ahead of the budget plan, it may become necessary to slow things down a bit through procurement delays and curtailment of travel and/or training. Theoretically, training should be the last thing cut, but because so much of the budget (often 90 percent or more) is for salaries, there is little left to cut!

Empowering the Company Officer with Money. Some departments have taken the rather radical approach of empowering you, as the company officer, with funds, based on historical data, to run your station. If you can plan and manage well, you will likely be able to save some funds to buy discretionary items that the station needs. On the other hand, if you make no effort to reduce costs, it is unlikely that the station will see any rewards from the process.

Acquiring Needed Resources

Even after ensuring that the department will have the necessary funds to operate throughout the budget year, the planning does not stop. The department must plan for major incidents within their jurisdictions. To accomplish this, departmental officers need to conduct a needs assessment for their districts. Conducting a needs assessment will outline the specific resources that will be needed for different types of incidents. Even the largest departmental budgets cannot accommodate for unforeseen incidents that require resources beyond its capabilities. Conducting a needs assessment for major incidents should include the following steps:

1. *Play the "what if" game*—Planning for target hazards within a jurisdiction can prove to be difficult. The process starts by mapping out the hazard and planning for any and all possibilities.

2. *Spawn ideas*—Once all of the possibilities have been mapped out, ideas must be generated on how to mitigate these hazards. In this step, the order of importance is not as significant as is generating all achievable ways that the hazards can be mitigated.

3. *Organize ideas*—After all of the ideas for mitigation have been generated, they must be placed in order of significance. This step should fall in line with the three fire service priorities: Life Safety, Incident Stabilization and Property Conservation.

4. *Identify sources for resources*—As discussed earlier, the normal operating budget does not plan for incidents that requires specialized resources for a catastrophic event. Therefore, within the needs assessment, an outline of needed resources must be developed ahead of time. Plan for the needed resources and devise a list of where these resources can be obtained.

5. *Paying for the resources*—Nothing is free; therefore, the plan should estimate the cost for each individual resource. The plan should also include an arrangement on how these resources are to be purchased.

6. *Construct the timeline*—When the resource is ordered, how long is it going to take to receive it? Most of the time ordering directly from the manufacturer is not the fastest way to receive a product. This was proved in the 9/11 attacks of 2001. The Fire Department of New York recognized early that their firefighters needed something other than turnout gear to work the rubble pile. Hundreds of sets of turnout gear were ruined while sifting through the pile. The plan was to order two sets of heavy duty coveralls for each member so that they could always have a clean set. Logistics personnel called various manufacturers to try to place the order. The problem was that the manufacturers could not turn out the coveralls fast enough. As a back-up plan, Logistics officers started calling all of the smaller vendors within a three-state area and were able to purchase all of the needed coveralls.

Whatever the method used, departments must have a plan. Departmental policies and procedures need to be developed to outline the steps for obtaining resources for any given situation. These procedures need to list where to obtain resources and how they will be funded.

GOOD FINANCIAL MANAGEMENT

The company officer should do the following:
- Keep track of the station's needs.
- Identify items that would improve efficiency.
- Involve everyone in the budget process.
- Talk with other company officers to learn how to effectively manage the station's resources.

You are encouraged to review the procurement process for your respective department and develop a budget for your station using the identified process. This may necessitate creating a budget line for each separate item your station is going to need for the entire budget year or it may mean grouping items within categories. If there is a need for items that you do not normally receive, complete the appropriate acquisition forms or prepare a budget request for those items, justifying and prioritizing the need. Readers are also encouraged to write a specification for one of these items and have others review the specification to see if it clearly and accurately defines what is to be purchased. What you will discover is that developing budgets and specifications is a time-consuming and laborious process. However, they are the lifeblood of your organization and require devoted attention and detail.

Managing Time

Of the three resources on this short list, time may be the most precious for everyone. As previously stated, in most paid departments, salaries for your and the other members' time accounts for 90 percent or more of the budget. So time management is essential for running an efficient department. Yet, some members are content to come in, put in their hours, and go home. What they do on their days off is their business. What they do when they are on duty is your business. Wasting the day away is not acceptable and should not be tolerated by them, you, or the customer.

PARKINSON'S LAW

Have you ever noticed that activities take about as long as the time allowed? This is called Parkinson's Law, a concept named for C. Northcote Parkinson and first published in 1957. According to Parkinson, work expands to fill the time available for its completion.

You should manage your time just as you manage any other resource. You should find new ways of doing things so that you are not just putting in your time, but you are making good things happen, all the time.

A lot of attention has been given to time management in recent years. The popularity of this topic suggests that many people have problems managing their time. Like management in general, time management starts by setting goals and objectives. In this case, we are probably setting personal goals and objectives.

Next comes personal planning and scheduling. This activity is more than just making a list of things that need to be done. Good time management correlates the management of our time with our previously set goals and objectives. Planning allows us to

TIME MANAGEMENT

- Plan and prioritize.
- Delegate as appropriate.
- Establish goals for yourself.
- Be organized and reduce clutter.
- Make a schedule and stick to it.
- Fit other things around these events.
- Identify and work to eliminate time wasters.
- Give first attention to difficult and complex tasks.
- Announce meetings, start them on time, and have an agenda.

SAVE TIME BY PLANNING

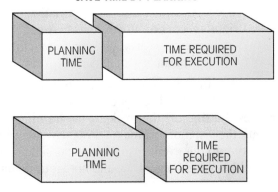

FIGURE 5-4 Good planning will save time in the long run.

prepare for events, rather than react to them. When you are in a reaction mode, you are inefficient and unproductive.

Some people seem to enjoy this mode. They operate in a crisis mode, and run from one crisis to another, never being able to break out of the rat race and get ahead of the events of their life. People who wait until the last moment to do their Christmas shopping or prepare their income tax reports are good examples. Most would benefit on focusing their time and energy on fewer things and concentrate on doing them well.

Of course, there are many things in your life that you cannot control. In the fire station, you should be available to your personnel, to your boss, and to the public. You also need to be available to respond to alarms. But there is still a lot of time you can manage. Set up a program that makes the day become a routine, and where your precious time is used effectively (see **Figure 5-4**). Here are several suggestions:

- If meetings are needed, have them at a regular time and have an agenda. Make the meeting brief but effective.
- If you want time to work in the office, let your people know when that is. Ask your people not to disturb you unless they have an urgent need.
- Avoid putting things off. Taking time to decide to put something off simply wastes time and gets nothing done.
- Distinguish between important and urgent matters. Important things need to be done, but not right away. We know about most important events well in advance of any deadline. Most important events take time to plan and execute. A training program for the coming year is an important event.
- Urgent events usually need to be done right away. Some urgent events are unexpected. You cannot anticipate emergency responses; you have to deal

with them when they arise. Likewise, a call from your supervisor asking for some information that is needed right away is an urgent event. A call from home that indicates that your spouse or child is seriously ill is also an urgent event.

People allow important events to become urgent events through the lack of good time management techniques. Putting off the work on the training program until the last moment makes it become an urgent event. Avoid urgent events as much as possible. Your goal should be to manage time, and not let time manage you.

Protect your time by setting goals and objectives. Starting with goals and objectives helps you decide what really must be done by you, what can be delegated to others, and what can be omitted. Remember, without a goal, you will not know your destination.

Time management requires organization and some method of record keeping. Being organized allows you to readily access information and prevents you from wasting valuable time looking for documentation and/or records. If you take the time to establish and maintain an effective filing system, it will save time in the long run. This also includes electronic files. An *electronic filing system* refers to a computer-based system that allows you to receive and store documents in electronic form. An electronic filing system also capitalizes on the use of automated technologies and promotes cost savings. Some departments have a network of computers with a central server where the files are stored in a centralized location and always backed up. Others use stand-alone computers, relying on the local hard drives to store the information. The problem with the latter scenario is that if the local computer "crashes," you can lose all of your stored information. This is why it is important that the computer be protected with an antivirus program and that your files be backed up in some fashion.

It is not only important how and where you store your files, but it is also important how you name your files. Some name electronic files by the last name of the person who sent the information; others name and organize files by topic. In most organizations, there are no policies for how to name or keep an electronic file; hence, the format is left to the individual. Just remember that the file name is the principal identifier for a record and should place some context of what is in the record or how the record is related to other files, records, and/or projects. If your electronic records relate to a particular subject or function or result from the same activity, they should be similarly named and kept within the same set of folders.

When naming your files, you may wish to include some of the following common elements: name of intended audience (e.g., school groups), name of

group associated with the record (e.g., Haz-Mat team), budget or project number (e.g., line item 7021), project name (e.g., rescue truck specifications), department name (e.g., Public Education), version number (e.g., Version 1.1), date of creation (document 9-21-08), name of creator (e.g., Clinton), description of content (equipment inventory), or a combination of any of these (e.g., Meeting Minutes 9-21-08). Consistently named records foster organization and collaboration based on a mutual understanding of how to name files and use electronic media.

Subfiles can be kept within the main menu files, but remember that the main thing is to manage your electronic files so you can effectively find what you want. If you get too many subfolders or subfiles, you can spend a lot of time opening the main folders looking for what you want. Likewise, if you get too many files on the main menu, you can still spend a lot of time searching.

Files can also be stored on other portable electronic media such as computer disks (CDs) or flash drives. Portable hard drives are also available. The advantage of the portable media is that you can take it with you and access it on another computer.

Electronic files are also typically more secure than paper files, since most computers are set up for multiple users, requiring that you have a username and password to access them. However, the files are susceptible to being accessed by the public if the computer is connected to the Internet and you do not have proper firewall protection,

One should remember that electronic files, just like emails, are subject to the open records act, and as a result, anything you have stored or sent on a department computer can be requested for public review and used as evidence in a court of law. If you delete documents subsequent to any litigation, you can be guilty of destroying or spoliation of evidence. So keep all you do professional and above board.

Most organizations have some type of policy that identifies how long department-related files should be maintained. Absent of any policy, there are no hard-and-fast rules concerning record retention. Files or documents related to personnel issues usually have to be maintained for at least 5 years. If a person has an accident or a medical-related issue, files or documentation related to these issues usually has to be kept for the term of employment, if not longer. When in doubt, use the 10-year mark as a good rule of thumb.

When you dispose of electronic media with sensitive information, make sure it is deleted from all in boxes, out boxes, deleted items, and trash bins. Otherwise, deleting a file from a single location often just moves it to another, where it can be accessed by others. When destroying paper files, make sure you shred them. If you leave personal information for others to find, you can be at fault for disclosing personal information. It is just a lot easier to do things right the first time.

One final note on records—understand the laws and department policies that govern the use, access, and release of information. While you should try to meet the needs of your citizens, and make your organization a transparent one, there are certain laws pertaining to what information the public and you have access to. If a citizen comes in and wants to know who got hurt in a wreck you just got back from, you cannot release that information. You may be able to tell them that a 45-year-old man suffered back injuries, but you cannot release a name or any identifiers because that violates the Health Insurance Portability and Accountability Act of 1996 (HIPAA, Title II). A good tactic to use in these situations is take the opportunity to talk about preventative issues related to the event. If the event was related to a fire, take the time to discuss the home evacuation plan and how it can help if such an event was to happen to them.

Department policies related to the use and release of information may also be applicable. Some departments have a policy not to release anything but rather refer them to the administrative section or public information officer. There are also laws related to access, use, and release of personnel information. For example, you should not discuss discipline issues with anyone other than your supervisors and the involved employee. Some agencies terminate employees who access personal patient or employee information without proper authorization and/or justification.

Using a personal planner also assists with organization. Many planning tools are available, ranging from the traditional paper-based planner to computer-based

FIGURE 5-5 Keeping track of your personal time and commitments is also important.

devices such as the personal data assistant (PDA) (see **Figure 5-5**). Whatever form is used, it is important to keep the planner up-to-date. Use it to help set priorities and to identify items that can be delegated to others.

TEN PROVEN TIME EATERS

1. *Lack of personal goals and objectives.* Where are you headed?

2. *Lack of planning.* Set priorities, set a schedule to meet your goals, and follow through on your priorities. Manage your time; do not let time manage you.

3. *Procrastination* is putting off things that need to be done. It is poor time management. Making the decision to put off the activity takes time. It is often more effective to just bite the bullet and do it.

4. *Reacting to urgent events, often the result of procrastination.* These events force you to rearrange your schedule to meet the need of the organization or other individuals.

5. *Telephone interruptions.* They are frequently unwanted and frequently unnecessary. Learn to conduct business in a polite, businesslike way and get off the phone.

6. *Drop-in visitors. Same comments as the preceding telephone item.* Be polite, do your business, and move along.

7. *Trying to do too much yourself.* Learn to focus on what is important, and learn to delegate what you can.

8. *Ineffective delegation.* When delegating, learn the steps for doing it effectively.

9. *Personal disorganization.* In its milder forms, disorganization presents problems due to scheduling conflicts, the inability to find things when needed, and so forth. In a more serious form, it means that you have not set goals and objectives and that you lack focus.

10. *Inability to say "no."* For some, this item may be the hardest on the list. Tie your time to your goals, and focus on priorities. If the request supports your

goals, go for it. If it does not, politely say that you are sorry, but you are unable to take on that activity. Obviously, this may not set well with your supervisor, but even here, supervisors sometimes can overload your plate. Show your supervisor a list of what you are working on, and ask for help in setting priorities. Your supervisor will give you the guidance that you need and may also find (possibly as a surprise) that you really are overloaded.

Of those items that remain on your list, number them to identify their priority. Try to avoid the temptation to clear up the little things first or simply do the things you enjoy the most. Do the important things first. The little things can be fitted in later or held over until another day.

Planning your time helps you to control your time. When you are managing your time, you are effective and productive. When you let time control you, you will become frustrated and stressed as work piles up around you.

Planning can be done at the start of the day and at the end of the day. At the start of the day, your mind is fresh and you probably have a pretty good idea of what the day will bring. On the other hand, planning at the end of the day allows you to focus on the accomplishments of the day and helps you identify what remains to be accomplished. A listing of these items is the starting point for the next day of work. An advantage of this practice is that you and your team members can subconsciously think about the work, even while you are off duty.

Give attention to the vital few tasks that are really important (see **Figure 5-6**). Research has shown that

we often spend inappropriate amounts of time dealing with trivial issues, thus saving little of our time for the really important items. Give priority to the important management tasks.

Most fire officers have come up through the ranks, and some spend an inappropriate amount of the time doing firefighter stuff rather than managing firefighter activities (**Figure 5-7**). Many feel more comfortable doing the job themselves. They are proficient and enjoy these activities. Quite often, this is how they got promoted (see Leadership Tetrahedron in Chapter 6). These officers may also have some anxieties about delegating. They may have some concern about their work being evaluated and are unwilling to let that evaluation rest on the accomplishments of a team member. And many would rather perform physical work than sit at a desk and perform strictly mental work. As we indicated earlier, the management functions, such as planning and organizing, require considerable mental effort.

As you rise through the ranks, you will find that you will become increasingly concerned with management activities. At the lower levels in our organization, technical skills are important. In the middle

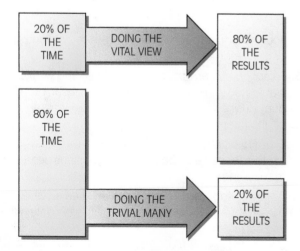

FIGURE 5-6 Effective time management increases productivity and requires management give appropriate attention to the tasks that are really important.

FIGURE 5-7 Planning the day involves all affected personnel.

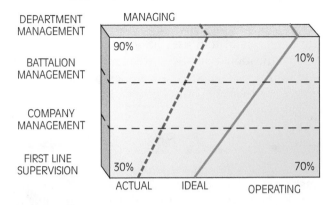

GIVE PRIORITY TO MANAGEMENT FUNCTIONS

FIGURE 5-8 Give proper attention to management functions.

ranks, interpersonal skills are critical. In the upper ranks, management skills become most important (see **Figure 5-8**).

We defer commanding and coordinating for now. That leaves us with controlling. When controlling, you are looking for problems. Solving problems is one of the manager's jobs.

SOLVING PROBLEMS

From time to time, you will face problems that will require a bit of thinking for their solution. When you have a problem, you should follow a proven, logical approach to find the solution. The first step may be simply recognizing that you have a problem. Many managers do not see problems occurring, or, if they do, they fail to take any corrective action. Here are the generally accepted steps for solving problems:

1. *Accurately define the problem.* That may seem simple, but in many cases it is the most difficult step. And in many cases, it is the most critical step, for if you do not correctly identify the root of the problem, the rest of the process is doomed.

2. *Gather information.* Get ALL of the facts and hear ALL sides of the issue. Review the laws, policy statements, and regulations from federal, state, and local sources that may address the issue.

3. *Analyze the information.* See if you have everything you need.

4. *Look for alternative solutions and list all possible solutions.* At this point in the process, you want to be open minded. Invite others to offer ideas.

5. *Select one or more of the best solutions.* At this point, you want to start ranking the solutions in terms of expected outcomes. Some solutions may cost too much or take too long. There are always

compromises. If the answer were easy, you would not be in this process.

6. *Take action.* Usually at this point, you must get approval to implement your solution. You have done all the work—identified the problem, gathered data, considered alternative solutions, and listed one or more solutions you are recommending. This information is usually conveyed to the supervisor in a short memo. Your supervisor should be able to quickly review what you have done and give approval.

7. *Monitor the results.* This step is also important but often overlooked. You want to be sure that you follow the process and see what happens. Most changes bring about some negative consequences. You want to be sure that the good consequences outweigh any bad consequences the change brings with it.

8. *Take corrective action if necessary.* At this point, you may have to simply start all over.

Notice how the five management functions show up here: In steps 1 and 2 you are planning, in step 3 you are organizing your information, in step 6 you are commanding and coordinating, and in steps 7 and 8 you are controlling.

When the answer to the problem comes too easily, make sure that you correctly identified the real problem and that you found the correct solution. When you solve your problem, you may have to change a procedure or policy within your organization. Policy changes should be carefully managed and not the result of a knee-jerk reaction to a one-time event.

This method is applicable, regardless of how the problem is recognized. As an officer, you could simply recognize the problem yourself or it could be brought to you by a supervisor, subordinate, or citizen. Some departments have structured ways in which to handle each; however, the methodology to solving them is basically the same. If you recognize the problem yourself, you need to analyze the situation, accurately identify the problem, look for solutions to that problem, take action, and monitor the results. This process may or may not involve others or require feedback be given. If the problem is reported to you from an outside source or a subordinate, this usually requires some sort of notification process to your supervisor.

Notification could be in verbal or written form. The form of notification is usually dependent upon the severity of the reported problem. For example, if a subordinate reports that he or she recognizes a problem with the facility and/or equipment, it typically is reported to the appropriate personnel and handled at the officer level. This situation would typically require the officer to only verbally report the issue to his or her respective supervisor and note it within the station log. If the subordinate's reported problem was related to some form of harassment, then this rises to a higher level and should be immediately reported to the officer's supervisor in verbal fashion to make him or her aware of the issue and then followed up with written documentation that states the specifics related to the complaint. The same basic guidelines are true for a citizen's complaint. If a citizen is complaining about trash in the yard of the station, this issue would be addressed at the officer's level and noted in the station log. The defining, analyzing, and solution process would be very brief. If the citizens reports that a fire truck caused damage to their personal property, this would require a more formal handling of the complaint, requiring extensive investigation, fact finding, and documentation. The solution process for this type of complaint usually takes place above the level of the company officer and often involves the department's administrative staff, risk manager, and others. Therefore, it is critical that the officer's information be as detailed and accurate as possible.

Some departments have very specific guidelines for certain types of complaints. It is the officer's responsibility to know how to properly handle a complaint and use the appropriate forms and communications channels. Readers are encouraged to examine the reporting policies of their respective department.

OFFICER ADVICE

When you feel the answer to a problem has comes too easily, review the problem-solving steps and make sure that you correctly identified the real problem and that you found the correct solution.

MANAGING CHANGE

As a manager, you should make change a requirement in your organization. Identify the desired outcomes, sell the benefits, involve your people, and reward the participants. Some individuals and organizations deal

TECHNOLOGY ADAPTATION LIFE CYCLE

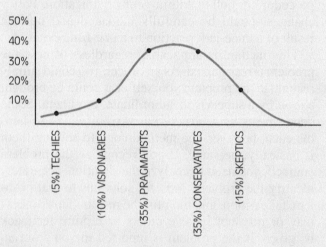

FIGURE 5-9 The Technology Adaptation Curve indicates an approximate distribution of members ability to deal with the changes brought by new technology. Where do you fit in?

When change is ongoing, some accept it more rapidly than do others. In today's fast-paced world, change often comes with the development of new technology.

Personnel adapt to new technology in various ways (see **Figure 5-9**). These individuals can usually be identified and placed into one of five categories:

1. *Techies or Innovators:* This small minority is the first to the store to buy any new product. New technology is their main interest. They like the challenge of being in the forefront. Their opinions are important as they often influence those who follow.

2. *Visionaries or Early Adopters:* Like the pioneers before them, these individuals are ahead of the pack. They are interested in using technology to improve productivity. They are more challenging and more demanding of new technology, for they are looking at the results, not the means.

3. *Pragmatists:* Pragmatists are not pioneers. They are pushed into the adoption of technology by the need to increase productivity, rather than being pulled by technology itself. They need support, understanding, and interaction with peers to survive.

4. *Conservatives:* Conservatives are not comfortable with technology. Tradition is strong with them, and they join only when there is no real alternative. When they do adapt to new technology, they tend to adopt popular concepts and proven products.

5. *Skeptics:* Skeptics resist new technology. They will cost the organization more than the results are worth. Best leave them alone.

with change quite well. They carefully analyze their problems, rapidly adapt new ideas, and go on with business. Others tend to be more resistive. In these cases, change often passes them by.

Change breeds resistance. With change, conflict may become more apparent. We should convert the energy of conflict into positive values: Recognize that competition is healthy and that we can use the results of the conflict to further positive outcomes. With conflict, we tend to take a position, and as a result, there are winners and losers.

NOTE

With change, conflict may become more apparent.

Change that is easily accomplished, or that has the highest visibility, or that makes you look good to your boss may not be what is really needed for the organization.

We discuss how to introduce change in Chapter 7.

MANAGING CUSTOMER SERVICE

Most fire service organizations enjoy a noncompetitive environment when it comes to the delivery of services. However, most fire departments, and certainly most public fire departments, are in a very competitive environment when it comes to obtaining financial resources. Fire departments must compete for their share of the community's tax dollars. Their competition comes from the other agencies: the police department, the school board, the public works department, and all of the other agencies that contribute to the security and welfare of the citizens. Today, communities are faced with increasing demands for their services. At the same time, many are also faced with funding these demands with limited resources. Increasing taxes is one answer, but from a citizen's viewpoint as well as a politically viewpoint, this is an unattractive option. For much the same reasons, funds from state and federal resources have diminished recently, all of which has been aggravated by a slow economy.

Given this situation, one would wonder why a fire department would be concerned about its relationship with the recipients of its services, its customers. Think about yourself as a customer for a moment. If we had a choice between several well-known fast-food restaurants, two that offer similar menus and nearly identical pricing, we would likely patronize the one that offered the better customer service. The same would apply whether we were buying tools or shoes. What would provide that satisfaction? Well, courtesy and prompt, personal attention from the employees would be a good start. And maybe, having them stay with us long enough to be sure that we received everything we expected, maybe even everything we wanted, would give us satisfaction. The ultimate test is having them do more than just what is required, having them exceed our expectations.

Today, many fire departments are doing just that: They are trying to provide the customer with the best possible service. During emergency situations, you are often placed in situations with very uncertain outcomes, accompanied by significant dangers where you have to offer the best possible service. You work in burning buildings or in the middle of a dangerous highway. You work at all hours of the day or night, often in the worst weather. Whenever and wherever you go, you go because people are having a problem, and they need your help. They call you when there is a fire or a leak, when someone gets sick, or when someone is injured. In nonemergency situations, we should be still trying to provide the best possible service as well. We should continually see and treat these people as our customers.

Three Major Components of Customer Service

First: Always be nice. Treat people with respect and consideration. Show patience and consideration, even if it is three o'clock in the morning. Your customers may be impressed by your quick arrival but what really matters, and what they will remember the longest is what you do after you get there. Fire trucks do not treat medical problems or put out fires; people do. People also show compassion. Remember to treat the customer, every customer, the way you would want your mother to be treated. Provide the best service, and show compassion.

Second: Regard everyone as a customer. Many times the problem is straightforward and simple for us, but for the victim's family and friends, things are very complicated. Factor these folks into the customer equation as well. What you do and say has a lasting impression on everyone present. Regardless of whether you do well, or not so well, they will tell others what happened. Which story do you want them to tell?

There are others who wander into our workplace. Most of what you do attracts spectators. While you must control them and keep them out of harm's way, you should be aware that they have a natural and genuine concern about what is going on. You, as a provider, have the advantage of a front row seat at every show. But that does not mean that they cannot stay and watch from a safe distance. Good Samaritans also stop to help and must sometimes be relieved with an informal and polite transfer of command. Imagine finding that the apparently uninformed foreign-born female civilian who is helping at the scene of an accident involving a child with a broken leg is actually an orthopedic trauma surgeon.

Our customers are everywhere. They walk into fire stations looking for directions to the post office across the street or directions to a street that is located across town. There are school kids and scouts who come to visit and high school students working on a science project. Always remember you are part of a bigger picture. Your department is one organization of many that are set up within the community to protect and SERVE the citizens of that community. Consequently, your job is to meet the needs of the citizens to the best of your capabilities. This is exactly what Brunacini was referring to in his customer service book. If they are looking for resources, help them find that resource, even if it is not related to what the department does. For example, a citizen may come into the fire station looking for a location where he can pay a water bill or wanting to find out how he can get a permit to hold a garage sale. Your job is to help the citizen locate that resource. If you don't know, take the time to find out!

The fire department is not a silo independent of other city, county, and state organizations. Just think how much better off the community would be if all departments within the community worked with the common goal of meeting the needs of the citizens. It starts at the company officer level. The company officer is typically the highest ranking officer in contact with the citizens. If he or she does not initiate this service, who will? Take a little time to learn about your city, county, or state organization and what services they have to offer so that you will be able refer a citizen's concern to the appropriate authority. Learn about your organization's policies and procedures related to public inquiries.

Firefighters are heroes to the citizens, but do not forget all those folks behind the scenes who seldom get the glory—the mechanics, payroll clerks, and dozens of others who provide support to the front-line firefighters and make it possible for us to do our job. They are customers, too. As company officer, your firefighters are also your customers. Your job is to make them successful. That means providing them with the tools, knowledge, and skills to get the

FIGURE 5-10 *Over maintenance of vehicles and equipment should be performed as a personal and professional labor of love.*

job done. How well they perform is a direct reflection on you.

Part of your customer service is how you look. Uniforms are a big deal for most members of the fire service. They indicate pride and identity. They connect us to our local government and the long and glorious tradition of the fire service. The uniform should be complete, neat, and worn properly. As company officer, it is up to you to make sure your crew is in the appropriate uniform and that the uniform is maintained and presentable. This also relates to your turnout gear; make sure it is properly maintained and that you and your crew wear it when you should.

Your vehicles are another highly visible clue that you are out and about (see **Figure 5-10**). They should be over maintained with a personal and professional labor of love. The tools you use are also impressive. They range from primitive to space-age, but they all have a function, and all should work when needed. Anyone watching you will be quickly able to tell if your vehicle, your tools, and yes, even you, are sharp and ready for the task at hand. A well-maintained crew and a well-maintained vehicle send a positive action-packed message to the world: You are ready, willing, and able to handle your customer's needs.

Third: Constantly raise the bar, raise the performance standard, and look for ways to improve, to offer more and better service to your customers. You should be constantly working to improve yourself and your organization (see **Figure 5-11**). We have to do that just to keep up with the not-so-perfect but constantly changing world around us. Remember, you want your customers to be delighted with your work.

FIGURE 5-11 The citizens paid for this equipment. They let you manage it for them. Show pride in the way it is maintained and operated. (*Photo courtesy of Fairfax County Fire and Rescue Department*)

ASSESS THE NEEDS OF YOUR CUSTOMERS

- Seek feedback from customers about their needs.
- Seek feedback from customers (including your crew) about their degree of satisfaction.
- Define requirements of the workplace to meet these needs.
- Establish goals.
- Define a management philosophy.
- Develop a self-directed team.
- Provide feedback to the team regarding performance.
- Incorporate new ideas and technologies.
- Seek to make continuing improvements to the process and the product.
- Fine-tune the mission.

BUILDING A MANAGEMENT TEAM

The following are characteristics of good management in a committee or group activity, including your company[3]:

- Provide an open, relaxed working atmosphere in which people are involved and engaged (see **Figure 5-12**).

- Provide an environment that encourages discussion in which virtually everyone participates. There can be no fear of retribution.

- Provide an environment in which the objective is well understood and accepted by the members.

- Provide an environment in which the members listen to each other, in which the discussion does not jump from one idea to another. Every idea is given

FIGURE 5-12 Successful company officers create an environment where there is open discussion and collaboration.

a hearing, and people are not afraid of being foolish by putting forth a creative idea, even if it seems extreme.

■ Provide an environment in which disagreement is allowed. The group is comfortable with this concept and shows no signs of having to avoid conflict.

■ Provide an environment in which agreements are not suppressed or overridden by premature group action. On the other hand, there is no tyranny in the minority. Individuals who disagree do not appear to be trying to dominate the group or to express hostility. Their disagreement is an expression of a genuine difference of opinion, and they can expect a hearing in order that a solution may be found.

■ Provide an environment in which most decisions are reached by a kind of consensus, where it is clear that everybody is in general agreement and willing to go along.

■ Provide an environment in which criticism is allowed but which there is little evidence of personal attack, either openly or in a hidden fashion. The criticism has a constructive flavor in that it is oriented toward removing an obstacle that faces the group and prevents it from getting the job done.

■ Provide an environment in which people are free to expressing their feelings as well as their ideas on a problem and on the group's operation.

■ Provide an environment in which clear assignments are made and accepted.

■ As the leader, provide an environment in which you lead but do not dominate. In fact, allow the leadership to shift from time to time, depending on the circumstances. Different members, because of their knowledge or experience, are in a position at various times to act as resources for the group. The members utilize themselves in this fashion, and they occupy leadership roles while they are thus being used.

To be a good manager, help make these conditions happen at your workplace.

ONGOING AND EMERGING MANAGEMENT ISSUES FOR THE FIRE SERVICE

Before we close out this two-chapter discussion of management, we should look at where the fire and emergency services are headed and give some thought to the major issues that face management in these organizations.[4] Your department may already be involved in one or more of these issues. As a member, you may be involved in helping deal with some of these as well. Most of these topics are addressed in this book.

Ongoing Issues of National Importance

■ *Leadership:* To move successfully into the future, the fire service needs leaders capable of developing and managing their organizations in dramatically changed environments.

■ *Prevention and public education:* The fire service must continue to expand the resources allocated to prevention and health and safety education activities.

■ *Training and education:* Fire service managers must increase their professional standing in order to remain credible to community policy makers and to the public. This professionalism should be grounded firmly in an integrated system of nationally recognized and/or certified education and training.

■ *Fire and life safety systems:* The fire service must support adoption of codes and standards that mandate the use of detection, alarm, and automatic fire sprinklers with special focus on residential properties.

■ *Strategic partnerships:* The fire service must reach out to others to expand the circle of support to assure reaching the goals of public fire protection and other support activities.

■ *Data:* To successfully measure service delivery and achievement of goals, the fire service must have relevant data and should support and participate in the revised National Fire Incident Data System. Likewise, NFIRS should provide the local fire service relevant analysis of data collected.

■ *Environmental issues:* The fire service must comply with the same federal, state, and local ordinances that apply to general industry and which regulate response to and mitigation of incidents, personnel safety, and related training activities.

Emerging Issues of National Importance

■ *Customer service:* The fire service must broaden its focus from the traditional emphasis on suppression to a focus on discovering and meeting the needs of its customers. This may include preparing for and being able to deliver an effective response to all forms of local disasters, be they natural or human-made.

- *Managed care:* Managed care may have the potential to reduce or control health care costs. It also will have profound impact on the delivery and quality of emergency medical care.

- *Competition and marketing:* In order to survive, the fire service must market itself and the service it provides, demonstrating to its customers the necessity and value of what it does.

- *Service delivery:* The fire service must have a universally applicable standard that defines the functional organization, resources in terms of service objectives (types and level of service), operation, deployment, and evaluation of public fire protection and emergency medical services.

- *Wellness:* The fire service must develop holistic wellness programs to ensure that firefighters are physically, mentally, and emotionally healthy and that they receive the support they needed to remain healthy.

- *Political realities:* Fire service organizations operate in local government arenas. Good labor–management and customer service relations are crucial to ensuring that fire departments have maximum impact on decisions that affect their future.

- *Interagency cooperation:* Today's environment requires that the fire service create and maintain effective working relationships with all elements of the community before a disaster occurs.

EXTREME RESOURCE MANAGEMENT

Lieutenant Marty Rutledge, Antarctic Fire Department, McMurdo Station, Antarctica

I think that company officers tend to take resource management for granted—it is a skill that often comes easily, especially to those who have been in the area and in the fire department for any length of time. I am no exception. I certainly understood making changes in the way the fire department does business under certain circumstances, like in ice and snow and freezing temperatures, but it was a very small part of my thought process. Then I took a job as a firefighter in Antarctica, and resource management became critical. Once I was promoted to lieutenant, I realized that I would have to incorporate issues related to resource management into every decision I make—not only to get the job done, but also to protect firefighters.

If one word was given to describe the biggest challenge it would most likely be the word "cold." The environment presents the greatest challenges, and these challenges go much further than one might think. The extreme cold is very hard on all mechanical equipment. Even new vehicles are as prone to weather-related problems as are much older vehicles. There is always the question in my mind, "Will the equipment function as it should?" The cold affects the physical and mental abilities of the firefighters and rescuers as well as general firefighter attitude. Summer season here can present some mild temperatures (mild relative to the continent), but these are short-lived and usually there are about 4 to 6 weeks of temperatures in the mid 30s ∫ F. This presents the possibility of outdoor training scenarios, annual hose testing, and driver/operator training, but all of this has to be accomplished during this short season; once the

opportunity is gone, it will be up to a year before it presents itself again.

How else does this environment impact everyday firefighting decisions?

- Rescue is a unique challenge with the unusual question, "Where do I send the victims once they are evacuated and/or rescued?" Victims not attired in extreme cold weather gear must be quickly evacuated to a safe warm area; minutes, or sometimes even seconds, in the environment could render an unprotected person dead or incapacitated.

- Fire attack could be no more than a total exposure protection scenario. The air in Antarctica is very dry and it is usually windy. Any wooden building could be a loss by the time an engine company arrives, even if the response time is quick! Fire attack almost always involves advancing dry hoselines as any hoseline not flowing water will quickly freeze. Ice on the fireground presents a hazard that will remain for months and must be a consideration in the total strategy and tactics plan. Wet turnout gear will quickly freeze and render firefighters handicapped while attempting normal duties. Firefighters can also quickly become hypothermic on a fire scene, and provisions must be made for firefighter rehabilitation.

- Salvage operations must be quick and coordinated. Anything usable that is not damaged by fire or water must be protected and moved, if at all possible. Products of combustion remaining must also be cleaned thoroughly to comply with the Antarctic Treaty. No waste can remain on the continent—it all must be packaged and shipped off continent for disposal.

Training, tabletop exercises, and constant fire tactical discussions are a must for Antarctica's firefighters and company officers. The things that many people are used to in firefighting back in the United States can be totally different here in this land of ice and cold. The basis for the firefighting work we do here shares many similarities with the training and firefighting done in the United States, but a constant focus on resource management in the conditions we face in Antarctica is what keeps our firefighters safe and effective.

Questions

1. Why is a focus on resource management so important for Lieutenant Rutledge?

2. What similarities do you see between resource management in your fire department and resource management in the fire department in Antarctica described in this case study?

3. What differences do you see between managing company resources within your fire department and how Lieutenant Rutledge's must manage his?

4. Provide an example of how resource management can affect the safety of firefighters in this case study.

5. Now, provide an example of how resource management can affect the safety of firefighters in your own department. Although this case study provides an extreme example, can you see the critical impact that management of resources has on your own department?

LESSONS LEARNED

This chapter contains practical tools to assist you in managing your resources. Learn to use your resources well, especially your own time.

These chapters only scratch the surface. To learn more about management, check for the newest management books in your library or bookstore.

KEY TERMS

agency shop   an arrangement whereby employees are not required to join a union, but are required to pay a service charge for representation

arbitration   the process by which contract disputes are resolved with a decision by a third party

authority   the right and power to command

budget   a financial plan for an individual or organization

capital budget   a financial plan to purchase high-dollar items that have a life expectancy of more than 1 year

closed shop   a term in labor relations denoting that union membership is a condition of employment

fact finding   a collective bargaining process; the fact finder gathers information and makes recommendations

Formal Bid   a formal request for cost of products and/or services that are over a specified dollar amount

goal   a target or other object by which achievement can be measured; in the context of management, a goal helps define purpose and mission

ISO   a family of standards for quality management systems. *ISO 9000* is maintained by ISO, the International Organization for Standardization, and is administered by accreditation and certification bodies

line-item budgeting   collecting similar items into a single account and presenting them on one line in a budget document

mediation   the process by which contract disputes are resolved with a facilitator

Memorandum of Understanding (MOU)   a pre-established contract to provide goods or services from a vendor prior to an emergent event

mission statement   a formal document indicating the focus and values for an organization

objective   something that one's efforts are intended to accomplish

open shop   a labor arrangement in which there is no requirement for the employee to join a union

operating budget   a financial plan to acquire the goods and services needed to run an organization for a specific period of time, usually 1 year

program budget   the expenses and possible income related to the delivery of a specific program within an organization

Request for Bid   a formal request for cost of products and/or services that are over a specified dollar amount

Request for Proposal (RFP)   a solicitation typically used when you need design, consultant, products, or typically services

Request for Quote (RFQ)   a non binding solicitation typically to determine cost for goods or services from pre-established vendors or contractors

union shop   a term used to describe the situation in which an employee must agree to join the union after a specified period of time, usually 30 days after employment

vision   an imaginary concept, usually favorable, of the result of an effort

REVIEW QUESTIONS

1. What is a mission statement?
2. What is meant by labor relations?
3. What are some of the typical issues covered by union contracts?
4. What are the major differences between management and member rights?
5. What is the process for implementing project management principles?
6. What is a budget?
7. What is the difference between a capital budget and an operating budget?
8. What is the difference between a formal bid and an RFQ?
9. What are the benefits of writing well-defined specifications?
10. How can you, as a company officer, improve the use of the department's financial resources?
11. How can you, as company officers, improve the use of the department's human resources?
12. How can you, as company officers, improve the use of the department's time resources?
13. How can you, as company officers, improve the use of your personal time?
14. What is your role, as a company officer, in managing customer service?

DISCUSSION QUESTIONS

1. What are the benefits of having a mission statement?
2. What is the mission statement of your organization? Do you agree with what it says? If you could make changes to your department's mission statement, what would you suggest?
3. What are some ways to assess the performance of your department relative to its stated mission and goals?
4. Why is customer service an import issue for emergency response organizations?
5. What are the major differences between management and member rights?
6. What is your department's procedure for data collection, recording, retention, and disposition?
7. Why is delegating difficult?
8. Why do some supervisors have problems delegating work?
9. Much of an officer's time is spent doing administrative work. What administrative duties would be appropriate for delegating? What administrative duties would not be appropriate for delegation?
10. How does your department manage routine expenditures?
11. How does your department manage capital expenditures?
12. What does your department purchase that is let out for bid and what is purchased directly?
13. What are the benefits of writing well-defined specifications?
14. What are the applicable purchasing laws and who is responsible for ensuring your department follows them?
15. What required policies does your agency have in place for acquiring resources?
16. Who within can your agency can legally expedite purchases?
17. What are the guidelines needed to submit a request for purchasing equipment?
18. How well do you manage your own time? What could you do to improve your time management?
19. What is the operating cost for your station and what percentage is for staffing?
20. What is the importance of accurate data entry and what are the effects of inaccurate data entry?
21. What are the ongoing and emerging management issues in your department?
22. Why is it important to develop and maintain working relationships before a disaster with other elements of the community?

ENDNOTES

1. Unions are discussed here for several reasons, even though we realize that not all firefighters are union members or that they even benefit from union representation. This section discusses a process that while formalized by a labor–management relationship, can bring benefits to the employees of any organization.

2. Unfair labor practices include interference with employees in their right to organize, interference with a labor organization, interference with employees who do not want to participate in union activities, discrimination in employment because of union participation, refusing to bargain in good faith, asking the employer to discriminate against an individual, and featherbedding provisions in the contract.

3. Adapted from Douglas McGregor, "The Managerial Team," in *The Human Side of Enterprise* (New York: McGraw-Hill, 1960).

4. Conference Report, Wingspread IV (Fairfax, VA: International Association of Fire Chiefs, 1996).

ADDITIONAL RESOURCES

Anthony, Derrick. "10 Common Mistakes New Company Officers Make." *FireRescue*, November 2007.

Carlson, Richard. *Don't Sweat the Small Stuff*. New York: Hyperion, 1997.

Cascio, Wayne .F. Managing Human Resources. 5th ed. Boston: Irwin McGraw-Hill, 1998.

Cavette, Chris. "Runs per Gallon." *Fire Chief,* December 2007.

Coleman, Ronny. "To Pursue Progress, Ask Yourself 'What If?'" *Fire Chief,* March 2007.

Cooper, Frank. *The Customer Signs Your Paycheck*. Everett, WA: Frank Cooper Publishing Company, 1984.

Covey, Stephen R. *First Things First*. New York: Simon and Schuster, 1995.

Johnson, Spencer, *Who Moved My Cheese?* New York: G.P. Putnam's Sons, 1998.

Lasky, Rick. "Customer Service that Serves Your Department." *Fire Engineering,* March 2004.

Lasky, Rick. "Pride and Ownership: The Love of Our Job—Our Mission." *Fire Engineering,* May 2004.

Lasky, Rick. "Pride and Ownership: The Love of Our Job—The Firefighter." *Fire Engineering,* June 2004.

Lasky, Rick. "Pride and Ownership: The Love of Our Job—The Company Officer." *Fire Engineering,* July 2004.

McGregor, Douglas. *The Human Side of Enterprise*. New York: McGraw-Hill, 1960.

Okray, Randy and Thomas Lubnau. *Crew Resource Management for the Fire Service*. Saddle Brook, NJ: Fire Engineering, 2004.

Poulin, Thomas. "Managed Expectations." *Fire Chief,* November 2008.

Project Management Institute, *A Guide to the Project Management Body of Knowledge* (PMBOK® Guide. 3rd ed.). PMI Institute, October 2004

Ritvo, Roger A., et al. *Managing in the Age of Change*. Burr Ridge, IL: Irwin Professional Publishing, 1995.

Serio, Gregory. "Re-evaluating Risk." *Fire Chief,* June 2008.

Urbano, Paul. "Company Officer Time Management." *Fire Engineering,* July 2008.

Wallace, Mark. *Fire Department Strategic Planning*. Saddle Brook, NJ: PennWell, 1998.

Wilmoth, Janet. "Union United." *Fire Chief,* February 2004.

6

The Company Officer's Role: Principles of Leadership

I have the unique and very satisfying opportunity to be a paid and a volunteer officer at the same time in two separate organizations. The opportunities I have in these two positions offer unique challenges to me in developing both as a manager and, more importantly, as a leader.

I believe the most important factor in successful leadership at any job is attitude. A person must be responsible for his or her actions, must be educated and trained, and must be able to communicate well..Firefighting is a very special job. Does it really matter if you are volunteer or paid? The job has to be done, and it has to be done right—every time. The foundation of attitude helps me to cope with the challenges of being a leader. Focusing on what to do, or what is wrong, is easy if you leave the personalities and the issue of "paid versus volunteer" out of the equation. The problem is just that, the PROBLEM, usually not the people.

Expectations are another part of the foundation for a positive attitude. "Good enough" simply does not cut it. If a company officer is solid, makes intelligent decisions, and continually acts and performs as a leader, then this lays the groundwork for the success of the entire team. The right attitude and defined expectations remove many of the obstacles to success and allow everyone to play by the same rules. If a leader can be consistent, fair, accept team input, and attempt to remove the personalities from the issues at hand, the team can thrive, no matter whether they are paid for their work or they provide service as volunteers.

Many firefighters are quick to place blame and foster negativity based simply on the paid versus volunteer issue. I firmly believe that the mark of a great fire service leader is that he or she can lead any type of firefighters, paycheck or no paycheck. Remembering that we are a service to our citizens and that we must be trained and make intelligent decisions helps to build a bridge between the paid and volunteer services—and get over it with success!

—*Street Story by Scott Windisch, Lieutenant, The Woodlands Fire Department,*
The Woodlands, Texas, and Deputy Chief,
Ponderosa Volunteer Fire Department, Houston, Texas

LEARNING OBJECTIVES

After completing this chapter the reader should be able to:

6-1 Define *interpersonal* and *group dynamics*.

6-2 Describe the principles (elements) of interpersonal and group dynamics used to obtain group cooperation.

6-3 Explain how understanding the human behavior of employees assists the company officer in performing their daily duties.

6-4 Describe the elements that determine leadership style.

6-5 Discuss ways to use leadership to build positive teamwork.

6-6 Identify the uses of leadership power.

6-7 Compare the principles (elements) used by supervisors to obtain cooperation of subordinates.

6-8 Define the different types of leadership styles available to the company officer.

6-9 Define the term *power* as it relates to company officer.

6-10 Describe the positive and negative aspects of power.

6-11 Identify the sources of leadership power.

6-12 Identify the types of leadership power.

6-13 Differentiate between the various types of leadership.

6-14 Determine which type of power is appropriate for a given situation.

6-15 Interpret when it is appropriate to use different types of leadership styles.

6-16 Describe cultural diversity.

6-17 Analyze the effect of cultural diversity on the fire service.

6-18 Explain the human resource policies and procedures that apply to the department.

*The FO I and II levels, as defined by the NFPA 1021, the *Standard for Fire Officer Professional Qualifications*, are identified in different colors: FO I = black, FO II = red.

INTRODUCTION

We discussed organizations in Chapter 3. In Chapters 4 and 5, we discussed management in organizations. We now look at the human side of our organizations, the people who work in the organization.

To be effective, the people in those organizations need leadership. Leadership focuses on people—people working together in organizations. In this chapter, we look at some principles of leadership. In Chapter 7, we look at the applications of these principles in your workplace.

> **NOTE**
>
> Leadership focuses on people working together in organizations.

WHAT IS THE DIFFERENCE BETWEEN MANAGEMENT AND LEADERSHIP?

Management is a set of processes that can keep a complicated system of people and technology running smoothly. The most important aspects of management include planning, budgeting, organizing, staff-

ing, controlling, and problem solving. Leadership is a set of processes that creates organizations in the first place or adopts them to significantly changing circumstances. Leadership defines what the future should look like, aligns people with that vision, and inspires them to make it happen in spite of the obstacles.[1]

ORGANIZATIONS AND GROUPS

An **organization** is a group of people working together to accomplish a task. There are many types of organizations. Fire departments are formal organizations. A fire department bowling league represents an informal organization. Both meet the definition: a group of people working together for a common purpose. In spite of the varied nature of organizations, they share several common characteristics:

All organizations have a purpose. The purpose may be to make a profit, protect the environment, provide an essential service to the community, or just have fun.

Second, all organizations are made up of people. Although a purpose statement is essential to provide the organization with goals and objectives, it takes people to make these things happen.

Third, these people will arrange themselves into some type of organizational structure. They will define the duties and responsibilities of the organization's members and establish rules and procedures.

A group is two or more people who come together for a common purpose. Therefore, **group dynamics** can be defined as how those individuals interact and influence one another. Groups will tend to develop numerous dynamic practices that separate them from other individuals. These practices can be referred to as **interpersonal dynamics**. In a group environment, the two major factors that are considered are the task at hand and the relationships within the group. Groups must be able to resolve conflict, build trust, and, above all, communicate. Interpersonal dynamics molds these groups into effective team units.

By their very nature, groups represent people who share common interests and goals. They may share a vision of what their group can do. For many groups, there is also a sense of permanence, a desire to see that the group flourishes and survives. As a result of these common interests and visions, the members work together to satisfy these interests. These shared values are what bind the members together as a group.

As we have already indicated, most of us like to be affiliated with groups. The members of a fire department are a group, and they are part of at least one group at work. There are also informal groups at work, groups of individuals who share common interests. Their interests may be bowling, softball, golf, fishing, photography, or other pursuits. Even here, we see some variation in how groups work. Softball requires a group just to field a team, and at least two teams are needed to play a game. Clearly, there is organizational structure here, for the team members must be able to play the various positions and must be able to work together to be successful. On the other hand, individuals in a group that goes fishing together may be more autonomous and unstructured. In these informal situations, we have the right to select what we want to do.

We may also belong to groups away from work. We may be active in a church, civic association, or neighborhood resident association. We may be involved with activities that support the development of our children, such as scouting or soccer leagues. All of these are groups. Within all groups, there are positions to be filled. Someone has to be the leader. That person may be appointed or elected. In most informal groups, that person may be the one who comes forward with an idea. The group accepts the idea and responds to the requests for help.

WE ALL BELONG TO INFORMAL GROUPS

Suppose that one of your neighbors suggests a party. If the weather is nice, it can be held outdoors. Someone starts with an idea; others add their ideas and agree to participate. The group comes to a consensus on the time and place that work best. Everyone agrees to make great food and to make purchases to support the gathering. The weather cooperates and everyone has fun.

Is it work? Yes, of course it is work. Does someone have to take the lead? Yes, at least initially. Will there be some who do not participate? Of course. But for most, the benefits far outweigh any efforts. All who attend go home feeling great and agreeing to do this again next year.

Who elected the leaders? Who made the rules? Clearly they were there, but they were very informal. All agreed to the idea and in so doing accepted some responsibility to the others for the overall success of the event. A group of people with a purpose. In this case, the purpose was just to have fun.

WHAT IS LEADERSHIP?

What is leadership? **Leadership** is defined as the personal actions needed by managers to get team member's to carry out certain activities. Leadership is achieving the organization's goals through others. Simply put, leadership is getting other people to do what you want done. Management, leadership, supervision, and administration: we hear these words used often, and we may confuse them. Indeed, they are often used interchangeably.

Leadership is one's influence over others. This definition infers that leadership is a management tool. If you think of the management of the resources of the organization, you must include the human resources. Managing human resources deals with personnel administration, the administrative activities associated with staffing an organization. These activities include hiring, promoting, and so forth, which are usually the job of the human resource or personnel department.

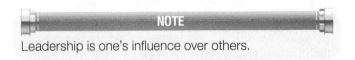

NOTE

Leadership is one's influence over others.

For the supervisors in the organization, managing human resources is more about leadership, the human relationships in the organization. Effective leadership deals with changing the personal conduct of others. Leadership skills are based on the feelings and attitudes that have grown out of both the relationship with a team member and the sum of all of the experiences that have occurred during the supervisor's life.

NOTE

President Harry Truman said, "A leader is a [person] who has the ability to get other people to do what they don't want to do, and like it."

We let our leaders use their experiences to attempt to influence our actions. How much tolerance we have for this control over us is determined both by our personality and the personality of the leader. Some are more tolerant than others. We all know people who are tolerant of their organizational situation and stick around, regardless of the conditions. We also know people who frequently change jobs for one reason or another. In some cases, they may lack the tolerance needed to survive in a difficult work environment.

NOTE

We let our leaders use their experiences to attempt to influence our actions.

Several things generally help make the work environment a better place:

- Sound organizational objectives
- Clear policies and guidelines
- Consistent management
- Clear definition of duties
- Open lines of communications
- Individuals well matched with jobs
- Recognition of good work

We have discussed several of these important issues in previous chapters. Where these characteristics are present, it is easier to be a good leader. When members can see their place in the overall organization, when they are given information and allowed to contribute their ideas, and when their services are appreciated, they will be satisfied and productive. Textbooks, magazine articles, and professional development seminars all address the positive values of these important concepts. Yet we still see organizations where these basic qualities are missing. In these cases, effective leadership is more challenging. Many of the members as well as the leaders become frustrated. The organization is likely to be operating at less than full potential.

When poor work conditions exist in a fire department, everyone loses. Most firefighters join the fire service because of a love for the job. Most look for opportunities for professional growth and for serving their community; they like working with others, and they are generally willing to take on any reasonable task to help the citizens.

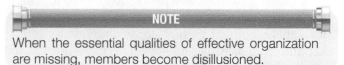

NOTE

When the essential qualities of effective organization are missing, members become disillusioned.

When the essential qualities of effective organization listed previously are missing, these members become disillusioned. Some members quit, while others just hang around, doing as little as they can. In these cases, everyone loses, and the department is probably not really serving the community as well as it could. As company officer you may not be able to do much that will have a department wide impact, but you should do what you can to ensure these qualities exist within your crew.

When we looked at simple organizational structure in Chapter 3, represented by a triangle, we saw that there were four major categories of employees: employees, supervisors, middle managers, and top managers. Applying these generic labels to the fire service, we usually call these same layers firefighters, company officers, senior staff, and the fire chief (as shown in **Figure 6-1**). In this chapter, we are focusing on the supervisory role of the company officer.

Company officers supervise firefighters. Because you are the first supervisory rank in the organization, and because you usually do not supervise other managers, we call you **first-level supervisors**. The leadership component of the supervisor's job is a large part of your overall responsibility.

The perception of the supervisor's job has changed over the last few years. Before World War II, and even

FIGURE 6-1 Company officers represent the first supervisory rank in the organizational structure of the fire service.

for a period afterward, the supervisor's role was simply to see that the work got done. This was where management interfaced with the members. The supervisor had the authority to define the job, monitor the workers' performance, discipline those who needed it, and fire those who did not respond.

Today, the role of the supervisor is much expanded. You might think of some other titles that help define the leader's job. Today's team leader is also a coach, foreman/forewoman, and supervisor. Regardless of the exact title or the perception it implies, the job is to lead the organization's human resources. An important part of that job is to develop and maintain the organization's human resources to their fullest potential. The role of the supervisor is important today in any organization and it appears that the importance of the supervisor's role will continue to grow.

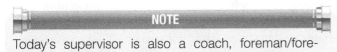

> **NOTE**
>
> Today's supervisor is also a coach, foreman/forewoman, and team leader.

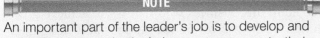

> **NOTE**
>
> An important part of the leader's job is to develop and maintain the organization's human resources to their fullest potential.

Increasing Responsibilities for the Company Officer

In many communities, the cost of providing government services is under close scrutiny. At the same time, fire departments are expected to provide increasing services. Increasing services while holding the line on budgets means that fire departments must become more efficient. In organizations of every kind, part of this cost reduction is being accomplished through the streamlining of the organization. In particular, middle management is a common target. For the fire service, this means that there are fewer layers of supervision between the fire chief and the station officers. More responsibility for the day-to-day operation of the organization is being passed to you, as the company officer.

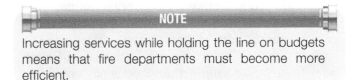

> **NOTE**
>
> Increasing services while holding the line on budgets means that fire departments must become more efficient.

For many fire service organizations, the greatest opportunities for further improvements in efficiency will come at the company level. While the organization is getting smaller, the job at the company level is becoming more complicated. We just noted that more work is being passed down to the company officers, and at the same time, most fire departments are responding to an increasing variety of emergency situations. As company officers, you must personally remain proficient in the skills needed to respond to the myriad of emergency situations that can be expected, and at the same time, you must see that your personnel are proficient and ready to perform their respective duties (see **Figure 6-2**).

In addition to the duties required at the emergency scene, many other supervisory tasks crowd the day. Preplanning, company inspections, and physical

FIGURE 6-2 As company officers, you are expected to provide leadership in the fire station as well as at the scene of emergency events.

training all compete for the company's time. Other issues, such as diversity training and improved customer service beg for attention. All of these require good leadership.

Where Do We Get Good Leaders?

It has been said that there are only two positions in a fire department to which outsiders can apply. One is the fire chief, and the other is as an entry-level firefighter. With few exceptions, all other positions are filled by promoting people from within the organization.

There are many benefits of such a promotion process. Certainly, you, as the company officer, need to understand the firefighter's job, and you will even be a firefighter during certain emergency activities. You, by the nature of your experience, also know the organization, understand the rules, and in most cases know the people with whom you work.

Emergency service organizations usually promote our personnel based on some combination of past performance, test scores, and evidence of relevant training and education. But new supervisors soon find that their technical competence does not mean as much at the supervisory level; they find greater challenges in people issues, administrative duties, and the responsibility for managing others.

MODELING HUMAN BEHAVIOR

To be a good leader, you should understand human behavior. Most of us in the fire service have had thousands of hours of training on the technical aspects of the job. You have been trained in fire suppression, emergency medical procedures, and how to deal with hazardous materials. Most fire companies spend about 10 percent of their time dealing with emergency activities. They spend most of the rest of the time they are together relating as humans. Few of us have had much formal training in human relations. Most of us would work better together if we could better understand how others react in various situations.

OFFICER ADVICE

Most of us would work better together if we could better understand how others react in various situations.

Working together is all about group behavior. Studying group behavior is a part of the science of sociology. Sociology includes the study of groups ranging from families and small workplace groups to entire communities and ethnic groups. We look briefly at several of the significant contributions to this field.

It is hard to take a photograph of human behavior. It would be even harder to show a picture of an attitude. Yet, we all have some concept of these terms. To help us understand these ideas, many of the writers on human behavior have conceived models to represent some aspect of human behavior. A diagram or chart usually represents these models.

Models are used to represent real things. When we model human behavior, we attempt to represent real human behavior. We use diagrams or drawings to help convey our thoughts in books. Unfortunately, the models do not always give a complete or precise picture.

Using models of human behavior helps us understand people. We do not need to be behavioral psychologists to realize that not all people are alike. But we can all benefit from the available information that enables us to better understand people. Understanding helps us get along better with others. These basic concepts are important to everyone, especially leaders.

To truly understand people a good leader must first evaluate themselves. They must decide if they are in it for the team or for themselves. There are a lot of people in this business who will try to achieve great things for themselves for the purpose of feeding their own ego. Good leaders develop the team along their journey, their personal agendas support the team's vision, and they continually try and improve the team. By evaluating human behavior, a leader can provide values, improve work ethic, and ultimately lead the team to success.

Maslow's Hierarchy of Needs

Abraham Maslow is best known for establishing the theory of a hierarchy of needs. Maslow suggested that human beings are motivated by unsatisfied needs and that certain lower needs need to be satisfied before higher needs can be satisfied. According to Maslow, there are four general types of needs (physiological, safety, love, and esteem) that must be satisfied before a person can act unselfishly. He called these needs "deficiency needs," and, once they are fulfilled, other (higher) needs emerge, and these, rather than physiological hungers, tend to dominate. When these are satisfied, again new (and still higher) needs emerge, and so on. As one desire is satisfied, another pops up to take its place. The layers in **Maslow's Hierarchy of Needs** can be described as follows (see **Figure 6-3**):

- *Physiological needs:* Physiological needs are the very basic needs such as air, water, food, sleep, sex, and so on. When these are not satisfied, we may feel sickness, irritation, pain, discomfort,

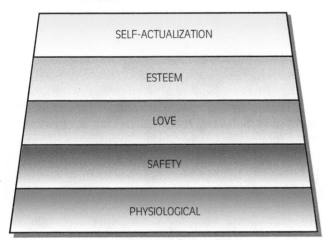

FIGURE 6-3 Maslow's Hierarchy of Human Needs.

and so on. These feelings motivate us to alleviate them as soon as possible to establish homeostasis. Once they are alleviated, we may think about other things.

- *Safety needs:* Safety needs have to do with establishing stability and consistency in a chaotic world. These needs are mostly psychological in nature. We need the security of a home and family. Many in our society cry out for law and order because they do not feel safe enough to go for a walk in their neighborhoods.

- *Social needs:* Love and belongingness are next on the ladder. Humans have a desire to belong to groups: clubs, work groups, religious groups, family, gangs, and so on. We need to feel loved (nonsexual) by others and to be accepted by others. Performers appreciate applause. Beer commercials, in addition to playing on sex, also often show how beer makes for camaraderie.

- *Esteem needs:* There are two types of esteem needs. First is self-esteem, which results from competence or mastery of a task. Second, there is the attention and recognition that comes from others. This is similar to the belongingness level; however, wanting admiration also has to do with the need for power. People who have all of their lower needs satisfied may drive very expensive cars because doing so raises their level of esteem. "Hey, look what I have!"

- *Self-actualization needs:* The need for self-actualization is "the desire to become more and more of what one is, to become everything that one is capable of becoming." People who have everything can maximize their potential. They can seek knowledge, peace, aesthetic experiences, and self-fulfillment.

Life is not as easy as Maslow's model might suggest. One does not suddenly meet the needs on one level and step up to the next level; we move in small increments up and down life's road, gradually moving from one level to another. There are no road signs along the way; in fact, it is unlikely we will know when we have moved from one level to another.

To some, Maslow's concept suggests one-way movement, that we are always moving up the steps in the hierarchy. Unfortunately, that is not true, as we move up and down the steps of the hierarchy during life's journey. We have good days and we have setbacks: We lose our job, our home is destroyed by fire or flood, our spouse or a child is injured. We are all at motion in this system, sometimes moving so slowly that we cannot perceive the motion. At other times, we move through the system rather quickly, especially when descending.

Although you may think of human behavior theory as being removed from our purpose here, today's fire officers should understand and apply Maslow's ideas. They should understand where their personnel are at on the scale and be able to help them move up the scale when they are interested in doing so. As company officers, you should realize that as you attempt to motivate people, you must find motivational tools that are consistent with your members' needs. Fame and glory are nice, but for the young firefighter who has a sick child or who is struggling to pay the rent, the basic needs are more important.

In the case of Maslow's Hierarchy of Needs, we are looking at a useful model, or tool, to help us understand and motivate others. As a company officer, you should recognize that your workers have needs on one or more levels, and you should try to determine which level they are on. You should also realize that once those needs have been satisfied, your workers will have greater needs. In other words, as workers' needs change, so do the factors that are effective in motivating them. Understanding human needs helps you better motivate others to improve themselves and the organization.

NOTE

As members' needs change, so do the factors that are effective in motivating them.

Herzberg's Perspective on Motivation

Many consider money to be the ultimate motivator in the workplace. But experience has shown that when you provide workers with a raise, the euphoria lasts about one payday; after that, most people settle back into their same old level of attitude and performance.

Frederick Herzberg described two factors that act on personnel at work. He called the first set **hygiene factors,** factors that are needed just to get people to come to work and to prevent dissatisfaction. Hygiene factors include company policies and administration, supervision, salary, interpersonal relations, and working conditions. Herzberg suggested that if these are done well, the workers will be satisfied. If they are done poorly or not at all, the workers will be dissatisfied.

Just as good physical hygiene helps prevent dissatisfaction with our health and fitness, so do the hygiene factors at work help prevent worker dissatisfaction. Good hygiene does not cure disease, but it does help prevent it. By the same token, Herzberg's hygiene factors do not motivate workers; they prevent workers from having bad feelings about the workplace.

Dr. Herzberg called his second set of factors **motivators.** Motivators encourage workers to rise above the satisfactory level and do excellent work. As he defined them, motivators include achievement, recognition, the work itself, responsibility, and advancement. Keep Herzberg's ideas in mind when you are motivating others.

When you really stop to think about it, both Maslow and Herzberg suggest a similar hierarchy of needs. Maslow had five layers; Herzberg has only two. In the case of Herzberg's model, remember that everything below the line will just keep workers from feeling bad. As supervisors, we have to add the items above the line to get positive results.

HERZBERG'S MODEL

Motivators

- Responsibility
- Job with a purpose
- Recognition of good work
- Opportunity for promotion
- Opportunity for achievement
- Opportunity for personal growth
- Threshold of satisfaction

Hygiene Factors

- Good working relationships
- Considerate supervisors
- Good working conditions
- Good pay and benefits
- Job security

DETERMINING THE BEST LEADERSHIP STYLE FOR MOTIVATING OTHERS

We have been discussing what motivates others and how understanding their motivations will make us more effective in motivating them. Let us look at other factors that might determine leadership style. These factors are more a function of the leader's needs and feelings, rather than those of the worker.

Leaders have various styles based on their own needs. We mentioned in Chapter 4 that, according to Douglas McGregor, there were two generally held management styles, determined by the leader's belief about the workers. McGregor's Theory X and Theory Y are called management style, but they have a great deal to do with the way you supervise. How you supervise is how you motivate others to do what you want to get done.

While having concern for people is admirable, in the real world, we are all expected to get the job done. Some might see the needs of the people and the need for getting the work done as a conflict. Being nice, as suggested by the Theory Y style of leadership, might not give you the productivity you want. On the other hand, being a taskmaster, as possibly represented by the Theory X style of leader, will not be well accepted in most situations by your teammates. It would appear therefore that these two forces, namely a concern for people and a concern for production, are in opposition, or at least in competition with one another, as you determine your own leadership style.

Needs Theory[2]

There is yet another way of looking at motivational factors. We have already seen that affiliation and recognition show up consistently as motivators. Research has also shown that some individuals like to have some influence over the group as well. So we add status, or **power,** to our list of motivators as we look at the way we can best motivate others.

Consider persons with high-status or **power needs.** They like to be in charge and to have influence over others. They like the structure of organizations and they like to get work done. Remember these three important characteristics: Persons with high-power needs like to be in charge, persons with high-achievement needs like challenges, and persons with high-affiliation needs like to work in groups.

People who have **high-achievement needs** typically take personal responsibility for their efforts, set their own goals, and take on new and demanding challenges. They tend to be creative, and for many with achievement needs, the strongest characteristic may be their need for feedback. These individuals usually set high goals for themselves, and they look for ways to get feedback on their performance.

> **NOTE**
>
> Persons with high-power needs like to be in charge, persons with high-achievement needs like challenges, and persons with high-affiliation needs like to work in groups.

The third group are those who have **high-affiliation needs.** We have already discussed group dynamics to some extent. These people thrive on it. They desire to belong and be accepted by the group. Although they like to work, they prefer to work as part of a group. In many ways, these individuals are on Maslow's third layer—they have a need to belong.

These three characteristics, the need to have status, to achieve, and to affiliate with others, are present in all of us. In many people, one of these needs characteristics may be far more dominant. When that is the case, you may be able to find work that fits their needs well. When you can match their needs with the task or the environment, they will be more than just satisfied, they will be motivated to work well. However, you should realize that you cannot always put people into the roles that they would most like. Frequently the organization's needs must also be considered.

> **NOTE**
>
> While having concern for people is admirable, in the real world we are all expected to get the job done.

The Managerial Grid

The grid represents a field in which the leader's concern for people and production are plotted (see **Figure 6-4**). The vertical axis represents the leader's concern for people, and the horizontal axis represents the leader's concern for production. The Managerial Grid was introduced in 1964 by Drs. Robert R. Blake and Jane S. Mounton. Their model of managerial behavior in an organizational setting suggested that morale (the people factor) and productivity are independent of one another. Thus, either one could flourish, or fail, in spite of the other.

There are infinite positions on the grid, but five positions are noted here for purposes of explanation:

1. In the lower right-hand corner, we see position A. Position A represents a leader with a passion for production while having a relatively low concern

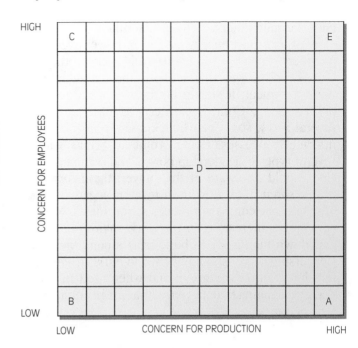

FIGURE 6-4 The Managerial Grid.

for people. Position A may be similar in some ways to Theory X. Persons having this style of leadership use their power to control people, telling them what to do and how it is to be done.

2. At position B, we see a minimum regard for both people and production. Here, the leader provides the minimum amount of supervision needed to survive within the system.

3. At position C, the leader has a strong bias for interaction with people, while having a relatively mild concern about production. In some ways, this person is a strong advocate of McGregor's Theory Y.

4. In the center of the grid we have a leadership style that is represented by a middle-of-the-road approach on both scales. While position D may be somewhat better than the person who operates from position A, B, or C, we see that neither people nor production gets full attention and that either can be compromised to accomplish the other.

5. At position E, we have the supervisor who has a high regard for both the workers and for production. These leaders use their own talents to integrate all the positive leadership qualities to bring out the full potential of all of their workers.

UNDERSTANDING POWER

The word *power* has been used several times in this chapter. How do you define *power* in the context of leadership? Leadership is the exercise of power, and we said that leadership is achieving the organization's goals through others. To have this influence, the company officer must have some sort of power over team members. *Power* is the ability to influence others in a constructive way that supports the individual and the department; it should not be seen as a negative or bad thing. Company officers acquire power in several ways, some of which comes from formal or informal processes. Being promoted brings with it several types of leadership power, and company officers should be aware of this power; these forms of power should never be used to gain negative influence over the company officer's subordinates. Misuse of power can create a hostile work environment and tear down the ability to build trust among the crew. Leadership power should be used to build strong relationships within the group and when used in a fair and encouraging way power will assist in completing assigned task.

LEADERSHIP POWER

Type of Power	Source	Use
Legitimate	Bestowed	Official position
Reward	Comes with the badge	Power over others
Punishment	Comes with the badge	Power over others
Identification	Individually earned	Charisma
Expert	Individually earned	Unique knowledge

Company officers work in a socially complex organization where they will need the support of subordinates, peers, supervisors, and external parties such as the community to achieve organizational goals. Building an effective influence strategy using a persuasive power base will be a challenge for new company officers.

For fire service personnel, the first type of leadership power comes with the badge of the office. Some refer to this as **legitimate power,** the power that is bestowed upon you as an officer in the organization. As an officer, you wear insignia that indicates you are a representative of local government. The name of the governing agency is usually prominently displayed on that insignia. So, some power comes with the position.

With that legitimate power, two additional types of power are implied. The first of these is the power

to reward people, or **reward power.** As a supervisor, you have the power to approve requests, recommend individuals for special assignments, and write recommendations and evaluations that will help the personnel in your company attain their personal goals.

Reward power is more than giving someone a medal or some time off. Reward power may be nothing more than name recognition, a smile, or an acknowledgment of effort. Reward power can be taking a moment to help someone who is having a bad day. Reward power is not taking your bad day out on someone else.

EMPOWERING OTHERS

Kimberly Alyn, a popular author and speaker on fire-service training and development, says, "Empowering people to think and make decisions wherever possible accomplishes many things:

- Building better teams
- Improves morale
- An increased sense of ownership
- A greater feeling of trust
- Higher productivity
- Better service to the customers
- A stronger public image
- A more efficient fire department
- Better succession planning."[3]

The other implied power that goes with the job is the power to punish, or **punishment power.** We usually think of punishment power as the authority to administer discipline, although, fortunately, this is not really an issue for most personnel.

Far more often, we punish people by simply failing to give recognition for a job well done or by withholding information that might be useful. This is not done on purpose; it is just a failure on the part of the supervisor to do things that are important in the job of being a good leader. Legitimate, reward, and punishment power come with the badge.

There are additional forms of power that one earns. While the legitimate, reward, and punishment power come with the position, this added power is earned by individuals through their personal actions. It may be influenced by the way they treat people, by their ability to communicate, and by their knowledge of the job. These qualities can be illustrated with terms like charisma and knowledge. Clearly these qualities are much more subjective than the legitimate power evidenced by the gold badge, but they are just as real. Terms such as **identification** power and **expert power** are often used to describe the qualities of role models and knowledgeable individuals.

NOTE

Legitimate, reward, and punishment power come with the badge.

HOW TO DETERMINE THE LEADERSHIP STYLE THAT IS RIGHT FOR YOU

Effective leaders work at achieving the organization's goals through the efficient labor of others. They strive to do this as well as possible, and at the same time, they are helping members reach their full potential.

But these goals are seldom easily attained in real life. We are frequently faced with the old debate: Which is more important—the task or the people? Sometimes, the task is critical. At other times, the people are far more important. Sometimes, both are important; sometimes, neither requires much attention.

Three factors can help determine your own leadership style: the team member, the leader, and the situation. Let us look at how each of these impacts on a leader's style.

Elements that Determine Leadership Style

- The Team Member
  - The team member's experience
  - The team member's maturity
  - The team member's motivation
- The Leader
  - The leader's self-confidence
  - The leader's confidence in the team members
  - The leader's feeling of security in the organization
  - The leader's perception of the organization's value system
- The Situation
  - Risk factors involved
  - Time constraints imposed
  - Nature of the particular problem
  - The organizational risk climate
  - The ability of the individuals to work as a team

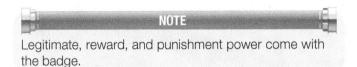

OFFICER ADVICE

Many leadership styles can be used. Dynamic and effective leaders make their style fit the situation.

Many leadership styles can be used. Dynamic and effective leaders make their style fit the situation. Certainly, the factors we have been discussing here should be integrated into the process. Four representative styles follow:

1. **Directing.** This style is characterized by lots of direction and mostly one-way communications. The supervisor tells what has to be done, provides direction, and monitors the results. The task gets more attention than the people. The personnel have little input into the decision-making process.

2. **Consulting.** Here, there is some discussion in which the supervisor seeks ideas, explains the needs and decisions, and "sells" the idea. The supervisor still gives lots of direction but maintains close presence, providing encouragement and reassurance. In this mode, the supervisor is available to provide support to both the task and the people.

3. **Supporting.** The leader encourages participation at all levels and shares responsibility for the process. Two-way communications are encouraged on a continuing basis. There is a sharing of the decision-making process. The leader facilitates growth by sharing information and asking questions that will enhance the workers' understanding of the situation. In this case, the leader has taken a supporting role, supporting the workers. There is little direction, but the leader continues to provide support, encouragement, and recognition.

4. **Delegating.** The supervisor essentially turns the management of the task over to team members. Direction is limited to setting the goal and defining the parameters. Communication may be limited, but when it occurs, it will be a cordial two-way process.

In these four definitions, you see a gradual change from the strong directive-type of leadership behavior to one in which the leader is more of a supporting player (see **Figure 6-5**). The first type is mostly supervisor-centered leadership, and communications are mostly one-way. As the process moves toward a more member-centered leadership style, you see two-way communications that are more effective, more freedom of thought and expression for the members,

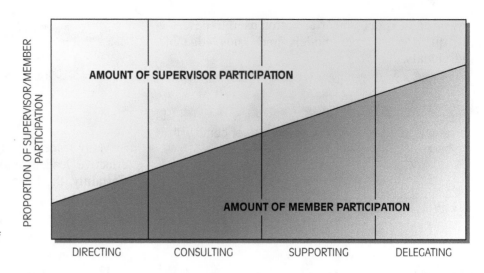

FIGURE 6-5 Leadership style varies depending upon the amount of member input allowed.

and members that are more involved with the decisions and the outcome.

Many of us tend to be too directive. Ease up a bit, and learn how and when to use each of the four leadership styles effectively. During a normal day, there are situations in which each is appropriate. Use the one that is most productive.

OFFICER ADVICE

"Never tell people how to do things. Tell them what to do and they will surprise you with their ingenuity."—George S. Patton Jr.

GOOD LEADERSHIP IN TODAY'S WORKPLACE

Several major social issues have confronted our society in recent times. These issues deal with diversity and harassment in the workplace. Diversity and harassment in the workplace are generally considered to be management issues, but we consider them here in the context of good leadership. Indeed, the organizational tone for policies regarding these important issues must come from the top of the organization. But this is not a book for fire chiefs, nor is it a book about what fire chiefs should do. The focus of this

THE LEADERSHIP TETRAHEDRON

Anyone who has any familiarity with the basics of fire behavior and combustion should be familiar with the Fire Tetrahedron. For decades, the Fire Triangle with the components of heat, fuel, and oxygen was used to describe the chemical structure of fire. It was then discovered that a fourth component was necessary for combustion to occur; Chemical Chain Reaction was added, and the Fire Tetrahedron was born.

The same is true of the promotional processes used within the fire service. For many years, we have operated under the assumption that seniority, written tests, and assessment exercises identified and measured the key components required to be a fire officer. However, just like the Fire Triangle, we may not be identifying all of the key components necessary to identify the best person to promote. As a direct result, the fire service often struggles with identifying and promoting the right people.

It is not due to a lack of trying, for the fire service typically has one of the most stringent testing systems of any profession. In fact, most of us are a product of this very system or are involved in a similar system right now. But are these systems evaluating ALL of the attributes required to be a good officer? To that end, the Leadership Tetrahedron was developed adding another facet that we should consider incorporating into our promotional processes (see **Figure 6-6**).

The Leadership Tetrahedron simply places traits required to be a fire officer, in place of the components essential for a fire. At the top of the Leadership Tetrahedron is "Desire." Good officers will always have the desire to aspire as well as to inspire, and possess the desire to excel at everything they do. They continually seek ways to improve themselves and others. They attend training sessions, seek feedback from their subordinates and their superiors on how they are doing, and strive to pass on to others what they know and have learned. Training is a top priority for these individuals,

FIGURE 6-6 According to the Leadership Tetrahedron a well rounded officer must be assessed in all areas.

because they do not operate under the assumption they know it all or have nothing left to learn. Persons who have lost this desire typically accept the status quo and merely perform duties expected of them, without striving to excel. Typically, chief-level officers, especially battalion-level officers, are probably the best and worst examples of this. The rank of battalion chief is as high as many wish to attain within the fire service. Anything higher typically requires a transition from shift work to an 8-to-5/Monday-through-Friday schedule, and anyone who has worked both realizes the 24-hour shift is a much more attractive schedule that affords a person a lot more time off, to pursue other ventures. Therefore, typically when a person attains chief, he or she has "made it." With no aspiration to go any higher, one subsequently loses the desire to learn.

Additional training classes are not required to retain the position, so why should they attend any? It is very likely they will earn the same amount of money whether they spend every day off attending a class or whether they attend none. Besides, submitting a new idea or policy or anything outside of status quo could potentially cause more work, so status quo it is.

We all have either worked for or know someone like this. They usually operate by a seat-of-pants management style. They are not up to date on the latest technical or management information, and while they may do their best, they are lacking the necessary skills because they are operating within a limited bubble of information. What they gain, they gain from mandatory in-service training sessions, or what they pick up from others who have not lost their desire and took the time to attend outside training sessions. It is interesting to note that these same individuals are by and large the biggest complainers and spend most of their time tearing down the department and its administration, while those who have the desire normally possess a positive attitude toward the department. The second component of the Leadership Tetrahedron is "Ability." No matter how much desire one possesses to become an officer, he or she may lack the required abilities or skills to do the job. It may simply be a matter of more training, or it may be

something that they will never possess. There was once a firefighter who had a desire to become a pump operator. Countless hours were spent attempting to train this individual, but he could never fully grasp the mechanics of pumping and would fall to pieces when placed under pressure. He had all the desire in the world; however, he lacked the skills needed to be an efficient pump operator and likely will never possess what it takes. The same can be true with an officer candidate. To be an effective officer, one needs to learn, practice, and be effective in certain skills. Ability can be broken down into a series of different skills sets, including but not limited to Standard Operational Guidelines, ICS system, code enforcement, building construction, basic firefighting tactics, leadership, etc. Typically, these are areas tested; evaluating how well a candidate knows operational guidelines and other basic fire-related information. If an assessment process is used, then it may attempt to place a value on a candidate's ability to put this information to use. Dependent on the process, it may be a small portion of the total score, or it may be the only score used in evaluating a candidate. The point being, it only tests the knowledge the candidate possesses at that time in a play situation, and it does not test all four components of the Leadership Tetrahedron. The third component of the Leadership Tetrahedron is "Courage," but not the run-into-a-burning-building-and-save-the-baby kind of courage. While it is necessary for an officer to be able to perform this task, this act should be done based on a risk/benefit analysis of the situation and not an abandonment of all common sense because someone is screaming, "Save my baby." The courage spoken of here is the ability of an officer to stand on his or her own two feet, assume responsibility for his or her actions, and make decisions based on his or her judgment and, when necessary, step outside the box and take risks. Officers who want a policy for everything, so they do not have to assume risks, will never be good officers. He or she typically operates this way because he or she believes they cannot get in trouble if they follow policy. Officers are needed who use discretion and common sense and who are not afraid to make decisions and do the right thing, even if occasionally it does not follow policy. If you do not believe this concept, try to write Standard Operational Guidelines for every situation you are likely to encounter, taking away an officer's discretionary decision making. If you are successful, then you will not need officers, only robots that can be programmed with the volumes of operational guidelines you develop. Furthermore, a good officer possesses the courage to assume his or her duties and not dodge those responsibilities just so they can remain one of the troops. He or she also possesses the courage to discipline when required. Additionally, a good officer should also have the courage to stand up for the troops when a new policy is unjust or unfair, or a new procedure is unsafe, and not roll over or look the other way because it is good for his or her career. If you do not believe these things take courage, then you have never been an officer.

The final chain reaction component of the Leadership Tetrahedron is "Human Relations." For no matter how much desire an officer has to inspire his or her subordinates, if that officer cannot relate to them, or if he or she does it in a manner that generates resentment, then the officer will not be effective.

No matter how many classes he or she may have attended, or skills sets an officer has perfected, or ability he or she has attained, the officer will be ineffective if he or she cannot relate to his or her subordinates how to get the job done.

No matter how much courage an individual possesses, if he or she has continually bullied his or her subordinates to get his or her way, then he or she will be ineffective.

A good officer needs to be able to communicate with others orally. A good officer needs to be able to relate to subordinates and occasionally express empathy for what he or she, as an individual, may be experiencing. If an officer is cold and removed, always goes by the book, and does not communicative with subordinates, it is unlikely that he or she will ever bond with the subordinates. The atmosphere will become and likely remain an us-versus-them atmosphere. These are not the types of officers whom the troops will follow. A good officer, while remaining an authoritative figure of the group, bonds with the group and comes to know who they are, what motivates them, and what their interests are. He or she also learns their strengths and weaknesses and then, if he or she possesses the desire, begins to make use of their knowledge and skills, capitalizing on the strengths of each individual while diminishing their weaknesses.

If you remove one of the components from the Fire Tetrahedron, you will not have a fire. If an officer does not possess all the components of the Leadership Tetrahedron, he or she will not be an effective fire officer.

Therefore, officers should be developed and evaluated through a process that evaluates all four components of the Leadership Tetrahedron. An officer candidate needs to be evaluated in all of these factors in real-world situations throughout his or her career, not just within the last few months or 15 minutes of an assessment exercise.

To that end, utilization of a promotional process that uses a written test, assessment exercises, and a peer evaluation to measure all of the Tetrahedron components is recommended. A written test will evaluate a candidate's basic knowledge as well as reduce the number of candidates to a manageable size for the other processes. A written test will also, to a limited degree, measure a candidate's desire, assuming the amount of time one puts into preparing for a test determines how well he or she will perform. An assessment exercise will evaluate a candidate's desire and ability. If properly designed, an assessment exercise can also determine a candidate's human relations skill; however, one must realize it will typically be limited, as most candidates are placed in the role of dealing with a member of the public or with a personnel matter one on one. The leaderless group exercise can evaluate a person's ability to work within a group setting; however, it, too, is limited due to the short time a candidate is evaluated and the dynamics of a small group. Despite these limitations, an assessment exercise is a valuable testing tool and should be used. The third and final evaluation tool recommend is a peer evaluation. A peer evaluation can measure the components of the tetrahedron that are missed with the written and assessment exercises. If candidate were evaluated by their peers, with whom they work on a regular basis and who have nothing to gain (or lose) in the process, then they will truly and accurately be evaluated in all the components of the tetrahedron. Promotional candidates would be assessed on their ability to perform in the real world and their ability to make decisions and think on their feet. Additionally, candidates would be evaluated on their courage and the capacity to do the right thing. Candidates would also be evaluated on their desire for self-improvement and their desire to work with others and pass on their knowledge. This could be done by rating the candidates in different characteristics like responsibility, job performance, communication skills, leadership abilities, attitude/demeanor, judgment, team player, integrity, human relations skills, problem analysis, handles stress/problems/adaptability, and initiative/motivation.

Unfortunately, seniority does not fit anywhere within the Tetrahedron. Seniority, while easy to calculate, only tells us how long a person has been on the job, not how good of a job that person has been doing during that time. A potential candidate may have 10 years of experience, but is it 1 year 10 times over, where they continue to make the same mistakes, or is it truly 10 years of wisdom and knowledge? Was it 10 years at a busy station, or was it 10 years at the slowest station in the department? Did that person have the desire to be part of a committee during those 10 years, or did he or she simply show up for the shift and go home at the end, forgetting he or she even worked for the department? Perhaps this person showed up for work on a regular basis but spent half the time conducting business for his or her part-time job. Seniority is a very poor indicator of whom is the best candidate for promotion and should not be used.

If one believes in this theory, then we hope that you can understand why there is a need to change the way we develop and identify the best officers.

text is the company officer, and in that regard, the leadership issues that company officers face.

FIREGROUND FACT

The fire service has had significant problems in the area of diversity and harassment.

The fire service has had significant problems in the area of diversity and harassment. We still see headlines in the trade journals and local papers suggesting that personal prejudice and a lack of good leadership on the part of a few have cast a cloud over many. Although much progress has been made, there is still plenty of evidence to suggest that the fire service needs to continue to train officers and other personnel in dealing effectively with diversity and harassment issues.

NOTE

Both diversity and harassment deal with personal attitudes toward others.

Both **diversity** and harassment deal with personal attitudes toward others. Those attitudes may be the product of a lifetime of family and community values over which we have little control. Regardless of our past culture and present beliefs, we must realize that within the workplace, laws and regulations determine the legal boundaries of our actions. To avoid problems, we should understand these laws and regulations.

Many federal laws prohibit discrimination. Starting with the U.S. Constitution, the Fifth Amendment provides that "No person shall . . . be deprived of life, liberty, or property without due process of law." The Fourteenth Amendment provides that "No state shall . . . deny any persons within its jurisdiction the equal protection of the law."

Given these constitutional guarantees, one has to wonder how these problems have occurred. Some suggest that the government failed to address the sociological problems in the communities and workplace with quick and positive leadership action and that, if it had, these problems would not have happened. Others suggest that employers were not fair in providing an equitable workplace for their personnel; that if their action had been more effective, we would not see the headlines in the newspapers nearly every day regarding diversity and harassment issues in our government, our military, and our major corporations.

If we lived in a perfect world, and if all government officials and all employers were good and considerate human beings, these things would not happen. But our world is not perfect, and people sometimes fail to understand the problem or the solution. Where good judgment and morality failed, the laws now provide incentives for proper action.

In 1964, Congress passed the Civil Rights Act. Title VI of the act deals with programs and activities that receive federal funding and prohibits discrimination on the basis of race, color, national origin, or sex. Title VII of the Civil Rights Act deals with employment and prohibits discrimination by employers on the basis of race, color, religion, sex, or national origin. The act also authorized the establishment of the **Equal Employment Opportunity Commission (EEOC)** to enforce the act. Of these two, Title VII has the greater impact on our workplace.

EQUAL EMPLOYMENT OPPORTUNITY

A condition in which all employees, and potential employees, are treated fairly. Equal employment opportunity is more than a legal right; it is a human right.

Several other significant laws bolster the power of the Civil Rights Act. The Equal Pay Act of 1963 prohibits discrimination in compensation on the basis of sex. The Equal Employment Opportunity Act of 1972 strengthened the authority of the Civil Rights Act of 1964 and expanded the power of the EEOC. The Civil Rights Act of 1991 provided that individuals may be personally liable (as well as their organizations) for discrimination in the workplace. It authorized jury trials and increased the limits on financial settlements. As a result, we have seen increasingly larger punishments, especially in cases involving sexual harassment.

FEDERAL LAWS PROHIBITING JOB DISCRIMINATION

- Title VII of the Civil Rights Act of 1964 prohibits employment discrimination based on race, sex, color, religion, or national origin.
- The Equal Pay Act of 1963 prohibits sex-based wage discrimination when men and women perform substantially equal work in the same establishment.
- The Age Discrimination Employment Act of 1967 protects individuals who are 40 years of age and older.
- Title I and Title V of the Americans with Disabilities Act of 1990 (ADA) prohibit employment discrimination against qualified individuals with disabilities in the private sector and in state and local governments.
- The Civil Rights Act of 1991 provides monetary damages in cases of intentional employment discrimination.

The EEOC enforces all of these laws. EEOC also provides oversight and coordination of all federal equal employment opportunity regulations, practices, and policies.

FIGURE 6-7   Today's students are tomorrow's firefighters. Tomorrow's firefighters will be more diverse than those in the fire service today.

THE CIVIL RIGHTS ACT OF 1991

The Civil Rights Act of 1991 made major changes in the federal laws against employment discrimination enforced by EEOC. Enacted in part to reverse several Supreme Court decisions that limited the rights of persons protected by these laws, the act also provides additional protections. The act authorizes compensatory and punitive damages in cases of intentional discrimination and provides for obtaining attorneys' fees and the possibility of jury trials. It also directs the EEOC to expand its technical assistance and outreach activities.

What Is EEOC and How Does It Operate?

The EEOC is an independent federal agency originally created by Congress in 1964 to enforce Title VII of the Civil Rights Act of 1964. The commission is composed of five commissioners and a general counsel appointed by the president and confirmed by the Senate. Commissioners are appointed for 5-year staggered terms; the general counsel's term is 4 years. The president designates a chair and vice-chair. The chair is the chief executive officer of the commission. The commission has authority to establish equal employment policy and to approve litigation. The general counsel is responsible for conducting litigation. The EEOC carries out its enforcement, education, and technical assistance activities through fifty field offices serving every part of the nation.

Finally, the Americans with Disabilities Act (ADA) of 1990 prohibits discrimination on the basis of disabilities, and covers employment, public accommodations, transportation, and telecommunications. The ADA is a complex law that applies to employment issues, our 911 systems, and providing proper access for citizens and employees in public places such as fire stations. Questions regarding local application of the ADA should be referred to higher management levels of your department or local government.

Diversity

Our communities are becoming more diversified. Likewise, most government and private business activities are, or at least should be, becoming more diversified. Until recently, white males dominated the workplace in nearly every industry, but this is changing, and in the very near future, the white male will be a minority member of the workforce!

Changes are taking place in the fire service as well. For the most part, the fire service has long been dominated by white males. However, that is changing, and women and minorities are entering the fire service in increasing numbers (see **Figure 6-7**).

Women deserve special mention here. They represent slightly over half the total population, but they represent a very small percentage of the fire service. Most of those who have entered the fire service have proved themselves to be quite capable of performing all the tasks associated with firefighting. In spite of some formative barriers that they had to overcome, many of the women who entered the fire service have done well. Some of those pioneers have proved themselves over and over and are serving capably as chiefs and senior officers in large, progressive fire departments.

Think about America's great history. Nearly all of our ancestors were immigrants, and at one time, they too were minorities. Half of them were female. Yet our parents and their parents were able to become successful. Many have given great service in public office, academia, and industry. Today's generation and tomorrow's generation should have the opportunity to accomplish their personal goals.

Public and private organizations are responding to these changes by seeking ways to hire, train, promote, and retain the best talent available, regardless of the gender, race, or ethnic background. As a result, we are seeing a more diverse workforce. Effectively supervising a diverse workforce is what good leadership is all about.

Benefits of Diversity

We should all be committed to making our organization the best it can be. We should be looking at the total community and all of its citizens as a source of new members for our organization. As new members enter our organizations, we must be aware that they may have different values. As standing members of the organization, we should look on these new values as an opportunity, rather than a problem. The new members may have new perspectives on existing social problems, may be able to speak a second language, and may provide many other benefits that we lack. But all too often we close the door on these opportunities because these members are different.

All of us want highly motivated individuals working for us. Opening the doors of the fire station to groups that have been traditionally excluded allows a greater number of interested and qualified persons to apply to our organization. A greater applicant pool means that departments can be more selective about the people they hire. Many large metropolitan fire departments have the luxury of selecting one candidate from 20 or more applicants. If done carefully, the selection process provides highly qualified candidates who are likely to survive the recruit training process and become well-motivated, long-term members of the organization. Their gender, race, or ethnic background should not be an issue.

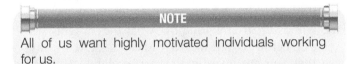

NOTE

All of us want highly motivated individuals working for us.

Maintaining good working relationships for this increasingly diverse group of members is another matter. The new members enter a culture that has been traditionally a white male organization. They may be uncomfortable. The more senior members of the fire service have had little experience in dealing with diversity issues, and some do not know how to deal with these diverse groups or how to handle the problems that arise.

For many women and minorities, the path to acceptance and the opportunities for advancement are a little harder than for the traditional member. In some cases these are merely perceptions, and in some cases, there are real barriers that prevent them from having an equal opportunity.

Many of the organization's rules are unwritten. We must be sure that the same information is presented to all of our members, regardless of our personal feelings about them.

ETHICS

Many of the organization's rules are unwritten. We must be sure that the same information is presented to all of our members, regardless of our personal feelings about them.

How to Supervise Diversity

Understanding and effectively managing diversity must start at the top of any organization. But we quickly see that top management can have little impact without the understanding and cooperation of everyone in the organization. We all need to proactively embrace the positive values of diversity and work hard to give all of our personnel the very best opportunity to grow to their full potential and satisfy their own goals.

Supervisors should take the time to learn about all new personnel and their personal values. The fact that these values may not be identical to our own should not be considered as a barrier but rather as an opportunity for learning more about ourselves and our personnel.

Supervisors should also be sure that women and minorities are fairly represented and involved with all of the department's activities. Put them on committees. Listen to their opinions with an open mind, and give their ideas a fair try. Give recognition and support when appropriate. If these actions are not completely comfortable and natural, supervisors should be honestly assessing their own weaknesses in this area and looking for opportunities to improve their dealings with others. Many of the problems facing supervisors today are the result of inappropriate actions in the past.

ETHICS

Supervisors should be sure that women and minorities are fairly represented and involved with all of the department's activities.

Harassment

To *harass* means to annoy, tease, or torment. **Harassment** of any kind should not be allowed in the workplace. This ban includes harassment of new members or any other efforts to make any individual feel uncomfortable. Where this harassment has sexual overtones, it is a clear violation of federal law. The law deals with both those who harass and those who permit harassing.

Title VII of the Civil Rights Act of 1964 covers sexual harassment in the workplace.[4] The act states that the behavior must be unwelcome sexually oriented conduct of a verbal or physical nature. There are two situations in which sexual harassment generally occurs. The first of these is considered a hostile work environment. A hostile work environment occurs when there are pictures, comments, and other offensive acts that inhibit a worker's performance. A second situation occurs when one asks for sexual favors as a condition of employment, promotion, or transfer. This implies an act of power by one over another in which a favor is requested in exchange for some personal action. Such harassment is referred to as *quid pro quo,* a Latin phrase meaning "this for that."

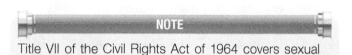

NOTE

Title VII of the Civil Rights Act of 1964 covers sexual harassment in the workplace.

We usually think of harassment situations in which there is a male–female relationship between two employees, with the male being the aggressor. In this stereotypical male–female harassment situation, the female is often the victim. Because of physical size and organizational position, men are generally thought to be more powerful. Sexual harassment is often about power.

While such circumstances do occur, you should realize that many other gender combinations can occur as well. The roles might be reversed—the female may be the aggressor. And it is possible that both parties could be of the same sex. Likewise, you should realize that not all of the parties have to be members. Harassment may come from a visitor, a vendor, a contractor, or a service provider in your workplace. Finally, sexual harassment need not be about sex; it may be a general derogatory comment about a person because of that person's gender.

What to Do when Harassment Occurs

Supervisors have a special obligation to protect personnel from the consequences of sexual harassment. When a person comes to you, as a supervisor, with a complaint about sexual harassment, you must take action promptly. Encourage the complainant to provide you with specific information. You may need to help the process by asking open-ended questions like the following:

- Where did this occur?
- When did this occur?
- Who was involved?
- Were there any witnesses?

- Have you talked to anyone else about this?
- Has this happened before?
- How long has this been happening?
- Did you try to stop (this conduct) on your own?
- What was the reaction?
- What do you want me to do?

Notice especially the last question in the series. In some cases, no corrective action is desired. The complainant may just want you to know that the event occurred, to sort of "get it on the record." Be sure that you follow your organization's policy. Many organizations require at least some notification be made, even in these situations.

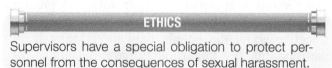

ETHICS

Supervisors have a special obligation to protect personnel from the consequences of sexual harassment.

In some cases, you will have to take some action. Be sure to follow your department's policies. Protocol may require the reporting and the presence of human resources personnel in the investigation, interviews, and corrective actions. Some departments direct the company officer to make the initial investigation, whereas others expressly preclude such action. If the policy directs you to conduct an investigation, you should follow the initial interview by checking with the witnesses, if any. Find out what they might have seen or heard. If talking to witnesses, start by saying that you are investigating a complaint of sexual harassment and that they have been named as a witness to the act. Do not provide them with specific information regarding either the name of the accused or the accuser; just focus on their observations.

Finally, after you have a good understanding of what has happened, talk to the accused. Again, do not reveal the names of the accuser or of any witnesses you have interviewed. Be serious and to the point, but at the same time, be open minded and fair. Here are some guidelines on asking questions:

When talking to witnesses, start by saying, "I am investigating a complaint of sexual harassment. I would like to ask you a few questions regarding what you may have seen or heard earlier today...."

When talking to the accused, start by saying, "I am investigating a complaint of sexual harassment by you. I would like to ask you a few questions regarding your actions (or conversations) this morning...."

This approach makes the process a little easier for both you and the person you are questioning.

In talking to the accused, ask the individual to respond to the allegation(s). In most cases, the accused will acknowledge the action and will offer an excuse that the actions were misunderstood. Instruct the person on the seriousness of the accusation, and advise the person regarding what is and is not acceptable workplace conduct. Advise the accused that there are laws and regulations that deal with these matters and those violations can have serious consequences.

GENERAL COMMENTS REGARDING ANY HARASSMENT ISSUE

- Move quickly.
- Let everyone you talk to understand that you are concerned, that you take these matters seriously, and that you will make every effort to be fair and just.
- Meet in private.
- Make notes of the meetings and safeguard them.
- Provide feedback to the accuser as to what you found.
- Follow your organization's policies to the letter.

NOTE

Most organizations have established policies for dealing with harassment. Know and follow your organization's policies.

RETALIATION

The EEOC is charged with enforcing Title VII of the Civil Rights Act, which prohibits employment discrimination based on race, color, religion, sex (including sexual harassment), or national origin, and protects personnel who complain about such offenses from retaliation.

In 2003, the EEOC heard a case in which an employee did the right thing and got fired! In this case, a nursing supervisor in a large hospital was terminated after she took complaints of an operating room employee, complaining about inappropriate behavior of a sexually harassing nature by a nurse, to her supervisors, the hospital's top management. As a result of the supervisor's action, the nurse complained to several doctors who, in turn, complained to top management. One prominent physician even threatened to take his surgical business elsewhere. Instead of responding to the supervisor and dealing with the matter appropriately, the hospital took no action on the behavior or the initial complaint and terminated the supervisor who brought the charges forward. She appropriately asked the EEOC for help. A 4-day jury trial resulted in a $4 million award against the hospital.

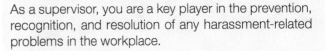

NOTE

As a supervisor, you are a key player in the prevention, recognition, and resolution of any harassment-related problems in the workplace.

The laws provide clear guidelines for organizations to follow in these matters. Most organizations have established policies for dealing with these issues. Know and follow your organization's policies. You may be personally at risk if you violate the law or your organization's policies in this important area. As a supervisor, you are a key player in the prevention, recognition, and resolution of any harassment-related problems in the workplace.

EEOC's Guidelines for Employer Liability for Sexual Harassment by Supervisors

- *When does harassment violate federal law?* Harassment violates federal law if it involves discriminatory treatment based on race, color, sex (with or without sexual conduct), religion, national origin, age, or disability or because the employee opposed job discrimination or participated in an investigation or complaint proceeding under the EEOC statutes. Federal law does not prohibit simple teasing, offhand comments, or isolated incidents that are not extremely serious. The conduct must be sufficiently frequent or severe to create a hostile work environment or result in a "tangible employment action," such as hiring, firing, promotion, or demotion.

- *Does the guidance apply only to sexual harassment?* No, it applies to all types of unlawful harassment.

- *When is an employer legally responsible for harassment by a supervisor?* An employer is always responsible for harassment by a supervisor that culminates in a tangible employment action. If the harassment did not lead to a tangible employment action, the employer is liable unless it proves that (1) it exercised reasonable care to prevent and promptly correct any harassment; and (2) the employee unreasonably failed to complain to management or to avoid harm otherwise.

- *Who qualifies as a "supervisor" for purposes of employer liability?* An individual qualifies as an employee's "supervisor" if the individual has the authority to recommend tangible employment decisions affecting the employee or if the individual has the authority to direct the employee's daily work activities.

SEXUAL HARASSMENT IS STILL A PROBLEM

In 1998, the U.S. Supreme Court heard four cases regarding sexual harassment. The number of cases reaching the Court suggests that sexual harassment is still a significant problem in the workplace. In writing for the Court, Justice David Souter noted, "It is well recognized that sexual harassment by supervisors is a persistent problem in the workplace. When a person with supervisory authority discriminates, his actions necessarily draw upon his superior position over the people who report to him." Justice Souter's comment is supported by the fact that the number of complaints filed with the EEOC more than doubled (from 6,500 to 16,000) between 1991 and 1998.[5]

Two of the 1998 cases dealt with the employer's liability. In *Burlington Industries v. Ellerth* and *Faragher v. Boca Raton,* the Court held that an employer could be held liable for its supervisors' sexual harassment under Title VI of the Civil Rights Act of 1964, even if it did not know of the misconduct. In some cases, the employer can defend itself by showing it took steps to prevent or correct harassment. In these cases, the Court redefined the employer's liability, not sexual harassment itself.

In another case, *Oncale v. Sundowner Offshore Services,* the Court held that the federal laws against sexual harassment on the job cover misconduct, even when the victim and the aggressor are the same sex. The fourth case, involving a school district, is not germane to this discussion. All of these decisions (except the school issue) were against the employer.

In past decisions, the Supreme Court defined sexual harassment and made clear that it was covered under the Civil Rights Act. The case of *Meritor Savings Bank v. Vinson* (1986) was the first significant sexual harassment to reach the Supreme Court. The Court held that an employee might sue the employer when discrimination based on sex has created a hostile work environment. The critical issue in this case was whether members of one sex were treated differently or discriminated against because of their sex. Under Title VII of the Civil Rights Act, it is unlawful for an employer to discriminate against an employee with respect to compensation, terms, conditions, or privileges of employment because of the employee's race, color, religion, sex, or national origin. In this case, the Court also held that a plaintiff need not show economic or other tangible injury as the result of such action.

In *Harris v. Forklift Systems, Inc.* (1993), the Court held that employees who sue for sexual harassment are not required to show that they suffered loss of their psychological well-being, much less any tangible injury, as the result of a hostile workplace environment—only that the condition existed and that they were subject to the condition. The test is whether a reasonable person subjected to the same discriminatory conduct would find that the harassment altered the working conditions and made it more difficult to do the job.

What does all of this have to do with your organization? These Supreme Court decisions support the employer who is proactive in taking action to stop sexual harassment in the workplace. They also suggest your organization not only needs to have policies but you need to enforce them.

In that regard, the following policies are recommended:

- Provide sexual harassment training for workers and supervisors.
- Have a policy of no tolerance, and make sure all personnel understand the rules.
- Create a system in which workers have an opportunity to complain without fear of retaliation.
- Promote supervisors carefully by recognizing leadership qualities as well as technical competency.
- Train supervisors to identify and investigate complaints.
- Take complaints seriously.

- *What is a tangible employment action?* A "tangible employment action" means a significant change in employment status. Examples include hiring, firing, promotion, demotion, undesirable reassignment, a decision causing a significant change in benefits, compensation decisions, and work assignment.

- *How might harassment culminate in a tangible employment action?* This might occur if a supervisor fires or demotes a subordinate because the subordinate rejects the supervisor's sexual demands, or promotes the subordinate because the subordinate submits to the supervisor's sexual demands.

- *What should employers do to prevent and correct harassment?* Employers should establish, distribute to all employees, and enforce a policy prohibiting harassment and setting out a procedure for making complaints. In most cases, the policy and procedure should be in writing.

- *What should an antiharassment policy say?* An employer's antiharassment policy should make clear that the employer will not tolerate harassment based on race, sex, religion, national origin, age, or disability, or harassment based on opposition to discrimination on participation in complaint proceedings. The policy should also state that the employer will not tolerate retaliation against anyone who complains of harassment or who participates in an investigation.

■ *What are important elements of a complaint procedure?* The employer should encourage employees to report harassment to management before it becomes severe or pervasive. The employer should designate more than one individual to take complaints and should ensure that these individuals are in accessible locations. The employer also should instruct all of its supervisors to report complaints of harassment to appropriate officials. The employer should assure employees that it will protect the confidentiality of harassment complaints to the extent possible.

■ *Is a complaint procedure adequate if employees are instructed to report harassment to their immediate supervisors?* No, because the supervisor may be the one committing harassment or may not be impartial. It is advisable for an employer to designate at least one official outside an employee's chain of command to take complaints, to assure that the complaint will be handled impartially.

■ *How should an employer investigate a harassment complaint?* An employer should conduct a prompt, thorough, and impartial investigation. The alleged harasser should not have any direct or indirect control over the investigation. The investigator should interview the employee who complained of harassment, the alleged harasser, and others who could reasonably be expected to have relevant information.

■ *How should an employer correct harassment?* If an employer determines that harassment occurred, it should take immediate measures to stop the harassment and ensure that it does not recur. Disciplinary measures should be proportional to the seriousness of the offense. The employer also should correct the effects of the harassment by, for example, restoring leave taken because of the harassment and expunging negative evaluations in the employee's personnel file that arose from the harassment.

■ *Are there other measures that employers should take to prevent and correct harassment?* An employer should correct harassment that is clearly unwelcome regardless of whether a complaint is filed. For example, if there is graffiti in the workplace containing racial or sexual epithets, management should not wait for a complaint before erasing it.

☐ An employer should ensure that its supervisors and managers understand their responsibilities under the organization's antiharassment policy and complaint procedures.

☐ An employer should screen applicants for supervisory jobs to see if they have a history of engaging in harassment. If so, and the employer hires such a candidate, it must take steps to monitor actions taken by that individual in order to prevent harassment.

☐ An employer should keep records of harassment complaints and check those records when a complaint of harassment is made to reveal any patterns of harassment by the same individuals (see the "Managing Time" section in Chapter 5 for more on keeping records).

■ *Does an employee who is harassed by his or her supervisor have any responsibilities?* Yes. The employee must take reasonable steps to avoid harm from the harassment. Usually, the employee will exercise this responsibility by using the employer's complaint procedure.

■ *Is an employer legally responsible for its supervisor's harassment if the employee failed to use the employer's complaint procedure?* No, unless the harassment resulted in a tangible employment action or unless it was reasonable for the employee not to complain to management. An employee's failure to complain would be reasonable, for example, if the employee had a legitimate fear of retaliation. The employer must prove that the employee acted unreasonably.

■ *If an employee complains to management about harassment, should the employee wait for management to complete the investigation before filing a charge with the EEOC?* It may make sense to wait to see if management corrects the harassment before filing a charge. However, if management does not act promptly to investigate the complaint and undertake corrective action, then it may be appropriate to file a charge. The deadline for filing an EEOC charge is either 180 or 300 days after the last date of alleged harassment, depending on the state in which the allegation arises. This deadline is not extended because of an employer's internal investigation of the complaint.

Further guidance on harassment can be found on the EEOC's Web site: www.eeoc.gov.

Although a general statement elsewhere in this book indicates that the terms "employees" and "subordinates" include volunteer members of the fire and rescue services, for purposes of this text, let us be very clear that the laws regard members of volunteer organizations in much the same way as they do employees and employers. Volunteer organizations are expected to provide a fair and equitable workplace for their members and for the public they serve. Although the threat of liability may be present, the risk of adverse publicity should be an even greater motivator to do the right thing. The loss of the reputation of a volunteer organization could lead to the loss of community support; most volunteer organizations need strong community support to exist.

Good leaders can prevent harassment in their workplace. Take a proactive stance on harassment. Should you become aware of any form of harassment, take quick, firm action to let people know that the action is inappropriate. Failure to do so could be embarrassing and expensive for both you and your organization.

ETHICS

Volunteer organizations are expected to provide a fair and equitable workplace for their members and for the public they serve.

LEADING OTHERS

Steve Mormino, Lieutenant, FDNY, New York

Lieutenant Steve Mormino says arguments between his firefighters are not uncommon—when you are living in the close quarters of a firehouse, there are going to be disagreements—but for a good leader, they are also controllable.

Recently, an issue came up between firefighters over a policy confusion. In their company, the most senior firefighter on a call has the first choice of position. One firefighter with seniority, who was on duty for someone else, thought he should have had first choice. Another firefighter of less standing knew the company policy—the rule only applies if you are working your own shift. The two started arguing about who should have gotten the senior position. At first, Mormino let them argue it out, but when their chests started to rise and one pushed another, he got involved, getting in between them before things got out of control.

Then, he sat down with both of them together, reiterated the policy, and let them cool down. The battalion chief, who had been upstairs in his office, heard everything—so he called Mormino later to make sure everything was OK. Mormino assured him it was. Typically, Mormino will not bring minor arguments to a higher level if they are resolved.

If they are to a point where the issue could continue—racial slurs or sexual harassment—or if a serious infraction had been made or a crime committed, he would have taken the matter to a higher level. But Mormino does not see that too often because the policies on expected behavior are clearly posted in the firehouse and stated in the regulations manual, which everyone is required to read annually. The firefighters, he says, also know that punishment could have involved loss of mutual privileges or vacation days; the knowledge that these rewards could be taken away usually keeps problems from escalating.

Things between the two firefighters ended up fine within minutes. One of them was on the lieutenants' list, and because things were calmed down, that possibility for promotion has not changed.

Questions

1. Why do adults act like this?
2. What do you think of Lieutenant Mormino's approach to this situation?
3. What would have likely happened if Lieutenant Mormino had not intervened?
4. A mark of a good leader is the ability to anticipate problems like these and head them off as early as possible. Do you agree with that statement? How can it be accomplished?
5. What approach would you take in a situation like this?

LESSONS LEARNED

One of your most important roles as a company officer is leading others. Leading involves working with others and trying to influence others to accomplish the organization's goals. This can be done in many ways, but when done effectively, productivity is increased and both supervisors and workers seem happier and more satisfied. Part of working with others involves accepting proper organizational behavior where everyone is treated fairly and with respect. We continue this discussion on leadership and provide some practical tools for effective leadership in Chapter 7.

KEY TERMS

consulting   seeking advice or getting information from another; as a leadership style it implies that the leader seeks ideas and allows contributions to the decision-making process

delegating   sharing work, authority, and responsibility with another; as a leadership style, delegating implies the most generous sharing of the officer's leadership role

directing   controlling a course of action; as a leadership style, it is characterized by an authoritarian approach

diversity   a quality of being diverse, different, or not all alike

Equal Employment Opportunity Commission (EEOC)   federal government agency charged with administering laws related to nondiscrimination on the basis of race, color, religion, sex, age, or national origin

expert power   a recognition of authority by virtue of an individual's skill or knowledge

first-level supervisors   the first supervisory rank in an organization

group dynamics   how individuals interact and influence one another

harassment   to disturb, torment, or pester

high-achievement needs   according to the needs theory of motivation, individuals with high-achievement needs accept challenges and work diligently

high-affiliation needs   according to the needs theory of motivation, individuals with high-affiliation needs desire to be accepted by others

hygiene factors   as used by Frederick Herzberg, hygiene factors keep people satisfied with their work environment

identification power   a recognition of authority by virtue of the other individual's character or trust

interpersonal dynamics   dynamic practices developed by groups that separate them from other individuals

leadership   the personal actions of managers and supervisors to get team members to carry out certain actions

legitimate power   a recognition of authority derived from the government or other appointing agency

Maslow's Hierarchy of Needs   a five-tiered representation of human needs developed by Abraham Maslow

model   representation or example of something

motivators   factors that are regarded as work incentives such as recognition and the opportunity to achieve personal goals

organization   a group of people working together to accomplish a task

power   the command or control over others, status

power needs   a recognized motivator to be in charge or have group affiliation

punishment power   a recognition of authority by virtue of the supervisor's ability to administer punishment

reward power   a recognition of authority by virtue of the supervisor's ability to give recognition

supporting   as a leadership style, the supporting process involves open and continuous communications and a sharing in the decision-making process

REVIEW QUESTIONS

1. What is a group? What are some of the characteristics of a group?
2. Why do people like to work in groups?
3. What is leadership?
4. What is leadership power?
5. How does Maslow's theory apply to the leader's job in the fire service?
6. Why should the demographics of the fire department mirror those of the community it serves?
7. How does diversity help the fire service better serve its community?
8. Name and briefly describe the principal laws that provide guidelines to encourage diversity in the workplace.
9. Name and briefly describe the federal laws that prohibit sexual harassment in the workplace.
10. What action should you take, as a company officer, when a member comes to you with an allegation of sexual harassment?

DISCUSSION QUESTIONS

1. Is leadership a learned or a natural skill?
2. Is there any benefit of studying the leadership traits of others?
3. How are leaders selected in your organization? Is this a valid process? Might there be a better solution?
4. Which leadership style is best for dealing with routine activity at the fire station?
5. Which leadership style is best for dealing with implementing a new procedure?
6. Which leadership style is best for dealing with the conditions at the scene of an emergency?
7. Discuss the five types of power discussed in the chapter.
8. Which type of power is most effective for company officers?
9. How does one determine the leadership style that is best for them?
10. The text suggests that diversity and harassment have been issues in the fire service. Would you agree? How did the fire service in your community respond to these issues? What has or is being done in the fire service in your community in response to these issues?

ENDNOTES

1. Frederick Taylor, *The Principles of Scientific Management.* New York: Harper, 1911.
2. Not to be confused with "needs" as used earlier by Maslow.
3. Kimberly Alyn. "The Power of Empowerment in Leadership." *Firehouse* (October 2008).
4. The U.S. Code defines sexual harassment as follows: "Unwelcome sexual advances, requests for sexual favors, and other verbal or physical conduct of a sexual nature constitute sexual harassment when (1) submission to such conduct is made either explicitly or implicitly a term or condition of an individual's employment, (2) submission to or rejection of such conduct by an individual is used as the basis for employment decisions affecting such individuals, or (3) such conduct has the purpose or effect of unreasonably interfering with an individual's work performance or creating an intimidating, hostile, or offensive working environment."
5. "For Employers, a Blunt Warning," *Washington Post,* June 27, 1998.

ADDITIONAL RESOURCES

Alyn, Kimberly. "The Power of Empowerment in Leadership." *Firehouse,* October 2008.

Bell, A. Fleming. *Ethics in Public Life.* Chapel Hill, NC: The Institute of Government, 1998.

Bendrick, Mark. "A Fair Shake." *Fire Chief,* April 2008.

Certo, Samuel. *Supervision: Concepts and Skill Building.* 3rd ed. New York: McGraw-Hill, 2000.

Covey, Stephen R. *Principle-Centered Leadership.* New York: Simon and Schuster, 1992.

Crane, Bennie and Julian Williasm. *Personnel Empowerment.* Saddle Brook, NJ: Fire Engineering, 2002.

Edwards, Steven. *Fire Service Personnel Management.* Upper Saddle River, NJ: Prentice Hall, 2000.

Hogan, Lawrence J. *Legal Aspects of the Fire Service,* 3rd ed. Frederick, MD: Amlex, Inc., 2003.

Kanterman, Ronald. "Leadership Excellence: Balancing Management with Leadership." *Fire Engineering,* October 2008.

Maxwell, John. *There Is No Such Thing as Business Ethics.* New York: Warner Business Books, 2003.

Schneid, Thomas D. *Fire and Emergency Law Casebook.* Albany: Delmar Publishers, Inc., 1997.

Schrage, Doug. "Officer Mentoring Preserves Fire Service Heritage." *Fire Engineering,* July 2007.

Swinhart, Dominick. "360-Degree Performance Evaluations." *Fire Engineering,* August 2008.

Wilmoth, Janet. "Where Are the Leaders?" *Fire Chief,* June 2006.

Wilmoth, Janet. "A Leader's Book Club." *Fire Chief,* February 2008.

7

The Company Officer's Role in Leading Others

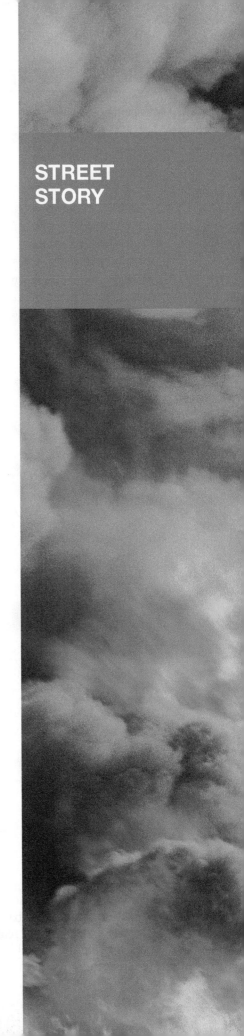

I 've been with the London, Ontario Fire Department for more than 20 years now, working in a range of different areas. At one point I became a team coordinator for our special teams units, and then I became an officer. After working as an officer for a while, I moved on to training, where I have been for 5 years. Within the past year, I became the assistant director. I am responsible for all aspects of training for our department of about 400 people, along with the five other people in this division.

When I was working on the teams, I built up a lot of experience, and it was hard to find a way to turn around and give back that experience. You can give it to a couple of people on your team, but I wanted to turn it into something that I could invest in and create a wider audience. So much of what we do as company officers is on-the-job learning, and I think it is important for company officer candidates to see the action models. We can learn details and statistics in a classroom, but that is not the same as it is in a real world situation. I always say that it is critical when there is a new crew or group for them to see everything in action. So basic competency that you can learn in the classroom is the foundation, and then you have to get out there in the real world, and sometimes that is not as easy, and sometimes discipline is necessary. We get skilled people here all the time who just need to develop their qualities.

Sometimes a difficult situation can arise when you have talented groups of people all working together, and one or two people may be causing a problem within the group. One such thing happened about 2 years ago. We were working with a group from many different municipalities. They were all experienced before they came to us. We ran them through some basic training exercises, and I saw there was no challenge for them in that particular training. So we compounded the situation, and what we found was that the increase in pressure caused them to make mistakes. When I found that there were individuals who were able to adapt to those changes, I put them in charge of their team. One of these teams was not performing up to par. There was one individual on the team who was not performing as well as the others. So I gave the group a new, complex scenario; left it open ended for them to figure out how to solve it; and then had an immediate debriefing with them after they tried. Then I got on the offensive and asked them if they were aware that they were not performing the way they should be. One of the positive things I saw was that the group told me what they knew about their performance. I pointed out some other things they could do to improve their performance they might not have otherwise noticed, and the team leader took responsibility. I then gave them time to go away with what they had learned. They had their own discussion, and we saw their cohesion go way up. I think the notion of giving people the space to go off and digest everything on their own and then bring it back is important when it comes to corrective action.

Supervising teamwork can be hard, but one of the big things that I'm a fan of is that in a team, everyone brings something to the table. As a leader, you have to be a model of that behavior and enable others to bring their strength to the table, and there's a trust and understanding that get built, and the leader has to foster that.

—*Shawn Smith, Assistant Director of Training,*
London Fire Department, London, Ontario, Canada

LEARNING OBJECTIVES

After completing this chapter the reader should be able to:

7-1   Describe how a job description impacts a unit member's evaluation.

7-2   Identify the company officer's role in the professional development of agency members.

7-3   Distinguish between coaching and counseling of employees.

7-4   Analyze a given situation and apply the appropriate coaching or counseling techniques.

7-5   Identify the responsibilities of the supervisor in recognizing those behaviors requiring referrals to a member assistance program.

7-6   Identify the appropriate use of coaching and counseling.

7-7   Explain the purpose of unit member evaluations.

7-8   Explain the evaluation process as it pertains to job performance.

7-9   Identify the standard operating procedures pertaining to unit member evaluations.

7-10   Discuss how each of the common evaluator errors affects a member's evaluation.

7-11   List the common evaluator errors that influence a member's evaluation.

7-12   Identify forms required for conducting a member evaluation.

7-13   Explain the agency's policy for documenting an individual's disagreement with the content of the evaluation.

7-14   Describe how a job description impacts a unit member's evaluation.

7-15   Explain how the components found in a succession plan interact to produce a comprehensive career development guide for fire personnel (e.g., FESHE, Emergency Services Professional Development Model).

7-16   Describe the role of discipline as it relates to employee evaluation.

7-17   Identify the characteristics of a group.

7-18   Analyze the various phases of group dynamics within a work unit.

7-19   List individual behaviors that constitute a problem within the department.

7-20   Identify the generally accepted steps for resolving problems.

7-21   Explain how individual behavior can affect organizational culture.

7-22   Describe the services available to a member in need of assistance.

7-23   Explain how work-related stress can impact the performance of unit members.

7-24   Explain how to encourage unit members to seek professional assistance.

7-25   Explain how to follow up the unit member's request.

7-26   Describe the policies and procedures that define the supervisor's role in the member assistance referral process.

7-27   Identify the standard operating procedures pertaining to federal, state, municipal, and departmental human resource management.

7-28   Interpret federal, state, municipal, and departmental human resource management policies and procedures and how they impact members of the organization.

7-29   Explain how change can affect organizational culture and behavior.

7-30   Identify the process used by the agency to develop a new policy or procedure.

7-31   Describe how to implement change in a manner that minimizes organizational and individual conflict.

* The FO I and II levels, as defined by the NFPA 1021, the *Standard for Fire Officer Professional Qualifications*, are identified in different colors: FO I = black, FO II = red.

INTRODUCTION

This chapter continues our discussion on leadership. In Chapter 6, we discussed the basic concepts of leadership. Here, we fit them into the reality of the fire service. Remember that the focus of this book is the company officer, the first-line supervisor in the fire service.

BEING A SUPERVISOR IS DIFFERENT

When people are promoted to supervisors, they quickly find that the work is different and in some ways, more difficult than when they were one of the workers. The typical firefighter has developed an impressive array of technical skills, knowledge, and abilities that make the firefighter quite competent in that role. The new job as a supervisor is quite different.

NOTE

When people are promoted to supervisors, they quickly find that the work is different and in some ways, more difficult than when they were one of the workers.

Administrative duties, understanding the organization's rules and regulations, and supervising others are demanding tasks. Communications skills are vital. As a supervisor, you have to communicate with your team members. As a company officer, you are expected to be able to communicate with the rest of the organization and the public.

Few organizations, including most fire departments, prepare prospective supervisors for the position. Fire departments promote good firefighters and expect them to become good officers. Some fire departments provide excellent training as new supervisors enter the officer ranks, whereas others provide a continuing program of professional development to provide the supervisor, leader, and manager with the leadership tools needed for the increasing responsibilities.

No one can anticipate every situation that may occur in the role as a new supervisor. The fact that you are reading a book on the topic is a positive step, but even here, what is learned from the book or in a classroom will only begin the process of personal professional development. You are encouraged to continue the development process: Read every article and take every class on leadership you can.

OFFICER ADVICE

Continue the development process: Read every article and take every class on leadership you can.

The selection of new firefighters into your organization and the new firefighters' entry-level training are normally not the province of the company officer. However, once these new firefighters have survived the selection and initial training process, they are usually handed to you—the company officer. They will likely work in that environment until they too become company officers.

NOTE

For most of those entering the fire service, the person who has the greatest impact on them in their entire career is probably their first company officer.

As supervisors and as company officers, it is your job to develop the new firefighter into a true professional. This is a critical time for both you and the new firefighter. For most of those entering the fire service, the person who has the greatest impact on them in their entire career is probably their first company officer. You have an awesome opportunity and responsibility here!

You might look upon new firefighters as simply resources. Indeed, they are now a part of the company team. At the same time you should look on new firefighters as part of your team and your profession, and help them move along in a continuing process of development, just as you have done. In fact, a significant part of any supervisor's time should be devoted to the development of his or her subordinates. The task of developing subordinates is one of the most challenging tasks a company officer faces, and at the same time is one that can provide great personal satisfaction. There is nothing more satisfying than watching a subordinate develop to full potential.

ON LEADERSHIP

Leaders take risks. Leaders have to have the confidence to try new things. Leaders motivate people to work as a team and accept responsibility, even in failure. Leaders understand that failure is only an event that can be a learning tool. Like the best of sports teams, good leaders sometimes lose, but like good teams, they come back and are willing to try again.

Many famous Americans were great leaders in their time. George Washington was a great general in the Revolutionary War. He went on to become president of the United States. But he was not always successful: Washington lost eight engagements early in the war. Ulysses S. Grant was a great Civil War general. He too went on to become president of the United States. Before the Civil War, Grant was a failure in business and in military service.

As you work to become a leader, try to demonstrate your leadership skills at every opportunity. Let the people you work with know that you are not afraid to take reasonable risks. Let them know that you believe in sharing the praise when things go well and personally accepting the responsibility when they do not.

BUILDING A RELATIONSHIP WITH YOUR TEAM MEMBERS

The First Meeting

The first meeting with a new firefighter should define the job, establish positive relationships, and build a friendly atmosphere. Although new firefighters are qualified to serve in that position, they may not have actually worked in this particular environment.

You may have to show them some pretty basic things, like where their lockers and bunks are located, where to hang their turnout gear, and where you want them to sit on the apparatus. You will have to pick up where the recruit academy left off in terms of training. Many fire departments have a structured program that allows the new firefighter to review what was covered in recruit school while learning new material that is needed on the job. And the new firefighter should be expected to become familiar with the buildings and streets in the company's first-due response area. Much of this can be done using a senior firefighter as a mentor.

NOTE

Many fire departments have a structured program that allows the new firefighter to review what was covered in recruit school while learning new material that is needed on the job.

THINGS THAT YOU CAN DO FOR YOUR TEAM MEMBERS

- Share information about the job and the organization.
- Share knowledge about the role of the department and how they fit in.
- Share knowledge about trends in growth, promotions, and opportunities.
- Identify alternative career plans and the implications of each.

OFFICER ADVICE

The supervisor should establish a clear definition of the job.

The supervisor should establish a clear definition of the job. This seems rather elementary to the established firefighter, but think back to the first days when you were on the job: Think about how nice it would have been if someone had taken a few minutes a day to explain the job in some detail to you.

When you are talking to a new team member, try to avoid information overload. Take just a few minutes a day to cover the needed items, rather than trying to do it all at once. When we get too much information, it is difficult to sort out all that we have heard, and it is certainly difficult to remember all the details, especially when we are entering a new profession. When you are explaining the tasks that need to be done,

clearly define the members' role and show them how they fit into the overall organization. Give the team member a chance to talk and to ask questions.

OFFICER ADVICE

Give the team member a chance to talk and to ask questions.

Teaching individual small activities may seem trivial to the seasoned professional, but we all know that proficiency in the little details is what keeps us prepared for the big emergencies. For example, if the new firefighter is responsible for maintaining some piece of equipment on the apparatus, explain what the tool does and why the maintenance is important, both in terms of being able to use the tool effectively and the personal safety of the firefighter using the tool.

Set well-defined job standards with new firefighters. There should be plenty of information available to help you determine what these job standards are. Your organization should have a detailed position description of the firefighter's duties. That position description may refer to additional requirements, such as those listed in NFPA 1001, *Standard for Firefighter Professional Qualifications*. Make sure that your firefighters understand the way the job should be done, which may not necessarily be the way that it is done now. Having this set of standards and expectations for the job provides a gauge for firefighter proficiency and ultimately evaluation of this proficiency.

OFFICER ADVICE

Set well-defined job standards with new team members.

For most new members, sorting through all of this information will be a bit overwhelming. As a supervisor, you should help them by assigning some priorities. If appropriate, put some deadlines on the first few items on the priority list to give the new member some specific goals to attain in the first few days and weeks on the job.

An important part of this process should include a regular review of the new member's progress. Although a formal review will occur after some specific time of service, typically 6 months or a year, you should not wait that long with a new member. Hold a mini review to help track the new member's accomplishments, provide positive reinforcement for the good work the member is doing, make adjustments where necessary, and set new goals for the short-term future (see **Figure 7-1**). Such meetings should be held once or twice a month.

FIGURE 7-1 Encourage your team members to openly discuss their performance.

OFFICER ADVICE

An important part of setting standards should include a regular review of the new member's progress.

Determining the Team Members' Competency and Commitment

The team members' performance is determined by their competence and their commitment to the work. Their competence is determined by their knowledge, skills, and abilities; their commitment is determined more by their attitude, motivation, and confidence in themselves. Together these two dimensions present four possible situations:

- *Type 1: High competency/high commitment = good performer.* This member has the skill and self-confidence needed and is ready to move to areas of greater responsibility. Keep this member challenged.

- *Type 2: Low competency/high commitment = good student.* This characterizes many of us while we are on the learning curve. The novice member is learning new skills but now realizes that this process is a lot more complicated than initially thought. As a supervisor, your role is to provide assurance and help this novice gain self-confidence.

- *Type 3: High competency/low commitment = poor attitude.* In this case the member is learning but lacks the self-confidence or motivation needed to go it alone. Work on attitude and self-confidence.

- *Type 4: Low competency/low commitment = unwilling and unable.* You may need to review the individual's goals. Look for learning disabilities. Take shorter steps and have patience. Provide positive motivation and evidence of your confidence.

TIPS FOR GOOD LEADERSHIP

- Plan the assignment of responsibilities.
- Patiently guide team members toward effective performance.
- Recognize that there may be more than one way to do something. Be flexible.
- Avoid constant interference so that team members can develop and become more independent.
- Use effective communications techniques.
- Document examples of noteworthy performance, both good and bad.

FOUR EFFECTIVE TOOLS FOR TEAM MEMBER DEVELOPMENT

We have discussed at some length the leadership role of the supervisor in the company environment. Although the role is critical to the collective productivity and safety of the members assigned to that company, we should also look at the role of the supervisor as it relates to each member as an individual. Effective supervision promotes an environment that encourages each member to develop to their full potential, which not only benefits the department but also provides the very kind of intangible rewards that separate the good workplaces from the great workplaces. This interaction is accomplished in four ways: mentoring, coaching, counseling, and performance evaluations. Later in this chapter, we will discuss a fifth effective tool—discipline.

Mentoring

In some departments, the new member spends up to 20 weeks in recruit school learning all the skills needed to be a good firefighter and EMT. Upon assignment to the company, the new recruits find that the only skills they use are related to cleaning the station and fire trucks.

On the other hand, some departments immediately start the new recruit through a mentoring program—a formal, structured process that continues the training that was a part of the recruit school. (This is even more important when a member does not go through a full recruit school process.) This is more than just pairing the new member with a seasoned one: That process has some benefits, but in many cases the new member becomes an underling for the older one.

A good mentoring process must start with a formal decision from a senior officer. The next step is to get the training division to develop a mentoring program—a formal program that includes goals, objectives, topics to be covered, evaluation instruments, a definition of the mentor's role, and a method for final evaluation of both the process and the new member's knowledge at the end of the program. Even if your department does not have a formal program should not stop you from instituting one at your station. Just be sure to put in benchmarks or some way to monitor progress. Remember; you can not manage what you can not measure!

A key to the process is the mentor. A "mentor" is a trusted counselor, guide, tutor, or coach. Mentors should have a positive attitude about the department and the mentoring process. They should have the positive qualities of good character and set a good positive personal example.

Good communications skills are also important in a mentor. As with any instructor, the mentor should be able not only to convey the information but also to offer constructive criticism to help the new member. Recruits should be encouraged to ask lots of questions. The mentor must have lots of answers. This is a time to review the material covered in recruit school at a slower pace. The instruction is individualized, and emphasis can be put upon the individual's needs without having to be concerned about a class schedule. It is an especially good time to cover the material required in NFPA 1001 for Firefighter II.

Ultimately, you, as the company officer, are responsible for the firefighter's performance. You have to oversee the mentor and the mentoring process to assure its successful outcome.

MENTORING

The word *mentor* comes to us from Greek mythology. Mentor was the name of Odysseus' wise and faithful advisor. Today, we use the term to describe a friend and role model, a person who lends support in many ways to help one pursue a goal.

The mentoring process can be briefly summarized with the words "Lead, follow, and get out of the way." Lead, as used here, is showing the way by being a teacher and role model. Follow involves watching, advising, and counseling the student or new member. And by get out of the way, we mean letting go, delegating, or withdrawing. The process is not linear. It involves concurrent activity in all three areas. As a new member masters one skill, the mentor can move on to another area and start the process all over. A good mentor will continue to observe, advise, and counsel when appropriate.

The process should be done at all levels. As a good company officer, you should be the mentor for that senior firefighter who looks like promotion material. And that senior firefighter should be the mentor for the new firefighter who just joined the company.

Who is the company officer's mentor? One would like to think that a good battalion chief would take on that task, and that same process should work all the way up the chain of command to the very top. The fire chief should be mentoring one or more of the senior staff so that they are prepared to take on the additional responsibilities as the chief of the department.

We might draw an analogy here. In the U.S. Navy and Coast Guard, there is a definite path to the top, especially for those in operational commands such as ships and air stations. The assignment process facilitates that activity, but the real value lies in the personal values of the individuals involved in the chain of command. Good commanding officers will take their executive officers under their wing and teach them everything needed to become a good commanding officer. This is an operational necessity in wartime, for the commanding officer could conceivably be a casualty. But the process has everyday value—good executive officers should be groomed to be good commanding officers. No book or school can cover all that is needed. Likewise, the process should be working down the chain of command so that the newest and most junior person in the unit has a designated mentor.

What motivates this process? In the fire service as in the military, the senior officer (fire chief) is responsible for the performance of the entire organization. To bring out and develop the best talent in any organization, good supervisors should be facilitating the development of all of those who report to them. This means being a good mentor, coach, counselor, and advisor. All of these are part of the job of a good supervisor. The goals are to improve performance and to prepare the team member for moving to positions of greater responsibility.

In addition to helping the member master today's responsibilities and prepare for those of tomorrow, a good mentor provides information on survival skills: how to dress, what to say, and how to get ideas approved.

Remember the three steps:

1. **Lead**—pass on knowledge and skills allowing the student or new member to safely and effectively function in the new environment.

2. **Follow**—advise and counsel, providing a continuing program of development as the new member actually functions in the new environment.

3. **Get out of the way**—initiate the process of handing off as the student or member gains the skills and ability to do the job alone.

THE MASTER FIREFIGHTER AS A MENTOR

One excellent example of this program provides a special position called master firefighter. In this department, there is one (and only one) master firefighter in each company. The master firefighter is a seasoned member and could be compared with the first sergeant in an army unit—although neither is an officer, both have an abundance of expertise. This expertise is acknowledged with respect; the master firefighter has both the expert and identification power discussed in the previous chapter. The master firefighter has demonstrated proficiency as a firefighter, and if the department has technicians serving in specialty areas (for example, hazmat or rescue), the master firefighter should be currently qualified in at least one of these. The master firefighter should also be a certified instructor. Finally, a competitive process will help identify the best-qualified individuals to serve as master firefighters. Ideally, there will be a pay increase for the individual to make this all worthwhile. The master firefighter is the designated mentor who guides the new members successfully through the mentoring process.

Coaching

Coaching is a term we borrow from the sports world, but it fits well here. When coaching, you are helping a team member improve knowledge, skills, and abilities. Coaching usually focuses on one aspect of the job at a time, and the **coach** works with the student until the desired level of competence is demonstrated (see **Figure 7-2**). We do this with our children. We should do it with our team members, too. It is an easy and satisfying process.

NOTE

When coaching, you are helping a team member improve knowledge, skills, and abilities.

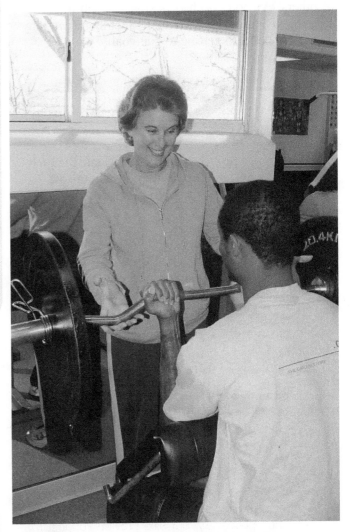

FIGURE 7-2 The traditional coach helps people develop physical skills.

Coaching is usually initiated by the leader. It is an informal process that requires little preparation. The leader already knows the technical aspects of the task. Teaching the task and having the patience to work with the member may require some self-control, but the process is really quite easy. Take time to describe the steps that make up the job, demonstrate the proper method, and ask the member to demonstrate the same sequence. Ask the member to talk through the task, just as you did when you were demonstrating the task. This does two things: First, it reinforces the learning process, and second, it verifies that the member's thinking, as well as the actions, are correct. While watching and listening to your team member, give feedback and reassurance.

Coaching is an informational process that helps members improve their skills and abilities. Coaching implies a one-on-one relationship that treats members as full partners, that communicates a sense of mutual respect and trust, a willingness to listen, and even a willingness to compromise. An effective coach has self-confidence and the ability to win the confidence of others. A good coach is flexible enough to recognize the vision of success and the fact that there may be several paths to getting to that goal.

NOTE

Coaching is an informational process that helps members improve their skills and abilities.

An effective coach has that vision, the self-image of success, and the ability to communicate that vision effectively to others. An effective coach can see beyond the obvious.

For a new person entering a workplace, this means that the coach or supervisor can see beyond the present and envision the full potential of the new member (see **Figure 7-3**). There is much to be learned from the way coaches treat their players and from how the players react to that treatment.

TIPS FOR EFFECTIVE COACHING

- You are a role model. The example you set is one of the most important elements in your effort.
- Gain the agreement of your members regarding the goals that are set. They must be high, mutually agreeable, and attainable.
- Progress toward these goals must be measured against reasonable standards and on a regular basis.
- Feedback is essential. Be as positive as possible. Criticize when necessary, but also provide suggestions for improvement.

FIGURE 7-3 Company officers coach people to help them develop professional skills.

Performance is often a function of how members are treated and the expectations that are indicated. Where success is expected, success will likely follow. Where failure is expected, it will surely follow. Successful leaders have the ability to transmit high expectations to their members. Members, more often than not, will perform to these expectations. High expectations lead to high performance. This, in turn, reinforces the high expectations, and even higher performance is expected next time. The converse is also true.

Think about what happens when you have a good working relationship with good members. You tend to be comfortable with them and are likely to spend

AN EXAMPLE OF COACHING

Let us use teaching knot tying as an example of coaching. Knot tying is a basic skill, but some firefighters have a hard time remembering the basic knots, especially when they are not often used. The officer or a more senior firefighter can take the new firefighter to a quiet corner of the station and work on just one knot. Help the student, and then watch the student tie it a few times. A few hours later, ask the student to do it again. Next day, have this student tie the knot again and then try another knot. Within a few days, the firefighter will have the skill and self-confidence to tie every knot on the list. This same idea works with just about any lesson being taught. Coaching takes time, but the satisfaction for both parties is huge!

more time with them. You find favor with their work, and, therefore, you find it easier to talk to them and compliment them on their work. As a result, they will probably continue to get better. But what happens when you do not have a good working relationship with a member? You probably are not comfortable with that person. Therefore, you do not spend time with this person, and you do not see this person's work, good or bad. As a result, this member probably does not improve. Whose fault is that?

Counseling

Counseling is the second leadership tool available to help members. **Counseling** may focus on some specific aspect of the job, but more likely, the focus is on general attitude or behavior. Counseling should always be done in private. Keep the counseling session business like but friendly. In fact, counseling can be very informal. After all, the purpose is to help the member improve. To be effective, it should be accomplished shortly after the unsatisfactory behavior was observed.

NOTE

Counseling is the second leadership tool available to help members. Counseling may focus on some specific aspect of the job, but more likely, the focus is on general attitude or behavior.

For example, a new firefighter while on an emergency medical services (EMS) call shows some rather obvious signs of displeasure at having to treat a patient at the scene of the incident. Although no one said anything, other firefighters and bystanders saw what was happening. The time to correct this problem is back at the fire station (see **Figure 7-4**). As the company officer, you should call the firefighter into your office and ask the firefighter to discuss relevant feelings. Counsel the individual that while these feelings are understandable, they should be suppressed, at least in public. If this is the first time this situation has been noted, just point out to the member the action at issue, why it is unacceptable, and define what is acceptable. Most members will get the message.

The counseling session should be as constructive as possible under the circumstances. Good leaders are able to counsel members without either one becoming emotional. If you and the member have a good relationship, and if the communication process is working well, the information can be quickly and effectively conveyed without having to make a major production of the event. Ridiculing, threatening, or verbally attacking a member is generally ineffective.

OFFICER ADVICE

To be effective, counseling should be accomplished shortly after the unsatisfactory behavior was observed.

FIGURE 7-4 Counseling focuses on improving some specific aspect of member performance, attitude, or behavior.

Counseling is a tool for member improvement, but it also is a problem-solving process. The problem should be clearly identified, and an explanation should be offered as to why it is a problem. Have agreement with the member about the problem and the corrective action that must be taken. Give the member an opportunity to explain the actions at issue or express relevant feelings on the matter. Close the session by agreeing on the next step. A follow-up meeting may be needed in some cases. Pick a specific time and date that are mutually convenient for the next meeting, and make a point to log that meeting in your personal activity planner.

CHECKLIST OF UNSATISFACTORY JOB PERFORMANCE IDENTIFYING THE TROUBLED TEAM MEMBER

Do any of the team members you supervise have any of the following characteristics?

Absenteeism
- Excessive sick leave
- Frequent Monday, Friday and/or weekend absences

High accident rate
- Accidents on the job
- Accidents off the job (but affecting job performance)

Difficulty in concentration
- Work requires greater effort
- Jobs take more time

Confusion
- Difficulty in recalling instructions, details, and so forth
- Increasing difficulty in handling complex assignments

Spasmodic work patterns
- Alternate periods of high and low productivity
- Missed deadlines

Generally lowered job efficiency
- Mistakes due to inattention or poor judgment
- Improbable excuses for poor job performance

Poor team member relationships on the job
- Overreaction to real or imagined criticism
- Complaints from coworkers

If you have checked any of the items, you may have identified a team member whose work performance is affected by a personal or emotional problem. Document each occurrence in your supervisory records, and when the situation warrants action,

conduct appropriate supervisory intervention. When the performance deficit cannot be attributed to a "management problem," your intervention should include a referral to the Employee Assistance Program (EAP).

One final comment: Remember that the member will naturally be feeling down after the discussion. Once the counseling session is over, consider the option of rebuilding the relationship. Go and get a cup of coffee. Together!

THE FIVE PROFESSIONAL DEVELOPMENT ROLES OF A LEADER

Mentor	Helps members learn the job
Coach	Helps members with self-assessment
Appraiser	Provides members with a reality check
Adviser	Helps members identify their goals
Referral agent	Connects members with their own future goals

Performance Evaluations

Performance evaluations are another important member development tool. Many members and even some supervisors look upon performance evaluation as a necessary but undesirable part of the job. If we look upon evaluations as a way of improving performance, things might be a lot easier.

NOTE

Performance evaluations are another important member development tool.

Compared to coaching and counseling, evaluations are usually quite structured. Evaluations are structured in that they are usually done on a regular schedule, and they follow a standard procedure, usually dictated by a reporting form.

If coaching and counseling have been taking place as they should, there should be no surprises for members in the evaluation interview. In fact, the members should be able to write the evaluation themselves. Although the written record is important, the performance evaluation interview is important too, especially for newer members of the organization (see **Figure 7-5**).

In most organizations, formal personnel evaluations occur once or twice a year, but in reality, the evaluation procedure should be a continuous cycle.

FIGURE 7-5 Performance evaluations are an important member development tool.

The cycle should start with goal setting, then move to encouragement, self-assessment, watching performance and providing feedback during the period, and a formal performance review and goal setting. The cycle is continuous (see **Figure 7-6**).

THE 360-DEGREE ASSESSMENT

Typically, performance appraisal has been limited to a feedback process between employees and supervisors. However, with the increased focus on teamwork,

employee development, and customer service, the emphasis has shifted to employee feedback from the full circle of sources depicted in the diagram below. This multiple-input approach to performance feedback is sometimes called "360-degree assessment" to connote that full circle.

There are no prohibitions in law or regulation to using a variety of rating sources, in addition to the employee's supervisor, for assessing performance. Research has shown assessment approaches with multiple rating sources provide more accurate, reliable, and credible information. For this reason, the U.S. Office of Personnel Management supports the

CYCLE OF PERFORMANCE MANAGEMENT

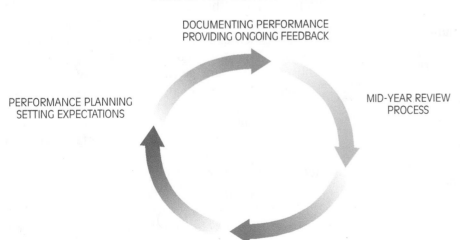

DOCUMENTING PERFORMANCE
PROVIDING ONGOING FEEDBACK

MID-YEAR REVIEW
PROCESS

PERFORMANCE PLANNING
SETTING EXPECTATIONS

PERFORMANCE EVALUATION
INTERVIEW

FIGURE 7-6 The cycle of performance management represents the continuous process of goal setting, observation, and performance evaluation.

use of multiple rating sources as an effective method of assessing performance for formal appraisal and other evaluative and developmental purposes.

The circle or, perhaps more accurately, the sphere of feedback sources consists of supervisors, peers, subordinates, customers, and oneself. It is not necessary, or always appropriate, to include all of the feedback sources in a particular appraisal program. The organizational culture and mission must be considered, and the purpose of feedback will differ with each source. For example, subordinate assessments of a supervisor's performance can provide valuable developmental guidance, peer feedback can be the heart of excellence in teamwork, and customer service feedback focuses on the quality of the team's or agency's results. The objectives of performance appraisal and the particular aspects of performance that are to be assessed must be established before determining which sources are appropriate. The following information discusses the contributions of each source of ratings and feedback.

> **NOTE**
>
> In addition to the positive aspects of the system, there are cautions that need to be considered. For the full version of the 360-degree assessment, visit: http://www.opm.gov/perform/wppdf/360asess.pdf.

Superiors

Evaluations by superiors are the most traditional source of employee feedback. This form of evaluation includes both the ratings of individuals by supervisors on elements in an employee's performance plan and the evaluation of programs and teams by senior managers.

What does this rating source contribute?

The first-line supervisor is often in the best position to effectively carry out the full cycle of performance management: Planning, Monitoring, Developing, Appraising, and Rewarding. The supervisor may also have the broadest perspective on the work requirements and be able to take into account shifts in those requirements.

The superiors (both the first-line supervisor and the senior managers) have the authority to redesign and reassign an employee's work based on their assessment of individual and team performance.

Most federal employees (about 90 percent in a large, government-wide survey) believe that the greatest contribution to their performance feedback should come from their first-level supervisors.

Self-Assessment

This form of performance information is actually quite common but usually is used only as an informal part of the supervisor-employee appraisal feedback session. Supervisors frequently open the discussion with:

"How do you feel you have performed?" In a somewhat more formal approach, supervisors ask employees to identify the key accomplishments they believe best represent their performance in critical and noncritical performance elements. In a 360-degree approach, if self-ratings are going to be included, structured forms and formal procedures are recommended.

What does this rating source contribute?

The most significant contribution of self-ratings is the improved communication that results between supervisors and subordinates.

Self-ratings are particularly useful if the entire cycle of performance management involves the employee in a self-assessment. For example, the employee should keep notes of task accomplishments and failures throughout the performance monitoring period.

The developmental focus of self-assessment is a key factor. The self-assessment instrument (in a paper or computer software format) should be structured around the performance plan but can emphasize training needs and the potential for the employee to advance in the organization.

The value of self-ratings is widely accepted. Approximately half of the federal employees in a large survey thought that self-ratings would contribute "to a great or very great extent" to a fair and well-rounded performance appraisal. (Of the survey respondents who received ratings below "Fully Successful," more than 75 percent thought self-ratings should be used.)

Self-appraisals should not simply be viewed as a comparative or validation process but also as a critical source of performance information. Self-appraisals are particularly valuable in situations where the supervisor cannot readily observe the work behaviors and task outcomes.

Peers

With downsizing and reduced hierarchies in organizations, as well as the increasing use of teams and group accountability, peers are often the most relevant evaluators of their colleagues' performance. Peers have a unique perspective on a coworker's job performance, and employees are generally very receptive to the concept of rating each other. Peer ratings can be used when the employee's expertise is known or the performance and results can be observed. There are both significant contributions and serious pitfalls that must be carefully considered before including this type of feedback in a multifaceted appraisal program.

What does this rating source contribute?

Peer influence through peer approval and peer pressure is often more effective than the traditional emphasis to please the boss. Employees report resentment when they believe that their extra efforts are required to "make the boss look good" as opposed to meeting the unit's goals.

Peer ratings have proved to be excellent predictors of future performance. Therefore, they are particularly useful as input for employee development.

Peer ratings are remarkably valid and reliable in rating behaviors and "manner of performance" but may be limited in rating outcomes that often require the perspective of the supervisor.

The use of multiple raters in the peer dimension of 360-degree assessment programs tends to average the possible biases of any one member of the group of raters. (Some agencies eliminate the highest and lowest ratings and average the rest.)

The increased use of self-directed teams makes the contribution of peer evaluations the central input to the formal appraisal because by definition the supervisor is not directly involved in the day-to-day activities of the team.

The addition of peer feedback can help move the supervisor into a coaching role rather than a purely judging role.

Subordinates

An upward-appraisal process or feedback survey (sometimes referred to as a SAM, for "Subordinates Appraising Managers") is among the most significant yet controversial features of a "full circle" performance evaluation program. Both managers being appraised and their own superiors agree that subordinates have a unique, often essential, perspective. The subordinate ratings provide particularly valuable data on performance elements concerning managerial and supervisory behaviors. However, there is usually great reluctance, even fear, concerning implementation of this rating dimension. On balance, the contributions can outweigh the concerns if the precautions noted next are addressed.

What does this rating source contribute?

A formalized subordinate feedback program will give supervisors a more comprehensive picture of employee issues and needs. Managers and supervisors who assume they will sufficiently stay in touch with their employees' needs by relying solely on an "open door" policy get very inconsistent feedback at best.

Employees feel they have a greater voice in organizational decision making and, in fact, they do. Through managerial action plans and changes in work processes, the employees can see the direct results of the feedback they have provided.

The feedback from subordinates is particularly effective in evaluating the supervisor's interpersonal skills. However, it may not be as appropriate or valid for evaluating task-oriented skills.

Combining subordinate ratings, like peer ratings, can provide the advantage of creating a composite appraisal from the averaged ratings of several subordinates. This averaging adds validity and reliability to the feedback because the aberrant ratings get averaged out and/or the high and low ratings are dropped from the summary calculations.

Customers

Executive Order 12862, Setting Customer Service Standards, requires agencies to survey internal and external customers, publish customer service standards, and measure agency performance against these standards. Internal customers are defined as users of products or services supplied by another employee or group within the agency or organization. External customers are outside the organization and include, but are not limited to, the general public.

What does this rating source contribute?

Customer feedback should serve as an "anchor" for almost all other performance factors. Combined with peer evaluations, these data literally "round out" the performance feedback program and focus attention beyond what could be a somewhat self-serving hierarchy of feedback limited to the formal "chain of command."

Including a range of customers in the 360-degree performance assessment program expands the focus of performance feedback in a manner considered absolutely critical to reinventing government. Employees, typically, only concentrate on satisfying the standards and expectations of the person who has the most control over their work conditions and compensation. This person is generally their supervisor. Service to the broader range of customers often suffers if it is neglected in the feedback process.

PERFORMANCE EVALUATION SYSTEMS

Performance evaluations are clearly an important supervisory activity. The real purpose of performance evaluations is to document the member's performance. In reality, we are looking at the performance levels of three sets of people: those who are above average, those who are well below average, and all the rest.

Regardless of the official system that is in place, all members should be receiving constant feedback on their performance. A word of acknowledgement or praise is feedback. A need for coaching or counseling is feedback. These are informal tools, whereas the performance evaluation process is a formal tool. In reality, we have two appraisal systems working side by side. The first, the informal appraisal system, is based on the interpersonal relationship between the supervisor and the member. The second, the formal appraisal system, provides a formal method of documenting the member's performance on a regular and systematic basis.

Employee-Supervisor Evaluation

Supervisor:_____ Position:_____

Evaluation Period:_____ To_____

1	2	3	4	5
Unsatisfactory	**Marginal**	**Satisfactory**	**Superior**	**Distinctive**
Does not meet Expectation	Comes close to Meeting Expectation	Meets Expectation	Surpasses Expectation	Greatly Surpasses

1-Encourages Teamwork and Cooperation.................................. 1 2 3 4 5

Comment:_____

2-Exhibits a Positive Attitude Toward Subordinates.................... 1 2 3 4 5

Comment:_____

3-Ability to Control Work Environment.................................... 1 2 3 4 5

Comment:_____

4-Works Well With Others.. 1 2 3 4 5

Comment:_____

5-Utilizes Positive Problem Solving... 1 2 3 4 5

Comment:_____

6-Maintains Professional and Calm Environment In Stressful Situations.......... 1 2 3 4 5

Comment:_____

7-Motivates others to Increase Their Own Personnel Skills & Development...... 1 2 3 4 5

Comment:_____

8-Improves Management & People Skills................................... 1 2 3 4 5

Comment:_____

9-Utilizes Good Communication Skills...................................... 1 2 3 4 5

Comment:_____

10-Keeps Skyland FD Best Interest As Top Priority........................ 1 2 3 4 5

Comment:_____

11-Reflects Positive Image to Subordinates and Community.............. 1 2 3 4 5

Comment:_____

12-Maintains Trust & Respect of Subordinates........................... 1 2 3 4 5

Comment:_____

13-Demonstrates Good Decision Making Ability.......................... 1 2 3 4 5

Comment:_____

14-Exhibits Self Control... 1 2 3 4 5

Comment:_____

15-Open To Suggestions & Constructive Criticizem...................... 1 2 3 4 5

Comment:_____

16-Stays Current on New Technology in The Fire Service................ 1 2 3 4 5

Comment:_____

17-Maintains a Learning Work Environment.............................. 1 2 3 4 5

Comment:_____

18-Maintains Positive Work Environment................................ 1 2 3 4 5

Comment:_____

19-Allows for Team Management with Subordinates in Proper Environment....... 1 2 3 4 5

Comment:_____

20-Always Places Highest Priority On Safety Of His/Her Crew........... 1 2 3 4 5

Comment:_____

Employee Summary

_____

_____

_____

_____

_____

Summary of Evaluation # of 1's_____ # of 2's_____ # of 3's_____ # of 4's_____ # of 5's_____

Scale: 1-20 = Unsatisfactory 21-40 = Marginal 41-60 = Satisfactory 61-80 = Superior 81-100 = Distinctive

(A)

I. EMPLOYEE INFORMATION

Name: Title:

Department: Division:

The employee should submit the self-evaluation to the supervisor by January 30th.

II. ASSESSMENT OF EMPLOYEE PERFORMANCE AND GOAL ATTAINMENT

A. PERFORMANCE EXPECTATIONS - *Review the essential functions on the job description. Then consider how well the expectations listed below are met within the context of those essential functions. If "meets expectations" is checked, supporting comments are welcome, but not required. If you feel that any of the expectations are consistently exceeded, provide specific details to support this conclusion. If "needs improvement" is checked, provide a specific explanation. Identify your comments with your initials.*

If any item is not applicable to the position, put "NA"

Complete numbers 1 – 9 for all full-time employees.

CATEGORY / *EXPECTATIONS*	SELF-EVALUATION	SUPERVISOR'S EVALUATION
1. Knowledge of Work - *Demonstrates understanding of the work assignment, mastery of job skills.*	☐ Meets Expectation ☐ Needs Improvement	☐ Meets Expectation ☐ Needs Improvement
Explanation/Support:		
2. Dependability - *Can be relied upon to meet work schedules, job responsibilities, and commitments.*	☐ Meets Expectation ☐ Needs Improvement	☐ Meets Expectation ☐ Needs Improvement
Explanation/Support: Meet deadlines. Prepared and ready to start classes on time. cs		
3. Efficiency - *Prioritizes work responsibilities and utilizes time accordingly.*	☐ Meets Expectation ☐ Needs Improvement	☐ Meets Expectation ☐ Needs Improvement
Explanation/Support:		
4. Quality of Work - *Work is thorough, accurate, and meets specified standards.*	☐ Meets Expectation ☐ Needs Improvement	☐ Meets Expectation ☐ Needs Improvement
Explanation/Support:		

CATEGORY / *EXPECTATIONS*	SELF-EVALUATION	SUPERVISOR'S EVALUATION
5. Works with Others - *Contributes to the success of the department and division through effective teamwork.*	☐ Meets Expectation ☐ Needs Improvement (Note: If you are a supervisor, you can skip this item and refer to # 3 on the Supervisor Supplement)	☐ Meets Expectation ☐ Needs Improvement (Note: If this employee is a supervisor, you can skip this item and refer to # 3 on the Supervisor Supplement)
Explanation/Support:		
6. Initiative - *Displays energy and determination in overcoming obstacles, solving problems, and keeping the work flowing.*	☐ Meets Expectation ☐ Needs Improvement	☐ Meets Expectation ☐ Needs Improvement
Explanation/Support:		
7. Planning - *Systematically plans work assignments resulting in minimal delays, waste and duplication of effort.*	☐ Meets Expectation ☐ Needs Improvement (Note: If you are a supervisor, skip this item and refer to # 2 on the Supervisor Supplement)	☐ Meets Expectation ☐ Needs Improvement (Note: If this employee is a supervisor, skip this item and refer to # 2 on the Supervisor Supplement)
Explanation/Support:		
8. Judgment - *Actions and decisions are appropriate, based on sound reasoning and common sense.*	☐ Meets Expectation ☐ Needs Improvement	☐ Meets Expectation ☐ Needs Improvement
Explanation/Support:		
9. Safety - *Through example, emphasizes the importance of a workplace that is safe and secure. Follows established safety practices while working and, when applicable, wears personal protective equipment. Attends safety training programs related to the position.*	☐ Meets Expectation ☐ Needs Improvement	☐ Meets Expectation ☐ Needs Improvement
Explanation/Support:		

Complete numbers 10, 11, and 12 for Curriculum Faculty, Continuing Education Instructors, and Staff with Instructional Responsibility ONLY.		
10. Knowledge of Subject(s) - *Demonstrates expertise in the assigned subject matter and stays current in the field.*	☐ Meets Expectations ☐ Needs Improvement	☐ Meets Expectations ☐ Needs Improvement
Explanation/Support:		

(B)

FIGURE 7-7 Evaluation forms from three organizations are shown.

ANNUAL COMPANY OFFICER PROFESSIONAL DEVELOPMENT REVIEW (PDR)

PART B – SECTION 1 – SUBJECTIVE SUMMARY SHEET

SUBJECTIVE CORE COMPETENCY	RATING	MULTIPLIER	TOTAL
1. Courteous and helpful to the public	_____	1	_____
2. Cooperates with co-workers, good team player	_____	1	_____
3. Willing and able to communicate openly	_____	1	_____
4. Respectful and tolerant of others	_____	.5	_____
5. Remains positive and optimistic	_____	.5	_____
6. Possesses required education, training and skills	_____	1.5	_____
7. Emphasizes quality in work activities	_____	1	_____
8. Completes assignments in a timely manner	_____	1	_____
9. Exercises good judgment	_____	3	_____
10. Complies with adopted City and Department policies	_____	.5	_____
11. Remains calm and in control during stressful situations	_____	1.5	_____
12. Careful with City property/equipment	_____	1	_____
13. Complies with regular work schedule	_____	.5	_____
14. Stays on task without being distracted	_____	.5	_____
15. Makes good, productive use of time	_____	.5	_____

"Our Quality of Service – Your Quality of Life"-   *FY 2003-2004*

ANNUAL COMPANY OFFICER PROFESSIONAL DEVELOPMENT REVIEW (PDR)

	RATING	MULTIPLIER	TOTAL
16. Performs assigned tasks with minimal supervision	_____	.5	_____
17. Responds to requests from supervisor	_____	1.5	_____
18. Seeks additional duties and responsibilities	_____	.5	_____
19. Willing to offer suggestions or new ideas	_____	1	_____
20. Operational ability	_____	1.5	_____
		TOTAL	_____

DIVIDED BY 20 = _____ =
TOTAL CREDIT EARNED FOR THE SUBJECTIVE RATING SECTION

All support documentation for ratings other than "2"
in each category shall be provided and approved.

"Our Quality of Service – Your Quality of Life"-   *FY 2003-2004*

(C)

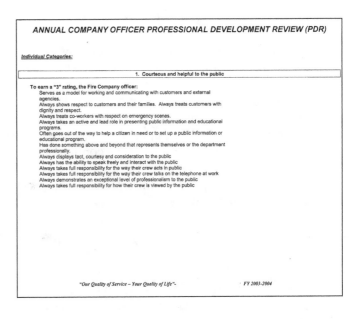

ANNUAL COMPANY OFFICER PROFESSIONAL DEVELOPMENT REVIEW (PDR)

Individual Categories:

1. Courteous and helpful to the public

To earn a "3" rating, the Fire Company officer:
   Serves as a model for working and communicating with customers and external agencies.
   Always shows respect to customers and their families. Always treats customers with dignity and respect.
   Always treats co-workers with respect on emergency scenes.
   Always takes an active and lead role in presenting public information and educational programs.
   Often goes out of the way to help a citizen in need or to set up a public information or educational program.
   Has done something above and beyond that represents themselves or the department professionally.
   Always displays tact, courtesy and consideration to the public
   Always has the ability to speak freely and interact with the public
   Always takes full responsibility for the way their crew acts in public
   Always takes full responsibility for the way their crew talks on the telephone at work
   Always demonstrates an exceptional level of professionalism to the public
   Always takes full responsibility for how their crew is viewed by the public

"Our Quality of Service – Your Quality of Life"-   *FY 2003-2004*

FIGURE 7-7 Continued.

Objectives of member appraisal systems:

- To provide feedback on the member's performance
- To serve as a foundation for member guidance to continue the development of the member
- To document the work of members with talent for positions of greater responsibility
- To justify adjustments in compensation and position classifications

There are a number of systems for evaluating members: Some are very complicated; others are easy to use. Some have a series of questions addressing various traits of the member's performance. Others simply ask for a pass/fail assessment. Whatever system is used, all employee evaluations should follow both the department's guidelines and the city's or town's guidelines. Let us briefly examine several of the more common systems (see **Figure 7-7**).

Graphic Rating System

In this case, the evaluator marks the box that best describes the member's quality of work, attitude, cooperation, and so forth, not unlike a report card. A problem with this system lies in the evaluator's interpretation of the words. This problem is compounded by the fact that many new evaluators lack experience in the system and tend to rate people with excessive leniency or strictness, based on their own standards. Even when the evaluation system suggests a distribution of marks, for example, 10 percent are outstanding, 20 percent are above average, and so forth, the average company officer is evaluating three or four people, all of whom are at different levels of professional growth and maturity. These factors make this system very difficult to use.

Rank Ordering

Another commonly used system is a process of rank ordering. In this system, supervisors are asked to rank their members from good to bad. This can be difficult, but most supervisors can pick out the top 10 percent and the bottom 10 percent, and that may be all that we are interested in. Again, because of the relatively few number of members under the command of any one supervisor, and because these members are most likely at different points in their professions, this process is also difficult to use.

COMMON RATER ERRORS

- *Halo effect.* The halo effect is based on a rater's tendency to overweight certain factors among the overall information available. This results in a score that is distorted either too high or too low.
- *Central tendency.* The opposite of the halo effect, here, the rater tends to put scores in the center of the scale. This type of error defeats the purpose of the process.
- *Contamination.* A rating score that is based on information that is not relevant to the performance standard. This raises questions about the integrity of the entire rating process and could well lead to scores that are not indicative of the member's worth.
- *"Like me" syndrome.* In this situation, the raters use themselves as the benchmark, rather than using a performance standard. Raters should be measuring performance against common standards, using job descriptions and similar objective criteria.

Behaviorally Anchored System

Another common system is the behaviorally anchored system. In this case, brief descriptions define the level of competence in a range of performance measures.

There are several advantages of this system. First, it asks the evaluator to assess the performance against the standard, something that should be taking place in all systems. In this case, that is reinforced because the levels of performance are clearly stated. Second, this clear listing of performance parameters helps reduce errors introduced by rater bias, making the system more creditable.

Most of the systems ultimately reduce the words to numbers, resulting in a final score. Many members and many evaluators see the score as the overall outcome of the process and lose sight of the fact that the performance appraisal system is meant to be a performance improvement tool.

In many cases, developing scores is necessary. For example, in the military, performance, as documented on semiannual performance evaluation reports, is a major criterion for selection for promotion, and for officers is the major criterion for selection for promotions or selection for graduate school.

Another approach is to eliminate the score concept all together. In many cases, the members are not in competition for promotion, and even if they are, their past performance is but a small part of the overall selection process. In a majority of the cases, the real purpose is to identify those who are well above average and all who are well below average. The well-above-average members are the rising stars of the organization. They should be identified and given an opportunity to strive for positions of greater responsibility. Those who are well below average must also be identified; their deficiencies must be pinpointed, and they must be informed that their work is not up to standard. Additional training, coaching, and counseling may be indicated.

Some would argue that this is ultimately the role of the member evaluation process. Identify those who are not doing their fair share, address the problem, and take appropriate corrective action. When such action fails to bring improvement, it may be necessary to terminate that member. We think that a better approach would be for supervisors to do their job and address any deficiencies in performance early on, long before it surfaces in the member performance evaluation system.

- Performance appraisal should be a continuous process.
- Continually observe and monitor members' performance.
- If the member is doing a good job, let the member know that.
- If problems develop, let members know that, too. In this case, it is even more important to discuss with the member the nature of the problem. Allowing the inadequate activity or attitude to continue could lead to a significant performance failure during an emergency situation. Such action could lead to an accident, with injury or worse.
- Likewise, do not save up negative performance observations for the appraisal interview, for all the reasons listed in the preceding paragraph.
- Implement a regular process of systematic performance review so that there are no surprises at the appraisal interview. This is a very simple step, and yet it is the one that leads to the greatest anxiety. Many members (and many supervisors) dread dealing with the surprise bit of bad news at the interview. Deal with issues as they arise.

The performance appraisal interview is the primary formal feedback members receive on their work. In many organizations, it is (unfortunately) the primary means by which members are told what (and how) they need to improve. Regardless of their role, performance appraisal interviews are an important step in the overall process of member development.

The interview should include both positive and negative feedback. Negative feedback does not mean that one has to look for something wrong in every member. Many are doing superior work in all areas. But almost everyone has some area of weakness, especially when compared to the relative strengths of other performance areas. In such cases, the supervisor should help guide the member to continued improvements that will produce even greater amounts of excellent work.

Many members complain that the interview is really one-sided, because the supervisor does all the talking. Generally, the better approach is for the supervisor to be less dominant, talking less and listening more. This means that the supervisor has to let the member do a lot of the talking. For this to be effective, the member has to have adequate notice of the pending interview. Another issue is that the member may not agree with the contents of the evaluation. If this is the case, the fire officer must follow the department's guidelines for documenting the disagreement. The next step is for the evaluation to be reviewed by the next link in the chain of command. The outcomes from this process are not healthy for the organization because in essence someone is going to lose. The fire officer's evaluation points could be overruled, which can cause conflict, or the member's disagreement could be overruled, and that can also cause conflict. It is better served for the fire officer and the member to discuss the issues and work it out among themselves.

Ideally, interviews should be scheduled a week or two in advance. (In some organizations, members are given the option of picking their own time within an established framework. For example, members may sign up for an interview at any time during the afternoons of the following week.) At such interviews, the supervisor should ask members to evaluate themselves. Then, the member and the supervisor can compare their results. This makes the member a more equal player in the interview process and reinforces the supervisor's role as a coach and member developer.

Regardless of the approach, all such interviews are important tools. As such, both the member and the supervisor should be well prepared. The interviews should be scheduled so that there is complete privacy, free of interruptions, and that there is no sense of hurry. Instituting these systems and rules properly would help relieve the sense of dread that arises during member evaluation times.

OFFICER ADVICE

Make a habit of noting something about the performance of every member on a regular basis.

Writing Member Evaluations

Many find that writing the member evaluation is an arduous task. Here are several suggestions:

First, make a habit of noting something about the performance of every member on a regular basis (see **Figure 7-8**). During each work cycle or work week, a supervisor should have seen something good, or bad, about each member. There are many different types of evaluation forms; therefore, it is important for the fire officer to research the forms used by the department.

Remember the lessons of *The One Minute Manager:* Look for people doing something right. We often remember to record shortcomings; make it a habit to record accomplishments, too. And remember that when you see something that requires corrective action, you should counsel the individual right away. There should be no surprises at the evaluation session. Unlike most records within the fire department, employee evaluation records are not available to the public. All employee evaluations should be recorded and placed within their

LOG FOR OBSERVING PERFORMANCE		
Employee's Name:		
Supervisor's Name:		
Date	*Behavior Observed*	*Discussed (Yes or No)*

FIGURE 7-8 Good supervisors have an organized system that simplifies the recording of observations of member performance.

STEPS FOR AN EFFECTIVE EVALUATION REVIEW

A good evaluation interview starts with a review of the significant accomplishments of the period. It should focus on the positive accomplishments. Good news is always easier to deliver and receive. Every counseling session, even for the best members, should identify areas where improvement is desired. It may be a matter of improving performance levels in existing skills, or it may be a matter of motivating the member to move on to new areas. Give your members a specific target to aim for. Your expectations become their goals for the following reporting period.

When the interview is over, the member should have a clear understanding of the completion to the following sentences:

- My leader wants me to do more _____ _____
- My leader wants me to do less _____ _____
- The part of my job that matters most is _____
- Excellent performance in my job requires _____
- I should learn to take care of _____ _____
- I think that my leader would say that my performance is _____

The member should be motivated to set some goals, too. Examples might include the following:

- I want to do more _____ _____
- I want to do less _____ _____

When the member and the supervisor concur in this process, they are almost certain to see some progress. It may be necessary to compromise on a few of the items, but remember that your job as supervisor is to motivate your members to their full potential.

To motivate the member, it is recommended to come up with a succession plan for him or her. A succession plan takes key individuals and starts training them for the next level within the fire department. Most departments today use some form of development model for their employees. The U.S. Fire Administration has developed the Fire and Emergency Services Higher Education (FESHE) Program. This program establishes an organization of post–secondary institutions to promote higher education and to enhance the recognition of the fire and emergency services as a profession to reduce loss of life and property from fire and other hazards. Fire officers should always keep in mind that part of their job is to train the people who work with them for the next level within the department. Train them to do your job.

During a good evaluation interview, it is important to let the subordinate talk too. Many members, especially younger ones, are silent during these interviews, but they should be encouraged to express their feelings and views on matters pertaining to their own performance and goals as well. Frequently members are their severest critic; they overestimate their shortcomings while underestimating their own accomplishments.

At every counseling session or evaluation interview, you should be sure to reinforce your member's contributions with compliments about what was done well. Show confidence and express personal pleasure in the member's abilities and contributions. Always give the member an opportunity to ask questions or speak freely before the meeting is ended.

respective personnel files. The law requires that all evaluations must be maintained and documented.

OFFICER ADVICE

Measure your members against a fixed and known standard.

Through these recorded observations, you will likely have plenty of quality material to use on the evaluation itself. In fact, you will probably have too much. That is okay; you can select the best and most representative material and write great evaluations filled with specific details. And you will not have to try to remember 6 month's or a year's worth of accomplishments as you try to write the report. Once you

get in the habit of doing this recording, the process becomes easy. Several examples of the forms that you will have to fill out are provided (see **Figure 7-7**).

OFFICER ADVICE

Let your members know what you expect, and when they give a little bit more, give them some recognition for their effort.

Measure your employees against a fixed and known standard. Several tools are helpful in benchmarking performance levels. First, there are national performance standards for firefighters, emergency medical technicians (EMTs), and the various technical specialties in the fire service. Use these as a measurement gauge in determining competence. Most departments add some additional performance requirements. Keep these in mind and let your subordinates know that you are using these as the criteria for performance measurement.

OFFICER ADVICE

A good evaluation interview starts with a review of the significant accomplishments of the period.

In addition, many departments have professional development programs for their members. These programs usually involve both training and educational requirements to help members increase their skills and abilities in the workplace. Successful accomplishment and application of these professional development requirements should be encouraged and formally recognized.

OFFICER ADVICE

Every counseling session, even for the best members, should identify areas where improvement is desired.

PUT PEP IN YOUR PERFORMANCE EVALUATIONS

Sometimes a new name helps make an old problem easier to manage. We know of one organization that calls its annual performance review process the Performance Enhancement Plan, or PEP. The process is similar to what we are describing here but is structured so that everyone shares the work more equally.

The members start the process by evaluating their own work and completing their portion of the report form, including both the evaluation of specific performance items and the narrative. The form is then forwarded (sometimes electronically) to the supervisor, who reviews what the member has prepared. Supervisors can concur or differ with (but not change) the marks and remarks, and then supervisors add their own comments. There are only two choices in marking: meets expectations or does not meet expectations. During the performance interview, any differences are reconciled, overall performance is discussed, and goals for the coming year are identified and recorded.

GIVING RECOGNITION

We may differ in the way we want to be recognized for our work, but recognition is one of our basic needs. Maslow, Herzberg, and others have identified the need for recognition. If you only talk to your members when their performance is below acceptable standards, you lose the value that comes from positive recognition. Let your members know what you expect, and when they give a little bit more, give them some recognition for their efforts (see **Figure 7-9**).

Giving recognition is easy; however, there are several pitfalls. Do not be overly zealous in giving praise. If you compliment every act, you eventually dilute the recognition process. On the other hand, it is also easy to fail to give recognition when it is due or to acknowledge that the work is worthy of special recognition.

There will be times when exceptional performance should be recognized with an award from the department. The award may be for valor at the scene of an emergency, or it may be for exceptionally meritorious service during nonemergency activities. The recognition may come in the form of a medal, citation, or other formal presentation (see **Figure 7-10**). As the company officer, you should remember that the awards process starts with you; if you do not take the initiative to recommend the award, it is unlikely that anyone else will. There will be plenty of opportunity for the department's senior staff members to pass judgment on the merits of the award. They will never have that opportunity if the member's act or performance is not brought to their attention by you—the company officer.

Recognition does not have to be a medal or pay raise. It might be just a comment, a word of appreciation, or only a nod. At times, such gestures are as good as any tangible reward you could provide. Whatever form the recognition may be in, make sure that it is sincere. One of the basic tenants of the Dale Carnegie program is "Show sincere, genuine appreciation."

Recognition should be personal. It should identify the individual, the act, and that you are personally grateful for the work. We usually teach students to try to avoid starting a sentence with "I." This is one of the exceptions. It means a lot when you say something like, "I was very pleased with the way that you. . . ." Think about how your team member will feel!

As we have already stated, recognition fulfills a basic human need. Recognition motivates members to want to do even more and to be even more effective.

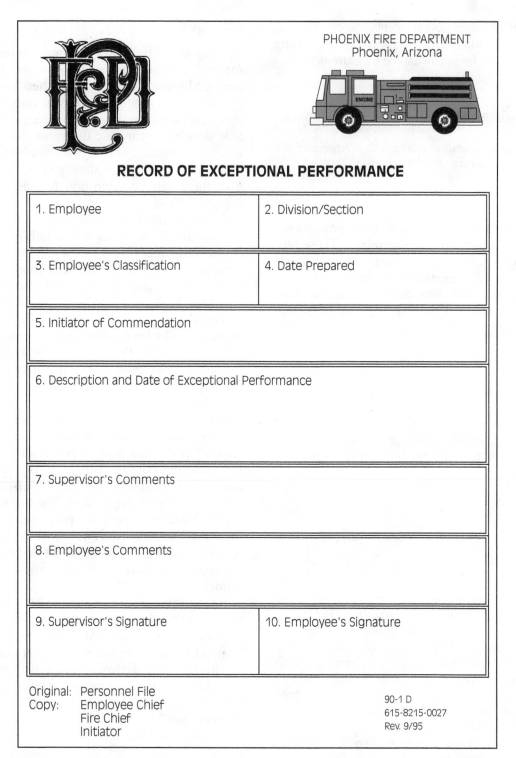

PHOENIX FIRE DEPARTMENT
Phoenix, Arizona

ENGINE

RECORD OF EXCEPTIONAL PERFORMANCE

1. Employee	2. Division/Section
3. Employee's Classification	4. Date Prepared

5. Initiator of Commendation

6. Description and Date of Exceptional Performance

7. Supervisor's Comments

8. Employee's Comments

9. Supervisor's Signature	10. Employee's Signature

Original: Personnel File
Copy: Employee Chief
           Fire Chief
           Initiator

90-1 D
615-8215-0027
Rev. 9/95

FIGURE 7-9 Awarding certificates of accomplishment or providing notes of exceptional performance are an important part of the recognition process.

FIGURE 7-10 A formal award provides highly visible recognition of a member's accomplishments. *(Photo courtesy of Fairfax County Fire and Rescue Department)*

QUALIFICATIONS OF A FIRE OFFICER

It is by no means enough that officers of the fire service be capable firefighters. They must be that, of course, but also a great deal more. They must have a liberal education, refined manners, punctilious courtesy, and the nicest sense of personal honor.

They should be the soul of tact, patience, justice, firmness, and charity. No meritorious act of a team member should escape their attention or be left to pass without its reward, even if the reward is only a word of approval. Conversely, they should not be blind to a single fault in any team member, though at the same time, they should be quick and unfailing to distinguish error from malice, thoughtlessness from incompetence, and well-meant shortcomings from stupid or headless blunders.

With apologies to John Paul Jones, USN.

TAKING DISCIPLINARY ACTION

Having talked with a member about a performance problem and having seen no change, your next step may be to take some form of **disciplinary action**. If you do not take such action, the problem will likely continue. In addition, if you do not take action, you will lose part of your credibility and part of your authority as a leader.

When taking disciplinary action, think of it as another way of improving member performance and behavior. This is usually far more comfortable for both you and the member than thinking about disciplinary action as a form of punishment.

When counseling or administering punishment, remember the following:

- Focus on the behavior, not the individual.
- Help the member maintain self-esteem.
- Work to maintain a constructive relationship.

OFFICER ADVICE

When taking disciplinary action, think of it as another process for improving member performance and behavior.

Make an effort to be consistent and fair in the treatment of members over a period of time. This is often difficult. No two members are the same, and no two offenses are quite the same. However, you should consider several factors when deciding whether to use discipline. First, the discipline should have some relation to the offense. Stealing and dishonesty are far more serious than arriving 5 minutes late for duty. You should also take into consideration members' recent work, their past disciplinary record, and any mitigating circumstances. Finally, some organizations require that the company officers inform their supervisors of the action taken. Find out what is appropriate.

ETHICS

Make an effort to be consistent and fair in the treatment of members over a period of time.

Discipline Procedures

Most organizations have an established discipline procedure. In most cases, there is a progressive system that generally follows the steps of oral reprimand, written reprimand, suspension, demotion, and finally, termination. Let us briefly look at each of the steps and see how they might be effectively used to help improve a member's performance.

Oral Reprimand

An **oral reprimand** is much like a private counseling session between the supervisor and the member, except that it is likely that the unsatisfactory conduct in question has been discussed previously. You should let the member know that the conversation constitutes an oral reprimand and that the next step in the progressive disciplinary process is a written reprimand. The nature of the unsatisfactory behavior problem and the corrective action that is desired should be clearly identified during this session. Try to get the member to understand the nature and consequences of the actions at issue. Ask the member to help spell out the corrective action. Although no formal report is required, it is a good idea to make notes of the conversation.

Written Reprimand

This second step in the disciplinary process indicates that the unsatisfactory behavior has continued. As the company officer, you should administer the **written reprimand.** Lacking any established procedure to the contrary, a brief memo to the member that identifies the nature of the unsatisfactory behavior, the fact that previous discussion has been had, and the fact that the unsatisfactory behavior has continued should be spelled out, along with the desired corrective action. A sentence or two on each of these topics is adequate, often in memo form. The member should be asked to sign a copy of the reprimand acknowledging its receipt. Give the original to the member. Company officers should keep signed formal documentation for oral or written conversations.

Transfer

The next step might be a possible **transfer.** A transfer may provide a solution to the problem, not just shifting the problem to someone else. Transferred members are entitled to the same fair treatment you would give any other member, but at the same time, the new supervisor should be aware of the prior discipline problem.

Suspension

At this point, we are well beyond the punishment authority of most company officers in the fire service. **Suspension** is a serious punishment and is usually administered by a senior fire department officer. Disciplinary procedures vary, but most departments provide for a formal hearing, usually in the more senior officer's office, with the company officer and the member each having an opportunity to present their version of the story. Most of the real evidence will be in the form of written records showing that the unsatisfactory conduct occurred and that previous disciplinary action has not been effective.

Again, we should emphasize that the purpose of the disciplinary process is to improve the member's performance or conduct. If that occurs, the punishment is effective. There are plenty of senior fire officers on duty today who have been on the carpet at some time during the early part of their profession. The punishment, coupled with a natural maturing process, helped them to correct the problems they may have had in their younger days.

NOTE

The purpose of the disciplinary process is to improve the member's performance or conduct.

Demotion

The punishment of **demotion** has a lasting impact. It takes away both pay and power. Obviously demotion is not very effective in the case of firefighters, because they are already in an entry-level position. As with suspension, the disciplinary procedure is usually conducted by a senior fire department officer. Make a detailed record of the procedure.

Termination

Termination is used when all other steps have failed. Termination is costly for both the member and the organization. A terminated member must be replaced. A replacement must be recruited, selected, trained, and brought into the organization. However, in spite of these considerations, termination may be appropriate in some situations.

For serious offenses, one or more of these steps may be omitted.

WHEN USING DISCIPLINE

- Focus on taking corrective action, not on administering punishment.
- Be sure to coach and counsel before using discipline.
- When disciplining, be fair, be firm, and DO care.

Fundamental to the process is the fact that before any disciplinary action can be implemented, the supervisor must first give a warning. Members must be informed clearly that certain actions will result in disciplinary actions. This is a very important step. It is not the members' fault if they do not know the company's rules, ethics, and standards. It is the supervisor's responsibility to educate and inform. It is easier to accept discipline if the rules and standards are clearly stated beforehand and understood.

In order for this to be productive, the supervisor must take timely disciplinary action. The sooner the discipline is imposed, the closer it is connected to the violation. This enables the member to associate the discipline with the offense, rather than with the dispenser of the discipline.

This is important, because you do not want your team members to associate the punishment with you; you want them to know that they broke the organization's rules, and as the supervisor, it is your duty to punish. You are there to help them and guide them to perform to the best of their abilities to the betterment of the association. Your discipline has nothing to do with the supervisory relationship; it merely establishes standards to work by. As the dispenser of the discipline, you should react as soon as possible to the violation.

APPLYING DISCIPLINE

Punishment should have the following characteristics:
- Be associated with a specific unsatisfactory conduct or behavior
- Occur soon after the offense
- Be firm and fast
- Be uniform among all persons
- Be consistent over a period of time
- Have informational value
- Clearly identify the problem and the desired corrective action
- When possible, be administered by the same person who can reward good performance, usually the company officer

TAKING CARE OF YOUR TEAMMATES

Fairness for One and All

Throughout the process of developing your team, it is important to remember that everyone should be treated fairly. People have different needs and goals. Some have a high need for recognition. Others have a greater need for affiliation. Recognize these needs in your teammates. Finding the best way to motivate each of your teammates is one of the most important things you do in being an effective leader. Use the most effective approach on each teammate.

OFFICER ADVICE

Finding the best way to motivate each of your teammates is one of the most important things you do in being an effective leader.

We have discussed the importance of setting goals. Be sure that the goals are reasonable and fair for each teammate. Although all firefighters should have the same skills, some will be better in some things than others will. Some learn some skills faster than others do. However, it should all balance out in the final analysis. Work with individuals to help them overcome their weaknesses, and recognize their strengths by letting them do that which they do best, at least some of the time.

Encourage all of your teammates to participate in the goal-setting process for themselves and for their company. Get them involved in working together to help the company reach its full potential. Getting teammates involved is called **empowerment.** Empowerment allows members to have a feeling of ownership in the organization.

Maintaining fairness has taken on new dimensions with the increasingly diverse workforce we see in today's fire service. Diversity comes in many dimensions. Variations in race, gender, age, experience, and other factors tend to make each member unique. Leading such a diverse group presents it own special challenges. The key to understanding and leading a diverse workforce lies in understanding and flexibility. Take time to understand each teammate's needs. Try to be flexible in accommodating those needs.

Along with being fair, the supervisor must take the lead in making sure that the workplace is comfortable for all who work there. Set a personal example and encourage others to imitate your lead in the following:

- Avoid disrespectful slang terms and names.
- Eliminate offensive pictures, cartoons, and other printed matter from the workplace.
- Avoid ethnic and sexist jokes.
- Speak calmly and slowly.
- Listen equally to each teammate.
- Spell out those unwritten rules.

Dealing with Conflicts

With all the demands upon them, your people are under considerable pressure. In the fire service, one may see more of life's raw edges in a day than most citizens see in a lifetime. We expect our firefighters to treat people and to go places that most citizens would like to avoid. We expect our drivers to operate on busy streets, getting us safely and quickly to the scene, without the risk of an accident. And we ask our teammates for increasing efforts around the fire station and in the community. It is no wonder that our people are under pressure.

Pressure can create **conflict.** Often times, these conflicts stem from personnel behavior problems. When people work together under pressure, conflict is natural. Conflict can be healthy. It implies that people are working hard and trying to be successful in meeting their goals, both their personal goals and those of the organization. Conflict can be unhealthy when it involves personality clashes. When this type of conflict occurs, it needs to be quickly resolved—another job for you, as the company officer.

NOTE

When people work together under pressure, conflict is natural.

As an effective company officer, you should be able to detect a conflict arising while it is still in its incipient stages. Although it is not always easy, you should work with both parties to understand what is happening and resolve the problem. Here are some suggestions:

- It takes two (or more) to have a conflict. Get the parties together, tell them that you think "we have a problem," and ask the parties to help you find a solution.
- Try to get the parties to acknowledge that there is a problem and that the conflict is causing a problem in the workplace.
- Work with the involved parties to calm the situation and to solve the problem.
- When a solution is found, get a verbal commitment from each party on how each is going to support the solution. Make sure that everyone understands the other's position.
- Set a specific time to review progress on the situation.

Employee Complaints and Grievances

It is natural for team members to complain about things in the workplace, especially in fire stations. When you take four or five bright and active young people and lock them up together in a building away from their own homes for 24 hours, and complaints are inevitable. Some of this is just ordinary grousing, but sometimes the issues and the people involved can get quite serious.

A first expression of dissatisfaction might be considered a **gripe.** Prolonged or repeated dissatisfaction over the same topic might lead to a **complaint.** The difference is not important. However, the next step, a **grievance**, is serious and can occur when the complaining becomes significant or does not bring relief. All three deserve your attention as the company officer.

If someone complains about the weather, there is not much you can do about it. If someone complains about the coffee pot not working, that may require some action. You need not resolve every gripe or complaint personally, but you should provide a healthy atmosphere where perceived problems can be dealt with in a healthy and respectable manner. Understand that perception is reality to those who come to you. You may not perceive the issue to be a problem, but if they come to you, obviously they think it is and it should be taken it seriously.

Complaints should be channeled into positive action. You should expect adults to solve some of their own problems and to try to get help from others when this does not work. When neither of these works, members may turn to their supervisor for help. You should set some ground rules here. If a team member complains about some work-related item at the dinner table or while the crew is engaged in recreational activity, you can say, "I would like to hear more about that. Please come see me about that when we can discuss this in

greater detail in private." You are politely saying; lets discuss this at the appropriate time and place. When members have a problem, they should take the initiative to discuss the issue with the officer in a private setting, not in front of the crew, especially newer firefighters.

NOTE

Complaints should be channeled into positive action.

Sometimes, just letting the member talk over the problem helps to resolve the issue. At other times, you may suggest some form of action to help members to help themselves. And at times, you will have to take action to help resolve the problem. Listening to and acting on material complaints will improve the workplace environment. At the same time, you do not want to pamper our people by listening to and attempting to accommodate their every whim.

Solving problems can be difficult if there is no outlined plan. Although there are many types of problem-solving processes, the steps listed next contain the basic elements for solving most problems that occur in and around the firehouse.

1. *Problem definition:* What exactly is the problem? Writing down the issue will validate that you are focusing on the problem at hand and makes sure that nothing is overlooked. In the problem definition, the fire officer should also list what he or she wants to achieve through this process.

2. *Problem analysis:* Break down the problem into the following:
   - Where are we?
   - What is the current situation?
   - Who or what is involved?

   By breaking down the problem, a fire officer can focus on the root causes and determine if the problem is affecting the organization as a whole or if it is an individual or isolated problem.

3. *Analyze the solutions:* This step turns into a positive-versus-negative competition. The fire officer should brainstorm and write down all of the possible solutions to the problem. Then the officer should take each proposed solution and write down the positive and negative points that are relative to each solution.

4. *Select the best solution(s):* Look through all of the factors, both positive and negative. While this process is often a "gut instinct" choice, it is often found that this choice is not the most obvious solution. After this process is over, fire officers may find that they have numerous solutions or they may find that they do not have a solution at all. In the case of numerous solutions, rank them in order of preference and then proceed with the top ranked. If no solution is found,

then the process starts back over. Make sure that the problem has been correctly defined.

5. *Move forward:* Institute the solution and monitor the outcomes.

NOTE

Sometimes just letting the member talk over the problem helps to resolve the issue.

Company officers must be familiar with their organization's **grievance procedure**. Where members are represented in a bargaining unit by union, the grievance procedure is part of the union contract, and contracts usually allow the members to have someone from the union with them during any action that may be taken.

If member frustration has escalated to the grievance level, it indicates that there is a problem that members cannot resolve, that they are unhappy with the situation, and that they have decided to make a formal issue of their complaint. It may also indicate that there is a communications problem in the organization. Most organizations have a formal grievance procedure; the procedure forces the issue to be heard in a timely manner.

Dealing with the Disenchanted Firefighter

Some fire companies have a seasoned firefighter who has developed a negative attitude. This firefighter may be sullen or even openly hostile. When he or she does speak, he or she complains a lot, but offers few suggestions and seldom takes the initiative to solve problems. He or she also tries to do as little work as he or she can. Not only does he or she set a bad example, his or her negative attitude can become contagious if left unchecked.

As a supervisor, you should take the lead and try to gradually restore the disenchanted firefighter. You may want to start with improving his or her attitude. Give him or her a specific task that he or she can accomplish. When it is successfully accomplished, give him or her some recognition for effort. Repeat the process. By taking the rehabilitation process in small increments, you may be able to turn the firefighter with a negative attitude into a positive performer.

Employee Assistance—When Is It Warranted?

Employee Assistance Programs, or EAPs, were brought up earlier in this chapter. While mentioned previously, it is important to take a broader look at when these types of programs should be recommended. EAPs were developed to help employees deal with personal or work-related issues that would otherwise adversely affect their work performance and/or health. Many of the issues that personnel face deal directly or indirectly with stress. Narrowing down all of the indicators of stress would prove to be impossible; therefore, we will focus on work-related stress indicators. Work-related stress is a destructive physical reaction that occurs when the job demands do not match the ability or capabilities of the worker. Work-related stress often results from the demands of the job as well as the working conditions that firefighters face. There are many different perspectives on how to deal with work-related stress problems. One such perspective says that the differences in individual characteristics are important in predicting whether certain job conditions will result in stress. In looking at the characteristics of a firefighter and the job conditions that he or she will face, we can see a rapid road to stress overload. The problem is that the majority of firefighters bottle-up these feelings for fear of being looked down on by their peers. Therefore, it is imperative that fire officers learn the signs of stress so that they can encourage the member to seek professional assistance. Some of the more obvious signs include mood swings, sleep disorders, eating problems, and relationship issues. Company officers will probably be the first people within the department to notice these signs. This is why it is important for the company officer to let subordinates know that they can always come and talk no matter what the circumstances.

Ideally, members will refer themselves to an EAP long before their work is affected. However, this is rarely the case when dealing with emergency service workers. If job performance has suffered and the employee does not seek help on his or her own, it is suggested that the company officer take the initiative and offer to arrange an appointment with an EAP. This can prove to be a difficult discussion. Therefore, here are some tips on how to handle these types of discussions:

1. Make sure that all performance issues have been properly documented.
2. Tell the employee that you have noticed that his or her performance has been suffering and refer the employee to your documentation.
3. Let the employee know the circumstances for continuing performance problems.
4. DO NOT try to diagnose the problem.
5. Let the employee give his or her view on the issue.
6. DO NOT engage in discussions that deal with personal issues. Focus on work performance issues only.
7. Review the EAP procedures with the employee.

Whether the employee seeks assistance on his or her own or the company officer offers to help, it is important to follow up on all requests for assistance. Documentation must be placed within the members personnel file. These steps should be listed within departmental as well as the city's or town's policies and procedures.

WHAT GOOD LEADERS DO

John Buckman, chief of the Germantown Township (Indiana) Volunteer Fire Department and recent past president of the International Association of Fire Chiefs, notes that good leaders in the fire service, and elsewhere, demonstrate several common and predictable characteristics (there are always exceptions of course):[1]

Good leaders are bright, mentally agile, and alert. Intelligence is the quality that helps one gather information, sort through the information, and make accurate critical decisions in a timely manner.

Good leaders seek out responsibility. First, there are a few dominant people who have the necessary skills to be in charge, the courage to make things happen, and the willingness to accept responsibility when things go wrong. Chief Buckman says these are the qualities of a natural leader. Second, and more commonly, we see a pattern in people who spiral upward into leadership roles. They start early and gradually ascend the leadership ladder with increasingly important positions, more challenging assignments, larger spans of control, more people to supervise, and more power. When departments have professional development programs, planned professional progression provides a succession of gradual and critical development activities that prepare leaders for subsequent positions.

Good leaders are informed. Skills and knowledge create leadership power. Good leaders are constantly renewing their knowledge base in both administrative and operational skills. They are up-to-date on what others are doing. In turn, they are setting a pace that others will try to follow.

Good leaders are good communicators—good speakers, good writers, and good listeners. Since so much of good leadership involves informing and persuading others, those who cannot communicate cannot succeed.

Good leaders also have a sense of humor. Humor is an enormous asset. Fire officers endure trauma on the street and boredom in meetings. Anger and hostility can be present in either setting. Humor, appropriate and timely of course, often provides a much needed relief.

HUMAN RESOURCE MANAGEMENT AT A GLANCE

All of the information in this chapter thus far can be bundled into the human resource management realm. Leading members, developing members, evaluating members, and even taking disciplinary action are all parts of human resource management. To become an efficient leader and manager of human resources, it is important to understand the federal, state, municipal, and departmental human resource management policies, procedures, and laws. One such law that affects the majority of fire service members is the Fair Labor Standards Act (FLSA). Also known as the Wages and Hours Bill, the FLSA enforces the rights and wages of nonexempt employees. It is recommended that all fire officers study the FLSA laws so that they can be ready for questions from members. There have been numerous lawsuits filed between fire departments and municipalities that deal with these laws. The FLSA applies to "any individual employed by an employer" but not to independent contractors or volunteers because they are not considered to be "employees" under FLSA. While there are some exceptions to the FLSA laws, the one that the fire service most often deals with is the "white collar" exemption. This exemption is applicable to those employees deemed "exempt" by the employer. However, courts have overturned this ruling by the employers because they could not prove that the employee fit "plainly and unmistakably" within the exemption terms.

The Occupational Safety and Health Administration, or OSHA, is tasked with preventing work-related injuries, illnesses, and deaths by issuing and enforcing standards for workplace safety and health. The Occupational Safety and Health Act, which created OSHA, also created the National Institute for Occupational Safety and Health (NIOSH), which is the lead research agency for occupational health and safety.

While there are many different laws and standards that deal with human resource management, fire officers are encouraged to study those that pertain directly to themselves and their subordinates. Firefighters are notorious for finding presumed loopholes in just about any law or standard; therefore, it becomes necessary for their officer to have an understanding of these laws and standards before the questions arise. Understand the local policies and procedures at the federal level as well as the state and municipal levels. Know the department's policies and procedures on how to deal with questions and/or complaints.

INTRODUCING CHANGE

It is important that an organization be able to change. If it cannot, it may not survive. Likewise, it is important for individuals to be able to change.

Most change is good; it allows us to keep up-to-date and to use modern ideas and technology to get the job done. As the speed with which we share information and technology increases, so will the rate of change.

FIGURE 7-11 New equipment may break from tradition, and require changes in attitude and procedure.

As one sage put it, "If you think we have had a lot of change in the last 10 years, wait until you see what happens in the next 10 years!" (See **Figure 7-11**.)

We all change during our lifetime, but accepting change is an individual thing: Some do it better than others do. For us as individuals, accepting change is something we can plan, control, and monitor on our own. When it becomes too painful, we can slow down or reverse our decision.

NOTE

Most change is good; it allows us to keep up-to-date and to use modern ideas and technology to get the job done.

Changes in organizations are another matter. Although the change will impact upon us as individuals, just as with our own initiatives, with organizational change, we probably do not have much opportunity to set the pace. We often have little or no knowledge of the change in advance of its implementation, and we have little opportunity to plan, control, and assess its impact. As a result, changes in organizations are much more difficult on the team members.

It is natural to resist change. Our resistance may come from the fact that we are comfortable with the present process. We may not have the training or education to deal with the change or the new process that comes as a result of the change. We are intimidated by the unknown or by the rumors we have heard. All are valid reasons why people resist change. We all want to share the benefits of change, but few of us want to pay the price of changing from the present.

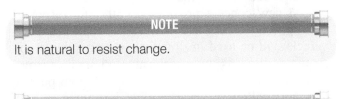

NOTE

It is natural to resist change.

NOTE

Introducing change and managing its effective implementation is a leadership skill.

Although things may be running well at your fire station, there will come a time when you have to make a change. Make sure there is a valid reason for the change. Change that is easy to accomplish or that has the greatest visibility may not be what is really needed. Change may be the result of one of your own ideas, or it may be the result of an initiative that has come down the chain of command. Regardless of its source, it has to be implemented at the station level.

Introducing change and managing its effective implementation is a leadership skill.

Good leadership provides a culture where change is not only welcomed but also required. With good leadership, you should have open lines of communications. You should be encouraging members to offer new ideas about their work and the service they provide to the community they serve. As a supervisor, you should provide a climate where a member's ideas are given serious consideration.

Changes in fire apparatus design are among the most conspicuous changes in the fire service. When one of these vehicles arrives in your community, citizens, as well as firefighters, will quickly note the differences in appearance.

Introducing changes to an organization may be evident to the outsider and may have a profound impact on the way the department serves its community. In Richmond, Virginia, the fire chief was directed to reduce the cost of services. After a careful analysis, he determined that the city needed a new fleet of fire trucks. In May 1997, the city placed an order for over $12 million in new equipment. Included in the package are new quints and new rapid-response rescue vehicles. One might wonder how spending $12 million would save money, but the chief was able to convince everyone, including the firefighters, that these new vehicles could provide better service to the citizens of the city. At the same time, the initial expense would allow for a modest and gradual reduction in personnel, the retiring of an old fleet of vehicles that were expensive to maintain, and the closing of several no-longer-needed fire stations. In the long run (15 years), the department expects to save more than the acquisition cost of the new equipment.

The new concept provides an all-quint response concept for working fire alarms. The rapid-response vehicle will be used in situations when a quint is not required, such as auto accidents and smaller fires, since the rapid-response units have 750-gpm pumps and 500 gallons of water. What is unique is that the quints and rapid-response units will be located together and staffed by the same personnel. The company will determine which unit to take, based on the dispatch information.

Introducing a change of this magnitude requires a great amount of work. You need to have a good concept, to do your homework and to get all the answers, and then to work to win the support of the members and the community.

Let us look at several key ingredients for providing the environment where change is accepted. First, keep your members involved in the process. Let them know that you want to hear about their ideas.

Let them know that you are open to other ways to do things, even where the present system may be of your own creation. We are surrounded with hard working teammates. Some take the job more seriously than do others, but all have the ability to offer ideas on how to do things better.

NOTE

Change introduces uncertainty into our lives and often introduces new events over which we have no control.

Work to overcome the obstacles of change. We usually fear the unknown. Change introduces uncertainty into our lives and often introduces new events over which we have no control. We often do not know all the details when changes occur. We do not know if or how they will affect us. We do not understand why the change has occurred and what future change may be about to occur. All of these anxieties can be overcome with good communications.

When introducing change, keep your team members informed of what is going on, and let them be a part of the process, rather than a victim of its passing. Let them know the time frame and the depth of the impact it will bring. Depending upon the time, circumstances, and conditions involved, you may be able to use one of the leadership styles described in the last chapter to sell the idea, rather than ramming the idea down your members' throats. Let them ask questions. Try to provide honest answers to their questions. Be positive and supportive of both them and the change (see **Figure 7-12**).

You can expect to hear some negative comments. Let those comments be heard and try to respond with honest and positive answers if you can. If you need information from elsewhere in the organization, say so. One of the main components of successfully introducing change is to keep the communications open.

OFFICER ADVICE

One of the main components of successfully introducing change is to keep communications open.

The greatest fear occurs when change involves personnel action. We all like to know what is happening in our lives, especially when we are at work. When people get moved around, when they have their

FIGURE 7-12 New ideas are not always welcomed.

working hours changed, and when they have to take on new duties, they want to know about these things in advance. Try to keep these events to a minimum, but at the same time, let your people know about the necessary events well in advance.

Others you work with—the police, other fire departments, hospitals, and of course the public—may be affected by the change. Let them know what is going on. For most organizations, this is an excellent time to build some good public relations.

Once a change has been introduced, it is always a good idea to follow up, to remain in touch with what is happening in the organization. Review the change and the impact it has brought to the organization. Change seldom comes without a price. With each good event, there are usually some negative side effects. The goal is to have the positive effects outweigh the negative ones. Monitoring and fine-tuning the results of the change will increase the net results (see **Figure 7-13**).

When change is successfully implemented, it is a good time to recognize those who made it happen. Recognize the individual and the team that led the change. Also give some recognition to those who were able to successfully accommodate the change. Use these steps to update, change, or even create a new policy or procedure. The following are some key steps for successfully introducing change:

■ Create a climate where change is welcomed and not feared.

■ Let team members be involved in finding new ways and in implementing new ideas from others.

■ Reduce stress and anxiety by keeping people informed.

■ When introducing change, be positive.

■ Consider the impact of the timing of the change on your team members and introduce the changes as gradually as possible.

■ Follow up and monitor the total impact of the change.

FIGURE 7-13 New ideas should always be given fair consideration. This foam unit is an effective urban interface vehicle.

WHAT'S NEW?

Any list of "new" ideas will bring laughter from some readers. For some, these ideas have been around for so long they are no longer considered new. Others may not even have heard of these ideas or realized their value as applied to the fire service.

A Needs Assessment of the U.S. Fire Service reported the result of a survey of departments on using new technology. They found the following:[2]

- Only 25 percent of all departments have thermal imaging cameras.

- Only 4 percent of the departments have the equipment to collect chemical or biological samples for remote analysis.
- Only 3 percent of the departments have mobile data terminals.
- Only 2 percent of the departments have advanced personnel location equipment.

In general, those departments that reported they did not have the equipment also reported they did not have any plans for getting the new equipment.

REFLECTIONS ON THE DEPARTURE OF OUR LEADER

Our leader is leaving us. I can recall when he arrived. We were concerned, as we always are when a new officer is assigned to the company. What will the new company officer be like? What things will he change? Well, we were soon to find out. First, we noticed how personable he seemed; he seemed comfortable with everyone. As the days passed, we noticed that he made it a point to spend some quality time with everyone. Sure, he spoke with everyone, but, more importantly, he listened to everyone, too. He did not make promises that he could not keep and did not try to solve all of our problems by himself. But he always listened and offered encouragement.

We also noticed that the new officer was refreshingly unassuming. He was not here on some ego trip. He was here because he wanted to be, because he had earned it, because he was qualified for the job, and because he enjoyed being here. He was as sharp as anyone around, but he sometimes admitted he did not know the answer. But he always knew where to go to find it.

He enjoyed the job and maintained a sense of humor. He worked us hard, but he also made sure we had fun. We recently won an award for being the best company in the department. At the award ceremony, our company officer thanked the chief for the award, but he also publicly thanked each of us and made sure that everyone present knew that we had done the work. We excelled because of his leadership; he did not mention that, however.

We also performed well because of his teaching. I guess you might call it mentoring or coaching. He was always concerned about our professional growth and gave us room and motivation to grow. He let us make decisions and take responsibility for our actions. He praised us when we did well. When we did not do as well as he thought we could, he let us know that too, but always calmly, tactfully, and fairly.

Our boss is getting a promotion and a transfer. We will miss him. But we are far better off for his having been here with us these past several years. He helped each of us grow to be better than we were before, and collectively, we are now a far better team—the best one around, in fact! I guess some of his influence will still be here. In many ways, he taught us to think and work as he does. And I think we are getting pretty good at it.

Do you suppose that he was preparing us to take his place?

STILL NOT CONVINCED?

Taking command of a fire department is much like taking command of a ship. As the chief, you are the boss, and you can make a difference. Some choose to be caretakers, and some choose to make improvements. In either case, the vision has to start at the top, but the actual work to implement the change takes place at all levels.

In a book entitled *It's Your Ship*, Michael Abrashoff, a retired navy captain and the former commanding officer of the *USS Benfold*, tells of the changes that occurred while he was aboard. When Abrashoff went aboard *Benfold*, he found a ship with all of the best technology but with an unmotivated crew. Their lack of motivation led to safety and readiness problems. He set about changing that by first working on his own leadership skills and then those of the crew. Within months, he created a crew of inspired problem solvers, eager to take the initiative and ready to take responsibility for their own actions.

The chapter headings read like an action plan for change:

- Take command
- Lead by example
- Listen aggressively
- Communicate with purpose and meaning
- Create a climate of trust
- Look for results, not salutes
- Take calculated risks

- Go beyond standard procedure
- Build up your people
- Improve your people's quality of life

Look at the chapter headings. Do these titles look like what we have been discussing in this text?

This is a book for leaders at all levels. Certainly, it is good reading for fire chiefs, but it is even better reading for fire officers on their way up: These "rising stars" are the ones who will make the changes today and in the future as their sphere of influence increases.

LEADING OTHERS

Tony Tricarico, Captain Squadron 252,
Special Operations Command,
FDNY, New York

As a captain in Special Operations for the FDNY, Tony Tricarico has had the opportunity to work with some of the most experienced firefighters in the country. And being part of that department affords him the opportunity to get cutting-edge training. But when he goes home to Long Island and serves as truck company captain at his volunteer fire department, it is a much different situation. That company is made up of everyone from lumbermen to insurance salesmen; Tricarico is quick to say they are a terrific group, but for as hard as they work, they do not necessarily have the same training opportunities as the FDNY. This requires that Tricarico supervise them a little more closely. That is why he feels it is his responsibility to take his own training experiences and disseminate them to his company.

Not long ago, he was taught an impressively fast and simple way to displace a dashboard, in which the "jaws of life" angled at such a degree that it did not block access to the passenger. Afterwards, he decided to hold a practice training session with his Long Island group and coach them through the process in a hands-on manner.

He saw the virtue in doing this not a few weeks later, when his company was called to the scene of a bad accident. The car had pulled out of an intersection into a 7-lane road—where drivers typically go 70 mph—and was T-boned. The car had spun out, hit a curb, and was in pretty bad shape when the company arrived.

From the size-up, it was apparent to Tricarico that the entire roof would have to be taken off the car, so EMS personnel could get access to the driver. The next step was taking off the driver's side door, and finally, he decided it was necessary to displace the dash, as the patient's legs were trapped underneath it. Supervising the scene, Tricarico guided his group in the action they had practiced, and he was pleased to see that they did exactly as they had been trained. Though the driver did not survive, there was no way they would have been able to free him any more quickly, and Tricarico was proud of his men. Later, he told them so, because he stresses the importance of letting your staff know they have done a terrific job, even when it seems most thankless.

Questions

1. What kind of leadership power does Captain Tricarico have at his volunteer company?

2. Does good leadership transcend the organization? In other words, can an effective leader in one organization take those leadership skills to another organization and be effective?

3. Captain Tricarico notes that he has to supervise his volunteer personnel more closely than he does at his Special Operations Unit at FDNY. Do you think that statement is correct? Why?

4. How would you feel about someone like Captain Tricarico bringing ideas like those in the story to your department?

5. Sometimes, in spite of the best of efforts, we lose the customer. How do you think the leader should deal with their personnel in this kind of situation?

LESSONS LEARNED

The following information summarizes what we have covered in Chapters 6 and 7 and should provide you with a list of practical tools to help you become an effective leader.

1. *Have an orientation session.* We usually have an orientation session for new firefighters. What do we do for experienced firefighters who are transferred to a new station? What do we do when the supervisor changes? In every case, there is a need for taking a few minutes to get acquainted. For the supervisors who do this regularly, it is a good idea to have a short checklist, not unlike a teaching outline, to help facilitate the transition and cover every issue with every member. This session is a good time to explain personal values as well as any policies that may be unique to this workplace. This session is a good time to listen to the members, hear about their goals and ambitions, and possibly their problems and needs. Remember—do not try to cover everything in one session. Give the new member information in installments that can be easily absorbed.

2. *Have regular meetings and share information with employees.* One effective meeting that should happen regularly is a planning meeting. The planning meeting may be nothing more than getting together with everyone after the equipment checks at the start of the shift for a cup of coffee and a brief discussion about what will happen during the upcoming shift. Some planning meetings may be longer, such as planning the training schedule for the next 3 months. Plan for future events, and let your teammates share the information. Let teammates be involved in the decision-making process whenever possible.

3. *Set goals for the company and for every member.* Some goals will be collaborative efforts that require group acceptance, whereas other goals may be for individual accomplishment. In any case, get the member(s) actively involved in goal setting. Use these goals as a benchmark for measuring progress during evaluations.

4. *Provide regular feedback on performance.* This does not require a formal evaluation with a complicated report. It may only take a minute or two, and usually focuses on some specific task or time frame. It is a lot easier for both the supervisor and the member to deal with these performance issues when they are addressed in small increments. When corrective action is desired, it will come a lot sooner.

5. *Empower your people.* Offer opportunities for your people to enhance themselves and provide oppor-
tunities for their growth and learning. Encourage them to seek training and education at every opportunity. Provide opportunity and incentive for individuals to increase their knowledge.

6. *Provide instruction and help,* be available, show interest and support, and recognize accomplishments. But do not supervise, unless necessary.

7. *Ask your members for ideas and suggestions.* Listen to the members' ideas and suggestions. Respond to all of them, and, when appropriate, see that the ideas are forwarded up the chain of command for action. Be sure that members get credit for their work and suggestions.

8. *Be a role model.* Do the right thing.

Throughout the chapters on management and leadership, we have tried to share with you some new ideas, ideas that are being used in modern organizations. Maybe some of these are ideas that you have seen used in your organization, and maybe they are some that you have not seen. All have been used in the fire service and elsewhere with good results.

Michael Abrashoff discusses ownership in the book *It's Your Ship.* Many modern managers, including fire chiefs, see this approach as a way to resuscitate tired organizations, providing better and more efficient service to citizens. These ideas are valid but face significant challenges from years of tranquility and status quo. Introducing change is a challenge in itself, but with something as bold as empowering people and suggesting that they have "ownership in the organization," things can be difficult.

The message for company officers—you can make changes in the way you do things a lot easier than your chief can: You are much closer to your people, and there are fewer of them to convince. Try some of the ideas discussed here, and see what happens. Do not be content to just do it the way we always have and then complain about the organizational climate. Take on small challenges first and then work your way up to larger undertakings. As you move to positions of greater responsibility and influence, continue to introduce new ideas. Read the trade journals. You will find that nearly every issue has something about these modern concepts in management, leadership, and supervision.

So, what is the company officer's role? You should aspire to be the best company officer in your department. That means you manage the best company in the department. You should have the cleanest fire station and fire apparatus in town. Your teammates should be enthusiastic about their job. They should be the best-trained, fittest, and most-qualified fire-fighters anywhere. They should all be competing for advancement, and you should have prepared them to take your place.

While coaching and counseling those around you, take time to work on your own development needs. Steve Covey, author of the best seller *Seven Habits of Highly Successful People,* calls this taking time off from your work to improve your skills as taking time to "sharpen the saw." Covey compares our learning process to working with tools, noting that more gets done with sharp tools, but that production is sacrificed in the process. He suggests that we also have to take time off to sharpen our thinking tools. His message: "Take time to keep your mental saw sharp!" There is a wealth of information on the topic of supervision and leadership. Every library and bookstore has books and tapes on the subject. You are encouraged to learn all that you can and to share your ideas and experiences with others.

KEY TERMS

coach   a person who helps another develop a skill

complaint an expression of discontent

conflict   a disagreement, quarrel, or struggle between two individuals or groups

counseling   one of several leadership tools that focuses on improving member performance

demotion   reduction of a member to a lower grade

disciplinary action   an administrative process whereby a member is punished for not conforming to the organizational rules or regulations

empowerment   to give authority or power to another

grievance   a formal dispute between member and employer over some condition of work

grievance procedure   formal process for handling disputed issues between member and employer; where a union contract is in place, the grievance procedure is part of the contract

gripe   the least severe form of discontent

oral reprimand   the first step in a formal disciplinary process

succession plan   the formal or informal training of subordinates to assume your position within the organization

suspension   disciplinary action in which the member is relieved from duties, possibly with partial or complete loss of pay

termination   the final step in the disciplinary process or the incident command process

transfer   a step in the disciplinary process that provides the member a fresh start in another venue

written reprimand   documents unsatisfactory performance and specifies the corrective action expected; usually follows an oral reprimand

REVIEW QUESTIONS

1. What is leadership?
2. What is coaching?
3. What is counseling?
4. The text suggests that the performance evaluation system is a continuous process. Explain.
5. Describe the disciplinary process in your department.
6. What are some general rules for effective punishment?
7. How should you, as company officers, deal with conflict?
8. How should you respond to complaints and grievances?
9. How should you introduce change?
10. How can you, as a company officer, become an effective leader?

DISCUSSION QUESTIONS

1. As a company officer, describe how you would treat a new firefighter joining your company.

2. Describe how you would integrate this new firefighter into your team.

3. Describe how you would encourage the new firefighter's professional development.

4. Describe something you would do at your fire station to make it more attractive to the first minority member to join your company.

5. As a company officer, how would you introduce a new procedure (or a change to an existing one) mandated by law or regulation?

6. As a company officer, how would you introduce a new procedure (or a change to an existing one) that is the result of your own initiative?

7. As a company officer, how would you deal with a new idea suggested by one of your teammates?

8. How is change accepted in your department?

9. Other than those mentioned in the text and in the Student Manual, what changes are occurring in our industry?

10. What changes do you foresee in our business over the next five years?

ENDNOTES

1. Jon Buckman, "What Good Leaders Do," *Fire Engineering,* August, 2003.

2. *A Needs Assessment of the U.S. Fire Service,* Emmitsburg, MD: U.S. Fire Administration, 2002.

ADDITIONAL RESOURCES

Anthony, Derrick. "To Enact Departmental Change, Company Officers Must Become Team Players." *FireRescue,* March 2008.

Bennis, Warren. *On Becoming a Leader.* Reading, MA: Addison-Wesley, 1994.

Blanchard, Kenneth and Spencer Johnson. *The One Minute Manager.* New York, NY: Berkley Books, 1982.

Calfee, Mica. "Leadership Do's and Don'ts." *Fire Engineering,* April 2008.

Compton, Dennis. *When in Doubt, Lead!* Still water, OK: Fire Protection Publications, 1999.

Effective Supervisory Practices. 2nd ed. Washington, DC: International City Management Association, 1984.

Emery, Mark. "How to Give Your Crew What They Need to Succeed." *FireRescue Magazine,* April 2004.

Emory, Mark. "The Fire Station Pyramid of Success." *Firehouse,* five part series starting Feb 2008.

Gayk, Ray. "Fire Officers Should Emulate Military Leaders to Improve Leadership Under Stress." *FireRescue,* October 2008.

Gayk, Ray. "What to Do When You Inherit a Personnel Problem." *FireRescue,* August, 2008.

Goswick, William. "Power and Influence: A Guide for Fire Officers." *Fire Engineering,* July 2007.

Humphrey, B., and J. Stokes. *The 21st Century Supervisor: Nine Essential Skills for Frontline Leaders.* Hoboken, NJ: John Wiley, 1999.

Krames, Jeffrey A. *The Rumsfeld Way: Leadership Wisdom of a Battle-Hardened Maverick.* New York: McGraw Hill, 2003.

Marinucci, Richard. "Counseling: Listen, Don't Judge." *Fire Engineering,* August 2007.

Martinette, C. V. "Surviving the Leadership of Others." *Fire Engineering,* April 2004.

Maxwell, John C. *Leadership 101.* Nashville, TN: Thomas Nelson, 2002.

Peters, Tom and Nancy Austin. *A Passion for Excellence.* New York, Random House, 1995.

Plunkett, Richard. *Supervision: Diversity and Teams in the Workplace.* 8th ed. Upper Saddle River, NJ: Prentice Hall, 1996.

Simpson, Jeff. "Is Your Leadership Bicycle Broken?" *Fire Engineering,* October 2006.

Shoebridge, Todd. "To Become a Respected Company Officer, Know Your People." *Fire Engineering,* July 2006.

Staley, Michael. *Igniting the Leader Within.* Saddle Brook, NJ: Fire Engineering, 1998.

Woodbury, Debbie. *Excellence in Supervision: Essential Skills for the New Supervisor.* Boston, MA: Crisp Learning, 2001.

Zeiss, Tony. *Nine Essential Laws for Becoming Influential.* Tulsa, OK: Triumphant Publishers International, 2000.

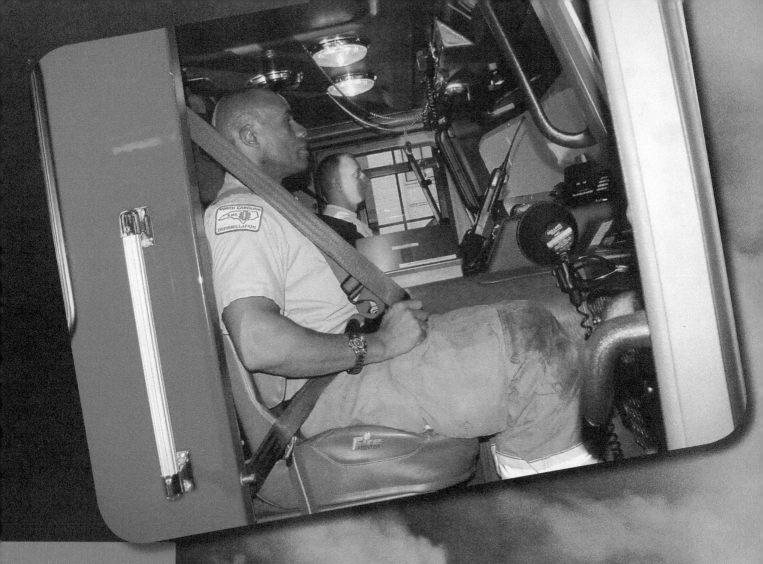

8

The Company Officer's Role in Personnel Safety

- Learning Objectives
- Introduction
- The Value of Safety
- Annual NFPA Reports on Firefighter Safety
- Theory of Accident Causes
- Regulatory and Standards Setting Bodies

- Fire Department Occupational Safety and Health Program
- Lessons Learned
- Key Terms
- Review Questions
- Discussion Questions
- Endnotes
- Additional Resources

After many years of serving as a firefighter, I was uncertain of what to think about a promotion to a captain's position managing the department's Health and Safety Program. I would be giving up the schedule and stations I was used to for a desk job at headquarters. Worst of all, I would be giving up the emergency responses and the excitement of working in the field.

Little did I know how valuable the experience would be in returning to the field several years later as a company officer. While managing the department's safety program, I had gained a wealth of knowledge about Self Contained Breathing Apparatus (SCBA) and the wide variety of regulations and equipment associated with respiratory protection for firefighters. When operating in the field, we typically do not realize all the complexities of the equipment and the issues we face. The SCBA is a tool that all firefighters use to protect themselves and to keep their respiratory systems safe while performing their jobs, right? But what about SCBA equipment that fails or is not properly maintained? What about the great importance of the fit testing program and its relationship to the overall effectiveness of the SCBA in making firefighters safe on the job? There are so many factors to consider beyond the daily checks and throwing the packs on before entering a structure.

In retrospect, I am thankful for my experience in managing my department's safety program as it has given me an invaluable perspective on how complex the issues are related to firefighter safety. It has also helped me realize the importance of training on safety issues, such as SCBA emergencies, and to place high priorities on the seemingly mundane tasks associated with safety, like fit testing, PPE gear checks and fittings, and SCBA maintenance. As a company officer, I believe that instilling the value and respect of the safety program in my team members is critical and is an important responsibility of my position.

—Dominic Depaolis, Captain,
Metropolitan Washington Airports Authority, Dulles, Virginia

LEARNING OBJECTIVES

After completing this chapter the reader should be able to:

8-1   Identify death and injury statistics for emergency incident-related duties.

8-2   List the common causes of accidents and injuries in the fire service.

8-3   Identify general safety policies and procedures.

8-4   List the common causes of accidents and injuries in the fire service from unsafe conditions or actions.

8-5   Describe fire department policies as they apply to firefighter wellness and fitness.

8-6   Describe the sixteen (16) Firefighter Life Safety Initiatives.

8-7   List the common causes of personal injuries at a selected emergency operation.

8-8   Identify departmental operating procedures pertaining to on-the-job safety.

8-9   Describe hazard mitigation measures for emergency and non-emergency situations.

8-10   Identify the theories of accident causes.

8-11   Review in-house and national reports addressing trends and conditions that result in accidents, injuries, and occupational illnesses and deaths.

8-12   Describe the safety policies and procedures of the agency.

8-13   Explain the need for accident records and reports.

8-14   List the steps for conducting an initial accident investigation.

8-15   Describe the procedures for responding, reporting, and processing line-of-duty deaths.

8-16   Describe the laws affecting the releasing of medical records and reports affecting the member.

8-17   List the types of health exposures with which fire personnel may come in contact.

8-18   Describe the components of an infectious disease control program as contained in the department standard operating procedures.

8-19   Describe safe procedures and/or practices that are applied during non-emergency operations.

8-20   Identify the standard operating procedures addressing safety for non-emergency operations.

8-21   Identify the stresses that may be encountered by unit members involved in emergency services.

8-22   Describe the supervisor's role in unit member crisis intervention.

8-23   Describe the confidentiality requirements pertaining to the unit member outlined in the human resources policy and procedures.

* The FO I and II levels, as defined by the NFPA 1021, the *Standard for Fire Officer Professional Qualifications*, are identified in different colors: FO I = black, FO II = red.

INTRODUCTION

An article in a fire service magazine stated that the leading causes of accidents among fire service personnel are managers who do not manage, supervisors who do not supervise, and firefighters who do dumb things.[1] Safety is an organization value. In addition to the deaths and injuries reported each year, a growing number of firefighters are being exposed to situations that lead to occupational illnesses and conditions that have debilitating and sometimes fatal consequences. Firefighters have long had a high rate of heart and respiratory disease and cancer. Firefighters providing emergency medical care may be exposed to a wide range of biological hazards.

FIREGROUND FACT

In the United States, approximately 100 firefighters are killed, and approximately 80,000 are injured each year while on duty.

Company officers should be intimately involved with the department's health and safety program. Most injuries occur to firefighters. Most firefighters have a company officer as their supervisor. Most fire departments, several NFPA standards, and the courts hold the supervisor responsible for the firefighter's safety.

Each year in the United States, approximately 100 firefighters are killed while on duty, and approximately

80,000 are injured. Although the number of fatalities and injuries has decreased by 15 percent over the past 20 years, the rate of these losses (the number of injuries per fire call) has increased by 25 percent. In the past decade, several high-profile incidents involving firefighter fatalities have brought the public's attention to the issue of firefighter mortality. Although the attention from the national media was fleeting, the awareness of these events has changed the fabric of the fire service and prompted many organizations and fire departments to initiate programs to protect firefighters.[2]

THE VALUE OF SAFETY

You cannot put a value on safety, but you sure can put a price on the consequences. Regardless of whether the firefighter is a paid or a volunteer member, injuries cost money. Anyone who has had recent contact with a health-care system knows the cost of good medical care. In the case of firefighters, someone is going to have to pay these costs, and ultimately, it is the citizens of the community. For paid departments dealing with tight budgets and minimum staffing issues, an injured firefighter brings an added cost in terms of providing a replacement. While the injured firefighter is in a "no duty" or "light duty" status, another firefighter is working for the injured person, probably at overtime rates. Thus, the injured $20-an-hour firefighter is actually costing the community $50 an hour. Additional costs to the department include medical insurance, workers' compensation insurance, and repairs for damaged equipment.

There are additional impacts—indirect costs—that cannot be measured in dollars. For the injured firefighter, there is the pain, suffering, possible permanent injury, and even the embarrassment that

comes from the injury. The firefighter's working colleagues also suffer, not only in working additional shifts for the injured teammate but also in the emotional costs associated with their sympathy for a hurt friend. And finally, the injured firefighter's family pays a cost for its anxiety, support, and concerns associated with the member's injury.

All of these concerns resulted in the adoption of NFPA 1500, *Standard on Fire Department Occupational Safety and Health Program*. We discuss NFPA 1500 later in this chapter.

Background of Safety Activity in the Workplace

Concerns for workplace safety and health issues are not new or unique to the fire service. In the United States, awareness of workplace safety and health issues date back to the 1600s. Textbooks on industrial safety usually include a history of workplace safety concerns dating back to the start of recorded history.

As technology advanced, so did the hazards. The demand for steam power increased with the arrival of the industrial age. With the increased need for power, there was an increased demand for coal. Most of the coal had to be mined, which was and still is dangerous work. Long hours, poor ventilation, and extremely demanding work conditions led to black lung disease and a host of other problems. Miners were essentially the slaves of the mine owners. Efforts to bring some balance to the mining industry started in the 1860s. It took 60 years and the full efforts of both the United Mine Workers of America and the U.S. Congress to increase the use of safety appliances in the coal mines.

AN OVERVIEW OF THE 118 FIREFIGHTERS WHO DIED WHILE ON DUTY IN 2007

- Sixty-eight volunteer firefighters and 50 career firefighters died while on duty.
- There were 7 firefighter fatality incidents where 2 or more firefighters were killed, claiming a total of 21 firefighters' lives.
- Eleven firefighters were killed during activities involving brush, grass, or wildland firefighting, the lowest in over a decade.
- Activities related to emergency incidents resulted in the deaths of 76 firefighters.
- Thirty-eight firefighters died while engaging in activities at the scene of a fire.
- Twenty-six firefighters died while responding to or returning from emergency incidents.
- Eleven firefighters died while they were engaged in training activities.
- Fifteen firefighters died after the conclusion of their on-duty activity.
- Heart attacks were the most frequent cause of death for 2007, resulting in 52 firefighter deaths.

Source: Emmitsburg, MD, USFA, News Release, USFA Releases Firefighter Fatality Report.

NOTE

Regardless of whether the firefighter is a paid or a volunteer member, injuries cost money.

The workplace changed as the United States became more industrialized and technology improved. Instead of a company focusing on production at any cost, worker safety was also considered. In many instances, a process change brought on by a concern for safety also brought increased productivity. For example, during the early years of the Industrial Revolution, gears and belts were unguarded, and workers were frequently required to place themselves in dangerous situations. When workers were injured, they were replaced with little or no compensation. The growth of organized labor and the development of laws dealing with workplace conditions started a process that continues today.

One industry that has made remarkable progress in the area of worker safety in recent times is the railroad industry. America's railroads, possibly prodded by the government, their labor unions, and the cost of workplace accidents, have made a dramatic change in their accident statistics over recent years. These improvements were accomplished by eliminating unsafe equipment and procedures, having and enforcing policies that provide for safety, and by having highly viable recognition programs at all levels that reinforce both the company's and the industry's commitment to worker healthand safety.

One of the best examples of this today is Norfolk Southern Corporation, one of the nation's largest railroad companies. Recognized by government and industry for its safety practices, Norfolk Southern started a process in 1987 of training top executives in the principles of industrial safety, including the concept that teamwork is required to reduce accidents. Since implementation, that teamwork has reduced injuries by 80 percent, and, as a result, the company has earned the industry's highest safety award each of the past 15 years.

Norfolk Southern's business is transportation, and its goal is to make a profit. However, its vision statement is quite revealing: "Our vision: to be the safest, most customer focused and successful transportation company in the world." As a part of its vision, safety gets top priority. Its goal is simple: zero accidents and zero injuries. At the core of Norfolk Southern's safety program are the following six tenants:

- All injuries can be prevented.
- All exposures can be safeguarded.
- Prevention of injuries and accidents are the responsibility of each employee.
- Training is essential for good safety performance.

- Safety is a condition of employment.
- Safety is good business.

What has all of this got to do with the fire service? We can draw several parallels between the railroad industry and the fire service. Both are dangerous professions, and safety and health issues are a concern for both industries. During the past 25 years, we have seen increased awareness for safety in both industries. Let us look at ourselves. Are you as safe as you can be? Are you looking for and sharing ways to improve health and safety for your firefighters? Does your fire department have programs to improve health and safety? Do you have programs that reward safety?

SAFETY

Safety comes from eliminating unsafe equipment and procedures, having and enforcing policies that provide for safety, and having viable recognition programs at all levels that reinforce both the company's and the industry's commitment to worker health and safety.

FIREGROUND FACT

Nearly half of all on-duty firefighter fatalities are the result of heart attack. Personal fitness and good health are within the realm of everyone's personal control.

Safety Activities in the Fire Service

By the nature of the occupation, emergency responders are at risk from accidents that can result in illness, injury, and death. Incident safety is a concern for all emergency response personnel. Firefighters are taught to deal with these problems. Why then, do we have the injury statistics we do?

Firefighters tend to be aggressive, action-oriented people, especially around other firefighters. At times, this behavior is desirable, but over aggressiveness can lead to a disregard for firefighter safety. As the company officer, you must learn to channel this enthusiasm and energy into productive activity. Although it is sometimes difficult to change attitudes about safety, some firefighters will respond to a discussion that focuses on their importance as a member of a team. In this context, every firefighter is dependent upon every other firefighter for success and survival. If one member of the team is injured, everyone is at risk. This logic can also be extended to how we serve the public. When we are not able to perform, we are not able to help our customers. For most firefighters, serving others is our mission.

SAFETY

Every firefighter is dependent upon every other fire-fighter for success and survival.

Many fatalities occur as a result of physiological and psychological stress, manifested as a heart attack or stroke, rather than some traumatic event (see **Figure 8-1** and **Figure 8-2**). The work can be very demanding at times and is often undertaken in unfamiliar settings with uncertain outcomes. These facts suggest that proper rest, good physical condition, proper health care, and proper lifestyle are just as important as protective clothing and an effective incident command system.

SAFETY

Proper rest, good physical condition, proper health care, and proper lifestyle are just as important as protective clothing and an effective incident command system.

Nature of injury	Deaths	Percentage
Sudden cardiac death	39	37.9%
Internal trauma	29	28.2%
Asphyxiation	23	22.3%
Crushing	1	1.0%
Stroke/aneurysm	2	1.9%
Drowning	2	1.9%
Electrocution	1	1.0%
Burns	4	3.9%
Projectile wounds	2	1.9%
Total	**103**	**100.0%**

FIGURE 8-1 Firefighter fatalities by nature of injury. (*Source: Rita F. Fahy, Paul R. LeBlanc, Joseph L. Molis, Firefighter Fatalities in the United States 2007, NFPA, July 2008.*)

Cause of injury	Deaths	Percentage
Exertion/Stress/Other	41	39.8%
Struck by/contact with object	32	31.1%
Caught or trapped	23	22.3%
Fell or jumped	6	5.8%
Electrocuted	1	1.0%
Total	**103**	**100.0%**

FIGURE 8-2 Firefighter fatalities by cause of injury. (*Source: Rita F. Fahy, Paul R. LeBlanc, Joseph L. Molis, Firefighter Fatalities in the United States 2007, NFPA, July 2008.*)

Many of these factors—attitude, conditioning, health care, and lifestyle—are individual concerns. Supervisors should set a good positive example, encourage fitness and good health practices, and even provide on-duty opportunities for improving every team member's fitness and health.

Knowledge of the job is also a matter of personal concern and initiative; it all comes down to the firefighter's attitude and awareness. In many cases the firefighter may need motivation and some help from the supervisor, and for most firefighters, the supervisor is the company officer.

Firefighting is a hazardous occupation. The action is fast paced and hot, and there are often problems with the structure itself. Many ask, "What reasonable person would go into a burning building while everyone else is running out?" Going into burning buildings is no longer what the fire service is all about. Firefighters are expected to understand how fire grows and to assess the fire situation. The conditions may suggest that not entering the building is the right action. Interior fire fighting involves advancing hose lines to the seat of the fire. Nearly half of those who are killed during structural firefighting operations were advancing hose lines at the time of their injury.[4] This task is typically assigned to the newest firefighters; you cannot expect those in their first year on the job to be aware of all the dangers involved.

FIREGROUND FACT

Nearly half of those killed during structural firefighting operations were advancing hose lines at the time of their injury.

Some accidents are considered surprises. A collapse might be an example; however, even in these circumstances, a better understanding of fire behavior and building construction should have suggested a different course of action. During a recent fire in a commercial building, the truss roof collapsed trapping two firefighters who had ventured inside this well-involved structure. The building had been there for years. One has to wonder if the department had been there to survey and preplan, or if anyone told the firefighters about the dangers of a truss roof. The collapse of a truss roof during a fire should not come as a surprise.

National Fallen Firefighter Foundation

Recognizing the need to do more to prevent line-of-duty deaths and injuries, the National Fallen Firefighters Foundation has launched a national initiative to bring prevention to the forefront.

In March 2004, the Firefighter Life Safety Summit was held in Tampa, Florida, to address the need for change within the fire and emergency services. Through this meeting, 16 Life Safety Initiatives were produced to ensure that *Everyone Goes Home*®.

The first major action was to produce 16 major initiatives that will give the fire service a blueprint for making changes.

The National Fallen Firefighters Foundation will play a major role in helping the U.S. Fire Administration meet its stated goal to reduce the number of preventable firefighter fatalities. The foundation sees fire service adoption of the summit's initiatives as a vital step in meeting this goal.

ANNUAL NFPA REPORTS ON FIREFIGHTER SAFETY

Each year, the NFPA publishes two reports on firefighter safety. One of the reports focuses on firefighter injuries; the other on firefighter fatalities. Both reports are published in the NFPA Journal. Although these topics are not pleasant, these reports provide valuable lessons learned and should be reviewed by all fire service professionals.

Safety professionals look at past statistics for several reasons. First, they provide considerable information about what causes the accidents that lead to firefighter deaths and injuries. The published safety statistics are indications of the activities that are causing the greatest problems for the fire service at the national level. However, these statistics can indicate problems that may not yet be apparent in a particular department. In such cases, these statistics should be a wake-up call and suggest changes in policy or the need for training.

Departments that do not enforce their policies (or do not have any policies) regarding the use of personal protective equipment or that do not have and routinely use a good incident command system at the scene of every event cannot expect positive results from members regarding occupational safety. How do your department's accident statistics compare with the published data?

THE U.S. FIRE ADMINISTRATION GOALS

- A 25 percent reduction in on-duty firefighter fatalities within 5 years
- A 50 percent reduction in on-duty firefighter fatalities within 10 years

SIXTEEN FIREFIGHTER LIFE SAFETY INITIATIVES

1. Define and advocate the need for a cultural change within the fire service relating to safety, incorporating leadership, management, supervision, accountability, and personal responsibility.

2. Enhance the personal and organizational accountability for health and safety throughout the fire service.

3. Focus greater attention on the integration of risk management with incident management at all levels, including strategic, tactical, and planning responsibilities.

4. All firefighters must be empowered to stop unsafe practices.

5. Develop and implement national standards for training, qualifications, and certification (including regular recertification) that are equally applicable to all firefighters based on the duties they are expected to perform.

6. Develop and implement national medical and physical fitness standards that are equally applicable to all firefighters, based on the duties they are expected to perform.

7. Create a national research agenda and data collection system that relates to the initiatives.

8. Utilize available technology wherever it can produce higher levels of health and safety.

9. Thoroughly investigate all firefighter fatalities, injuries, and near misses.

10. Grant programs should support the implementation of safe practices and/or mandate safe practices as an eligibility requirement.

11. National standards for emergency response policies and procedures should be developed and championed.

12. National protocols for response to violent incidents should be developed and championed.

13. Firefighters and their families must have access to counseling and psychological support.

14. Public education must receive more resources and be championed as a critical fire and life safety program.

15. Advocacy must be strengthened for the enforcement of codes and the installation of home fire sprinklers.

16. Safety must be a primary consideration in the design of apparatus and equipment.

Understanding the Nature of the Injuries

The largest percentage of firefighter injuries and deaths occur at the scene of fires. However, statistics show that every aspect of the firefighter's work routine has the potential for injury. Each year, a significant number of firefighters are killed and injured while responding to and returning from calls, as well as at the scene of emergencies (see **Figure 8-3**). Note that while firefighter fatalities have declined over the last 30 years, the numbers in all categories have remained fairly constant over the past decade.

Year	Total	Firefighting, fireground	Responding, returning	On scene at non-fire calls	Training	Other on-duty
1977	157	78 (49.7%)	41 (26.1%)	14 (8.9%)	10 (6.4%)	14 (8.9%)
1978	173	84 (48.6%)	48 (27.7%)	19 (11.0%)	5 (2.9%)	17 (9.8%)
1979	125	83 (66.4%)	20 (16.0%)	9 (7.2%)	3 (2.4%)	10 (8.0%)
1980	138	74 (53.6%)	37 (26.8%)	9 (6.5%)	4 (2.9%)	14 (10.1%)
1981	136	75 (55.1%)	35 (25.7%)	8 (5.9%)	7 (5.1%)	11 (8.1%)
1982	128	67 (52.3%)	40 (31.3%)	8 (6.3%)	6 (4.7%)	7 (5.5%)
1983	113	60 (53.1%)	26 (23.0%)	12 (10.6%)	4 (3.5%)	11 (9.7%)
1984	119	59 (49.6%)	37 (31.1%)	8 (6.7%)	3 (2.5%)	12 (10.1%)
1985	128	63 (49.2%)	33 (25.8%)	9 (7.0%)	6 (4.7%)	17 (13.3%)
1986	119	48 (40.3%)	42 (35.3%)	6 (5.0%)	6 (5.0%)	17 (14.3%)
1987	132	54 (40.9%)	37 (28.0%)	12 (9.1%)	17 (12.9%)	12 (9.1%)
1988	136	65 (47.8%)	39 (28.7%)	10 (7.4%)	11 (8.1%)	11 (8.1%)
1989	118	59 (50.0%)	29 (24.6%)	6 (5.1%)	12 (10.2%)	12 (10.2%)
1990	108	48 (44.4%)	24 (22.2%)	17 (15.7%)	8 (7.4%)	11 (10.2%)
1991	108	53 (49.1%)	26 (24.1%)	8 (7.4%)	14 (13.0%)	7 (6.5%)
1992	75	38 (50.7%)	21 (28.0%)	4 (5.3%)	5 (8.0%)	7 (8.0%)
1993	79	34 (43.0%)	21 (26.6%)	8 (10.1%)	7 (8.9%)	9 (11.4%)
1994	105	61 (58.1%)	19 (18.1%)	7 (6.7%)	7 (6.7%)	11 (10.5%)
1995	98	42 (42.9%)	32 (32.7%)	12 (12.2%)	4 (4.1%)	8 (8.2%)
1996	96	32 (33.3%)	32 (33.3%)	8 (8.3%)	8 (8.3%)	16 (16.7%)
1997	99	41 (41.4%)	27 (27.3%)	12 (12.1%)	7 (7.1%)	12 (12.1%)
1998	91	40 (44.0%)	17 (18.7%)	15 (16.5%)	11 (12.1%)	8 (8.8%)
1999	112	56 (50.0%)	32 (28.6%)	10 (8.9%)	4 (3.6%)	10 (8.9%)
2000	103	39 (37.9%)	24 (23.3%)	8 (7.8%)	14 (13.6%)	18 (17.5%)
2001*	443	38* (36.9%)	26 (25.2%)	2 (1.9%)	12 (11.7%)	25 (24.3%)
2002	97	46 (47.4%)	19 (19.6%)	10 (10.3%)	11 (11.3%)	11 (11.3%)
2003	106	29 (27.4%)	37 (34.9%)	9 (8.5%)	12 (11.3%)	19 (17.9%)
2004	105	30 (28.6%)	36 (34.3%)	9 (8.6%)	12 (11.4%)	18 (17.1%)
2005	87	25 (28.7%)	26 (29.9%)	4 (4.6%)	11 (12.6%)	21 (24.1%)
2006	89	38 (42.7%)	18 (20.2%)	6 (6.7%)	8 (9.0%)	19 (21.3%)
2007	103	36 (35.0%)	31 (30.1%)	7 (6.8%)	13 (12.6%)	16 (15.5%)

* Percentage breakdowns do not include the 340 fireground deaths at the World Trade Center, September 11, 2001.

Note: These figures are based on NFPA's annual tracking of all firefighter deaths. Some figures have been updated since the publication of the reports for those years.

FIGURE 8-3 Firefighter fatalities by type of duty. (*Source: Rita F. Fahy, Paul R. LeBlanc, Joseph L. Molis, Firefighter Fatalities in the United States 2007, NFPA, July 2008.*)

With departments providing a greater variety of service to their communities, firefighters are also being injured at the scene of non-fire emergencies. Many injuries occur at fire stations. Injuries and even a few deaths occur during training activities.

The most frequent injuries are sprains and strains (see **Figure 8-4**). These invisible injuries usually are the result of improper lifting. A second large category of injuries is wounds. Wounds include all kinds of skin injuries from cuts to the hands to stepping and falling on sharp objects. Given the nature of these injuries, it would seem logical that most of the injuries occur at the scene of fire-related emergencies (see **Figure 8-5**).

FIREGROUND FACT

The most frequent injuries are sprains and strains.

Check your own department's safety records. The statistics should suggest opportunities for corrective action to improve safety performance. If 50 percent of your department's injuries are back injuries and are occurring as the result of improper lifting, it makes sense to focus corrective actions in this area. Selecting the biggest target usually provides the greatest opportunity for improvement.

Reducing These Injuries

Sprains and strains can be reduced by proper physical conditioning and by using proper lifting methods. When a department is experiencing a large percentage of sprain-and-strain-type injury from a particular activity, some revision in procedure is indicated. Reducing cuts and other wound-type injuries may require providing and using proper protective equipment. Effective supervision and fire scene management can also help reduce the firefighter's injuries from falling and from exposure to sharp objects and other dangerous conditions that exist on the fireground.

THEORY OF ACCIDENT CAUSES

Accidents can be defined as unplanned occurrences that result in injuries, fatalities, and damage to or loss of equipment and property. Preventing accidents is extremely difficult in the absence of an understanding of the causes of accidents. Many attempts have been made to develop a prediction theory of accident causation. Here are several popular theories.

Domino Theory

Herbert Heinrich developed his widely held theory in the 1920s, while working for Travelers Insurance. He suggested a "five-factor accident sequence" in which each factor would actuate the next step in the manner of toppling dominoes lined up in a row. The sequence of accident factors is as follows:

- Ancestry and social environment
- Unsafe practice (personal fault)
- Unsafe actions or condition
- Accident
- Damage or injury (loss)

In the same way that the removal of a single domino in the row would interrupt the sequence of toppling, Heinrich suggested that removal of one of the factors would prevent the accident and resultant injury (see **Figure 8-6**). He noted that the third element, the unsafe act, led to nearly 90 percent of all accidents. For most of us, the greatest opportunity is preventing the unsafe act.

Human Factors Theory

Another popular theory is called the human factors theory. This theory also promotes the idea that most accidents are the result of human error and suggests that three broad factors lead to accidents:

- *Overload.* Overload might be affected by the environmental conditions, internal factors, and situation factors such as the level of risk or the lack of good instructions and supervision.
- *Inappropriate response.* Examples might include the removal of safety guards or ignoring safety instructions.
- *Inappropriate activity.* An example might be persons taking on a task for which they are not qualified, misjudging the degree of risk involved, and not telling their supervisor that they are not able to do the task.

Multiple Causation Theory

Multiple causation theory is an outgrowth of the domino theory, but it suggests that for a single accident there may be many contributory factors. According to this theory, the contributory factors can be grouped into two categories:

- *Behavioral.* This category includes factors pertaining to the worker, such as improper attitude, lack of knowledge, lack of skills, and inadequate physical and mental condition.

Year	Total	Sprain, strain	Wound, cut, dislocation, fracture	Smoke, gas inhalation or resp. distress (other)	Eye irritation	Burns and smoke inhalation	Fire or chemical burns	Heart attack, stroke	Thermal stress	Other
1981	67,510	16,530	20,455	12,485	5,315	n/a	7,545	300	n/a	4,880
1982	61,370	14,955	16,450	10,880	6,155	n/a	5,990	315	n/a	6,625
1983	61,740	15,415	17,470	11,470	5,025	n/a	6,470	370	n/a	5,520
1984	62,700	16,870	16,295	10,105	4,205	n/a	6,640	430	3,990	4,165
1985	61,255	16,545	15,435	9,625	4,580	n/a	6,215	230	4,520	4,105
1986	55,990	15,455	15,200	8,090	3,985	n/a	6,270	360	3,260	3,370
1987	57,755	16,565	14,770	8,040	4,905	n/a	5,770	240	4,260	3,205
1988	61,790	20,695	14,205	7,400	3,825	n/a	6,475	260	4,955	3,975
1989	58,250	22,360	13,625	7,220	4,025	n/a	4,815	255	3,040	2,910
1990	57,100	20,885	13,120	7,095	4,025	n/a	5,180	235	3,505	3,055
1991	55,830	19,655	11,285	7,525	3,645	n/a	4,960	325	4,630	3,805
1992	52,290	19,020	11,920	6,335	3,805	n/a	5,105	335	2,775	2,995
1993	52,885	18,810	11,910	5,540	3,145	n/a	5,990	295	3,430	3,765
1994	52,875	18,855	12,275	6,175	3,115	n/a	5,470	330	3,160	3,495
1995	50,640	19,280	11,680	5,700	3,140	n/a	4,890	345	2,935	2,670
1996	45,725	17,455	9,865	5,400	2,735	n/a	4,360	300	2,720	2,890
1997	40,920	15,590	9,710	3,820	2,265	n/a	3,755	205	2,840	2,735
1998	43,080	18,735	9,010	3,745	2,215	n/a	4,040	300	2,760	2,275
1999	45,550	17,925	9,880	4,435	2,615	n/a	4,060	395	3,570	2,670
2000	43,065	19,500	8,695	3,725	2,220	n/a	3,850	250	2,175	2,650
2001	41,395	16,410	10,355	3,925	n/a	1,190	3,255	310	2,315	3,635
2002	37,860	15,735	9,200	2,790	n/a	975	3,205	345	2,415	3,195
2003	38,045	16,830	9,195	2,890	n/a	980	2,765	235	2,145	3,005
2004	36,880	17,890	7,370	2,915	n/a	585	2,860	290	1,875	3,095
2005	41,950	18,620	8,570	3,390	n/a	750	2,930	315	2,480	4,895
2006	44,210	20,655	8,705	3,755	n/a	575	3,070	350	2,280	4,820
2007	38,340	17,280	8,195	2,710	n/a	695	2,650	395	2,410	4,005

Note: Prior to 1981, somewhat different categories were used. "Wound,cut" also includes bleeding and bruising. Dislocations and fractures can be provided separately from wounds and cuts. From 2001 onward, the category " eye irritation" was dropped and the category " burns and smoke inhalation" was added.

FIGURE 8-4 Firefighter injuries by nature of injury. (Source: Michael J. Karter, Jr., United States Firefighter Injuries 2007, NFPA, October 2008 and previous reports in series.)

Year	Total	Firefighting, fireground	Responding, returning	On scene at non-fire calls	Training	Other on-duty
1981	103,340	67,510 (65.3%)	4,945 (4.8%)	9,600 (9.3%)	7,090 (6.9%)	14,195 (13.7%)
1982	98,150	61,370 (62.5%)	5,320 (5.4%)	9,385 (9.6%)	6,125 (6.2%)	15,950 (16.3%)
1983	103,150	61,740 (59.9%)	5,865 (5.7%)	11,105 (10.8%)	6,755 (6.5%)	17,685 (17.1%)
1984	102,300	62,700 (61.3%)	5,845 (5.7%)	10,630 (10.4%)	6,840 (6.7%)	16,285 (15.9%)
1985	100,900	61,255 (60.7%)	5,280 (5.2%)	12,500 (12.4%)	6,050 (6.0%)	15,815 (15.7%)
1986	96,450	55,990 (58.1%)	4,665 (4.8%)	12,545 (13.0%)	6,395 (6.6%)	16,855 (17.5%)
1987	102,600	57,755 (56.3%)	5,075 (4.9%)	13,940 (13.6%)	6,075 (5.9%)	19,755 (19.3%)
1988	102,900	61,790 (60.0%)	5,080 (4.9%)	12,325 (12.0%)	5,840 (5.7%)	17,865 (17.4%)
1989	100,700	58,250 (57.8%)	6,000 (6.0%)	12,580 (12.5%)	6,010 (6.0%)	17,860 (17.7%)
1990	100,300	57,100 (56.9%)	6,115 (6.1%)	14,200 (14.2%)	6,630 (6.6%)	16,255 (16.2%)
1991	103,300	55,830 (54.0%)	5,355 (5.2%)	15,065 (14.6%)	6,600 (6.4%)	20,450 (19.8%)
1992	97,700	52,290 (53.5%)	5,580 (5.7%)	14,645 (15.0%)	7,045 (7.2%)	18,140 (18.6%)
1993	101,500	52,885 (52.1%)	5,595 (5.5%)	16,675 (16.4%)	6,545 (6.5%)	19,800 (19.5%)
1994	95,400	52,875 (55.4%)	5,930 (6.2%)	11,810 (12.4%)	6,780 (7.1%)	18,005 (18.9%)
1995	94,500	50,640 (53.6%)	5,230 (5.5%)	13,500 (14.3%)	7,275 (7.7%)	17,855 (18.9%)
1996	87,150	45,725 (52.5%)	6,315 (7.2%)	12,630 (14.5%)	6,200 (7.1%)	16,280 (18.7%)
1997	85,400	40,920 (47.9%)	5,410 (6.3%)	14,880 (17.4%)	6,510 (7.6%)	17,680 (20.7%)
1998	87,500	43,080 (49.2%)	7,070 (8.1%)	13,960 (16.0%)	7,055 (8.1%)	16,335 (18.7%)
1999	88,500	45,550 (51.5%)	5,890 (6.7%)	13,565 (15.5%)	7,705 (8.7%)	15,790 (17.8%)
2000	84,550	43,065 (51.0%)	4,700 (5.6%)	13,660 (16.2%)	7,400 (8.8%)	15,725 (18.6%)
2001	82,250	41,395 (50.3%)	4,640 (5.6%)	14,140 (17.2%)	6,915 (8.4%)	15,160 (18.4%)
2002	80,800	37,860 (46.9%)	5,805 (7.2%)	15,095 (18.7%)	7,600 (9.4%)	14,440 (17.9%)
2003	78,750	38,045 (48.3%)	5,200 (6.6%)	13,855 (17.6%)	7,100 (9.0%)	14,550 (18.5%)
2004	75,840	36,880 (48.6%)	4,840 (6.4%)	13,150 (17.3%)	6,720 (8.9%)	14,250 (18.8%)
2005	80,100	41,950 (52.4%)	5,455 (6.8%)	12,250 (15.3%)	7,120 (8.9%)	13,325 (16.6%)
2006	83,400	44,210 (53.0%)	4,745 (5.7%)	13,090 (15.7%)	7,665 (9.2%)	13,690 (16.4%)
2007	80,100	38,340 (47.9%)	4,925 (6.1%)	15,435 (19.3%)	7,735 (9.7%)	13,665 (17.1%)

Note: Prior to 1981, the categories were somewhat different.

FIGURE 8-5 Firefighter injuries by type of duty. (*Source: Michael J. Karter, Jr., United States Firefighter Injuries 2007, NFPA, October 2008 and previous reports in series.*)

■ *Environmental.* This category includes improper guarding of other hazardous work elements and degradation of equipment through careless use and unsafe procedures.

The major contribution of this theory is to show that rarely, if ever, is an accident the result of a single cause or act.

Pure Chance Theory

According to the pure chance theory, every one of any given set of workers has an equal chance of being involved in an accident. It further implies that there

is no single discernible pattern of events that leads to an accident. In this theory, all accidents are treated as acts of God, and there are no existing interventions to prevent them.

Symptoms versus Causes Theory

The symptoms versus causes theory is not so much a theory as an admonition to be heeded if accident causation is to be understood. Usually, when investigating accidents, we tend to fasten upon the obvious causes of the accident to the neglect of the root

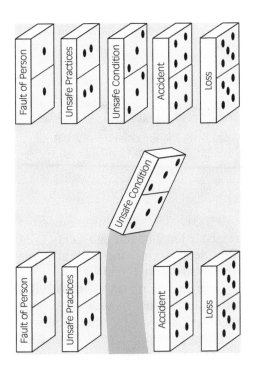

FIGURE 8-6 The Domino Theory. Removing any element breaks the chain and reduces the risk of an accident.

causes. Unsafe acts and unsafe conditions are often the symptoms—the proximate causes—and not the root causes of the accident.

Structure of Accidents

The belief that accidents are caused and can be prevented suggests that we study those factors that are likely to favor the occurrence of accidents. (A good example is the study that follows any large-loss fire.) By studying such factors, the root causes of accidents can be isolated, and necessary steps can be taken to prevent the recurrence of the accidents. These root causes can be grouped as "immediate" and "contributing." The immediate causes are unsafe acts of the worker and unsafe working conditions. The contributing causes could be management-related factors, the environment, and the physical and mental condition of the worker. In most cases, a combination of causes must converge in order to result in an accident.

Summary of Accident Causation Theory

Accident causation is a very complex field. No one theory dominates; in fact, many theories are commonly taught. From all of these, we can learn two things: Accidents do not just happen, and through good preventative action, we can prevent many accidents.

THE NATIONAL FIRE FIGHTER NEAR-MISS REPORTING SYSTEM

The National Fire Fighter Near-Miss Reporting System (www.firefighternearmiss.com) is a voluntary, confidential, nonpunitive, and secure reporting system with the goal of improving firefighter safety.

Submitted reports will be reviewed by fire service professionals. Identifying descriptions are removed to protect your identity. The report is then posted on this web site for other firefighters to use as a learning tool.

REGULATORY AND STANDARDS SETTING BODIES

There are several federal and nongovernmental entities that have established standards related to firefighting and firefighter safety.

The Occupational Safety and Health Administration (OSHA) is a regulatory agency under the U.S. Department of Labor that was created by Congress in 1970. Its role is to work with employers and workers to reduce workplace hazards and to implement new, or to improve existing, workplace health and safety programs. One of OSHA's functions is to develop and enforce mandatory job safety and health standards.

The National Institute of Occupational Safety and Health (NIOSH) is an agency of the Department of Health and Human Services that conducts research on various safety and health problems, provides technical assistance to OSHA, and recommends standards for OSHA's adoption. NIOSH establishes permissible exposure limits to chemicals and other workplace hazards, and tests and approves self-contained breathing apparatus.

The U.S. Environmental Protection Agency (EPA) has created several regulations that affect firefighters in the area of hazardous materials response. The Superfund Amendments and Reauthorization Act (SARA) of 1986 got the EPA involved in regulating response to hazardous material incidents. Title I of SARA made training and response planning mandatory for all fire departments that intend to respond to hazardous materials incidents. EPA regulations also help to support local efforts for hazardous materials safety for communities and the environment. Title II of SARA includes the Emergency Planning and Community Right-to-Know Act, which requires the EPA to work with states and localities to plan for hazardous materials incidents and to give communities the tools to obtain information about hazards in their localities. Under this section, businesses and citizens who store hazardous materials are required to inform the fire department about the types of materials, their quantities, and their hazards.

The U.S. Department of Transportation (DOT) is involved in firefighter health and safety issues in several ways. First, DOT regulates the construction, testing, and maintenance of compressed gas cylinders. Second, DOT has umbrella authority for developing hazardous materials transportation safety policy and helps provide sources of funding for hazardous materials training. For example, the Hazardous Materials Transportation Uniform Safety Act of 1990 directed DOT to collect registration fees from shippers and carriers of hazardous materials. These funds are then put into a national grants program to help train emergency responders to deal with hazardous materials incidents. DOT has also established requirements for drivers of certain motor vehicles (including fire trucks in some states) to obtain commercial drivers' licenses. Finally, DOT has established standards for training of emergency medical responders as well as standards for equipment carried on ambulances.

The foregoing are government agencies. In addition, we have the National Fire Protection Association (NFPA). Although it has no regulatory enforcement powers, it serves an extremely important function by bringing together experts to write and update standards that affect firefighting and fire protection. The NFPA works through a consensus process, which means that experts and interested parties representing different points of view are assigned to a technical committee where they present and debate ideas and ultimately agree on the most reasonable language for each standard.

FIRE DEPARTMENT OCCUPATIONAL SAFETY AND HEALTH PROGRAM

In 1987, the membership of NFPA adopted a new and somewhat controversial document entitled NFPA 1500, *Standard on Fire Department Occupational Safety and Health Program*. Although NFPA 1500 introduced many new concepts to the fire service, much of its content is taken from well-accepted practices of private industry. NFPA 1500 is the start of a major undertaking; it provides a framework for a series of new standards that would help fire departments plan, implement, and manage effective health and safety programs.

NFPA 1500 has had a significant impact on the fire service. NFPA 1500 was designed to help fire departments reduce the frequency and severity of accidents and injuries to their members. It has increased the awareness level about safety in every aspect of the job. Equipment has changed, procedures have changed, and new incident management systems and personnel accountability systems have been added.

But no law or standard can make you safe. In spite of all of these advances, personal safety is still a matter of your own attitude.

> **FIREGROUND FACT**
>
> NFPA 1500 was designed to help fire departments reduce the frequency and severity of accidents and injuries to their members.

NFPA 1500 contains the "minimum requirements" for a fire department health and safety program. The term "fire department" includes all organizations that provide fire suppression functions, including private, government, military, industrial, and public sector organizations. As used in the standard, the term "member" applies to volunteers, whether paid or unpaid, and paid personnel, whether full-time or part-time, of any such department. As company officers, you should be familiar with its contents. The following is a brief summary of its contents with some commentary.

Chapter 1. Administration

The first chapter of NFPA 1500 defines the scope, purpose, and application of the standard and defines the terms used in the following chapters.

Chapter 2. Referenced Publications

This chapter, new to the current NFPA 1500, provides an extensive list of publications referred to within the standard.

Chapter 3. Definitions

Chapter 3 is also new. It provides an extensive glossary of safety-related terms.

Chapter 4. Fire Department Administration

Chapter 4 of NFPA 1500 explains risk management, provides information for starting a health and safety program, and defines the role of the various officials. It requires a written health and safety policy and establishes goals to help reduce accidents and injuries. This chapter also discusses the concept of risk and risk management, the role of the department's **health and safety officer**, and the value of accident investigations.

Section 5.5 of NFPA 1521, the *Standard for Health and Safety Officer*, provides the following: The health and safety officer shall investigate, or cause to be investigated, all occupational injuries, illnesses, exposures, and fatalities, or other potentially hazardous conditions involving fire department members and all accidents involving fire department vehicles, fire apparatus, equipment, or fire department facilities, develop corrective recommendations that result from accident investigations, and shall submit such corrective recommendations to the fire chief or the fire chief's designated representative.

Source: Reproduced with permission from NFPA 1521, Fire Department Safety Officer, Copyright©2008, National Fire Protection Association. This reprinted material is not the complete and official position of the NFPA on the referenced subject, which is represented only by the standard in its entirety.

The Health and Safety Officer

According to NFPA 1500, every fire department should have an individual assigned to the duties of a health and safety officer. The **health and safety officer (HSO)** must be an officer from within the agency and should be formally appointed by the fire chief or the head of the agency. The health and safety officer must have access to the chief and other senior officers in the department. As such, the HSO also needs the personal skills to work with people and build strong collaborative relationships.

The health and safety officer's role varies, depending upon the direction of the chief and the support one gets from other members of the staff. The HSO should have a thorough knowledge of activities undertaken by the department and the hazards associated with these activities. The HSO should also be familiar with the department's policies as well as the laws, standards, and regulations that affect firefighter safety.

The HSO's most important job is the implementation and management of the department's safety program. In some corporations, this person is frequently called the organization's "risk manager." Although the ultimate responsibility for the safety of the department's personnel lies with the fire chief, the chief can delegate some of the daily management activity of the health and safety program to the HSO.

The HSO's most important job is the implementation and management of the department's safety program.

While the HSO's normal activity is administrative in nature, the HSO should also be available to assist at the scene of an emergency as the **incident safety officer (ISO)**. Such events allow the HSO to see firsthand the activities of the fire department's personnel at work. At times like these, training and procedural deficiencies become evident. The incident commander may not have time to notice or record these problems, but the ISO should. Corrective action can be implemented later.

Investigating the Cause of Accidents

One of the HSO's duties is to investigate accidents. Accidents should be investigated whenever someone is injured or when fire department equipment or property is damaged. Accident investigations tend to be clouded with a negative gloom, especially when a

DUTIES FOR THE HSO

- Developing standard operating procedures (SOPs) for activities that present risks to the organization, including procedures to deal with both routine and unusual events
- Developing and providing safety training programs
- Developing and managing injury and exposure reporting systems
- Reviewing draft directives to ensure that safety considerations are adequately addressed
- Developing and managing a viable fitness program with periodic testing for all personnel in the organization
- Reviewing regular and special reports that deal with firefighter health and safety issues

- Researching, testing, and evaluating planned purchases of such items as apparatus and protective clothing and equipment
- Keeping up-to-date on new laws, regulations, and standards that affect firefighter safety and health issues and keeping appropriate members of the department informed of the impact of these documents
- Investigating accidents that result in injury or property damage
- Respond to work incidents, including fires, hazardous material incidents, and technical rescue activities where firefighters are at risk, and identifying and mitigating such risks
- Attending critiques of such incidents
- Attending safety committee meetings

department uses the investigation to find fault with an individual rather than with the environment in which the accident occurred. As a result, many people get defensive when the department starts asking questions about their performance. Accident investigations play an important part in a department's overall safety program. Accident investigations should focus on finding ways to improve workplace safety and should become the basis for developing new procedures and training programs to help prevent similar circumstances from happening again. In many cases, the accident investigation and its report are required by law.

OFFICER ADVICE

Accident investigations should focus on finding new ways to improve workplace safety and become the basis for developing new procedures and training programs to help prevent similar circumstances from happening again.

Any fire officer may be asked to conduct an accident investigation. Asking officers to conduct accident investigations will increase their knowledge of the investigation process and increase their awareness of the causes and consequences of common accidents, as well as gather the information needed by the investigation process.

Why Accident Investigations and Reporting are Important

There are many good reasons for investigating accidents. Let us start with the legal requirements. For accidents involving injuries, department or local government policy commonly requires an accident investigation. OSHA and NFPA 1500 also require such investigations. In cases involving vehicles, accident investigations are required by the state department of motor vehicles (or similar agency).

The information gathered during an investigation—possibly required by procedure, policy, or regulation—provides a wealth of information that can be used for a variety of purposes:

- To identify sources of accidents by determining and documenting the materials and equipment frequently involved in accidents
- To compile information that will allow comparisons of accident data from other jurisdictions and agencies
- To indicate a need for a revision of an unsafe practice
- To indicate a need for the modification of the design of equipment or apparatus

- To indicate a need for revising a procedure or practice
- To help reduce or eliminate the conditions that led up to the accident
- To establish or revise policies that would prevent similar accidents
- To train and instruct personnel on ways to prevent the accident from reoccurring
- To serve legal purposes, either to establish a basis for filing a claim or to defend against a claim filed against the fire department

In all cases, the principal reasons for investigating and documenting accidents are to prevent similar situations in the future. The knowledge gained from the information gathered during such investigations should become the basis for training and for sharing the experience with others to inform them of the hazards involved. "Training shall address recommendations arising from the investigation of accidents, injuries, occupational deaths, illnesses, and exposures and the observation of incident scene activities" (Section 4.9.2 of NFPA 1500).

Usually, the investigation process involves filling out various forms. The forms help guide the investigation and help the officer conducting the investigation obtain and record all of the necessary information.

What Should Be Investigated?

NFPA 1521, Standard for Fire Department Safety Officer requires an investigation for all occupational injuries, illnesses, exposures, and fatalities, or other potentially hazardous conditions involving fire department members, and all accidents involving fire department vehicles, fire apparatus, equipment, or fire department facilities. Usually, department policies spell out the threshold for investigations, in terms of severity of injury or estimated property damage. In general, all accidents with injuries requiring medical care and all accidents in which property damage exceeds a certain value should be investigated.

Who Should Investigate?

Some departments appoint a health and safety officer and make that person the designated accident investigation officer. Some departments use command officers (for example, battalion chiefs) for accident investigations. Where vehicles are involved, some departments and some state and local government policies require that law enforcement personnel conduct the investigation.

As a general rule, the investigation should be conducted by someone not involved with the accident or in the chain of command of those who were involved in the accident. Using investigators with no vested interest in the outcome of the investigation will avoid even the appearance of a conflict of interest. Another consideration is the seriousness of the accident. The seniority and the experience of the investigating officer should also match the seriousness of the event: A fatality will require a much more detailed investigation than a $500 fender bender.

Responder Injuries

Depending upon the severity of the injury, the health and safety officer or other designated officer may conduct the investigation. Where equipment failure is known or suspected, the equipment should be impounded for further testing by appropriate authorities. Statements from the injured person(s) and witness(es) should be obtained as soon as possible. Today, it seems that everybody has a camera; if the accident was recorded on film, try to get a copy of the print or tape. In any case, take photographs of the accident scene.

Responder Fatalities

The death of a responder is a stressful event for any organization. While there is concern for family and teammates of the lost member, there are several agencies that will require information about the event, and many require prompt reporting. There are several places to go for assistance: Both the International Association of Fire Chief (IAFC) and the International Association of Firefighters (IAFF) provide information to their members, and the federal Public Safety Officers Benefit (PSOB) program is explained on the web at http://www.ojp.gov/BJA/grant/psob/psob_main.html. The National Fallen Firefighter Foundation provides the following information at www.firehero.org.

Here are important steps that a department needs to take to help the firefighter's family, members of the department, and the community.

1. Notify the family of the fallen firefighter.
2. Once you are sure the family has been notified, get information to members of the department, local and state officials, and the National Fallen Firefighters Foundation. Help is available from the Foundation's Local Assistance State Teams. To report a Line of Duty Death (LODD) please call toll free at 1-866-736-5868.
3. Contact the Department of Justice's Public Safety Officers' Benefits (PSOB) Program. When you

report a firefighter death, have basic information available on the incident, your department, and the fallen firefighter and his or her immediate next-of-kin. You may contact the PSOB Office at 1-888-744-6513.

PSOB offers a lump sum death benefit to survivors of public safety officers who die in the line of duty from a traumatic injury or heart attack. There are many procedures that need to be followed. Call PSOB even if you are not sure whether your firefighter's family will qualify for benefits under this program.

4. Use a checklist to determine what needs to be done immediately, before and during the funeral, and longer term. Be sure you know what the requirements are in your jurisdiction for conducting an autopsy.

If you would like to speak directly with another senior fire officer who has lost a firefighter in the line of duty and can offer some professional and personal support, please contact the Foundation.

5. Find out what benefits exist for survivors of fallen firefighters in your state. Then start contacting the state officials for each program. Benefits may include lump-sum death payments, workers' compensation, funeral benefits, pensions and retirement programs, scholarships, and nonprofit/private support.
6. At the family's request, begin preparations for a fire service funeral or memorial service. A comprehensive Funeral Guide will help you plan a fitting tribute. Download a copy or call the Foundation to have one sent immediately.

Accidents Involving Vehicles

Accidents involving vehicles (with or without fatalities and injuries) require a report because of the damage to government equipment and private property. Because of the complexities of such investigations, the health and safety officer or other designated officers may be needed for such investigations. In many jurisdictions, the police conduct the investigation. However, we must be careful here. Since both the fire department and police department are part of the same local government, many jurisdictions—to avoid any apparent conflict of interest by the police investigators—ask the state police to conduct the investigation where there is damage to civilian property.

If the fire department conducts the investigation, there are specific reports required for traffic accidents. In addition to completing the appropriate reports, the investigating officer should take photographs of the overall accident scene as well as close-up photographs showing damage. Such photographs should provide the following information:

- Position of the final resting places of all vehicles involved in the accident
- Accident damage to each vehicle including close-up photographs
- Accident damage to any objects relevant to the accident
- Evidence of the path of travel both before and after the impact
- Evidence of the point of impact (tire marks, debris, and so on)
- The view that each driver had when approaching the accident scene

When vehicle or equipment failure is suspected or known to be a factor in the collision, the vehicle or equipment should be impounded for further testing. When fire department vehicles are involved in an accident, they should be carefully inspected for damage before being placed back in service.

In such cases, department policy should require that all persons involved in the accident be tested for use of alcohol or drugs. Because of the urgency for such tests, all officers should be aware of any such requirements and comply with the organization's policies.

Benefits

No one wants an accident to occur, but when one does occur, we should try to learn something from the experience. After the accident, the following should be routine:

- Investigate the accident.
- Identify the causes.
- Report the findings.
- Develop a plan for corrective action.
- Implement the plan.
- Evaluate the effectiveness of the corrective action.
- Make changes for continuous improvement.

A safety program that routinely follows these steps will help reduce accidents (see **Figure 8-7**).

Health and Safety Committee

Important provision for improving the health and safety of members in any organization is the creation of a safety committee. To be effective, the committee should include members from management, member organizations, and the safety officer.

Safety committees have several functions: They recommend policy, respond to reports of unsafe conditions or practices, provide information for the organization and its members, and provide research and recommendations in matters pertaining to health and safety issues.

So Who Is Really Responsible for Safety?

The fire chief shall have the ultimate responsibility for the fire department occupational safety and health program as specified in NFPA 1500, *Standard on Fire Department Occupational Safety and Health*

FIGURE 8-7 Safety programs, utilization of PPE, and safety officers can have a significant positive impact on reducing injuries and fatalities.

Program. A health and safety officer (HSO) shall be assigned to manage the fire department occupational safety and health program. The HSO shall report directly to the fire chief or to the fire chief's designated representative.

Each fire department should have a designated HSO. The HSO can be assigned as a full-time or part-time position, depending on the size and character of the fire department. Additional assistant health and safety officers shall be appointed when the activities, size, or character of the fire department warrants extra safety personnel. If the health and safety officer is not available, additional assistant health and safety officers shall be appointed to ensure proper coverage.

The incident commander shall have ultimate responsibility for incident scene safety as specified in NFPA 1561, *Standard on Emergency Services Incident Management System*. An incident safety officer (ISO) shall be appointed when activities, size, or need occurs. The incident safety officer function can be pre-designated or appointed by the incident commander as needed. If the pre-designated incident safety officer is not available, the incident commander shall appoint an incident safety officer (see **Figure 8-8**).

The function of incident scene safety must be carried out at all incidents. It is the responsibility of the incident commander who cannot perform this function, due to the size or complexity of the incident, to assign or request response of an incident safety officer to do this function. There are, however, incidents that require immediate response or appointment of an incident safety officer, such as a hazardous materials incident or special operations incidents. These types of incidents should be defined in the fire department's response policy or procedure to ensure that the incident safety officer responds.

Because you, as a company officer, could well be assigned as an incident safety officer, let us review

FIGURE 8-8 The incident safety officer is just as important at training fires as at real incidents.

the duties of that position, as outlined in NFPA 1521, *Standard for Fire Department Safety Officer*:

At All Incidents

- The incident safety officer shall monitor conditions, activities, and operations to determine whether they fall within the criteria as defined in the fire department's risk management plan.

- The incident safety officer shall ensure that the incident commander establishes an incident scene rehabilitation tactical level management unit during emergency operations.

- The incident safety officer shall monitor the scene and report the status of conditions, hazards, and risks to the incident commander.

- The incident safety officer shall ensure that the fire department's personnel accountability system is being utilized.

- The incident commander shall provide the incident safety officer with the incident action plan. The incident safety officer shall provide the incident commander with a risk assessment of incident scene operations.

- The incident safety officer shall ensure that established safety zones, collapse zones, hot zones, and other designated hazard areas are communicated to all members present on scene.

- The incident safety officer shall evaluate motor vehicle scene traffic hazards and apparatus placement and take appropriate actions to mitigate hazards.

- The incident safety officer shall monitor radio transmissions and stay alert to transmission barriers that could result in missed, unclear, or incomplete communication.

- The incident safety officer shall communicate to the incident commander the need for assistant incident safety officers due to the size, complexity, or duration of the incident.

- The incident safety officer shall survey and evaluate the hazards associated with the designation of landing zone and working with helicopters.

During Fire Suppression Operations

- The incident safety officer shall ensure that a rapid intervention crew is available and ready for deployment.

- Where fire has involved a building or buildings, the incident safety officer shall advise the incident commander of hazards, collapse potential, and any fire extension in such building(s).

- The incident safety officer shall evaluate visible smoke and fire conditions and advise the incident commander, tactical level management units' officers, and company officers on the potential for flashover, backdraft, blow-up, or other fire events that could pose a threat to operating teams.

- The incident safety officer shall monitor the accessibility of entry and egress of structures and the effect it has on the safety of members conducting interior operations.

During Emergency Medical Service Operations

- The incident safety officer shall ensure compliance with the department's infection control plan and NFPA 1581, *Standard on Fire Department Infection Control Program, during EMS operations.*

- The incident safety officer shall ensure that incident scene rehabilitation and critical incident stress management are established as needed at EMS operations, especially mass casualty incidents (MCIs).

During Hazardous Materials Operations

- The hazardous materials incident safety officer shall meet the requirements of NFPA 472, *Standard for Professional Competence of Responders to Hazardous Materials Incidents.*

During Special Operations

- The individual who serves as the incident safety officer for special operations incidents shall have the appropriate education, training, and experience in special operations.

- The incident safety officer shall attend strategic and tactical planning sessions and provide input on risk assessment and member safety.

- The incident safety officer shall ensure that a safety briefing, including an incident action plan and an incident safety plan, is developed and made available to all members on the scene.

- The incident safety officer shall meet with the incident commander to determine rehabilitation, accountability, or rapid intervention needs. For long-term operations, the incident safety officer shall ensure that food, hygiene facilities, and any other special needs are provided for members.

Infectious Disease Control Program

Most fire departments have a designated infection control office; however, this can be included as one of the many duties of the HSO. Regardless, this person is responsible for developing policies and procedures referred to as the *infection control plan*. This plan is written to cover both the spread of infections and the investigation of suspected exposures to infections. The plan should include the following elements:

- *Records maintenance section*—Medical records are strictly confidential for both employees and citizens, so it is essential to have a section that lists how these records will be maintained. It should explain that the employee is allowed to review his or her own medical record. In turn, it should explain what is contained within those files. Often, the employee's medical file includes all memos referring to medical conditions, leave adjustment forms that pertain to medical comments, medical history forms, and immunization records. In addition, patient records cannot be released without the consent of the patient; this is included in the 1996 HIPAA the Health Insurance Portability and Accountability act. For more information on HIPAA visit the Center for Medicare & Medicaide Services (CMS) website at http://www.cms.hhs.gov/HIPAAGenInfo/. For fire departments to be legally covered, they should have an approved HIPAA release of information form that needs to be signed by the patient. This form needs to be approved by the department's legal representative. When in doubt, do not release any information. Under HIPAA regulations, the person who releases the information is liable. In addition to federal issues, the patient could also take civil action.

- *Education section*—This section should outline the steps for initial and annual refresher trainings. Personnel should be trained in the following areas:
  - Epidemiology and symptoms of bloodborne diseases
  - Transmission modes
  - The entire exposure control plan
  - Methods for recognizing activities that may involve exposures
  - Information on the types, proper uses, location, removal, handling, decontamination, and disposal of personal protective equipment (PPE)
  - Information on vaccinations
  - Actions to be taken in the event of an exposure

- *General sanitation section*—This section involves the steps for cleaning and sanitizing equipment and

PPE. At no time should kitchens, bathrooms, or living areas be used to disinfect or decontaminate any equipment or PPE.

- *PPE section*—This section should include departmental policies and procedures for the use of PPE. It should list the department and individual responsibilities regarding the use of PPE.

- *Preresponse and postresponse section*—This section will outline the steps for removal and replacement of equipment that has been contaminated.

- *Exposure section*—This section is a guide for employees and supervisors in the management of any and all exposures. It should outline all of the steps that need to take place in the event of an exposure.

- *Program evaluation section*—This section summarizes the quality monitoring of the program, the steps for inspecting station facilities, analysis of reported exposures, and a quality and compliance report.

Fire department personnel are susceptible to all types of health exposures both inside and outside the station walls. Fire officers should ensure that all personnel are trained on the hazards that they face as well as how to deal with them upon being exposed.

Chapter 5. Training and Education

Chapter 5 of NFPA 1500 provides direction for training programs at all levels. Firefighter safety is closely related to good firefighter training. When new recruits enter the fire service, they enter a new culture that includes training on a variety of basic procedures, many of which will remain with the firefighters for the rest of their lives. That training must include information about firefighter safety. While safety should be a topic of instruction itself, it is also important to incorporate safety into all training activities. For example, when teaching the proper use of power tools, safety considerations must always be included.

If firefighters are trained properly, they will know the correct actions to take during emergency activities. If the training is sufficient, they will have acquired good, safe working habits and, hopefully, will have forgotten any bad habits that they may have brought with them before entering the fire service. During times of stress, and sometimes when no one else is looking, firefighters may have the tendency to forget some of the good habits they learned during their recruit training. Many accidents occur during these moments. Safety practices and good habits learned during recruit training must be reinforced throughout the firefighter's work life. As the company officer, you have a leading role in this activity, both by setting a positive personal example and by reinforcing safety practices at all times. Correcting unsafe acts during training, and even during actual events, reduces accidents among firefighters. The approach should be positive and constructive.

SAFETY

Correcting unsafe acts during training, and even during actual events, reduces accidents among firefighters.

Firefighter Training Need Not Be Dangerous

It is important that fire departments train firefighters in conditions that are as realistic as possible, while at the same time protecting firefighters from harm. The challenge of accomplishing both is very real. Since 1987, reported training-related injuries have increased by nearly 21 percent. In 2001 alone, almost 7,000 training-related injuries were reported. That year, a firefighter died and two were injured while participating in a live-burn training situation.

Live-fire training buildings and simulators, fed by propane or natural gas, have many built-in safety features. Unfortunately, they may not provide the same quality of realism as live-fire training in acquired structures. The training buildings, for example, fail to teach students to react to the diverse conditions encountered in real fire operations. In addition, the need to provide training in situations other than fires, such as hazardous material incidents, increases the danger to participants. In 1997, for example, a member of a unit training to respond to a chemical incident was overcome by a nerve agent and needed to be administered an antidote to stop the seizures.

The U.S. Fire Administration (USFA), after reviewing the risks of firefighting training and identifying the risks, has recommended ways to reduce deaths and injuries associated with training activity. The USFA report underscores the inherent danger of such training, but it also reiterates the importance of experience gained in real, rather than closely controlled, training fires. The report also highlights the importance of following currently accepted procedures and standards to avoid training deaths and injuries and the need for instructors to avoid situations for which the students are not yet prepared. In that regard, students should be trained and certified to the Firefighter I level before being exposed to the hazard of live-fire training.

Lessons learned cited in the special report include the following:

- It is increasingly important that firefighters receive training in fire behavior and extinguishment methods for different types of buildings.

- Because modern protective equipment may make life-threatening fire conditions less obvious, firefighters must be trained to recognize the visual and physical clues to impending danger.
- During training, a firefighter's physical stress level should be monitored continuously, and departments should consider stronger physical screening programs and long-term health and wellness programs to reduce training-related heart attacks and strokes.

Where these training activities are conducted in acquired structures, extreme care must be given to the condition of the building. These structures are typically old and burn readily. Where these training activities are conducted in training facilities, care must be given to the type and size of fire that is built. Even the best live-fire training facility will fail when repeatedly subjected to intense fires and firefighting activity. NFPA 1403, *Standard on Live Fire Training Evolutions in Structures*, provides directions for safe training in either environment, and it must be understood and followed if training is to be conducted safely. In addition, it is important that the training be conducted in accordance with a plan and under the supervision of persons who have the experience needed to conduct such training activity.

SAFETY

The training should be conducted in accordance with a plan and under the supervision of persons who have the experience needed to conduct such training activity.

UNNECESSARY DANGERS CREATED DURING TRAINING

We continue to see evidence of training efforts that, although well-intended, result in dangerous situations, especially for the newest members of fire departments. Some departments look upon live-fire training for their recruits as an initiation ceremony or rite of passage in which the instructors build fires that will cause at least a few of the new members to retreat. (Instructors with melted visors should be considered suspect.) Every year we see reports of injuries and occasionally even deaths resulting from such reprehensible conduct.

Chapter 6. Fire Apparatus, Equipment, and Drivers/ Operators

Chapter 6 of NFPA 1500 provides directions for the safe design and operation of fire department vehicles, tools, and equipment. Probably the most visible impact of NFPA 1500 is apparent in the design of today's fire apparatus.

Vehicle Safety Is Important, Too

The second leading cause of firefighter fatalities is the result of motor vehicle accidents. Over the past 20 years, motor vehicle accidents have accounted for between 20 and 25 percent of all firefighter fatalities. Approximately 25 percent of these fatalities were firefighters who died while responding to an emergency in their own vehicle. The second largest category (about 20 percent) involved accidents with fire department tanker vehicles. In both categories, failure to use seat belts was a contributing factor. In many cases, these accidents were preventable.[5]

In many cases, the accident involved injuries and property damage to civilians as well as to the firefighters. This activity is inconsistent for an organization that has a mission of saving lives and protecting property.

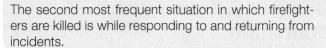

FIREGROUND FACT

The second most frequent situation in which firefighters are killed is while responding to and returning from incidents.

Although you may think of accidents involving apparatus and volunteers' privately owned vehicles racing to the scene as the source of these injuries, many of these injuries occur as the result of less spectacular acts. Firefighters sometimes slip and fall while running to board apparatus. Watching the firefighters' response at a well-disciplined fire station is almost a nonevent; the firefighters walk rapidly to their assigned apparatus, deliberately put on their gear, and mount up with little if any conversation. All of this is easily done in less than a minute. As a result, the crew leaves the station without gasping for air or pounding hearts. Their trip to the scene and their performance at the scene are certain to be safer.

While the driver of the emergency vehicle is responsible for the safe operation of the vehicle, you, as the company officer, are responsible for the safety of all the personnel in the company and the citizens with whom they come into contact. In the final analysis you, as the officer, are responsible for the safe operation of the vehicle.

Before the apparatus moves, all firefighters should be seated and their seat belts properly secured. This is usually routinely accomplished when the company is departing the station, but it is sometimes forgotten when the response starts from another site. Take that extra moment to make sure everyone is present and they are all properly secured. Once the response starts, you will be busy with traffic, both radio and vehicular.

Surprisingly, accidents also occur when crews are returning to quarters. Some of this is due to fatigue and the letdown that comes after the excitement has passed. You must keep your guard up for such events. Once the apparatus is back in quarters, all hands should work together to help restore the vehicle to a "ready" status for the next call.

We must remember that issues involving apparatus design and use were on everybody's mind in the mid-1980s when NFPA 1500 was developed and adopted. Chapter 4 addresses the role of the driver as well as the design of the vehicle. Provisions of NFPA 1500 indicate that the driver is responsible for the safe operation of the vehicle; for seeing that all firefighters are seated and secured before moving the vehicle; and for obeying all traffic laws, rules of the road, and all department policies regarding the use of emergency vehicles. The standard also indicates that the officer is responsible for the overall safety of assigned personnel and equipment, including the actions of the driver.

FIREGROUND FACT

According to NFPA there were an estimated 16,020 collisions in 2006 involving fire department emergency vehicles, where departments were responding to or returning from incidents. These collisions resulted in 1,250 fire fighter injuries.

Also, 1,070 collisions involving fire fighters' personal vehicles occurred in 2006 while responding to or returning from incidents. These collisions resulted in an estimated 210 injuries.

NFPA 1500 Requirements for Safe Operations of Fire Apparatus

■ Fire apparatus shall be operated only by members who have successfully completed an approved driver-training program or by trainee drivers who are under the supervision of a qualified driver.

■ Drivers of fire apparatus shall have a valid driver's license.

■ Vehicles shall be operated in compliance with all traffic laws, including sections pertaining to emergency vehicles, and any requirements of the authority having jurisdiction.

■ The fire department shall enact specific rules and regulations pertaining to the use of private vehicles for emergency response. These rules and regulations shall be at least equal to the provisions regulating the operation of fire department vehicles.

■ Drivers of fire apparatus shall be directly responsible for the safe and prudent operation of the vehicles under all conditions.

■ When the driver is under the direct supervision of an officer, that officer shall also assume responsibility for the driver's actions.

■ Drivers shall not move fire apparatus until all persons on the vehicle are seated and secured with seat belts in approved riding positions.

■ Drivers of fire apparatus shall obey all traffic control signals and signs and all laws and rules of the road of the jurisdiction for the operation of motor vehicles.

■ The fire department shall develop standard operating procedures for safely driving fire apparatus during nonemergency travel and emergency response and shall include specific criteria for vehicle speed, crossing intersections, traversing railroad grade crossings, and the use of emergency warning devices. Procedures for emergency response shall emphasize that the safe arrival of fire apparatus at the emergency scene is the first priority.

■ During emergency response, drivers of fire apparatus shall bring the vehicle to a complete stop under any of the following circumstances:

  □ When directed by a law enforcement officer
  □ Red traffic lights
  □ Stop signs
  □ Negative right-of-way intersections
  □ Blind intersections
  □ When the driver cannot account for all lanes of traffic in an intersection
  □ When other intersection hazards are present
  □ When encountering a stopped school bus with flashing warning lights

■ Drivers shall proceed through intersections only when the driver can account for all lanes of traffic in the intersection.

■ During emergency response or nonemergency travel, drivers of fire apparatus shall use caution when approaching and crossing any guarded railroad grade crossing, shall come to a complete stop at all unguarded railroad grade crossings, and shall ensure that it is safe to proceed before crossing any railroad track(s).

Using Seat Belts and Maintaining Safe Speed

Every year firefighters are killed in collisions involving their personal vehicles as they respond to emergencies. The majority of these firefighters are volunteers.

Two separate USFA studies of emergency vehicle crashes reveal that over 70 percent of those killed in these crashes were not wearing seat belts at the time of the incident. Ironically, firefighters preach the benefits of seat belts and child safety seats. They often fail to heed their own advice.

USFA suggests the following actions to help improve the safety of firefighters:

- Fire departments should adopt a policy that mandates the use of seat belts in all fire department vehicles and in firefighters' personal vehicles while responding to or returning from an incident.

- Fire departments should place warning signs in all department vehicles to remind firefighters of their responsibility to use seat belts.

- Drivers should limit response speed to the posted speed limit. The maximum speed may be lower than the posted speed in situations such as extreme weather, darkness, or congested traffic.

- Company and chief officers who observe unsafe driving should remind the offending firefighter of the department's policy. Firefighters who do not obey the policy should face disciplinary action.

Proper Use of Warning Lights Can Enhance Firefighter Safety

To enhance firefighter safety, recent editions of NFPA 1901, *Standard for Automotive Fire Apparatus*, have given increasing attention to the warning lights on fire apparatus. The standard now requires that apparatus have warning lights that are divided into two modes of operation: responding and blocking. The responding mode shall signal other drivers and pedestrians that the apparatus is responding to an emergency and is calling for the right-of-way. The second, the blocking mode, shall signal that the apparatus is stopped and blocking the right-of-way. Movement of the parking brake and/or movement of the automatic transmission gear selector automatically accomplishes the change-over from the response mode to the blocking mode.

Under ordinary conditions the specified warning lights provide an effective and balanced warning system. However, in some situations, turning off some warning devices can increase the safety of both civilians and emergency personnel. Consider the following examples:

- If other traffic needs to pass within close proximity to the parked apparatus, the possibility of distracting other drivers can be reduced if the headlights and lower-level warning lights are turned off.

- When responding in snow or fog, it could be desirable to turn off forward-facing strobes or oscillating lights to reduce visual disorientation of the apparatus driver.

- Flashing headlights are used in many areas as warning lights and provide an inexpensive way to obtain additional warning to the front of the apparatus. Daylight flashing of the high beam filaments is very effective and is generally considered safe. However, nighttime flashing can affect the vision of oncoming drivers as well as make driving the apparatus more difficult.

In some jurisdictions, headlight flashing is prohibited. However, in all areas, if flashing headlights are employed on fire apparatus, they should be turned off when the apparatus is responding at night and the regular headlights are needed. They should also be turned off, along with all other white warning lights, when the apparatus warning lights are in the blocking mode. Steady-burning headlights are not considered warning lights; consequently, they can be used in the blocking mode to illuminate the work area in front of the apparatus. However, in such situations, careful consideration should be given to avoid shining headlights into the eyes of oncoming drivers. (Additional information can be found in Chapter 13 of NFPA 1901, *Standard for Automotive Fire Apparatus, 2009.*)

These policies are the result of reports of serious accidents involving emergency apparatus while responding. For those who say, "We do not need to follow those rules because we have not adopted NFPA 1500," you might start thinking about the defense you will use after an accident in which you or one of your personnel was involved, and the attorney for the state or the injured party asks about your awareness of a nationally recognized standard on firefighter safety and health.

Firefighter Deaths from Tank Truck Rollovers

The National Institute for Occupational Safety and Health (NIOSH) routinely looks at systematic problems in the fire service. Heart attacks and vehicle accidents have been the subject of several of their recent studies.

NIOSH reports that mobile water supply vehicles, known as tankers or tenders, are widely used to transport water to areas beyond a water supply system or where the water supply is inadequate. Incidents involving motor vehicles account for approximately 20 percent of U.S. firefighter deaths each year; cases involving tankers are the most prevalent of these motor vehicle incidents. During 1977–1999, 73 deaths occurred in 63 crashes involving tankers. Of those deaths, 54 occurred in 49 crashes in which tankers rolled over (no collision), and 8 occurred in 6 crashes in which the tankers left the road (no collision). The other cases involved collision with another vehicle (10 deaths in 7 crashes) and collision with stationary object(s) (1 death) (NFPA 2000 statistics).

Tanker drivers may not be fully aware that tanker trucks are more difficult to control than passenger vehicles. A tanker truck requires a much greater distance to stop. Tankers weigh substantially more, and their air brake systems take more time to activate than the hydraulic/mechanical brake systems on smaller passenger cars. The effect is influenced by the amount of water the tanker is hauling and whether the tanker is baffled.

To reduce the risk of tanker truck rollovers, NIOSH recommends that fire departments take the following precautions:

■ Develop, implement, and enforce standard operating procedures (SOPs) for emergency vehicles, particularly with regard to the use of seat belts.

■ Ensure that drivers have necessary driving skills and experience and provide them with periodic refresher training.

■ Consider terrain, weather, and bridge and road conditions when purchasing a mobile water supply vehicle.

■ Adhere to the requirements of NFPA 1911 *Standard for the Inspection, Maintenance, Testing, and Retirement of In-Service Automotive Fire Apparatus,* for keeping a vehicle on a maintenance schedule and documenting the performance of the maintenance.

■ Inspect the complete vehicle at least once per year to comply with federal and state motor vehicle regulations.

■ Adhere to the requirements of NFPA 1901, *Standard for Automotive Fire Apparatus* for an approved mobile water supply vehicle.

■ Equip all vehicles with seat belts.

■ Ensure that water tank capacity is adequate and has proper tank mounting and sufficient front and rear weight distribution.

■ Ensure that the weight of the fully loaded vehicle does not exceed the gross axle weight rating of any axle and the gross vehicle weight rating of the chassis.

■ Ensure that the center of gravity of the vehicle does not exceed the chassis manufacturer's specified center of gravity.

■ Provide proper baffles to control water movement for all vehicles equipped with water tanks.

■ Verify that vehicles are of proper design and have adequate suspension, steering, and braking ability.

All drivers should do the following:

■ Recognize that they are responsible for the safe and prudent operation of the vehicle under all conditions.

■ Wear a seat belt when operating a vehicle.

■ Take training to meet the job performance requirements stated in NFPA 1002 *Standard for Fire Apparatus Driver/Operator Professional Qualifications* before driving and operating the vehicle.

■ Take refresher driver training at least twice per year.

■ Understand the vehicle characteristics, capabilities, and limitations.

■ Be aware of the potential for unpredictable driving by the public (excessive speed, failure to yield to emergency vehicles, and inattentiveness).

■ Adjust speed when driving on wet or icy roads, in darkness or fog, or under any other conditions that make emergency vehicle operation especially hazardous.

For more information on NIOSH's research of firefighter safety and health issues, go to http://www.cdc.gov/niosh/fire.

NOTE

NFPA 1500 clearly states that the officer (of the unit) is responsible for the overall safety of assigned personnel and equipment. This includes providing proper supervision of the driver.

Manual on Safe Operations of Fire Tankers Available from USFA

Noting that vehicle crashes are the second leading cause of all firefighter on-duty fatalities, the USFA published a report, entitled *Safe Operation of Fire Tankers.* This manual provides comprehensive information regarding the safety practices and principles of tanker vehicles for local fire departments. *Safe Operation of Fire Tankers* provides information related to human performance (driver training and operations) and technology (vehicle design) to enhance the safety of fire tanker operations. The manual also examines past incidents of crashes involving fire tankers that have killed firefighters, with a focus on how these fatalities could have been prevented. Fire departments will find *Safe Operation of Fire Tankers* a

ALCOHOL, SPEED, AND THE LACK OF SEAT BELTS

A 16-year-old Explorer firefighter riding as a passenger died as the result of injuries sustained when a fire department tanker rolled over. Following the accident, the driver of the vehicle was charged with operating while intoxicated and aggravated vehicular homicide.

The two firefighters were responding, along with four other units, to a rekindle of railroad ties. Since the tanker rolled over, questions about the vehicle's speed and the justification for an emergency response have surfaced and may be the subject of litigation. Many articles have been written about responding to "true emergencies." Several recent articles question if a rekindle of railroad ties constitutes a true emergency worthy of a five-vehicle emergency response.

It also appears that neither the Explorer firefighter nor the driver was wearing seat belts at the time of the accident. As shown elsewhere in this chapter, failure to use seat belts is a significant contributor to the fatalities associated with accidents involving fire apparatus.

valuable resource that provides information related to the current and applicable federal standards and regulations as well as national-level consensus standards and guidelines. The document can be downloaded from the USFA web site at *http://www.usfa.dhs.gov/downloads/pdf/publications/fa-248.pdf.*

Chapter 7. Protective Clothing and Protective Equipment

Chapter 7 of NFPA 1500 sets requirements for fire departments regarding providing and requiring the wearing of appropriate PPE (see **Figure 8-9**). Although the standard deals with some aspects of the design of protective clothing, it also states that members shall be provided with protective helmets, hoods, coats, trousers, gloves, and shoes to make a complete protective ensemble for the firefighter, and that departments shall establish policies requiring members to wear their protective clothing when exposed to the dangers of the job.

> **NOTE**
>
> With the arrival of NFPA 1500, the standards, for the first time, addressed human behavior. Other NFPA standards provide specifications regarding protective clothing, apparatus, and so on. In NFPA 1500, we see a requirement for departments to issue this equipment and for firefighters to use it.

NFPA 1500 declares that self-contained breathing apparatus (SCBA) must be worn (and used) when the atmosphere is hazardous, is suspected of being hazardous, or may rapidly become hazardous. To ensure a good seal, the standard prohibits beards, spectacles, and other objects that could interfere with the seal of SCBA against the face. The standard also provides for providing and using a personal alert safety system (PASS) device.

NFPA 1500 also provides that the equipment shall be cleaned at least once every 6 months. By design, turnout gear protects you from a host of hazards. In the process of providing you with that protection, the protective clothing becomes contaminated with all the products encountered during the emergencies you encounter, including carbon monoxide, hydrogen cyanide, asbestos, bodily fluids, gasoline, diesel fuel, and other contaminants. You

FIGURE 8-9 Well-dressed firefighters ready for business. To get the intended benefit from PPE, one must use it properly.

then expose yourselves, your colleagues, other citizens, and even your family to all of that material. NFPA 1581, *Fire Department Infection Control Program*, provides additional important information.

PROTECTIVE EQUIPMENT ONLY WORKS WHEN YOU USE IT

Firefighters are expected to perform normal human feats in superheated environments. To do this, firefighters are encased in a suit of protective clothing that, despite considerable technological advances, is considered by some to be about as comfortable as medieval body armor. That discomfort is a small price to pay for the protection from fire and other sources of injuries offered by today's modern PPE. But it only works if it is worn. Having an SCBA on your back is not good enough; you have to be breathing out of it to get the intended benefit!

And wearing a PASS device is not going to help others find you unless it is turned on. When firefighters repeatedly fail to properly use their SCBA or PASS device, and you, the company officer, do nothing to change the habit, who is responsible for injuries to the firefighters when they are caught in a situation where they lose their lives because the SCBA was not used or the PASS device not activated?

SAFETY

Wearing a PASS device is not going to help others find you unless it is turned on.

Outdated Equipment and Firefighter Safety

Over the past few years, the NFPA standards for PPE, fire apparatus, and other firefighter-related equipment have embraced modern technology and made the workplace safer for firefighters. The previously mentioned Needs Assessment discussed facilities, apparatus, and equipment owned and operated by America's fire service. With regard to firefighter safety, the report noted the following:

- Over half of all fire engines are over 15 years old, and over 10 percent are over 30 years old, meaning that they were designed and built before any of the current standards for fire apparatus were in place and are likely to be deficient in many safety-related features.

- Overall, fire departments do not have enough portable radios to equip more than about half of the emergency responders on a shift.

- The majority of fire service portable radios are not water-resistant, and more than three-fourths lack intrinsic safety in an explosive atmosphere.

- An estimated one-third of firefighters per shift are not equipped with SCBA equipment. Nearly half of the SCBA units are at least 10 years old.

- Nearly half of the emergency responders per shift are not equipped with PASS devices.

- An estimated 57,000 firefighters lack adequate personal protective clothing, and an estimated one-third of all protective clothing is over 10 years old.

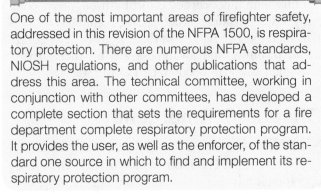

SAFETY

One of the most important areas of firefighter safety, addressed in this revision of the NFPA 1500, is respiratory protection. There are numerous NFPA standards, NIOSH regulations, and other publications that address this area. The technical committee, working in conjunction with other committees, has developed a complete section that sets the requirements for a fire department complete respiratory protection program. It provides the user, as well as the enforcer, of the standard one source in which to find and implement its respiratory protection program.

The technical committee can continue to develop and revise standards, but there must be a fundamental behavioral change in how firefighters and fire departments address fire service occupational safety. In turn, they must continue to educate their members and, most importantly, the administration, and the citizens to the hazards of the firefighting profession. The utilization and implementation of this standard can go a long way in reducing the staggering statistics involving firefighter fatalities and injuries, but only if given the training and resources to do so.

SCBA

The use of self-contained breathing apparatus (SCBA) would greatly reduce the number of firefighter asphyxiation injuries, which account for 20 percent of all firefighter injuries. The requirements for the use of SCBA in NFPA 1500 are very definitive:

SCBA should be provided and shall be used by all members working in areas where

- The atmosphere is hazardous
- The atmosphere is suspected of being hazardous
- The atmosphere may rapidly become hazardous

Many fire departments claim to have policies that mandate the use of SCBA, but in many cases, it is just a mandate to wear the cylinder. The pictures in many popular fire publications of firefighters wearing the SCBA cylinder but not using the facepiece clearly demonstrate the lack of enforcement efforts of their department's policies.

Enforcement of department policy starts at the highest level, but the real enforcer has to be you, the company officer. You should see that SCBAs are worn while doing interior firefighting, while on the roof of a burning structure, during defensive operations, and even after the fire is knocked down, because the produces of incomplete combustion become airborne during the overhaul process. All personnel exposed to these products should wear their SCBA until the air has been determined to be safe by a safety officer.

Only with a concerned effort by firefighters, company officers, safety officers, and command officers will there be a reduction in these injuries. The equipment, training, and awareness are all there. Every firefighter needs to become part of the solution.

Chapter 8. Emergency Operations

Chapter 8 of NFPA 1500 maintains that emergency operations and other situations, including training, shall be conducted with appropriate safety considerations, and it requires the use of incident management and personnel accountability systems. Most injuries occur while firefighters are at the scene of fires and other emergencies. Here, you, as the company officer, can have a great impact on the accident rate in the department. You must set a positive personal example in conduct, following the department's policies, wearing personal protective clothing, supporting the incident command system, and keeping your cool when the heat is on. Firefighters will see and remember a lot more from your example than they will from any lecture they may have heard in recruit school. Once the pattern is set, all members of the company must follow your example.

Firefighters should be continually reviewing the latest information regarding fire behavior, building construction, fireground tactics, the proper use of tools, and of course, the proper use of the incident command system (see **Figure 8-10**). Many fire departments get few opportunities to practice these activities in real fire situations, and as a result, some companies become weak on basic skills. As company officers, you should recognize this fact and use every opportunity for training and safely practicing basic skills.

Recognizing that emergency activities present the greatest threat to firefighters, NFPA 1500 provides policies for managing such events. These policies indicate that an incident management system must be used and the incident commander is responsible for the overall management of that system.

FIGURE 8-10 The incident commander is responsible for managing the overall incident and for the safety of the members at the scene.

RISK MANAGEMENT

The concept of risk management shall be utilized on the basis of the following principles:

- Activities that present a significant risk to the safety of firefighters shall be limited to situations where there is a potential to save endangered lives.
- Activities that are routinely employed to protect property shall be recognized as inherent risks to the safety of firefighters, and actions shall be taken to reduce or avoid these risks.
- No risk to the safety of firefighters shall be acceptable when there is no possibility of saving lives or property.

Risk a lot to save a lot; risk a little to save a little; but risk nothing to save nothing.

Personnel accountability is also addressed in NFPA 1500. **Personnel accountability** during emergency operations has become an important topic for the fire service in recent years, and much has been done to enhance firefighter safety through better tracking of personnel during emergency events. To enhance accountability and personnel safety, NFPA requires that firefighters work in teams of two or more. During the initial phases of an event, such as a structural fire, where a team of two may be required to enter the structure, a backup team of at least two individuals must be outside the structure, immediately ready to assist the members who are inside the hazardous area. This is the "two in, two out" policy. These personnel must be in constant contact with one another by radio or other means.

The emphasis on accountability and safety during emergency operations has led to the requirement for

NOTE

The requirements for two in, two out and for rapid intervention teams enhance firefighter safety by dedicating some of the on-scene resources for the rescue of firefighters who may need assistance.

rapid intervention teams (RIT). NFPA 1500 indicates that at least one team, consisting of at least two individuals, shall be kept immediately available for purposes of rescuing firefighters if the need arises.

While accounting for and ensuring the safety of firefighters during emergency operations are vital for the safety of our personnel, caring for firefighters also includes rotating crews out of action, offering them food and liquids, rendering medical examinations and care, and furnishing shelter from extreme weather conditions (see **Figure 8-11**). The U.S. Fire Administration (USFA), working with the International Association of Fire Fighters (IAFF) issued in 2008 an updated version of *Emergency Incident Rehabilitation*. The revised manual examines critical topics related to emergency incident **rehabilitation**, including operational issues, human physiology, weather issues, and technology and addresses ways to better protect firefighters and other emergency responders through the use of PPE and improved tactical procedures.

Emergency responder rehabilitation is designed to ensure that the physical and mental well-being of members operating at the scene of an emergency do not deteriorate to the point where it affects their safety. It can prevent serious and life-threatening conditions—such as heat stroke and heart attacks—from occurring. *Fireground rehab* is the term often used for the care given to the firefighters and other responders while performing their duties at an emergency scene. *Fireground rehab* includes monitoring vital signs, rehydration, nourishment, and rest for responders between assignments.

FIGURE 8-11 Fire departments should develop a systematic approach for providing rehabilitation for personnel at the scene. *(Photo courtesy of the Fairfax County Fire and Rescue Department.)*

Rapid Intervention Teams[6]

■ Fire departments shall provide personnel for the rescue of members operating at emergency incidents. Such personnel will be designated the rapid intervention team.

■ A rapid intervention crew/company shall consist of at least two members and shall be available for rescue of a member or a crew. When an incident escalates beyond an initial full alarm assignment or when significant risk is present to firefighters due to the magnitude of the incident, the incident commander shall upgrade the initial rapid intervention team to a full rapid intervention team that consists of four fully equipped and trained firefighters.

■ A rapid intervention team shall be fully equipped with the appropriate protective clothing, protective equipment, SCBA, and any specialized rescue equipment that could be needed given the specifics of the operation under way.

■ The composition and structure of a rapid intervention team shall be permitted to be flexible based on the type of incident and the size and complexity of operations.

■ The incident commander shall evaluate the situation and the risks to operating crews and shall provide one or more rapid intervention team commensurate with the needs of the situation.

■ In the early stages of an incident, which includes the deployment of a fire department's initial attack assignment, the rapid intervention team shall be in compliance with one of the following:

1. On-scene members designated and dedicated as rapid intervention crew/company

2. On-scene members performing other functions but ready to re-deploy to perform rapid intervention crew/company functions

■ The assignment of any personnel shall not be permitted as members of the rapid intervention team if abandoning their critical task(s) to perform rescue clearly jeopardizes the safety and health of any member operating at the incident.

■ As the incident expands in size or complexity, which includes an incident commander's requests for additional resources beyond a fire department's initial attack assignment, the dedicated rapid intervention team shall on arrival of these additional resources be either one of the following:

1. On-scene members designated and dedicated as rapid intervention team

2. On-scene personnel located for rapid deployment and dedicated as rapid intervention teams

- During firefighter rescue operations, each team shall remain intact.
- At least one dedicated rapid intervention team shall be standing by with equipment to provide for the rescue of members that are performing special operations or for members that are in positions that present an immediate danger of injury in the event of equipment failure or collapse.

Rapid Intervention Is NOT Rapid

The Phoenix Fire Department has a sobering message for firefighters operating in large buildings: "If you extend an attack line 150 feet, get 40 feet off the line, and run out of air, it will take 22 minutes to find and remove you from the structure."

This statement is based on 200 rapid-intervention drills conducted by the department following the loss of a firefighter in a fire in a supermarket. Clearly, the message is this: Do not let go of the hoseline. Several other messages came out of the experience:

- It takes 12 firefighters to rescue one.
- One in five rescuers will get in trouble themselves.
- A 3000-psi SCBA bottle lasts 18 minutes.

The study also looks at the roles of company and command officers at the fireground. A company officer's core competency is to command a fire company. A chief officer's core competency is to command fire companies. Assigning company officers as sector officers places them in a situation where they have to wear two hats: company officer and command officer. That is too much to expect of anyone.

Phoenix suggests that it would be more effective to send more command officers to a fire event to function as sector and division officers and to allow the company officers to command their companies, noting that it is a waste of resources to allow command officers to sit in their offices while a low-frequency, high-risk event, such as a fire in a commercial occupancy, is occurring in the city. As a result of this experience, Phoenix changed the way it operates its rapid-intervention teams by assigning more suppression companies to rapid intervention and by assigning more officers to the command process.

Who Is Leading When the Leader Is Working?

Helping your teammates and getting involved in the work are admirable qualities, but we have seen several instances when this activity leads to tragic outcomes.

U.S. Fire Administration data suggest that the most dangerous activity at the fireground is advancing hoselines. During interior fire attack, we expect you, as the company officer, to be close to the action. However, when you take the firefighter's place on the hoseline, we have to wonder who is supervising. Our guess is that nobody is supervising. In fact, that may be why this activity is so dangerous. Firefighters may need assistance at times, but your job is to supervise. As such, you are responsible for an ongoing size-up of the situation in your immediate area, for keeping an open link of communications back to the incident commander, and for scene safety. You cannot delegate these tasks.

When you get involved in the actual task at hand, you lose your ability to perform the other activities listed here. The activity that concerns us most is scene safety. You should be watching the progress of your teammates, watching for signs of building collapse and other structural problems, and watching for signs of rapidly changing fire conditions. You cannot do these tasks when you are down next to the nozzle, focusing your attention and energy on advancing the line.

The second situation where you might get involved and compromise the safety of your team is when you slide into the driver's seat of a fire apparatus. NFPA 1500 identifies clear and specific duties for the driver and for you with regard to safety. We would expect you to help by watching for traffic on your side and by locating street names. You also act as a safety valve by keeping the driver's actions in check.

Many accidents involving fire apparatus are the result of excessive speed. The speed may have been excessive for the road conditions (wet pavement, darkness, and traffic) or excessive for the nature of the call. Not all calls require a maximum effort. However, drivers (especially younger ones) frequently get caught up in that inevitable adrenaline rush that accompanies any response. Their zeal has to be tempered with your good judgment as a more mature officer.

But what happens when you move to the driver's seat? The safety responsibilities of the driver and the officer have now been placed on one individual, with no system of checks and balances. Few firefighters will know what constitutes a dangerous condition. Few will have the courage to tell you, their boss, "Slow down, you're going too fast!"

The message: Do not get so involved with the tasks at hand that you compromise the safety of all those around you.

Chapter 9. Facility Safety

Chapter 9 of NFPA 1500 requires that all fire department facilities must meet all applicable health, safety, building, and fire code requirements and that all facilities shall be inspected at least once a year to determine their compliance.

When you look at the injury statistics, you will notice that "other on-duty" is the second largest category after "fireground" as the place where firefighters are injured. Even in the busiest companies, firefighters spend most of their on-duty time at the fire station (see **Figure 8-12** and **Figure 8-13**). This area should be an easy target for improving occupational safety and health.

Most fire stations are reasonably safe and healthful. In many cities, however, older fire stations have suffered from years of neglect. Equipment does not work properly, floors and steps are smooth from years of wear, and many stations still have the traditional brass pole. Stations should be designed and maintained so that tripping hazards, areas that are poorly lit, and problems resulting from poor ventilation are eliminated.

Since firefighters live in this environment, they tend to become tolerant and take many of these hazards for granted. Firefighters spend a lot of time in the station, and it is their home, if you will, while working. The combination of increased exposure and decreased vigilance provides an environment in which accidents are likely to occur. Although everyone has a responsibility for helping keep the fire station safe, you, as the company officer, are clearly the person who will have the greatest impact. If you tolerate a sloppy and unsafe environment, few others will take the initiative to correct the problems. On the other hand, if you set a good positive example and take an active interest in keeping

IMPROVING SAFETY FOR OUR FIREFIGHTERS

NFPA 1500 indicates that a minimum of four individuals is required to start interior operations in a hazardous environment, which means that first-arriving units must assemble at least four members at the scene before entering a hazardous area. Although this policy may appear to limit the practices of some fire departments, it provides a significantly safer working environment for all members at all situations. The policy was established after careful consideration and considerable debate. In the final analysis, the purpose is to enhance firefighter safety.

A hazardous environment is defined here as an area where there is immediate danger to life and health (IDLH). It includes areas inside of a structure where SCBA is required for protection from smoke, the products of combustion, particulate matter given off by any materials, and oxygen-deficient atmospheres.

SAFETY

After the fireground, other on-duty injuries rank second. Because personnel spend a significant part of their duty time in the station, the station environment is an opportunity for improving firefighter safety.

(A)

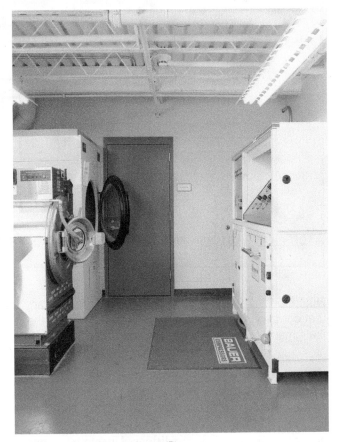

(B)

FIGURE 8-12 Interior views of two spaces of the station. (A) A typical bedroom. (B) The decontamination area of the apparatus bay. All dirty equipment is cleaned here before being returned to service or storage. (*Photographs courtesy of Stewart-Cooper-Newell Architects*)

(A)

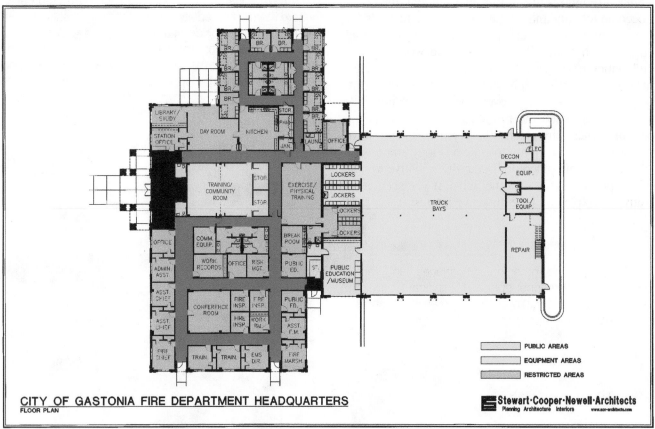

(B)

FIGURE 8-13 A modern fire station showcasing many safety features. The photograph (A) was taken from a position that corresponds to the lower right-hand corner of the drawing (B). This arrangement provides safety for the public and firefighters by controlling access and traffic flows. (*Photo and drawing courtesy of Stewart-Cooper-Newell Architects*)

the fire station safe, as well as presentable, others will be more likely to join in the effort.

As company officers, you should tour the station every day and inspect it once a month. Look for unsafe acts as well as unsafe conditions. Unsafe acts should be corrected immediately in a constructive manner, and many unsafe conditions can be fixed right away. Other problems may take longer.

OFFICER ADVICE

If you set a good positive example and take an active interest in keeping the fires station safe, as well as presentable, others will be more likely to join in the effort.

The fire department should also have a program of regular station visits conducted by the fire chief or other

PROVIDING A HEALTHY WORKPLACE

Chapter 8 of NFPA 1500 requires that all fire department facilities comply with all legally applicable health, safety, building, and fire code requirements, to wit:

- All fire department facilities shall be designated smoke free.
- Fire departments shall provide facilities for disinfecting, cleaning, and storage in accordance with NFPA 1581, *Standard on Fire Department Infection Control Program.*
- All existing and new fire stations shall be provided with smoke detectors in work, sleeping, and general storage areas. When activated, these detectors shall sound an alarm throughout the fire station.
- All existing and new fire department facilities shall have carbon monoxide detectors installed in sleeping and living areas.
- All fire stations and fire department facilities shall comply with NFPA 101, *Life Safety Code®.*
- The fire department shall prevent exposure to firefighters and contamination of living and sleeping areas from exhaust emissions.

The operation of a fire department requires the storage of, and indoor operation of, fire apparatus that are generally housed in an enclosed building. The need to keep the apparatus and other vehicles ready for immediate service and in good operating condition requires running vehicles indoors for response and routine service/pump checks. This makes storage in an enclosed area, such as an apparatus bay, necessary. The exhaust from all internal combustion engines, including diesel/gasoline-powered engines, contains over 100 individual hazardous chemical components that, when combined, can result in as many as 10,000 chemical compounds.

A large majority of these compounds are today listed by state and federal regulatory agencies as being cancer causing or suspected carcinogens. Over the past decade, it has been documented that fire department personnel exposed to vehicle exhaust emissions have had adverse health effects, including death, even in areas where only short-term exposure had taken place. Secondary effects of vehicle exhaust have been cited in the storage of sterilized medical equipment. The equipment is contaminated by exhaust emissions and handled by emergency services personnel while treating the public, thus creating a path for cross-contamination to the general public. In addition, there has been an effect on contamination to computers and emergency service electronics due to carbon deposits that lead to malfunction.

TWO OPPORTUNITIES FOR THE COMPANY OFFICER

During a rainstorm, personnel entering the station tracked the floor with their wet shoes because someone had removed the door mat in order to sweep it. After several hours of this, a member, ironically returning from a doctor's appointment, entered, slipped, fell, and was seriously injured.

In another station, the lighting in the apparatus area was either on or off. Turning all the lights on at night blinded those coming from the darkened quarters and left drivers momentarily blind as they entered traffic. Leaving the lights off presented dangers to firefighters moving rapidly to their apparatus.

Both cases had easy solutions, but in each case, someone had to take the initiative. In the first case, the problem could be easily fixed. In the second, several additional light fixtures may be needed and that may require some help. In both cases, someone, likely you—the company officer—should take the initiative. Once the idea catches on, others will offer ideas and take action to make the station a safer workplace.

SAFETY

Company officers, as part of their supervisory duties, should be taking an active role in the station's safety management activities to be sure that such accidents are eliminated.

senior officers. The department's health and safety officer should also be involved, helping locate and document problems, and helping find solutions to correct them (see **Figure 8-14**). The safety officer is the department's expert on station safety. Problems that are noted at one station are often present at others. Conducting these safety inspections becomes far more efficient and effective as you become experienced in doing so.

Ideally fire departments should provide appropriate training to enhance station safety. However, we all know time and money are usually in short supply. So fire departments and you, as company officers, should work together to obtain, maintain, and provide practical procedures regarding the safe use of all equipment, including the physical fitness equipment in the station. You and your department should also provide positive motivation to the firefighters concerning fire station safety by recognizing the safest companies and stations.

STATION INSPECTION RECORD

STATION _____   SHIFT _____   DATE _____

CAPTAIN _____   BATTALION CHIEF _____   DIVISION CHIEF _____

✓ - O.K
X - SEE REMARKS

I. PERSONNEL

A. Uniform . . . .
B. Protective Clothing . . . . .
C. Grooming & Cleanliness . .
D. Driver's License . . . .
E. I.D. Card . . . .
F. E.C. Card . . . .

5. SAFETY

A. Overhead Storage . . . .
B. Caution Signs . . . .
C. Safety Bulletin Board . . . .
D. Tool Storage . . . .
E. Combustibles Storage . . . .
F. Insecticide Chemical Stor. .
G. SCBA Tests . . . .
H. Extinguisher (s) . . . .
I. Smoke Detector (s) . . . . .
J. RADEF Equipment . . . . . .

6. EXTERIOR

A. Paint . . . .
B. Doors . . . . . .
C. Sidewalks . . . .
D. Ramps . . . . . .
E. Parking Area . .
F. Yard . . . . . .
G. Hose Tower . . .
H. Gas Pump . . . .
I. Flag Pole . . . .

2. OFFICE

A. Files . . . .
B. Maps . . . .
C. Log Book . . . . . .
D. Shift Change Log . .
E. Desk . . . .
F. Bulletin Board . . . .
G. Tourist Information .

3. ENGINE HOUSE

A. Floor . . . . . .
B. Work Bench . .
C. Hose Rack . . .
D. Storage . . . .
E. Turnouts . . . .
F. Apparatus . . .

7. INTERIOR

A. General Appearance . . . . . .
B. Cleanliness . . . .
C. Floors . . . .
D. Windows . . . .
E. Lights . . . .
F. Furniture . . . .
G. Storage Area (s) . . . .
H. Woodwork . . . .
I. Walls . . . .
J. Ceiling . . . .

Office / Kitchen / Day Room / Dormitory / Rest Room / Heater Room

4. PROGRAMS

A. Hazard Reduction . .
B. In Service Inspections
C. Pre Fire Plans . . . .
D. Water Systems . . . .

REMARKS _____
_____
_____
_____
_____

Fire 580 2415 016   Catalog # 9020 (Rev. 2/90)

FIGURE 8-14 Fire station inspection form.

Firefighters injured at fire stations are just as hurt as those who were injured at the fireground. These injuries result in the same lost time and medical costs as the injuries that occur during operational activities. Fire station injuries are often embarrassing and are almost always avoidable. You, as part of your management duties, should take an active role in improving the safety conditions in the station to be sure that such accidents are eliminated.

Chapter 10. Medical and Physical Requirements

Chapter 10 of NFPA 1500 requires that all members shall be given a physical examination before they start their recruit training and periodically thereafter; that departments shall establish and support programs that enable members to develop and maintain appropriate levels of physical fitness needed for the job; and that they shall establish physical performance standards for candidates and members who engage in emergency operations.

Reducing the Risk of Heart Attack

Approximately 40 to 50 percent of all on-duty firefighter fatalities are the result of heart attack. To reduce this toll, the USFA recommends the following:[7]

- *Having an annual physical examination.* Although especially important to older (40 or older) firefighters, all firefighters will benefit from having an annual physical. The physical exam should include checking for high cholesterol and cancer.

- *Modifying eating habits.* All firefighters should take steps to reduce fat intake and foods high in sodium and salt.

- *Considering supplements.* Although there is no clear consensus on the benefits of dietary supplements, a great deal of medical attention is given to the positive benefits of vitamin E, aspirin, multivitamins, and fish oil.

- *Taking a walk.* Good exercise is as important as what you eat. Walk or run every day and maintain muscular conditioning by regular workouts in the gym.
- *Avoiding the use of tobacco products.* For those who do use tobacco products, the greatest step you can take to enhance your present and future health is to stop.

Firefighting and many of the other tasks that firefighters perform are physically demanding. These activities require both good physical condition and strength. In many ways, the work is similar to that performed by professional athletes. Good professional athletes work out all year to maintain their excellent physical condition. They also have the advantage of having an opportunity to warm up before their workouts and games. Here we see a major distinction between the athlete and the firefighter: Firefighters are frequently called upon to perform at their best at the very start of an event, with no opportunity for mental or physical warm-up. No wonder strain and sprain injuries represent 49 percent of all injuries and heart attacks and strokes represent nearly 42 percent of all deaths in the fire service when firefighters do not have an opportunity to warm up (see **Figure 8-15**).

Having an opportunity to warm up would be beneficial, but other things can be done to reduce the risks firefighters face. All kinds of fitness programs are available, and although some are better than others, any program is likely to be better than none. Good physical condition does not eliminate the risk of injury, but it certainly reduces it, and good personal fitness

usually reduces the recovery time when injuries do occur (see **Figure 8-15**). Good body strength and flexibility reduce the likelihood for most of the types of accidents that lead to strain and sprain injuries. Good

FIGURE 8-15 Good physical conditioning is a part of good health.

WORKING TOGETHER TO IMPROVE FIREFIGHTER SAFETY

Heart attacks are the leading cause of fatalities among both paid and volunteer firefighters. The USFA and the National Volunteer Fire Council (NVFC) are continuing to work together to support a Volunteer Fire Service Fitness and Wellness Project—a partnership initiative to reduce loss of life among volunteer firefighters from heart attach and stress.

The USFA, working with the NVFC, issued in February 2009 a revised *Health and Wellness Guide* for the Volunteer Fire and Emergency Services. The *Health and Wellness Guide* now provides updated information on health and wellness issues, trends, and programs focused on the needs of the volunteer fire service. The document addresses fitness including aerobic exercise, flexibility, strength training, and diet; smoking cessation; and

other areas that will have a positive impact on volunteer firefighters.

"With heart attack, overexertion, and strain causing more firefighter deaths and injuries than any other cause, it is critically important for departments and personnel to focus on health and wellness," said NVFC Chairman Philip C. Stittleburg. The *Health and Wellness Guide* demonstrates ways to overcome these obstacles and provides direction for developing and implementing a department program.

This project also provides information on how volunteer fire departments can enhance compliance with appropriate National Fire Protection Association firefighter health and safety standards such as NFPA 1583, *Health Related Fitness Programs for Firefighters*.

This project complements existing USFA firefighter wellness and fitness partnerships with the IAFC and the IAFF to support the expansion of the *IAFF and IAFC Fire Service Joint Labor-Management Wellness Fitness Initiative* to additional fire departments.

aerobic conditioning, along with body strength, reduces the effort required for any task.

NFPA 1500 indicates that all fire departments shall have a fitness program "to enable members to develop and maintain an appropriate level of fitness to safely perform their assigned functions, and . . . shall require the structured participation of all members in the physical fitness program." Notice that the emphasis is on safety—"to enable members . . . to safely perform their assigned functions." This suggests a program that benefits the member. Many departments are showing the benefits of fitness to recruits when they join the department and are providing opportunities for them to develop to peak fitness and maintain that fitness until they retire.

Fitness testing should offer events that represent the tasks that firefighters are expected to perform on the job. Tests that combine such tasks as lifting objects, carrying something that represents an unconscious adult victim, dragging a hose, and simulating the motions of cutting a roof with an axe, are all fair and legal (see **Figure 8-16**). Every fire department should have this type of testing in place for all hands.

Along with a good fitness program, good nutrition can have a profound impact on how you feel and how well you perform. The firefighter's lifestyle has changed from one of sitting around drinking coffee and smoking cigarettes to a routine that allows little time for either. Many firefighters are also turning away from the traditional heavy meals that added inches around their waists (see **Figure 8-17**).

Company training programs should include information and activities to improve the firefighters' health and safety through exercise and proper diet. As a company officer, you can do a lot to improve your firefighters' health just by setting a positive personal example for your teammates to follow.

> **OFFICER ADVICE**
>
> As a company officer, you can have a significant impact on the health and safety of your personnel by just setting a positive example for them to follow.

Chapter 11. Member Assistance and Wellness Programs

Chapter 11 states that departments shall provide programs to help members and their families deal with problems associated with substance abuse, stress, and personal problems that might affect a member's on-the-job performance. Members should have access to member-assistance programs that provide the training and counseling needed to deal with the physical and mental problems associated with the job.

Chapter 12. Critical Incident Stress Program

Chapter 12 of NFPA 1500 deals with critical incident stress (CIS), stating that departments shall have a program for dealing with CIS.

We are seeing more and more about this topic in the trade journals. Only recently has the fire service recognized that mental condition, as well as physical condition, is a part of the firefighter's overall well-being. Many firefighters, especially those providing EMS care, tend to burn out from the routine aspects of the job. This gradual burnout is just as real and just as debilitating as the stress that comes from "the big one."

We usually recognize the stress that comes from large-scale events—the event that results in prolonged operations and multiple loss of life. For example, airplane crashes are probably near the top of the list, but everyday exposure to lesser stressors can have the same long-term impact. More typical situations may include fatalities of all kinds, but especially those involving children or fire department members, prolonged unsuccessful rescue efforts, and others. What are you and your fire department doing to protect members from the effects of these events and to enhance the psychological or physical well-being of your members?

FIGURE 8-16 Participating in a firefighter combat challenge can be one way of keeping fit.

FIGURE 8-17 Good nutrition will improve the way you feel, the way you look to others, and the way you perform.

NOTE

Critical incident stress cannot be avoided, but its effects can be lessened by having a support system in place.

Critical incident stress cannot be avoided, but its effects can be lessened by having a support system in place. Once again, you, as the company officer, have a pivotal role in helping firefighters cope with stress. As

FIVE GENERALLY ACCEPTED TOOLS FOR DEALING WITH CIS

1. *Training.* Fire departments have an obligation to prepare their personnel for what they will likely encounter. This does not need to be a course in the macabre, but rather a realistic presentation of what firefighters will see.

2. *Scene management.* At the scene of emergencies, you, as company officers, can reduce the impact of gruesome scenes by telling firefighters what they can expect to see, covering bodies, controlling entry, and otherwise reducing the unexpected and unwelcome sight of human carnage.

3. *Peer support.* Peer support comes from other members of the department. Following a critical incident, and sometimes even after several minor events, firefighters might be getting "near the edge." This is a critical point, one that can turn either way. As the company officer, you should recognize the signs and

take quick action to provide peer support for endangered firefighters.

4. *Debriefing.* As used here, a debriefing is frequently conducted after an incident where there is potential for a stressful impact on the members. The debriefing team may consist of fellow firefighters, trained mental health professionals, or others qualified in dealing with such situations. All members exposed to the stressful situation should go through this process.

5. *Counseling.* Counseling involves a mental health professional. Counseling may be needed for an especially stressful event or for the compounded effects of a series of events.

   As a supervisor, you should know that a firefighter who is having problems at home is less likely to be immune to stress in the workplace. Since you are responsible for the members' overall performance, health, and safety, you should be interested in helping members resolve their problems, regardless of the source.

supervisors, you must know and understand the mental makeup of your firefighters. You must be sensitive to the firefighters' needs and supportive of programs that will help the firefighters avoid or overcome mental stress. Discussion about the event and the emotions that follow and assurance that these feelings are normal help reduce more severe and long-term stress.

Company Officers Play a Key Role in the Physical and Mental Health of Their Teammates

Fire departments should have a written policy that establishes a program designed to relieve the stress generated by an incident that could adversely affect the psychological and physical well-being of fire department members. Firefighters frequently experience trauma, death, and sorrow. Critical incident stress is a normal reaction experienced by normal people following an event that is abnormal. Such stress should not be perceived as evidence of weakness, mental instability, or other abnormality. The emotional trauma can be serious. Symptoms can appear immediately after the incident, hours later, or sometimes even days or weeks later. The symptoms can last for a few days, weeks, or months so that the person can no longer function effectively. Occasionally, a professional counselor could be needed.

Critical incident stress is the inevitable result of trauma experienced by fire service personnel. It cannot be prevented, but it can be relieved. Knowing the signs and symptoms and how to respond to them after the occurrence of a critical incident can greatly reduce the chance of more severe and long-term stress. Rapid intervention, talking about the situation, and reassuring that these are normal reactions and feelings can help prevent more serious problems later on, such as family and marital problems.

To provide this intervention, the fire department should have access to a critical incident debriefing (CID) team. The main objective of the CID team is to lessen the impact of the critical incident, put it into the proper perspective, and help maintain a healthy outlook. The CID team should consist of other firefighters, support personnel, and mental health professionals specifically trained in stress-related counseling. The team should be well represented by all types of members whether volunteer, call, or paid, and by all ranks.

All members should have a minimum of a two-day training seminar with continuing education in stress-related training as an ongoing part of the team's regular meetings. (Monthly meetings are recommended for active departments, whereas quarterly meetings could be sufficient for less active departments.) All personnel should be able to initiate the debriefing procedure by contacting their supervisor or officer or the dispatch center. A contact list of the debriefing team members should be available in the dispatch center.

Debriefings should be held for incidents that have the potential for having a stressful impact on members. It is important to remember that an event is traumatic when experienced as such. Generally, debriefings should be held at a station within 1 to 3 hours after the incident. Debriefings should encourage brief discussions of the event, which in themselves help to alleviate a good deal of the stress. Debriefings are strictly confidential and are not a critique of the incident. Information should be given on stress reactions and steps that members can take to relieve the symptoms so that they can continue their normal activities as soon as the debriefing is over.

CHIEF ALAN BRUNACINI'S TWENTY-FIVE SAFETY TRUTHS

Phoenix Fire Chief Alan Brunacini was writing about fireground management and firefighter safety 25 years ago. His words are still true today.

1. Think.
2. Drive defensively.
3. Drive slower, rather than faster.
4. If you don't see, stop.
5. Don't run for a moving rig.
6. Always wear your seat belt.
7. Wear full turnouts and SCBA.
8. Attack with a sensible level of aggression.
9. Always work under sector command; no freelancing.
10. Don't ever breathe smoke.
11. Keep your crew intact.
12. Maintain a communications link to command.
13. Evaluate the hazard—know the risk you're taking.
14. Follow standard fireground procedures; know and be a part of the plan.
15. Vent early and vent often.
16. Use a hoseline that's big enough and long enough.
17. Don't ever go beyond your air supply.
18. Always have an escape route (hoseline or airline).
19. Provide lights in the work areas.
20. If it's heavy, get help.
21. Always watch your fireground position.
22. Look and listen for signs of collapse.
23. Rehab fatigued companies—assist stressed companies.
24. Pay attention all the time.
25. Everybody looks out for everybody.

Some common signs and symptoms of critical incident stress are fatigue, headaches, inability to concentrate, anxiety, depression, inappropriate emotional behavior, intense anger, irritability, withdrawal from the crew and/or family, change in appetite, increased alcohol consumption, and a change in sleeping patterns. To help alleviate some of the emotional pain, members can rest more; contact friends; maintain as normal a schedule as possible; eat well-balanced, schedu. led meals; keep a reasonable level of activity to fight boredom; express feelings; and talk to loved ones. Recent studies and research also indicate that exercise, especially soon after an event, can greatly reduce mental pain. Member assistance programs should always be available to members. The CID team is often the first step in providing the help that is needed and should be ready to serve to help minimize stress-related injury.[8]

Safety Will Always Be a Top Priority

The chapter on safety continues to be written. The second National Line-of-Duty Death Prevention Summit, held in 2007, reinforced the following topics:

Health and Wellness

- Every fire department should implement a no-tobacco use policy.
- Obtain vital signs on all crew members at the beginning of the shift.
- Every fire department should have a medical screening/fitness program in place.
- Firefighters' mental health should be monitored and counseling should be available.

Prevention

- Fire prevention should be included in fire department mission statement.
- Every firefighter should have training in public education and fire prevention.
- Personnel responsible for code enforcement should possess proper credentials.

Structural

- Stop rewarding unsafe behavior regardless of the outcome.
- Define responsibilities of every position.
- Teach firefighters the legal ramifications of engaging in unsafe practices.

Wildland

- Make sure all fire service personnel are familiar with all of the safety programs.

- Fitness programs should be established.
- Participate in the National Fire Fighter Near-Miss Reporting System.

Training and Research

- Create a national data collection center.
- Firefighter death and injuries should be discussed on the company level.
- Everyone should be held accountable for their actions, and the chief officers must lead by example.

Vehicles

- Make safety a core value by supporting LACK (Leadership, Accountability, Communication, and Knowledge).
- Require initial and refresher training for drivers.
- Champion the use of seatbelts.

ELEVEN SUGGESTIONS TO HELP COMPANY OFFICERS IMPROVE THEIR FIREFIGHTERS' SAFETY

1. Know the job. Company officers should constantly be working on improving their knowledge and skills.
2. Be observant of the actions of others.
3. Be a team player. Build team spirit and interdependence among all of the players.
4. Set a good personal example. Don't just talk the talk; walk the walk, too.
5. Be obviously concerned about safety at all times.
6. Provide constructive feedback to firefighters.
7. Communicate effectively in all your dealings, but especially in the area of safety.
8. Be assertive when necessary! Sometimes, it is the only way you can get compliance.
9. This is not a time to be timid about your feelings.
10. When all else fails, feel free to take appropriate disciplinary action.
11. Be truly positive about safety and health issues.

SOMETHING TO REMEMBER

- It takes 1 minute to write a safety rule.
- It takes 1 hour to hold a safety meeting.
- It takes 1 week to plan a safety program.
- It takes 1 month to put the plan into full operation.
- It takes 1 year to win the chief's safety award.
- It takes just 1 second to destroy it all with an accident!

Let us be careful out there!

LOSING FIREFIGHTER LIVES IS NOT OUR TRADITION

Jeff Pindelski: Battalion Chief, Downers Grove Fire Department, Downers Grove, Illinois; and Western Regional Director, Fire Department Safety Officer's Association.

In this day and age, firefighters are being provided with the most advanced tools and safety equipment that have ever been provided to the fire service, but we continue to lose more than 100 firefighters in the line of duty each year. More disturbing is the fact that we are responding to fewer and fewer actual working incidents each year, yet still losing the same amount of firefighters in line-of-duty deaths.

The fires faced by today's fire service are much different than 40 to 50 years ago. Fuel loads have increased and changed significantly affecting the fluctuation of the time-temperature curve as it relates to flashover. I remember learning in the fire academy nearly 20 years ago that flashover would normally take place within 12 to 16 minutes. Time frames to reach flashover conditions in today's environment have been demonstrated to take place in as little as 3 to 3½ minutes. Things once fabricated out of natural materials are now made of plastic. It has been demonstrated that 1 pound. of wood releases 8,000 BTUs when ignited; a small plastic shopping bag weighing much less than ½ of an ounce, however, has been shown to release as much as 19,900 BTUs.

Another tremendous challenge that faces the fire service is that of experience. Each day, the fire service is losing more and more members to retirement who have served in a busier era of the fire service. Rapid promotion is also becoming part of the overall challenge. As we lose the experienced members, we are being left with a generation of firefighters that may be the most highly educated generation ever, due to their access to information through avenues such as the Internet, but may come up short in other areas that are necessary to successfully lead our firefighters. It is realistically possible, in today's fire service, that some departments may actually promote a firefighter to the company officer level who has never gained practical hands-on experience at an actual fire. These should not be looked upon as shortcomings but rather as challenges that we will need to overcome. Fault does not fall onto any particular generation—each one only has to deal with the economic, political, and social environments that influence the time period in which they live.

The fire service environment requires fire officers to make life-and-death decisions in a quick manner based on incomplete information. When we are exposed to a situation, our brain stores that experience in its own file, almost like a photographic slide within a tray. When another situation is experienced, our brain will try to draw on another past experience or "slide" that closely resembles the situation being presented. From there, a solution to the situation will be developed. Obviously, the more files or "slides" in the bank will increase the odds of choosing the most correct solution to the problem. If we do not have exposure to a similar situation, there is a greater chance of failure.

Succession development and mentoring programs need to be placed at the forefront of our fire service. Tapping into the senior, experienced members before they retire needs to be a priority. Lessons learned over their career, as well as how the fireground environment has changed, can be invaluable to the newer members of the service. In addition, we must couple this information passing with scenario-based training that is repeated on a regular basis to help reinforce the principles being conveyed so that our future officer's "slide tray" has reliable information to draw upon. Situational awareness and risk/benefit analysis need to be understood by each and every person on the fireground and they must be empowered to speak up when they see potential problems.

Information added to these "slide trays," however, should not only focus on the positive or right ways to perform a task but should predominantly focus on things that may have gone wrong. It is important that we celebrate our successes, but if you think about our experiences throughout our lives, the most complete learning messages are taken from instances where we have not gotten the response or action that we desired. We have tremendous resources available to us to help share important lessons from firefighter fatalities and we talk about them when the case reports are issued but yet we continue to lose firefighters over and over for the same exact reasons. Just looking at the reports and "armchair quarterbacking" and placing blame is not going to lead to effective learning. We must take the reports and open our minds to what they are actually saying. How many times do we pick up a firefighter fatality or near-miss report and say, "I can see how this happened or I can see myself or my department falling into the same trap? It is from analysis such as this that we can learn and make the necessary changes at our department to reduce the chances of a similar event taking place in our own backyard.

It is an honor for every firefighter to be part of the rich fire service traditions that we partake; however, losing our life in the line of duty is not a tradition. Our attitude and culture are truly what lead us to be what we are. We must challenge the traditions and attitudes that lead us down the roads to failure. Just because we have done something a particular way for a long time does not necessarily mean that it is correct; it may just mean that we have been lucky. Look for the loopholes in your actions or your department's operations—if you think that something is good, try to find ways to break it and make it better.

Today's fire service operates and competes in a very volatile environment—both on and off the fireground. Being ready for these challenges related to personnel safety and overcoming them begins with the most integral position in the fire service—the company officer. Are you ready for this challenge?

Questions

1. What changes have occurred since Battalion Chief Pindelski entered the fire academy nearly 20 years ago?

2. Why are firefighter fatality statistics even more alarming than before?

3. What does Battalion Chief Pindelski think about the value of experience? How does he advocate we use the experience of others to improve new company officers?

4. Where can you find experience within your own department? How can you use that experience to help your own personal and professional development?

5. Has the fire service resigned itself to the fact that losing firefighters every year is part of our tradition? Do you agree with Battalion Chief Pindelski that this is not and cannot be considered fire service tradition? Why or why not?

Battalion Chief Pindelski is a co-author of another Delmar, Cengage Learning publication, Rapid Intervention Company Operations, *a great resource for company officers to learn more about firefighter emergencies, how to prevent them, and how to react to them.*

LESSONS LEARNED

Although individual firefighters have a responsibility for their own well-being, you, as company officers, also have a moral and legal responsibility for the safety and survivability of your firefighters. If an accident occurs, you must explain your action, or lack thereof.

NFPA 1500 compressively addresses a fire department health and safety program. As fire departments start to implement some or all of the procedures outlined in the standard, they will see improvements in the health and safety of their members. Many departments have already undertaken these steps, and the results are already apparent: There has been a significant reduction in the number of injuries and deaths in the fire service.

Providing a safe and healthful workplace should be the goal of every employer. The results of such an approach usually mean fewer accidents, higher morale, more productivity, and reduced liability resulting from accidents. This proactive approach can result in reduced risk to the members of the fire service as well as a reduced risk to their departments. Many fire departments tend to be reactive, rather than proactive, when it comes to firefighter safety. This position is unfortunate for all.

Accident, injury, and line-of-duty death statistics reinforce the need for comprehensive safety and health programs that address the needs of the fire service in both paid and volunteer organizations. Such programs must include policies regarding vehicle operations that address speed, the use of seat belts, the avoidance of alcohol, and obeying traffic laws. All are within your domain as company officers to manage.

Regardless of the fire department's organizational attitude about safety, you, as company officers, can have a significant impact on your own and your members' health and safety. Many of the requirements of NFPA 1500 can be introduced at the company level. As with many of the topics discussed in this book, setting the tone and enforcing the department's policies regarding safety is one of your many responsibilities. During emergency response, whether the call be a single-unit response for a dumpster fire, or a three-alarm event, you are responsible for the safety of the entire crew. At the scene of an emergency as well as in the fire station, firefighter safety is ultimately your responsibility.

KEY TERMS

health and safety officer   a person assigned as the manager of the department's health and safety program

personnel accountability   the tracking of personnel as to location and activity during an emergency event

rapid intervention team (RIT)   a team or company of emergency personnel kept immediately

available for the potential rescue of other emergency responders

rehabilitation   as applies to firefighting personnel, an opportunity to take a short break from firefighting duties to rest, cool off, and replenish liquids

REVIEW QUESTIONS

1. About how many firefighter injuries and fatalities occur each year?

2. Where do most firefighter injuries and fatalities occur?

3. List the most common causes of firefighter fatalities.

4. List the most common causes of firefighter injuries.

5. Under what conditions do these injuries occur?

6. What are some of the conditions that bring stress to firefighters? What are the signs and symptoms of stress?

7. What actions can you take as company officers to reduce injuries at the fire station?

8. What action can you take to reduce injuries while responding to emergencies?

9. What actions can you take to reduce injuries at the scene of emergencies?

10. What is your department's commitment to firefighter safety and health?

DISCUSSION QUESTIONS

1. What can you do as company officers to enhance the safety of your personnel?

2. What significant changes have occurred in the fire service within the last 10 years as a result of health and safety concerns?

3. Based on your own observations, what causes firefighters to get injured?

4. If you were assigned duties as the incident safety officer at the scene of a working structure fire in a residence, what would be your concerns? What would be your duties?

5. What are some of the signs and symptoms of stress? What role do you have as a company

officer in providing help for teammates showing signs or symptoms of stress?

6. What steps can you take to maintain a healthier body?

7. What steps can you take to maintain a healthier mind?

8. What steps can you take to personally maintain a healthier mind?

9. What steps can be taken by your company to improve firefighter safety and health?

10. What steps can be taken by your department to improve firefighter safety and health?

ENDNOTES

1. Frank C. Schaper and Gregg Gerner, "Why Firefighters Continue to Get Injured and Killed," *Firefighter's News*, March 1996, 1–2.

2. *Firefighter Fatality Retrospective Study, 1990–2000*. Emmitsburg, MD: U.S. Fire Administration, 2002, 13; and "2002 Firefighter Injuries," *NFPA Journal* (November–December 2002).

3. *Firefighter Fatality Retrospective Study, 1990–2000*. Emmitsburg, MD: U.S. Fire Administration, 2002.

4. *Firefighter Fatalities in the United States in 1998*. Emmitsburg, MD: United States Fire Administration, 1999, 26.

5. *Firefighter Fatality Retrospective Study, 1990–2000*, 21.

6. Reproduced with permission from NFPA 1500, *Fire Department Occupational Safety and Health Program*, Copyright©2007, National Fire Protection Association. This reprinted material is not the complete and official position of the NFPA on the referenced subject, which is represented only by the standard in its entirety.

7. For more information, see *Firefighter Fatalities in the United States for 2001*, published by the U.S. Fire Administration. This report is available on the Web at www.usfa.gov.

8. This material is taken from Chapter 12 of NFPA 1500.

ADDITIONAL RESOURCES

Alder, Mike and Mat Fratus. "The Impact of Department Culture on Fireground Safety." *Fire Engineering*, June 2007.

Arnold, Brian. "Who Wants to Go Home?" *Fire Engineering*, April 2007.

Byrne, Jarret. "Understanding Fireground LODDs." *FireRescue*, April 2007.

Coleman, John. "The Role of the Safety Officer." *Fire Engineering*, June 2008.

Comstock, "Safety and Liability." *Fire Engineering*, June 2007.

De Lisi, Steven. "Strategies for Safer Driving." *Fire Engineering*, February 2009.

Dittmar, Mary. "Firefighters and Heart Disease: Beyond the Statistics." *Fire Engineering*, December 2006.

Dunn, Vincent. "Does Aggressive Firefighting Cause Firefighters to Become Caught and Trapped?" *Firehouse*, July 2008.

Eisner, Harvey and Steve Kalman. "Hackensack: 20 Years Later." *Firehouse*, October 2008.

Ellis, Jeff and Martha Ellis. "Survival of the Fittest." *FireRescue*, April 2008.

USFA, *USFA Releases Firefighter Fatality Report*. Emmitsburg, MD, News Release. The statistics from USFA and NFPA may differ slightly due to their respective methods of gathering and analyzing the data.

Fahy, Rita F. *U.S. Firefighter Deaths Related to Training, 1996-2005*. June 2006, National Fire Protection Association.

Fahy, Rita F. *U.S. Firefighter Fatalities Due to Sudden Cardiac Death, 1995-2004*. June 2005, National Fire Protection Association.

Fahy, Rita F. *U.S. Fire Service Fatalities in Structures*. July 2002, National Fire Protection Association.

Firefighter Fatalities in the U.S. is published regularly by the USFA at Emmitsburg, MD. It is available at no cost by contacting them at usfa.dhs.gov.

Gerardi, Paul. "Fitness and Training Are Keys to Safety." *Fire Engineering*, June 2007.

Havner, J. M. "Having a Safety Committee Is Critical." *Fire Engineering*, June 2008.

Karter, Michael J. Jr. and Joseph L. Molis, *Firefighter Injuries in the United States*, October 2006, National Fire Protection Association.

McGrath, Mary. "Special Access." *Fire Chief*, April 2007.

Moore, Ron. "Safe Parking" (while working on the highway). *Firehouse*, October, 2003.

Routley, J. Gordon. "National Fallen Firefighters Foundation Firefighter Life Safety "Summit Report." *Firehouse*, June 2004.

Salka, John. "The Start of the Tour." *Firehouse*, June 2008.

Schneider, Jason. "Is Your Firehouse Making You Sick?" *Firehouse*, July 2007.

Shouldis, William. "Getting it Right: Clink it!" *Fire Engineering*, June 2006.

Stagnaro, Victor. "Firefighter Fatalities: The Preventable Loss." *Fire Engineering*, June 2006.

The *Health and Wellness Guide for the Volunteer Fire and Emergency Services* and *Emergency Incident Rehabilitation are both available without cost. Go to.*usfa.dhs.gov to order or download a copy

USFA. *Firefighter Fatalities in the United States in 1998*. Emmitsburg, MD: United States Fire Administration, 1999.

Wilson, Kevin. "Use SOGs to Prepare for a Mayday." *Fire Engineering*, February 2009.

The following NFPA standards pertaining to firefighter safety are available from the National Fire Protection Association, Quincy, MA: http://www.nfpa.org/index.asp

NFPA 1403. Live Fire Training Evolutions in Structures. latest ed.

NFPA 1451. Fire Service Vehicle Operations Training Program. latest ed.

NFPA 1500. Fire Department Health and Safety Program. latest ed.

NFPA 1521. Fire Department Safety Officer. latest ed.

NFPA 1561. Fire Department Incident Management System. latest ed.

NFPA 1581. Fire Department Infection Control Program. latest ed.

The National Fire Academy has two excellent two-day hands-off training programs that can help any fire department start and improve its safety program. These are entitled *Health and Safety Officer* and *Incident Safety Officer*. Contact your training officer or your state fire training director for additional information.

9

The Company Officer's Role in Fire Prevention

We took on a job of stupendous proportions with the idea of being honest, factual, and frank. We have been told by members of the fire service that the report was just that. The team dedicated all its efforts to the nine brave firefighters who so tragically lost their lives in that fire. I hope that the lessons we learned do not fall on deaf ears. I am fearful that they may be doing just that.

Task force members delivered an in-depth presentation to an extremely large audience at Fire-Rescue International in August. The presentation, like the report, is clearly divided into two equally important parts: fire prevention; and strategy, tactics, and risk-management. The first half of the presentation dealt with fire prevention. The audience was very polite, but facial expressions and body language showed their impatience. When we started the second part, about hoses, fire engines, and the like, the entire audience perked up and sat at the edge of their seats for 2 hours. This was a typical reaction at presentations we made around the country.

After the formal presentation, we opened the floor to questions. Hands flew up, but only two questions related to fire prevention. One of those came from a friend I asked to pose a question. This was despite the fact that one entire slide in the summary read, *"The fire could have been prevented if the property had been constructed and maintained in accordance with state and local codes."*

One man asked, "Charleston is one of many tragic events that have resulted in firefighter deaths. What can we learn that could help prevent this from continuing?" We discussed reactive measures like risk management, accountability, and incident command. These are highly important and we must be trained and retrained on them. The proactive elements of preventing fire and its spread are often an afterthought, if that. The three Es in fire prevention—*engineering, enforcement,* and *education*—are as important as strategy and tactics—*and they are directly related to saving firefighter's lives and preventing injuries.*

Code compliance issues—such as additions constructed without permits; lack of automatic sprinklers; inadequate fire walls; improperly stored flammable liquids, trash, and debris outside the loading dock (where the fire began); smoking; inadequate exits, nonworking fire doors; and locked or obstructed exits—were major factors in this tragedy. The building was last inspected in 1998. The inspector found exit obstructions, exit sign infractions, and storage practice issues. But there wasn't much follow-up. When the city began to follow the International Building Code, it discontinued the annual inspection program. At the same time the duty to perform fire inspections went from the fire department to the building department, with little or no communication between the two.

The fire could have been prevented if the property had been constructed and maintained in accordance with code. The additions would have been built to code or taken down. They would have either had 2-hour separations or been sprinklered. Flammable liquids would have been stored properly. In all likelihood, the fire would not have spread beyond the loading dock and, if the building had been sprinklered, would have been out when firefighters arrived.

In the wake of the fire, Charleston hired a fire inspection supervisor to be the liaison between the building and fire departments. The city also hired a plans reviewer, with extensive experience as a firefighter. Charleston is taking fire prevention seriously. I hope the fire service learns a message from the Sofa Super Store: Proactive emergency planning should be given top priority. This fire is a clear example of how fire prevention, engineering, enforcement, and education are directly related to saving firefighter lives and preventing injuries. It is worth a significant investment of time and budget. Do you have a building like the Sofa Super Store in your community? Are you putting firefighters at risk?

—*"Code Concerns" by Mike Chiaramonte, from
Fire Chief Magazine, November 2008*

LEARNING OBJECTIVES

After completing this chapter the reader should be able to:

9-1   List common causes of fire.

9-2   Identify a particular community need for which the fire department would need to be involved.

9-3   Identify where to find the data required for a report concerning trends, variances, or other given topic.

9-4   Describe the departmental information management and/or record keeping system(s).

9-5   Identify fire prevention education programs.

9-6   Describe the benefits of accurate collecting and reporting incident response data.

9-7   Identify the service organizations or other stakeholders within the community concerned with the specific need.

9-8   Select the appropriate policies and procedures in dealing with a public inquiry.

9-9   Identify the particular citizen's concern as it applies to the fire department.

9-10   Describe techniques (verbal and nonverbal) used to communicate with the public.

9-11   Describe policies and procedures pertinent to the citizen's concern.

9-12   Identify the particular public inquiry as it applies to the fire department.

9-13   Describe policies and procedures pertinent to the public inquiry.

9-14   Identify public relations practices when dealing with individuals and the public.

9-15   Describe how community demographics relate to the identified need.

9-16   Describe policies and procedures pertinent to the identified community need.

9-17   Describe the approach to use when requesting assistance from a service group or stakeholder within the community.

9-18   Describe the role and mission of the department relative to the identified need.

9-19   Explain the purpose of conducting a fire inspection.

9-20   Identify the model fire prevention and building codes available for adoption by municipalities.

9-21   List the steps necessary to the conduct a fire inspection.

9-22   Describe the permit model.

9-23   Explain the necessity of obtaining an inspection or administrative warrant prior to performing a fire inspection.

9-24   Describe the inspection model.

9-25   Describe the limits of the right of entry provisions contained in the model fire prevention codes.

9-26   Categorize the data collected into types of responses (e.g., EMS, fire, rescue, etc.).

9-27   Determine the response levels required by unit type to meet the needs of the response area.

9-28   Collect incident data history for the response area to be included in a report.

9-29   List the building occupancy classifications that are found in the current model building code.

9-30   Describe the characteristics of each occupancy classification found in current model building codes.

9-31   Identify the records-retention policies applicable to the organization.

9-32   Explain the procedure for data collection, recording, retention, and disposition.

9-33   Examine the mission and goals of the organization.

* The FO I and II levels, as defined by the NFPA 1021, the *Standard for Fire Officer Professional Qualifications*, are identified in different colors: FO I = black, FO II = red.

INTRODUCTION

Ask firefighters why they were drawn to the fire service. You probably won't hear "I felt a deep need to prevent fires, reduce property loss, make my city a more attractive place to live and do business; and to reduce the number of injuries and deaths." The mission of America's fire departments has dramatically expanded over the last century as public demands for emergency response have increased. Few could have envisioned fire departments performing urban search and rescue or swift water rescue or providing the level of emergency medical care that has become common in much of the United States today. When faced with new public threats, the fire service has always risen to the occasion, but it's important not to forget our primary mission: *the protection of lives and property from fire.*

NOTE

When faced with new public threats, the fire service has always risen to the occasion, but it's important not to forget our primary mission: *the protection of lives and property from fire.*

The primary mission of the fire service has always been the protection of people and property from fires. As public resources diminish, the role of the fire service will continue to expand even more.

THE U.S. FIRE PROBLEM

Between 1998 and 2007, there was an average of 1,664,500 fires resulting in an estimated $10,949,900,000 in direct dollar loss each year (see **Figure 9-1**). An average of 3,695 Americans lost their lives and another 19,405 were injured annually as the result of fire.

NOTE

While the United States has made progress in slightly reducing the number of fires, civilian fire deaths, and civilian fire injuries over the past decade, total property losses continue to climb.[1]

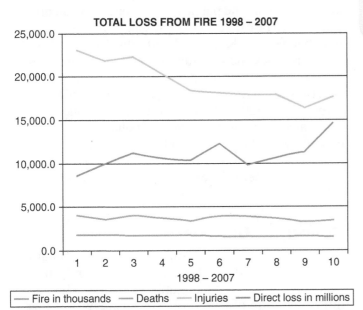

FIGURE 9-1 Statistics from Fire Loss in the U.S. 2007, National Fire Protection Association. (*Source: Michael J. Karter, Jr., United States Firefighter Injuries 2007, NFPA, August 2008 and previous reports in series*)

Fire Causes

Cooking is the leading cause of fires in the United States. Food preparation resulted in 28 percent of fires in the United States from 1995 to 2004. Incendiary or suspicious fires (arson) caused another 21 percent. The two leading causes of civilian deaths were arson, at 28 percent, and smoking, at 18 percent. The leading cause of injuries was cooking (24 percent), followed by open flame (18 percent) and arson (17 percent). Arson was, by far, the leading cause of property loss, at 26 percent.[2]

FIRE RESEARCH AND FIRE DATA

Data collection and analysis is a key element of your department's effect effort to protect your citizens from fire. It is of vital importance that fire data be timely and accurate, in order to identify trends and develop effective fire prevention strategies. Collecting no data is better than collecting inaccurate data. Consider your department's fire incident records as a key link in the chain that makes up the nation's fire statistics. Consider your department's information management system as one of the most important tools your department possesses. Without it, you're like a traveler without a map.

NOTE

It is of vital importance that fire data be timely and accurate, in order to identify trends and develop effective fire prevention strategies. Collecting no data is better than collecting inaccurate data.

In the fourteenth edition of *Fire in the United States* (August 2007) the editor noted: "Many records submitted to NFIRS by participating fire departments provide either incomplete or no information in some of the fields. Additionally, in preparing this report, it is assumed that participating fire departments have reported 100 percent of their fire incidents; however, this is not always the case. The completeness of all the information in the NFIRS modules will contribute to the refinement and confidence level of future analyses."[3] Unfortunately this is not new. It is important to follow national trends, as well as watch for local trends and conditions that may affect your community, and not necessarily the rest of your state or the nation.

The 1968 *Fire Research Safety Act*, established the National Commission on Fire Prevention and Control, a 24-member panel appointed by President Richard Nixon. Their report, *America Burning*, has proved to be one of the most significant forces for fire prevention and protection in United States history. Among the findings of the commission:[4]

- The need for more emphasis on fire prevention
- The need for fire departments to expend more effort on fire safety education, inspection, and code enforcement
- The need for better training and education for the fire service
- The need for improved built-in fire protection features in structures
- Increased involvement of the U.S. Consumer Product Safety Commission in regulation of materials and products affecting fire safety
- That important areas of research such as fire fighting, burn prevention and treatment, and protection of the built environment from combustion hazards were neglected

The commission called for the establishment of the U.S. Fire Administration (USFA) that would establish a national fire data system, monitor fire research, provide block grants to states and local governments for fire protection and prevention, and establish the National Fire Academy (NFA).[5]

USFA Data Collection, Analysis, and Dissemination

America Burning contained 90 recommendations. The first was for Congress to establish and fund the USFA to provide a national focus for fire protection and prevention issues. The second was that a national fire data system be established to "provide a continuing review and analysis of the entire fire problem."[6]

Lacking valid national statistics, the code development process must rely on anecdotal evidence that may or may not be valid. The National Fire Data Center studies and reports on the nation's fire problem, proposes solutions and priorities, and monitors proposed solutions. *Fire in the United States* is published by the USFA and distributed free of charge. The USFA Web site is a good place to start your data collection regarding a fire problem in your community. Your state fire marshal's office is another. The National Fire Protection Association (NFPA) collects fire data, and performs data analysis. Reports and studies are available at the USFA and NFPA Web sites. Many are free of charge.

Using Data to Attack a Problem

Every community has fire protection needs that can be addressed with the support of the public, elected officials, and the business community. In many communities, particular groups such as senior citizens or college students may be disproportionally affected by fire. Identifying the problem is the first step in addressing it. Gaining the cooperation of the affected group in recognizing that a problem exists and wanting to tackle it, and then enlisting the community to help solve the problem, is where the heavy lifting starts. Gaining the support of elected officials, the media and different stakeholder groups in the community should be the goal of every fire officer. Trust is slowly earned and can be quickly lost or destroyed.

The Stakeholders

People naturally form groups. Some are formal organizations with memberships. Some are merely neighborhood groups that form in response to a particular need or threat. Identifying these groups and enlisting their aid is a key element in addressing community needs. Reach out to these organizations. It is sometimes useful to contact them through intermediaries they trust such as the clergy, particularly if language is a communications barrier. Service and fraternal organizations, faith-based service groups, business associations and professional associations should all be enlisted to help develop and implement solutions. Form relationships with community organizations before the need is required.

Communication

Honesty and forthrightness is the key to effective communication with the public and the media. As an officer you will be expected to represent your department in public venues. Know and understand your department's policies and procedures for handling inquiries from the press and the public. Most state and local governments have specific policies for Freedom of Information Act (FOIA) requests that include fixed timeframes. Failure to follow the policies potentially exposes your department and you to judicial penalties. Ignorance of the applicable laws or departmental policies is never an excuse. Remember, you are an officer.

Concerned Citizens

No matter what the concern, company officers must act in a professional and considerate manner. Citizen concerns can run a wide gambit of issues. The first thing that a fire officer must do is distinguish if this particular concern can be handled by the fire department. Many concerns that are voiced by citizens have nothing to do with the fire department. However, it is important to let the citizen voice his or her concern before they are routed to another department. If the concern can be handled by the fire department it is important to receive all of the information. Fire officers are encouraged to document all citizen concerns for record keeping purposes. Let the citizen know that you understand the issue and that you will find an answer for him or her. Give them a time and date when they can expect a return call. This will give him or her a verbal confirmation that the concern has been understood. The next step is to be seen trying to handle the situation. For example, Mr. Smith calls into the fire station and voices his concern over his neighbor who is burning trash. The fire officer tells Mr. Smith that the department will look into it and call him back later on in the evening (verbal confirmation). The next step is to go and educate the neighbor on local burning ordinances. This allows Mr. Smith to see the fire truck and know the personnel are handling the concern (nonverbal confirmation).

Fire departments should develop policies and procedures for handling concerns raised by citizens. The procedures should outline what concerns should be handled at the company level and which concerns should be given to higher ranking officers. An example here would be a concern raised by Mr. Smith in regards to the fire inspection that was conducted at his business. This type of concern needs to be addressed by the fire inspector or the fire marshal. Policies and procedures should also outline the steps for documenting concerns. These concerns can be used when evaluating the community needs discussed later in this chapter.

Inquiries are not that different from concerns. These may be as simple as a citizen wanting to know the start time for the July 4th fireworks display or they could be as intense as a citizen wanting to know the cost of the new ladder truck. Either way, the fire officer must remain professional and handle the inquiry to the best of their abilities. If unknown, it is recommended that the officer take down all of the information and give the citizen a timeframe of when they can expect a return call from someone who can assist them. Once again, it is important for policies and procedures to outline the steps for handling public inquiries and it is important for you to know them. The policies should list steps for moving information up the chain of command as well as how they need to be documented.

Fireplace Ash Fires—A Local Problem

Fires involving the improper disposal of fireplace ashes are common in certain parts of the country, particularly around the Christmas holidays. In many homes, fireplaces are only used during the holidays, more for effect than for heat. Too often, these homeowners clean their fireplaces and remove the ashes before they have completely cooled. It is a national problem, but the greatest impact is in the Northeast and Mid-Atlantic states.

Fairfax County, Virginia, instituted a public education campaign after a rash of dwelling fires in the mid 1990s. Fires that involved ashes that reportedly had cooled for up to 24 hours accounted for 62 percent of the incidents. Fires that involved ashes cooled for up to 72 hours accounted for 8 percent.[7] The Fire and Rescue Department partnered with local retailers and had grocery bags printed with a warning against disposing of ashes in bags or plastic buckets.

Perception versus Reality

The public's perception of the fire department is often one that lacks knowledge. Most of the people within a community are satisfied that when they drive by the fire station and see fire trucks sitting in the bays. The perception that the job of a firefighter consists of washing trucks, play cards, eating, and getting cats out of trees is mostly the fault of the fire service. The fire service has often times failed to educate the public in what it really does. The reality is that the fire department must have the public's support. The public needs to know the reason that the fire truck is washed everyday is because it is one of our most important tools and we must keep in in top working order. Besides, the citizens paid for it and it should look good to show we are good stewards of community property. The public needs to know why the "big" fire truck is responding to medical calls in addition to an ambulance. By educating the public, a fire department earns respect. A fire department that is respected by the public will thrive when seeking new equipment, personnel, or even tax increases.

One sure way to earn the respect of the public is to identify their needs. This is an intricate process that involves the community's opinions and outlooks. To accomplish this, fire officers should use a community plan outline. When developing a community plan, there are some aspects that needed to be completed by the department and some that must be completed by members within the community. Community members should focus on the following elements:

- Description of community partners and committees
- Description of meetings and community, social services, and agency stakeholders
- Environment and natural resources, population, demographics, and socioeconomic data
- Residential density
- Community strategy for risk reduction
- Economic and ecological values
- Education steps for the entire community

Would some of these steps be easier for the agency to handle? Probably, but how else is the community going to fully understand the need if they do not do some studying for themselves?

The next elements of the plan need to be handled by the agency or agencies.

- Policies and procedures for all involved agencies
- Review of current studies and strategic planning
- Hazards and risks descriptions
- Materials and resources currently available
- Community safety education plans
- Outline policies and procedures based on the findings of the community

By putting together a community plan, the stakeholders will have a better understanding of the community needs. By involving the community, they will have a better understanding of the challenges that emergency service organizations face. The final outcome is respect for the fire department.

So—What's in It for Me?

The notion that fire departments are indispensable agencies, not subject to the same economic conditions that drive local government budgets, has always a been false one. For those who harbored any doubts, New York Mayor Michael Bloomberg announced the closing of sixteen fire companies in 2009. Facing a budget deficit of $4 billion, no city agency was considered too important to cut. The Big Apple isn't the only city grappling with the budget axe. When Philadelphia Mayor Michael Nutter ordered the closure of five engine and two ladder companies, IAFF Local 22 filed suit. The city's attorney's response was terse: "I think the fire fighters should acknowledge that they've already lost this issue. The current state of the law is that the city has the right to close fire companies. And it had done it, mindful of the safety of the fire fighters and the safety of the citizens."[8]

Economic downturns affect the tax base, resulting in less revenue for governments to provide services. But downturns aren't always caused by problems on Wall Street or Washington. When fires strike major employment centers, the impact might not have the same national implications, but for the communities involved, the blow is no less severe. When a welder's torch started an easily preventable fire at Farmland Food's processing center in Albert Lea, Minnesota, in 2001, 400 people lost their jobs overnight. The facility, which at the time was the largest employer in the region, never reopened. But the problems caused by the fire did not end there. Farmland Foods was the city sewer plant's biggest customer. Left with a sewer plant operating at

50 percent capacity, rates have risen 55 percent over the last seven years.[9] Twice the rates of Owatonna, 35 miles south on I-35. Businesses looking to expand or relocate began looking elsewhere. Most communities have factories, business parks, or tourist attractions that employ large numbers of people, generate sales and property taxes, and pump money into the economy by purchasing materials and services. The loss of a major employment center can change the economic climate within a jurisdiction overnight. The impact on government services including fire protection is usually swift and certain.

As a fire officer, the well-being of the public and your own people should be a primary concern. If you do not believe that preventing fires is a worthy effort for the good of the public, believe in it for your own job security and that of your people at the very least. As a fire officer, Chief Chiaramonte's words at the beginning of this chapter should come to mind every time you report for duty. Your first job as an officer is to safely bring all your people home.

OFFICER ADVICE

As a fire officer, Chief Chiaramonte's words at the beginning of this chapter should come to mind every time you report for duty. Your first job as an officer is to safely bring all your people home.

PREVENTING FIRES THROUGH REGULATION

On February 20, 2003, fire erupted at Station nightclub in West Warwick, Rhode Island, during a rock concert. Indoor pyrotechnic devices ignited foam plastic material that had been installed for acoustics purposes. The initial toll of 97 fatalities rose to 100 within three months, making it the fourth deadliest nightclub fire in U.S. history. State and local governments across the country scrambled to determine their vulnerability to a similar catastrophe. Fire chiefs and fire marshals were summoned to testify before elected officials, often in televised sessions.

The elected officials wanted reassurances that a system was in place to prevent such a large loss of life in their jurisdictions. Many fire officials were forced to admit that a comprehensive inspection program that included evening and weekend checks of public assembly occupancies did not exist. Few if any fire officials were given the opportunity to mention that the reason inspections were not being conducted was due to the lack of staff resources—resources that can only be provided by the same elected officials. Like the Sofa Super Store fire, the Station fire got the public's attention. It also got the attention of elected representatives on city and county councils. At least for a time, they demanded effective fire inspections programs, and they wanted them right now.

Demand for Inspections

The level of support from the public and elected officials for the inspection process is a constantly evolving one. **Acceptable risk,** which is a measure of the level of hazard potential that the public is willing to live with, constantly changes (see **Figure 9-2**). Incidents like the Station fire or the Oakland Fire Storm of 1991 result in a public demand for government officials to take some action that will provide assurance to the public. What elected officials want to hear is *why the catastrophic event that has aroused the public concern couldn't have happened in their community.* What they usually get is *what will be done in the future to keep the same thing from happening in their community.* In other words, has it been luck or an effective proactive prevention program that has prevented the catastrophe from happening? Most have to reply that their prevention programs are lacking and need to do better. We have been lucky up to know.

NOTE

What elected officials want to hear is *why the catastrophic event that has aroused the public concern couldn't have happened in their community.* What they usually get is *what will be done in the future to keep the same thing from happening in their community.*

FIGURE 9-2   The Oakland, California, fire in the fall of 1991 is one of the largest fire losses in American history. This picture was taken one year after the fire destroyed an entire community. *(Photo courtesy of Kim Smoke)*

Government programs are very often reactive not proactive. Government agencies have finite resources that must be allocated to achieve the greatest impact in support of their stated mission. Resources are often stretched thin.

Routine inspections by the fire service range from courtesy residential fire safety surveys by in-service companies, to specialized inspections for code compliance in business and industrial occupancies (see **Figure 9-3**). The preservation of property and protection of the public is the fundamental mission of the fire service. The reduction of hazards through the inspection process is one of the most effective means of accomplishing the mission. The inspection process is truly underestimated and often maligned by some firefighters who would rather fight fires, and by some fire

FIGURE 9-3 Company inspections afford an excellent opportunity to preplan while conducting safety inspections.

chiefs who can not see past the allocation of resources for training and staff, or the potential for political conflict caused by enforcement. Fire inspections reduce the rate and severity of hostile fires. To most people, the question "Just how many fires does an effective inspection program prevent?" is almost a rhetorical question and very difficult to answer. However, the effectiveness of inspections programs has been statistically proven.

In the 1970's the National Fire Protection Association (NFPA) and Urban Institute (UI) published a series of reports about the effectiveness of fire inspections. These studies are still cited as the leading resource for measuring inspection effectiveness. 1974: *Measuring Fire Protection Productivity in Local Government* 1976: *Improving the Measurement of Local Fire Protection Effectiveness* 1978: *Fire Code Inspections and Fire Prevention: What Methods Lead to Success?* The 1978 study was undertaken to determine, whether some fire code inspection practices *actually resulted in fewer fires, lower fire loss, and fewer civilian casualties.* The study focused on properties covered by fire codes, excluding one- and two-family dwellings.

The study used fire data from seventeen cities and one metropolitan county and grouped fires into three categories, based on the potential for intervention through the inspection process. The categories were those fires that were (1) due to visible hazards, (2) fires caused by foolish or careless actions, and (3) incendiary, suspicious, or natural-cause fires. The study determined that fires due to visible hazards that can be directly remedied by inspectors were responsible for only 4 to 8 percent of all fires. Fires due to carelessness or foolish actions or electrical or mechanical failures (sometimes due to lack of maintenance) accounted for 40 to 60 percent of all fires. The remaining fires (32 to 66 percent) were identified as incendiary, suspicious, or natural-cause fires, and were considered unpreventable by inspection.[10] When this information was compared to inspection practices, the findings underscored the effectiveness of routine fire inspections (see **Figure 9-4**):

> Fire rates appeared to be significantly lower in cities that annually inspected all, or nearly all public buildings. Jurisdictions that did *not* annually inspect most public buildings had rates of fires exceeding $5,000 that were more than *twice* that of those that *did* inspect them.

> Cities that used fire-suppression companies for a large share of their routine fire code inspections had substantially lower rates of fire, presumably because cities that exclusively used fire prevention bureau inspectors did not have sufficient personnel within the bureau to make annual inspections at all public buildings.

> Fires due to carelessness or foolish actions or electrical or mechanical failures (40 to 60 percent of all fires) are not preventable through direct action by inspectors, since these causes are not readily visible during an inspection.

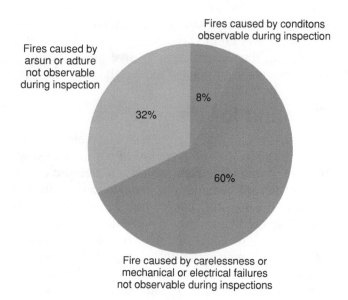

FIGURE 9-4 Only 4-8% of fires were caused by conditions that were observable during inspection, yet jurisdictions that inspected every building noted reductions in fires up to 50%.

Yet the rate at which these fires occurred was significantly lower in cities with regular inspection programs, *indicating that the department that inspects more frequently, has more opportunities to motivate occupants.* This also underscores the value of company inspections, recognizing the fact that although company inspections cannot be as in depth, due to the differences in training between inspectors in the bureau and those with other responsibilities, inspections by in-service companies have a significant impact on the incidence of fire within the jurisdiction.

In a nutshell, the study showed that although only 4 to 8 percent of fires were caused by conditions that inspectors could detect and take action to mitigate the hazard, yet the incidence of fire could be reduced by 50 percent by undertaking a comprehensive program in which every public building is inspected. Inspection is a positively motivating factor for the prevention of fire.

FIREGROUND FACT

The incidence of fire could be reduced by 50 percent by undertaking a comprehensive program in which every public building is inspected. Inspection is a positively motivating factor for the prevention of fire.

MODEL CODES

Fire prevention and construction regulation are complementary efforts of state and local governments to protect the public and preserve property. Both are driven by codes. **Codes** are merely systematically arranged bodies of laws or rules. Codes tell us what to do, or what not to do. Examples are the *United States Code,* the *Georgia Code,* and the *King County Code.* These are the codified laws of the United States, State of Georgia, and King County, Washington. The term *codified* refers to the fact that the laws are arranged according to a system.

Model codes are technical rules that are developed by organizations and are made available for governments to formally accept and put into effect within the jurisdiction. We refer to this formal acceptance as adoption of the code. Model codes are available for governments to adopt free of charge. In order to understand why private organizations invest thousands of hours of hard work into developing these codes, and why governments would adopt them, a look at the fire protection and construction regulation failures of the past is necessary.

The Development of Model Codes

Disastrous fires have been a staple of the American experience from our earliest days. The Jamestown Colony was almost totally destroyed by fire within a year of its founding. Fires ravaged our young cities, from coast to coast and border to border. Homes and businesses were destroyed, lives were lost, and insurance companies were bankrupted, leaving thousands of policyholders' claims unpaid. In the aftermath of every major fire, there was an outcry for action, followed by the passage of a new law or laws. These laws were overlooked or forgotten as soon as compliance became inconvenient or inexpedient. The political process is not necessarily the best medium for development of highly technical rules. Perhaps the most striking example was the City of Chicago, where within three years of the great fire that killed 300 and destroyed over 17,000 buildings, fire safety conditions were identified by insurance inspectors to be worse than before the fire![11]

If governments were not capable of technical code development, then who was? In its own self-defense and without the assistance of government, the fire insurance industry embarked on a method of ensuring profits and their own viability. The first attempts of the industry's technical arm, the National Board of Fire Underwriters, at developing standards involved setting specifications for "first class" woolen mills and sugarhouses, which would qualify for reduced rates by virtue of their class rating. When fires involving a whale oil substitute sold under the name of "kerosene oil" began to occur because there was no standard flashpoint for the lamp fuel, the board developed one.[12]

In 1892, a marked increase in fires at facilities considered to be better risks underscored a new hazard associated with economic success and innovation—electricity. An emergency meeting held in New York resulted in the formation of the Board's Underwriter's International Electrical Association, and development of the *National Board Electrical Code*, known today as the *National Electric Code* (NEC) NFPA 70. (NFPA 70). By 1901, the code was being enforced by 125 municipal governments. The National Board of Fire Underwriters transferred responsibility to the National Fire Protection Association (NFPA) in 1911. The *NEC* may well be the most widely used model code in the world.

By 1896, the National Board voted to expand the proposed building code, to be called the *National Board's Model Building Law.* In 1905 the first edition of the Board's *National Building Code* was distributed free of charge to all cities with population over 5,000 and to contractors, architects, fire marshals, and technical schools.[13] The National Board of Fire Underwriters later developed the *National Fire Prevention Code.* The Board published the *National Building Code* and *National Fire Prevention Code* through 1976.

THE EVOLUTION OF REGIONAL MODEL CODES

While several large cities had their own locally developed building and fire codes, many jurisdictions adopted the National Board's codes. In the twentieth century, three regional model code organizations emerged: Building Officials and Code Administrators International (BOCA) in the Northeast, Southern Building Code Congress International (SBCCI) in the Southeast, and International Conference of Building Officials (ICBO) in the Western states. Each published a complete complement of construction, fire prevention, and property maintenance codes for the better part of the twentieth century.

The International Codes

The regional code organizations formed the International Code Council in 1994, with the intention of cooperatively developing a single set of codes for the entire United States. Their efforts were so successful that they merged in 2003. BOCA, SBCCI, and ICBO no longer exist. The "I-Codes" have been widely accepted by state and local governments across the United States:

- The *International Building Code* (IBC) is adopted at the state or local level in 50 states and the District of Columbia (Washington, D.C.).
- The *International Residential Code* (IRC) is adopted at the state or local level in 48 states, District of Columbia, and the U.S. Virgin Islands.

- The *International Fire Code* (IFC) is adopted at the state or local level in 42 states and the District of Columbia.

International Fire Code Council (IFCC)

The International Fire Code Council (IFCC) was established by the ICC to represent the common interests of the fire service and the ICC, by providing leadership and direction on matters of fire and life safety, and to meet government, industry, and public needs. The IFCC was formed prior to the development of the *International Fire Code* (IFC). Fire service influence on the development of the IFC through the IFCC, is readily apparent. The IFC is a comprehensive, user-friendly document, clearly designed with the inspector in the field in mind.

The IFCC is comprised of 15 members, one each from the eight International Association of Fire Chiefs regions, four representatives from the National Association of State Fire Marshals, and three at-large members. Through the efforts of the IFCC, the International Association of Fire Chiefs and ICC are working toward a memorandum of understanding that would grant the fire service greater influence on fire-related matters, including the appointment of committee members.

Codes vs. Standards

The phrase "codes and **standards**" lumps together two important regulatory instruments in a way that makes it easy to conclude that they are the same thing—they are not. Both are key tools in fire prevention and construction regulation; both are developed through formal consensus processes, with public input; both are adopted by governments as regulatory tools. A code is systematically arranged body of rules—when and where to do or not to do something. A standard is a guide or rule to be followed. Codes tell you what to do. Standards tell you how. The IBC requires hospitals to be equipped with automatic sprinklers. The IBC references NFPA 13, *Standard for the Installation of Sprinkler Systems* for specifics on design, materials, installation, and performance.

NOTE

A code is systematically arranged body of rules—when and where to do or not to do something.

NOTE

A standard is a guide or rule to be followed. Codes tell you what to do. Standards tell you how.

CODE ADOPTION

The act of state or local governments to accept a model code and give it the effect of law is called adoption. Two basic methods are used, based on the laws in effect within the jurisdiction. **Adoption by reference** is simply the passage of legislation that states that a specific edition of a certain code will be enforced within the jurisdiction. Local governments typically do this by passing an **ordinance,** or local law. When model codes are adopted at the state level, the adopting legislation is generally in the form of a **statute.** Within the legislation, the code document is mentioned by reference only, and copies of the codes must be purchased from the model code groups. In some localities, adoption by reference is not legally possible. **Adoption by transcription** is a legal requirement, and the code is republished, usually with a numbering system for each section or article that complies with the requirements of the jurisdiction.

Model code groups permit adoption by reference for free—they make their money by selling books. Adoption by transcription is a different matter. A license to republish can be negotiated or, the model code organizations can be contracted to publish a special edition that meets the requirements of the adopting jurisdiction, and they get to sell the codebooks.

State and Local Adoption

Whether a code is adopted as a state minimum code that can be locally amended, adopted as a state **mini-maxi code** with no option of local amendment, or as simply a locally adopted code, there are legal requirements to ensure that adequate public notice is given, and citizens and special interest groups are given an opportunity to be heard regarding the proposed codes. Mini-maxi codes are favored by most business interests and by developers because they create uniformity within the state and all lobbying efforts can be directed at the state capital, not within each political subdivision. The fire service generally opposed mini-maxi codes since local control is reduced. One positive aspect of mini-maxi codes is that they promote training and political action on a statewide level, forcing fire service organizations to work together.

NOTE

The fire service generally opposed mini-maxi codes since local control is reduced. One positive aspect of mini-maxi codes is that they promote training and political action on a statewide level, forcing fire service organizations to work together.

WHAT CODES CANNOT DO

The cycle catastrophe followed by a public outcry of "there ought to be a law" isn't new, and won't be ending anytime soon. Sometimes, a legitimate need for new regulations is recognized, and new code provisions are adopted. In 1985, a fire at the Valley Parade Soccer Stadium in Bradford, England, claimed 86 lives and injured 200. Investigators identified the large volume of combustible materials beneath the grandstands as a significant factor in the fire spread. The fire, caused by discarded smoking materials that ignited trash and rubbish beneath the bleacher seats, engulfed a 290-foot-long grandstand in less than 5 minutes, and was televised live with the soccer match.[14]

The Bradford stadium fire identified a potential problem that was not clearly covered by fire code provisions. A provision regulating waste and combustible materials storage under grandstand seats was added to national model codes and exists today in both the UFC and IFC.

In the aftermath of the 2003 Station nightclub fire in West Warwick, Rhode Island, many called for new codes to ensure that such a catastrophe could never occur again. The lack of sprinklers was rightly identified as a significant factor in the fire that claimed 100 lives, but existing code provisions in place in the State of Rhode Island could have prevented the incident *had they been enforced*. The Rhode Island Fire Code, based on NFPA 1, *Uniform Fire Code* included provisions that addressed indoor pyrotechnic displays and foam plastic as an interior finish. All the codes or laws in the world will not prevent tragedies from occurring if they are not obeyed, and the largest fire prevention bureau in the nation cannot inspect every building every day.

THE NATIONAL FIRE PROTECTION ASSOCIATION (NFPA)

NFPA was organized in 1896 and incorporated in 1930. It was originally formed by the stock insurance industry to develop a single sprinkler standard that would be acceptable to all its member companies. In the late 1800s, there were nine different standards for sprinkler pipe size and head spacing within 100 miles of the city of Boston.[15] The group adopted the name of National Fire Protection Association and developed that standard. It lives on today as NFPA 13, *Standard for the Installation of Sprinkler Systems*.

NFPA publishes almost 300 codes, standards, and recommended practices developed by more than 205 technical committees. Many of the documents originated as National Board of Fire Underwriters pamphlets, and carried the subtitle "as recommended by the National Fire Protection Association." NFPA's recommended practices are one of the great, untapped resources for municipal fire officials. They are not designed to be enforceable documents, rather, a guide to good practice within given industries. Alerting business owner to the existence of these documents may do more for the prevention of fires than simply enforcing the code.

In 2003, NFPA partnered with the Western Fire Chiefs Association and incorporated the provisions of the Uniform Fire Code into NFPA 1, *Fire Prevention Code*. The new Document is titled NFPA 1, *Uniform Fire Code*. The *Uniform Fire Code* is adopted statewide in twenty states.

ESTABLISHING INSPECTION PRIORITIES

One of the initial steps in establishing an inspection program is the determination of which occupancies will be inspected, how often, and in what order. In the aftermath of a tragic fire in a particular type of facility, that group generally becomes the focus of public attention. Although understandable and not totally without merit, this system of "closing the barn door after the horse has run off" doesn't lend itself to success.

OFFICER ADVICE

A well-disciplined approach that prioritizes inspections based on risk and the actual hazards associated with certain occupancy classes is by far the better course.

A system of inspections based on the actual fire experience within the jurisdiction would appear to be the surest method of addressing the most pressing fire prevention and life safety issues. An accurate assessment of fire statistics for the jurisdiction should be a regular function of the fire prevention bureau. Unfortunately, the results of the study of a statistical sample as small as most cities or counties will not reveal a true picture of the relative risks associated with particular hazard classes.

For the purpose of discussion, let's assume you have been tasked with determining inspection priorities and in the course of your study of fire causes within your jurisdiction, you determine that welding and cutting were insignificant fire causes with negligible losses for the statistical period. Welding, cutting, and other hot work might go to the bottom of the list of inspections, if it made the list at all. In reality, it is a serious fire cause, with significant losses. The *2006 International Fire Code Commentary* described losses by a single insurance company within a five-year period of 290 hot-work related fires, with an average $1.4 million loss

FIGURE 9-5 Cutting torches use a mixture of acetylene and oxygen to generate a high-temperature flame. They cut metal by melting it.

per incident.[16] Sooner or later, welding and cutting would rightly assume its place in your jurisdiction's fire statistics, and would become an inspection priority—after the damage is done (see **Figure 9-5**). What method can be used to develop priorities that reflect actual hazards? The method is actually an integral part of the model fire codes, the *permit system.*

The Permit Model

The **permit model** of determining inspection priorities is simply inspecting those occupancies that are required to have a fire code permit. The 2006 *International Fire Code* contains forty-six operational permit categories and twelve construction permit categories; the UFC has seventy operational and nine construction permit categories. Operational permit categories include special amusement buildings, carnivals, compressed gases, covered mall buildings, cutting and welding, explosives storage and use, and operation of lumberyards and woodworking plants. Construction permits include installation or modification of fire alarm and detection systems, flammable and combustible liquid tank installation or modification, installation or modification of spray rooms or booths, and erection of tents and air-supported structures. Permit categories across both model fire codes are quite similar.

If you compare the list of permits from the model fire codes with large loss fires from the past, you will see a direct relationship. Permit categories are based on our national fire experience, and were established over the years in response to incidents involving large loss of life or significant financial loss from property damage.

The list is constantly evolving. The requirement in each of the model fire codes to operate a "special amusement building" can be traced directly back to the 1984 "Haunted Castle" fire at Six Flags Great Adventure Park in Jackson Township, New Jersey.

FIREGROUND FACT

If you compare the list of permits from the model fire codes with large loss fires from the past, you will see a direct relationship. Permit categories are based on our national fire experience and were established over the years in response to incidents involving large loss of life or significant financial loss from property damage.

Special amusement buildings are temporary or permanent structures for entertainment or amusement, in which the means of egress are intentionally confounded or not apparent due to visual or audio distractions, or are not readily available because of the nature of the attraction or mode of conveyance through the structure. Examples run from an attraction at Disney World on the high end, to a carnival fun house in a tractor-trailer, or even a Halloween maze put on by a local service club. In the Six Flags fire, eight of the approximately thirty visitors within the structure were killed when foam plastic material lining the interior of the structure was ignited. The combination of dense smoke, a rapidly spreading fire, and built-in confusion as to the location of exits within the structure was deadly.[17] In the aftermath, the memberships of the model code organizations voted to approve code changes regulating such structures and ensuring their inspection by requiring operational permits.

When faced with the prospect of inspecting a limited number of processes and occupancies, the decision to inspect those requiring permits under the fire code is a wise one. Not only are the facilities with the greatest risk for fire or life loss being afforded the attention of the fire department bureau first, the choice is based on the national fire experience.

A particular industry's cry of "why are you picking on us?" can be quickly answered by identifying the code requirement.

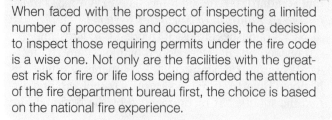

OFFICER ADVICE

When faced with the prospect of inspecting a limited number of processes and occupancies, the decision to inspect those requiring permits under the fire code is a wise one. Not only are the facilities with the greatest risk for fire or life loss being afforded the attention of the fire department bureau first, the choice is based on the national fire experience.

A significant benefit of the permit system that is often overlooked is the leverage that permits offer in gaining compliance and gaining entry inspect. Operation without the required permit is a violation of the model fire codes. Permits can be revoked for failure to comply with the code, or violation of permit conditions. Permit conditions include submission of the required application, payment of fees and permission for the fire inspect the facility.

The courts have consistently upheld the right of property owners, including business owners, to prohibit the entry of government officials without a search **warrant.** In other words, an owner of the local body shop owner has the constitutional right to refuse the entry of a fire inspector. What the owner doesn't have is the constitutional right to a fire code permit—no inspection, no permit. Spray painting cars in his body shop without the required fire code permit, for *spray application of flammable finishes*, is a violation of the fire code.

Another benefit of using the permit model is that permit requirements are based on the incidence of fire in the nation. The previous discussion of fire losses attributed to welding and cutting not being significant in a single jurisdiction, yet being a major fire cause nationally, points to the pitfalls of statistical analysis with too little data. Inspection priorities based on the permit model are the easiest to justify to elected officials, the public, and the courts.

The Inspection Model

The other method of determining inspection priorities is the **inspection model.** Occupancy classes and types of processes are selected for inspection based on perceived need. The need must be based on fire risk, or life hazard, and should be backed up with statistics of actual fire loss. The fire official must be able to justify why particular industries, processes or occupancies are selected, and why others are not. Inspecting by district or area immediately raises the question, "why were we selected first?" There may be a legitimate reason. The fire official must be ready to provide valid reasons to a skeptical press and public. The elected official who presses the fire official to "clean up" a particular part of the city for other than legitimate reasons will disavow any knowledge of the demand when the story hits the newspaper.

When prioritizing inspections using the inspection model, develop a written plan.

If the need does arise to secure an **inspection warrant,** the judicial officer will want to know why the particular occupancy was selected for inspection to start with. A good answer would be "Our standard operating procedure requires that all fuel storage facilities are inspected annually. We last inspected the facility twelve months ago," or "We inspect all fuel storage facilities every fall." The judicial officer represents the people

and is supposed to make certain that your request for an inspection warrant is not based on the fact that the business owner no longer gives firefighters a discount or that the mayor is a business competitor.

OFFICER ADVICE

When prioritizing inspections using the inspection model, develop a written plan.

INSPECTIONS OF EXISTING OCCUPANCIES

Within the big picture of inspection and enforcement of codes by state and local governments, the fire official probably has the broadest mandate. Most jurisdictions have a building official charged with the enforcement of a construction code that regulates new construction and renovations. Many jurisdictions have a property maintenance code that targets residential occupancies, particularly rental properties, to ensure that landlords meet minimum standards for safety and hygiene. Health inspectors are normally involved in food service and processing, health care, and other facilities that may have public health implications. In many states, assisted living facilities, foster care, and day care for both adults and children may be regulated by social service agencies. The fire official is generally the only official whose code applies to all the above, and should interact with all the other officials.

With very few exceptions, all buildings and structures, including residential occupancies, are within the scope of the model fire codes. What separates residences from other types of occupancies is that they are exempt from routine inspection.

NOTE

The **right of entry clauses** in the model codes, that permit the fire official to enter and inspect structures on a routine basis, exempt one and two-family dwellings and residential portions of multifamily structures. The code still applies, and must be enforced when conditions exist that pose a significant hazard.

What type of hazard reaches the level of *significant* is subjective, and has been debated among fire officials, civil libertarians, and the courts for years.

NOTE

With very few exceptions, all buildings and structures, including residential occupancies, are within the scope of the model fire codes. What separates residences from other types of occupancies is that they are exempt from routine inspection.

FIREGROUND FACT

The right of entry clauses in the model codes, which permit the fire official to enter and inspect structures on a routine basis, exempt one- and two-family dwellings and residential portions of multifamily structures. The code still applies and must be enforced when conditions exist that pose a significant hazard.

When Supreme Court Justice Potter Stewart wrote his famous opinion regarding obscenity in 1964, he stated, "I shall not today attempt to further define the kinds of material I understand to be embraced . . . *but I know it when I see it. . . .*"[18] Wrestling with just what constitutes a hazard that is significant enough to warrant government intrusion within someone's home is similarly perplexing. The hazard must be sufficient to justify what our forefathers wrote the Fourth Amendment to the Constitution to guard against. The condition of a 12-year-old's room probably does not reach the level of hazard, whereas 55 gallons of lacquer thinner, stored in the ground floor apartment of a painting contractor, certainly does. The fact that lacquer thinner is highly flammable, and a leak in such a large container would pose a threat to all the residents of the building, would probably lead the judicial officer to determine that the potential threat to the neighbors is greater than the possible harm caused by the intrusion into the painter's home. In most cases, inspection warrants can only be obtained after a request for entry to inspect has been denied by the tenant or property owner.

COMPANY INSPECTIONS

Most jurisdictions do not have sufficient staff within the fire prevention bureau to inspect every occupancy, yet studies have shown that jurisdictions that inspect every occupancy realize reductions in fire loss. The use of in-service companies to perform inspections has been successful in many jurisdictions.

Personnel should be trained to perform a basic safety inspection and directed to request assistance from the fire prevention bureau if they encounter imminent hazards.

The assignment of fire prevention bureau inspectors as a resource for specific occupancies improves communication. Checklist forms are a useful method of ensuring consistent inspections and provide a record of the visit for the department and the business owner. Company personnel should be informed that the inspection is a combination of a preplan visit, safety inspection, and chance to interact with the community they serve.

OFFICER ADVICE

Personnel should be trained to perform a basic safety inspection and directed to request assistance from the fire prevention bureau if they encounter imminent hazards.

The opportunity to preplan hazards, provide building familiarization, and establish cordial relations with the community can literally make in-service inspections the excuse that every chief wants to advertise the department (see **Figure 9-6**). By conducting thorough preplans, the department can assure that adequate resources will be available if an incident occurs. It is essential for preplans to devise a plan for each type of emergency that may occur. For example, if a school is being preplanned, then there should be a plan for a fire incident as well as a mass causality incident. Differences in incident types will dictate what resources are needed. Training for company inspections should be prefaced with the fact that the goals of the in-service inspection program are building familiarization to increase firefighter safety, fire and hazard reduction, and community relations. This information must be collected, documented, and distributed. Fire officers should train their firefighters on each of the preplans that are within their district.

NOTE

The goals of the in-service inspection program are building familiarization to increase firefighter safety, fire and hazard reduction, and community relations.

FIGURE 9-6 The chance to establish cordial relations with the community can literally make in-service inspections the excuse that every chief wants to advertise the department.

THE MODEL CODE SYSTEM OF SAFETY

The model building code system attempts to balance fire protection and safety features based on the needs of the occupants and the potential hazard level within a structure. This is achieved through the classification of buildings and areas within buildings based on use and occupancy. The first job of a building designer is to assign the appropriate classification to every room or space. The first job of every inspector for the life of the building is to ensure that the use or occupancy has not changed and, if it has, to take action.

The use of group classification is the first aspect that must be established in the design of a building. It should also be your first consideration during an inspection of a building.

The use and occupancy classification of a building is the most significant single design factor that will affect the safety of the occupants as well as the fire suppression forces that are called upon in the event of fire. The building's height and size, type of construction, type and capacity of exit facilities and fixed fire suppression systems are all dependent on this classification. The system of use group classification is the foundation for the building and fire prevention codes. A mistake in laying the foundation can prove to be a disaster later on. Failure to maintain the foundation can also prove to be a disaster.

OFFICER ADVICE

The use of group classification is the first aspect that must be established in the design of a building. It should also be your first consideration during an inspection of a building.

Use and Occupancy Classes

The model codes separate buildings and areas within into 10 general uses based on three factors:

1. Capacity of the occupants for self-preservation
2. Relative fire hazards associated with the activities inside the structure
3. Types and quantities of regulated materials within the structure

The classes are further subdivided into subgroups. A church, a museum, and a nightclub are all assemblies. The hazard is a life hazards, generally based on the large number of people. But the specific characteristics of their occupants and functions are drastically different, requiring different built-in levels of protection. The occupants of the church are probably very familiar with the building that they occupy. They have been there before and know the location of alternate exits. The occupants of the night club may not be so familiar. Dim lighting, loud music, and impairment by alcohol are all common features that may affect the ability of the occupants to identify a fire emergency and take appropriate measures to escape.

Group A: Assembly

Group A is subdivided by the IBC into five subcategories, based on function. The biggest hazard in Group A occupancies is the large number of people who are often unfamiliar with their surroundings. People generally feel safer exiting by the same route they entered. This can prove deadly if main exits are inadequate in size or are compromised. Additional hazards posed by large fuel loads such as sets, scenery or decorative materials, entertainment, and the effects of alcohol can increase the threat level. Building code requirements for fire protection and life safety features are among the highest.

Group B: Business

Group B includes physicians and other professional offices, civic administration, banks, barber shops, and post offices. Colleges and adult education rooms with occupant loads under 50 are Group B occupancies, as are dry cleaning pick-up and delivery. Group B is one of the least hazardous and has much less stringent requirements for fire protection and life safety features.

Group E: Educational

Group E occupancies include areas *not* used for business or vocational training (shop areas) for students up to and including the 12th grade. Colleges and universities are Business or Assembly areas depending on the number of occupants. Group E features a robust mix of life safety and fire protection features, and a requirement for 60-inch minimum corridors. The fire codes complement these built-in features with requirements for emergency planning and fire drills. *Day care facilities may be classed as Educational or Institutional depending on the number and age of the children.*

Group F: Factory and Industrial

Group F occupancies include buildings used for manufacturing and repair and are subdivided into moderate and low hazard facilities. High hazard factory and industrial areas are classified as Group H.

Group H: High Hazard

Group H occupancies are those in which the quantities of regulated materials exceed the permissible threshold established by the code. Group H is based on the quantity of materials in the building or area of the building, not what is being done with the materials. Group H is subdivided into five subgroups, based on the level of hazard.

Group I: Institutional

Group I occupancies are those in which persons whose capacity for self-preservation has been diminished, whether by age, illness, infirmity, or incarceration. Group I has the most restrictive fire protection and life safety requirements because in many cases, the occupants either cannot or should not evacuate the structure in a fire situation. Group I is subdivided, based on the ability of the occupants to escape or shelter in place. The fire and life safety features of a hospital are among the most stringent.

Group M: Mercantile

Group M occupancies include retail shops and stores, and areas displaying and selling stocks of retail goods. Automotive service stations that sell fuel and groceries are Group M. The difference between Group B-Business and Group M-Mercantile is the fuel load created by goods that are displayed and sold. Group B occupancies provide services. Group M occupancies display and sell merchandise.

Group R: Residential

Group R occupancies include hotels and motels, dormitories, boarding houses, apartments, town houses, and some one- and two-family dwellings (see **Figure 9-7**). Most single-family, duplex, and townhouses are now constructed using the *International Residential Code*. Group R occupancies are statistically the most hazardous. Consistently, more fire fatalities occur in Group R occupancies than any other.

Group S: Storage

Group S occupancies are used for the storage of goods and include warehouses, storehouses, and freight depots. Storage uses are separated into low and moderate hazard storage uses. Auto repair facilities that perform major repairs, including engine overhauls and body work or painting, are considered moderate hazard storage occupancies. Occupancies that store over the threshold quantities of regulated materials must be designed and constructed as Group H occupancies.

FIGURE 9-7 This apartment building is classified Group R-2. Group R occupancies are statistically the most hasardous. (*Photo by Jeff Payne*)

Group U: Miscellaneous

Group U occupancies are those that are not classified under any other specific use and include tall fences, cooling towers, retaining walls, tanks, etc.

Mixed Use Buildings

Buildings will often contain multiple occupancies, with different uses. A three-story office building might have a restaurant (Group A-2) and computer store (Group M) on the first floor and professional offices (Group B) throughout the rest of the building. Hotel/convention centers typically have Group R-1, Group B, and several Group A divisions (see **Figure 9-8**). The model building codes provide for these situations either by requiring that the whole building be constructed to all the requirements of the most restrictive use group or by separating the areas with fire rated assemblies or by separating the building with fire walls. Separating with fire walls actually allows the occupancies to be treated as separate buildings.

Changes in Use and Occupancy

Detecting and prohibiting illegal changes in use is a primary responsibility of the fire inspector. Some changes in use do not appear that drastic on the surface, while some pose immediate hazards to the occupants. Illegal changes in use are a violation of the fire prevention code. Changes in use require a complete analysis of a building prior to approval. It is as if the owner is building a new building on the site.

FIGURE 9-8 Hotel/convention centers typically have Group R-1 guest rooms, Group B offices, and several Group A divisions such as restaurants, meeting rooms and ausitoriums.

FIGURE 9-9 Inspectors should remember that sometimes people who work with dangerous materials or in potentionally hazardous environments may become complacent and neglect to comply with the basic rules of safety.

CONDUCTING INSPECTIONS

At times, the basic issues involving the establishment of an inspection program are obscured by the sheer volume of information that has to be considered. It is clear that the desired result of any fire inspection program is fewer fires, reduced fire loss from those fires that do occur, and fewer deaths and injuries. The 1975 American Insurance Association's Special Interest Bulletin No. 5, *The Value and Purpose of Fire Department Inspections*, listed seven objectives for an inspection program, which, if accomplished, would render a *substantial service* to the community they serve.[19] These points have value today.

1. To obtain proper life safety conditions—Life safety inspections warrant attention to the adequacy of exits, obstruction to adequate and orderly egress at the time of fire, the adequacy of building evacuation plans, and determination of the number of persons permitted in places of assembly.

2. To keep fires from starting—Persons who work among materials or situations that are hazardous often become negligent of their own safety, just as long periods without a fire overconfidence

and underestimation of the fire problem (see **Figure 9-9**).

3. To keep fires from spreading—The general public has little appreciation of the great value that structural features, such as stair and elevator enclosures, fire doors, and fire partitions, have in preventing the spread of fire.

4. To determine the adequacy and maintenance of fire protection systems.

5. To preplan fire fighting procedure—It is difficult to attack a fire intelligently without first knowing the building and its occupancy.

6. To stimulate cooperation between owners and occupants and the fire department—Nothing will assure closer cooperation between owners and occupants of buildings and the fire department than the interest of the department in not only preventing fires but also being better prepared to handle fires when they occur.

7. To assure compliance with fire protection laws, ordinances, and regulations—In many places, some or all of these matters will be under the jurisdiction of the fire department.

Whenever possible, review the records for previous inspections before you go to the site. If the same deficiencies and unsafe conditions are encountered, it may indicate the need for a frank discussion about conditions and ways to improve them. It might also indicate the need for assistance from the fire prevention bureau, especially if the deficiencies are particularly hazardous. This is the reason that departments must have record-retention policies in place. These policies safeguard the information so that it can be readily accessed and maintained.

Inspections and building surveys should be thorough, methodical and start on the outside of the building. Drive around the entire perimeter of a large building if you can, before you go in. You can inspect the exterior of the building on foot, as part of your inspection after checking in with the owner's representative. Cars or storage blocking hydrants, fire department connections, or exit discharges can be moved while you are touring the inside of the building if you notify the manager when you first arrive.

One final word about conducting company inspections. Remain professional. Make sure you and your crew looks professional, project a professional appearance and keep your crew together. Keep your behaviors professional; this is not the time to tell jokes or eye the opposite sex. If the property representative does not agree with what you are telling him or begins to get upset, do not argue. Politely excuse your self and call in the inspection bureau or your supervisor to complete the inspection, perhaps at a later time when emotions have had time to cool down.

NFPA 550, FIRE SAFETY CONCEPTS TREE

NFPA 550, *Fire Safety Concepts Tree* (see **Figure 9-10**) was first developed in 1974 and identifies fundamental strategies and lists objectives for achieving a fire-safe environment. The fault tree analysis system was first developed by H. A. Watson of Bell Telephone Labs in 1962 and made famous by Boeing Corporation in its application to the Minuteman Ballistic Missile Program.[20]

The goal of fire inspection programs is the safeguarding of life and property, and it is achieved through two basic strategies: *prevent fires from occurring*, and when they do occur through some failure, *manage the impact* on people and property.

Prevent the Ignition of Fires by Effectively Segregating the Three Elements of the Fire Triangle

- Control heat sources such as smoking, unsafe heating appliances, electrical hazards, open flames, and burning.

- Control fuels by regulating the storage and handling of specific materials.

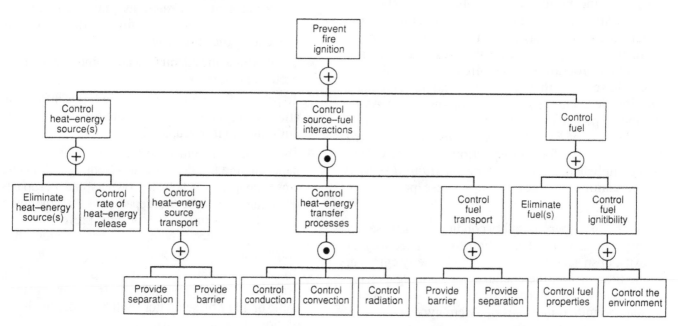

FIGURE 9-10 NFPA 550, *Fire Safety Concepts Tree* was first developed in 1974 and identifies fundamental strategies and lists objectives for achieving a fire safe environment. (*Source: Reproduced with permission from NFPA 550, Fire Safety Concepts Tree, Copyright © 2007, National Fire Protection Association. This reprinted material is not the complete and official position of the NFPA on the referenced subject, which is represented only by the standard in its entirety*)

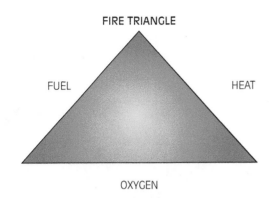

FIRE TRIANGLE

FUEL

HEAT

OXYGEN

FIGURE 9-11 The strategy used by the model fire codes to prevent fires is to segregate the three legs of the fire triangle.

■ Control the interaction of fuel, oxygen, and the heat source in storage, handling, transportation, and processing (see **Figure 9–11**).

Manage the Fire Impact by Effectively Regulating Construction, Use and Occupancy, and Occupant Training

■ Limit building height and area based on the combustibility and fire-resistance of construction elements.

■ Provide adequate fire-resistance for structural elements and separation assemblies.

■ Provide adequate emergency egress facilities.

■ Provide built in fire protection systems to detect and suppress fires.

■ Ensure that hazardous processes and occupancies are adequately separated from other occupancies.

■ Ensure that occupants and employees in critical occupancies receive appropriate training in evacuation and emergency response.

PREINCIDENT PLANS

In the first edition of *Building Construction for the Fire Service*, Frank Brannigan observed that:

A building does not drop from the sky as does a disabled aircraft. Neither is it transient, such as a ship, truck or a train. Very often it has been inexistence since long before any of the firefighters at the scene were born. Yet, sad to say, in many cases the building might as well be a space ship from Mars for all the fire department really knows about it.[21]

Company inspections are one of the best opportunities to develop or update pre-incident plans. Preplans typically consist of the following:

■ Occupancy classification(s)
■ Type of construction, height, area, and number of stories
■ Fixed fire protection systems
■ Building utilities
■ Nature of the activities conducted in the building
■ Life hazards and evacuation considerations
■ Water supply features
■ Access points to the site and building interior, both horizontal and vertical
■ Hazardous materials or processes
■ Exposure potential
■ Safety issues and features for civilians and fire personnel
■ Special resources that may be required
■ General firefighting concerns

NOTE

"A building does not drop from the sky as does a disabled aircraft. Neither is it transient, such as a ship, truck or a train. Very often it has been inexistence since long before any of the firefighters at the scene were born. Yet, sad to say, in many cases the building might as well be a space ship form Mars for all the fire department really knows about it."—Frank Brannigan[15]

Preincident Plans and the Codes

Often, the building and fire code are viewed as having little, if anything, to do with fire departments' operations. This is unfortunate, since the use classification is a key indictor of what the structure was designed for and the occupant's capacity for self-preservation. The occupancy classification listed on the certificate of occupancy must match the actual use, and structure should be consistent with the construction classification listed. Hazardous Materials Inventory Statements (HMIS) and Hazardous Materials Management Plans (HMMP) required by the fire code should also be included in preplans.

The model fire codes require buildings to employ the NFPA 704 marking system to identify areas where hazardous materials are manufactured, handled, used, or stored. The marking system is intended to alert first responders to the presence of hazardous materials (see **Figure 9-12**).

FIGURE 9-12   NFPA 704 marking system is intended to alert first responders to the presence of hazardous materials.

FIRE PREVENTION

Julie Miller, Captain, Inuvik Fire Department, Canada

Last April, in an isolated 190-person community in the Canadian Northwest Territory close to the Arctic Ocean, a park ranger's house burned to the ground. Witnesses had indicated kids were seen leaving the scene.

As a company officer for the Inuvik Fire Department, Captain Julie Miller had made it her goal to instill a district-wide prevention program, which included running the juvenile fire-setters intervention program. So several days after the house fire, she flew into the community and spoke to a 5-year-old and two 6-year-olds.

She interviewed the parents, and then each child separately to determine whether the fire was a curiosity or if psychological issues were to blame. It was found to be the former: The kids had no concept of the consequences of fire. They had entered the abandoned house and were playing with matches they had found, setting pieces of the couch on fire. One piece started to burn, and flames went up the wall, which caused the rapid fire spread. At that point, the kids were scared. By the time Miller got to them, they were terrified. Having worked with kids previously, she knew it was important to stress that they were not in trouble and

that she just wanted to prevent this from happening again. She discussed with them how fire works, using activity books that bolstered the talk and showing them videos of kids who had been burned.

After spending the day in the community, she gave the local fire department the resources to follow up if necessary. It is incidents like the one mentioned above that compels her to do Risk-Watch® public education programs in local schools several times a year and to train several other firefighters to do the same. According to Miller, there has been a direct correlation between public education and preventing fire occurrences within the community. An ounce of prevention . . .

Questions

Captain Miller made quite an effort to connect with the children involved in this incident.

1. Do you think her actions were appropriate?

2. Do you think her actions were beneficial to the children and others in that remote community?

3. What are the benefits of such activity?

4. Would you or your department have undertaken similar activities to prevent a recurrence of this type of situation?

5. Does your department have an organized juvenile fire-setter intervention program? If so, how is it used? If not, how would you start one?

LESSONS LEARNED

The primary mission of the fire service has always been the protection of people and property from fire. The role of the fire service continues to expand at the same time public resources diminish. It would be easy to make the false assumption that fire prevention regulation and inspection are aimed at the protection of property and are somehow less worthy than direct efforts aimed at life safety. If the loss of nine firefighters in the 2007 Charleston Sofa Super Store fire does not convince you that fire inspection and code enforcement, prefire planning, and firefighter safety are not necessary and complementary elements of firefighter safety, consider the economic impact of fire on the tax base. The loss of jobs, homes, careers, and the opportunity for a good education for our children are tied to the economic vitality of the community. Your first job as an officer is the protection of the public you serve and the safety and welfare of the firefighters that you lead.

KEY TERMS

acceptable risk the level of hazard potential that the public is willing to live with, that constantly changes based on public perception of the news and current events.

adoption to formally accept through legislative action and put into effect.

adoption by reference a method of code adoption in which the specific edition of a model code is referred to within the adopting ordinance.

adoption by transcription a method of code adoption in which the entire text of the code is published within the adopting ordinance.

code a body of law or rules which is systematically arranged. *When and where to do, or not to do something.*

inspection model a method of setting inspection priorities based on a jurisdictions actual fire experience or identified hazards.

inspection warrant an administrative warrant issued by a judicial officer to search or inspect a property, or perform a noncriminal investigation.

mini-maxi code a code developed and adopted at the state level for either mandatory or optional enforcement by local governments; that cannot be amended by the local governments.

model code technical rules developed and maintained by organizations and made available for governments to adopt and enforce. They are generally developed through a consensus through the use of technical committees.

ordinance a law of an authorized subdivision of a state, such as a county, city or town.

permit model a method of setting inspection priorities by inspecting those occupancies and processes that require fire code permits.

right-of-entry clause a code that permits the fire official to enter and inspect structures on a routine basis, exempt one and two-family dwellings and residential portions of multifamily structures.

standard a rule for measuring or a model to be followed. *How to do something, what materials to use.*

statute a law enacted by a state or the federal legislature.

warrant a legal writ issued by a judicial officer commanding an officer to arrest a person, seize property, or search a premises.

REVIEW QUESTIONS

1. List the three elements that determine use and occupancy classifications.
2. What is the difference between a code and a standard?
3. What is the key difference between Group B and Group M occupancies?
4. The NFPA 550, *Fire Safety Concepts Tree* identifies what two broad strategies for fire safety?
5. Which method of determining inspection priorities is the easiest to defend?
6. What is the principal hazard associated with Group A buildings?
7. Explain the term "acceptable risk."
8. Describe how a high-profile event can affect a community's tolerance for a particular hazard.
9. When is a warrant NOT necessary to conduct an inspection?
10. Define the terms *ordinance* and *statute*.

DISCUSSION QUESTIONS

1. Which model fire code is adopted within your jurisdiction, and is it adopted at the state or local level?

2. What percentage of buildings are inspected each year by the fire department in your jurisdiction? What steps would you take in order to inspect them all?

3. How are fire inspection priorities determined within your jurisdiction? Can you justify them?

4. What single building, facility, or complex would have the greatest impact on your local economy if it was destroyed by fire?

5. What single building, facility, or complex poses the greatest potential for loss of life in the event of a major fire?

6. What building code is in effect in your jurisdiction? Is it adopted on a local or state level?

7. What is the most hazardous occupancy classification in your jurisdiction?

8. Why can illegal changes in use create potentially hazardous conditions?

9. Why does the United States have such a high incidence of fire?

10. What are the potential public relations benefits of a company inspections program?

ENDNOTES

1. *Fire Loss in the U.S. 2007*, Boston, MA: National Fire Protection Association.

2. *Fire in the United States*, 14th ed. Emmitsburg, MD: US Fire Administration, National Fire Data Center, 2007, page 3.

3. Ibid, page 5.

4. *America Burning, the Report of the National Commission on Fire Prevention and Control.* Washington: National Commission on Fire Prevention and Control, 1973, page XI.

5. Ibid.

6. *America Burning, the Report of the National Commission on Fire Prevention and Control*, page 167.

7. *Can Your Ashes*, Fairfax County Fire and Rescue Department Press Release, January 2003.

8. Chris Brennan, "Firefighters Sue City to Stop Budget Cuts," *Philadelphia Daily News,* December 16, 2008.

9. Albert Lea Tribune, "City Sewer Rates at the Center of Company's Expansion," September 20, 2008.

10. John R. Hall, et al., *Fire Code Inspections and Fire Prevention: What Methods Lead to Success?* Boston, MA: National Fire Protection Association, 1978, page viii

11. Harry Chase Brearley, *Fifty Years of Civilizing Force.* New York, NY: Frederick A. Stokes, 1916, page 42.

12. Ibid, page 22.

13. National Board of Fire Underwriters, *Pioneers of Progress.* New York, NY: National Board of Fire Underwriters, 1941, page 125.

14. Tom Klem, "Investigation Report: Fifty-six Die in English Stadium Fire," *Fire Journal,* May 1986, page 126.

15. National Fire Protection Association, *NFPA 100 Years, a Fire Protection Overview.* Quincy, MA: National Fire Protection Association, 1996, page 9.

16. *2000 International Fire Code Commentary.* Falls Church, VA: International Code Council, 2002, page 577.

17. *International Building Code Commentary: Volume 1, 2000.* Falls Church, VA: International Code Council, 2001, page 4–63.

18. *Jabobellis v. Ohio,* 378 U.S. 184, 197, 1964.

19. Special Interest Bulletin No. 5, *The Value and Purpose of Fire Department Inspections.* New York: American Insurance Association, 1975, page 1.

20. NFPA 550 *Fire Safety Concepts Tree.* Quincy, MA: National Fire Protection Association, 1986, page 14.

21. Francis Brannigan, *Building Construction for the Fire Service,* 1st ed. Boston: National Fire Protection Association, 1971, page 1.

ADDITIONAL RESOURCES

Ahrens, Marty, et al. "Use of Fire Incident Data and Statistics." In *Fire Protection Handbook*. 19th ed. Quincy, MA: National Fire Protection Association, 2003.

America Burning, The Report of the National Commission on Fire Prevention and Control. Washington, DC: U.S. Government Printing Office, 1973.

America Burning Recommissioned. Emmitsburg, MD: U.S. Fire Administration, 2000.

America Burning Revisited. Washington, DC: U.S. Government Printing Office, 1990.

Bryan, John L. "Human Behavior in Fire." In *Fire Protection Handbook*. 19th ed. Quincy, MA: National Fire Protection Association, 2003.

Crawford, Jim. "Code Enforcement Never Goes According to Plan." *FireRescue*, April 2008.

Farr, Ronald R. and Steven F. Sawyer. "Fire Prevention and Code Enforcement." In *Fire Protection Handbook*. 19th ed. Quincy, MA: National Fire Protection Association, 2003.

Fire in the United States. 14th ed., is published regularly by the USFA at Emmitsburg, MD. It is available at no cost by contacting them at usfa.dhs.gov.

Hall, John R. "Using Data for Public Education Decision Making." In *Fire Protection Handbook*. 19th ed. Quincy, MA: National Fire Protection Association, 2003.

Hall, John R. and Arthur E. Cote. "An Overview of the Fire Problem and Fire Protection." In *Fire Protection Handbook*. 19th ed. Quincy, MA: National Fire Protection Association, 2003.

Kirtley, Edward. "Fire and Life Safety Education." In *Fire Protection Handbook*. 19th ed. Quincy, MA: National Fire Protection Association, 2003.

Robertson, James C. *Introduction to Fire Prevention*. 5th ed. Upper Saddle River, NJ: Prentice Hall, 2000.

Schaenman, Phillip. "U.K. Prevention Works." *Fire Chief* April 2008.

Regular reports regarding annual fire loss, loss fires, and firefighter injuries and fatalities appear in the *NFPA Journal*, the membership publication of the National Fire Protection Association.

Publications available from the U.S. Fire Administration: *Children and Fire in the United States, Fire Risks for the Older Adult, Heating Fires in Residential Properties, Protecting Your Family from Fire* (FA 130), *Socioeconomic Factors and the Incidence of Fire* (FA 170), and *The Rural Fire Problem in the United States* (FA 180).

These and many other reports are listed on the U.S. Fire Administration's Web site: www.usfa.gov or www.FEMA.gov. All can be ordered free of charge, and many can be downloaded.

Most of the reports from NFPA are available on the Web site: http:www.nfpa.org.

10

The Company Officer's Role in Understanding Building Construction and Fire Behavior

Some years back as a young firefighter, I had two close calls in a 2-week period. Through these near-death experiences, I learned a valuable lesson that I carried with me as I progressed to a company officer position.

On the occasion of the first close call, I was detailed to a busy truck company. It was a Sunday evening in January, and it had been a slow shift with only a few minor fires. As the structure fire alarm came in, the shift was relaxing and watching the Super Bowl. Our laid-back evening was about to change dramatically.

On arrival, we found a bowstring truss government building that measured approximately 65 feet by 125 feet. In the locked and darkened building, we could see heavy smoke conditions and the faint glow of fire in the rear. My partner and I used a K-12 saw, halligan bar, and a sledge to force a heavy steel door located in the rear of the structure. Then, with the use of 12-foot pike poles, we started bumping down the ceiling tiles, exposing a heavy volume of fire. An engine company led their 2½-inch line in the rear door and started hitting the fire.

A few minutes into our attack, one of the company officers started yelling for us to back out. Reluctantly, we backed out of the building and helped set up defensive operations. I was sure we were pulled out too early. I felt we were winning. Within a few minutes, I was proved wrong when, with a thunderous crash, the entire rear of the structure collapsed. The area where we had been working was now a debris pile with intense heat and flames rising some 60 feet into the air. Our lives were saved by a company officer who was savvy about building construction and the effects of fire on building components. His knowledge of this key fire service principle protected the safety of the firefighters under his command.

Just over a week later, I was in an attic apartment of a 2½-story balloon frame building with my rescue squad and an engine company. We came up a center stairway and were fighting a large fire in the rear of the structure. I crawled toward the front to search the bedroom and to vent the front window. Thick, dark smoke with intense heat banked down to the floor. The conditions were rapidly changing; superheated smoke now had me down into a belly crawl. Suddenly, the entire ceiling lit up in a flashover. I was almost melting. I quickly belly crawled in the direction of the screams of firefighters as they dove down the stairs. By the time I dove down the stairs behind the others, some of my gear was burning. A firefighter hit me with a fog line as I rolled to the bottom of the staircase. Once again, I was lucky and escaped with only partial-thickness burns, and again I learned a valuable lesson about the power of knowledge of the enemy when fighting fire. The conditions of the room I entered were ripe for flashover. Recognizing these conditions and taking alternate actions could have saved me painful burns.

Neither of these close calls had to happen. They could have been prevented if I had a better understanding of building construction and fire behavior and I had applied them to the situation. As a company officer, I strove to learn and apply these critical bodies of knowledge in everything I did, for the safety of my crew.

Today, firefighting has gotten even more dangerous than ever before. I believe this to be true for two reasons: Fires burn hotter due to the increase in synthetic fuels, and buildings are weaker due to lightweight construction. This is a dangerous combination. We cannot use 20-year-old knowledge and tactics in these changing times. Doing so could cost lives. Become a student of building construction and fire behavior. I sure did. Always remember to share your experience and knowledge with your brother and sister firefighters, as it is everyone's responsibility to make firefighting safer.

—*Steve Chikerotis, Battalion Chief/Assistant Director of Training,*
Chicago Fire Department, Chicago, Illinois

LEARNING OBJECTIVES

After completing this chapter the reader should be able to:

10-1   List the five general classes of building construction and alternative construction methods.

10-2   Describe the five general classes of building construction and alternative construction methods.

10-3   Describe fire dynamics within the five types of building construction.

10-4   Describe the impact of performance-based codes on building construction types and alternative construction methods.

10-5   Describe the effects of alternative construction methods on fire dynamics.

10-6   Describe the meaning of fuel load as it applies to the potential size and intensity of a fire.

10-7   Describe the effects of fuel load on the tactics and strategies employed during structural fire suppression.

10-8   Identify fire detection and suppression systems in the pre-incident plan.

10-9   Describe the function of the different types of fire protection systems currently in use.

10-10   Describe the importance of hazardous materials management plans and hazardous materials inventory statements to the inspections process.

10-11   Describe the NFPA 704 Hazard Identification System.

10-12   Describe the types of fire detection systems currently in use.

10-13   Describe the effects of fire protection systems on the tactics and strategies employed during structural fire suppression.

10-14   List the appropriate standards for the maintenance and inspection of sprinklers, standpipes, wet and dry chemical extinguishing systems, and fire alarm systems.

10-15   Explain the function of each fire detection system currently in use.

10-16   Identify the various types of alarm systems that are currently found in use.

10-17   Explain the operation of each alarm system that is currently found in use.

10-18   Describe NFPA 550, The Fire Safety Concepts Tree and its uses.

10-19   Explain each of the risk factors and community consequences related to structure fires within a response area.

10-20   Describe the effects of fire growth on the tactics and strategies employed during structural fire suppression.

10-21   Describe the effects of building construction on the tactics and strategies employed during structural fire suppression.

10-22   Describe the effects of water supply on the tactics and strategies employed during structural fire suppression.

* The FO I and II levels, as defined by the NFPA 1021, the *Standard for Fire Officer Professional Qualifications*, are identified in different colors: FO I = black, FO II = red.

INTRODUCTION

Many experienced firefighters have successfully moved from firefighter to fire officer with little or no formal training. To their credit, they were quite successful. But just think how much better they would have been if they had been trained and given a chance to prepare for that assignment.

This situation is especially true for new officers in operational situations. A new officer, riding on an engine or truck company arriving at the scene of a working fire or other emergency, may be overwhelmed. Some of this is due to a lack of experience, but some may be due to a lack of training.

At the scene of emergencies, the first-arriving officers are usually concerned with limited resources, time and many unknown factors. Working in these conditions is difficult and dangerous; leading others under such conditions can be extremely challenging. In such adverse conditions, you, as the company officer, are expected to be calm and decisive, to issue clear orders, and to keep track of all the activities. You are responsible for the safety of others. You are responsible for at least some aspect of the management of the incident and, in many situations, may be the incident commander.

SAFETY

Working in emergency conditions is difficult and dangerous; leading others under such conditions can be extremely challenging.

As you move to the officer's position in the organization, the roles change quite a bit. As an officer, you are now the manager and leader of the company. We have talked about management and leadership in previous chapters. All of that information applies in your role as a leader during emergency activities as well.

In the next four chapters, we discuss your role as the company officer in identifying and mitigating problems at the scene of emergencies. This chapter focuses on information needed for fighting fires in structures; we saw in Chapter 9 that structural fires are frequent and the loss is often high. This chapter deals with building construction and fire behavior. We will get to tactics in Chapter 13.

TODAY'S FIREFIGHTING REQUIRES EXPERIENCE, KNOWLEDGE, AND SKILL

Modern firefighting requires more than just experience; it requires knowledge and skill. With the gradual decline in the number of serious fires encountered, good officers will learn from every opportunity, including actual working incidents, participating in training and drills, attending seminars, and reading about the subject as much as possible.

During the 1870s, the chief of the London Fire Brigade, Massey Shaw, visited the United States. Chief Shaw later wrote the following:

When I was last in America, it struck me very forcibly that although most of the chiefs were intelligent and zealous in their work, not one that I met even made a pretense to the kind of professional knowledge that I consider so essential. Indeed one went so far as to say that the only way to learn the business of a fireman was to go to fires. A statement about as monstrous and as contrary to reason as if he had said that the only way to become a surgeon would be to commence cutting off limbs without any knowledge of the human body or of the implements required.

There is no such short cut to proficiency in any profession and the day will come when your fellow countrymen will be obligated to open their eyes to the fact, that, as a man learns the business of a fireman only by attending fires he must of necessity learn it badly. Even that which he does pick up and may seem to know, he will know imperfectly and be incapable of imparting to others.

I consider the business of a fireman a regular profession requiring previous study and training as other professions do. I am convinced that where training and study are omitted and men are pitchforked into the practical work without preparation, the fire department will never be capable of dealing satisfactorily with great emergencies.[1]

Chief Shaw's comments may seem disrespectful to some, but in the time frame in which they were offered, they may have been amazingly perceptive. Before you take sides here, we want to point out that the fire service has made a lot of progress in the development of the very professional knowledge that Chief Shaw found deficient.

Even the Best Will Be Challenged

Regardless of one's experience, training, and education, even the best officers will be challenged. Emanual Fried, in his classic text *Fireground Tactics*, offered the following advice:

Command on the fireground is a demanding task for the officer. To improve the ability to handle the situations we have to concede certain basic faults that we find in ourselves.

Admit these possibilities:

- You are going to get excited.
- You are going to yell.
- You are going to make mistakes.
- You are going to lose buildings.[2]

NOTE

Many factors contribute to the company being ready to deal with emergency situations, including your knowledge of the job.

In spite of these challenges, you and your fire department are in the business of saving lives and property during fires and other emergencies. You respond to a variety of events, using your skills and abilities to contribute something toward the successful outcome. Fortunately, most events do have successful outcomes. This does not happen by chance: Many factors contribute to the company being ready to deal with emergency situations, including your knowledge of the job.

Fighting Fires Is Still Our Business

Most textbooks and training programs still reflect the traditional nature of the fire service: They focus on fire-related emergencies. In reality, fire departments provide a host of other services to their community, including the delivery of emergency medical care, the mitigation of incidents involving hazardous materials, and various rescue capabilities usually tailored to the local community's needs (see **Figure 10-1**). As a result, we are seeing a decrease in the percentage of fire-related emergencies as a part of the number of total calls received by the fire department.

There is another factor that affects the department's firefighting activities. Where good fire prevention activities are in place, there are fewer fires. And where automatic fire detection and suppression equipment are present, timely fire notification and control lead to confined and smaller fires. This is good news for the fire service and for the community. However, it does present the fire service with a challenge: The basic fire department component, the company, is expected to be proficient in doing a greater number of things, while having fewer opportunities to practice the basic skills needed for its original task, effective fire suppression.

FIGURE 10-1   Firefighters are called upon to deal with many emergencies.

OFFICER ADVICE

Firefighters and fire officers should take advantage of every opportunity to plan and train for the various situations they may face.

All of this suggests that there is much to be done to prepare the company for the challenges it faces. Firefighters and fire officers should take advantage of every opportunity to plan and train for the various situations they may face. They must be familiar with the community they serve and the activities that regularly take place in that community. They must be familiar with the special hazards that exist, hazards that pose threats to themselves and the citizens they are obligated to help during an emergency.

UNDERSTANDING THE COMMUNITY'S NEEDS

Fire departments should be assessing their communities' risk factors. This is the first step in identifying the strengths and weaknesses of the department

in dealing with emergencies. The company has a role in this process, as well, and should have a significant interest in the outcome. Regardless of whether the community has one fire company or a hundred, some part of the community should be a particular fire station's primary concern. Members of the company responsible for that area should be interested in the risk factors in that area. They should look at specific situations where they may be called upon to save lives and protect property and should also be thinking about the overall **community consequences**.

We will review two items that were probably covered in your basic training—building construction and fire behavior. This information is not offered as an insult to your knowledge; however, it is possible that you may have forgotten some of this important material. It is also possible that some new information has been added, information that will make your job a little easier and a whole lot safer.

RISK FACTORS AND COMMUNITY CONSEQUENCES

Life risk factors are affected by the number of people at risk, their danger, and their ability to provide for their own safety.

Property risk factors are affected by the characteristics of building construction, exposures, occupancy, and available resources.

Community consequences are related to the impact of the event upon the community, both during and after the event is concluded. Some of these consequences are immediate; some last for generations.

For company officers, a general knowledge of your area of responsibilities is essential for effective operations. This knowledge includes the following:

- *Physical factors:* geographical size, population, valuation, response time, and topography of the community
- *Access factors:* access and barriers to all areas
- *Occupancy factors:* the nature of the businesses that occupy the buildings
- *Structural factors:* age, type, and density of structures
- *Resource factors:* fire department resources, and for firefighting purpose, water supply capabilities
- *Survival factors:* stairwells and other penetrations to allow for rescue, fire spread, and potential falling hazards for firefighters

BUILDING CONSTRUCTION

During the course of your fire service career, virtually every commercial building and all but a few residential buildings you encounter will have been constructed or renovated under a construction code.

NOTE

There is an axiom in the construction code business: Which of the five types of construction is preferred by design professionals for their projects? *The cheapest.*

There is a good reason. The design professional's client has a budget. Every construction project has a budget. Most buildings are built using the minimum type of construction that is permitted by the code.

The primary purpose of a building code is to prescribe standards that will keep buildings from falling down. Besides gravity, there are many forces that act against buildings. Snow loads, wind loads, and potential earthquake loads are all provided for the building codes. But the force that is addressed by the largest number of provisions is fire. The goals of the code are to:

- Maintain the integrity of the structure and prevent collapse
- Limit fire size by preventing fire spread
- Provide adequate time for occupants to escape or shelter in place

The goals can be met in a variety of ways, using fire resistance–rated construction, fire extinguishing systems, and noncombustible building materials and providing adequate fire separation distance. Two important terms to understanding construction types are **noncombustible** and **fire resistance.** In broad terms, a noncombutible building material will not burn, and fire resistance is the resistance of a building to collapse or to total involvement in fire. The late Frank Brannigan described it as *fire endurance.* It is measured by the length of time typical structural members and assemblies resist specified temperatures under test conditions.

How Construction Type Affects You

In his first edition of *Building Construction for the Fire Service,* the late Francis Brannigan noted: "In general it can be said the building makes the problem."[3] For the most part, fires that occur in buildings pose the greatest risk to the occupants and firefighting forces, and pose the greatest challenge. Brannigan made his point by describing a fire involving 50 upholstered chairs stacked in a barren field as one-unit response

using less than a tank of water. The same fire in a Hilton Hotel killed two people and injured 36 others. Fire intensity and behavior will, for the most part, be affected by the types of building materials used, the construction methods, and the build-in fire protection features installed during construction.

TYPES OF CONSTRUCTION

There are three key points to remember when dealing with building construction types:

1. All construction is either combustible (it will burn) or noncombustible (it won't).
2. The term **protected**, when applied to construction materials, means protected from the effects of fire by encasement. Concrete, gypsum and spray on coatings are all used to "protect" construction elements. Protected construction does not mean the building has a sprinkler system.
3. The code specifies the minimum requirements, but permits the use of materials that exceed those requirements. Making assumptions based on your view from the street is risky and potentially hazardous for you and your crew.

Some buildings with combustible components are actually more fire resistant than some noncombustible buildings. That is where protection of the structural elements comes into play. A lightweight steel building may be noncombustible but will fail quickly when exposed to a fire involving combustible contents. Conversely, buildings with masonry exterior walls and heavy timber joists have combustible components that hold up quite well compared to lightweight steel.

Determining Construction Type

A design professional lays out the building in a set of plans. The occupancy classification(s), building area, number of stories, location on the lot, distance from the lot lines, fire department access to the exterior of the structure, and whether the building will be equipped with sprinklers all have to be established before the minimum construction type can be determined. Occupancy classifications that are more hazardous, based on the vulnerability of the occupants (hospitals and nursing homes) or the relative hazards associated with the function of the building (repair garage versus an office building) or the materials handle or stored (manufacturing or processing hazardous materials), have more stringent construction requirements.

FIVE CONSTRUCTION TYPES

There are five basic construction types recognized by the *International Building Code* (IBC). The IBC's predecessor regional building codes used slightly different terms for the same basic types. Every jurisdiction has a large number of buildings constructed under one or more of the regional codes and fewer, relatively new buildings designed and constructed under the IBC. All of the codes use a combination letter/number system to identify the different types of construction. For simplicity, we will also include the common name associated with each type.

Type I: Fire Resistive

In Type I construction, the structural elements are noncombustible and are protected to the highest level. Type I is divided into two subtypes. The difference between them is the level of protection for the structural elements (expressed in hours). Only noncombustible materials are permitted for the structural elements. Structural steel is encased in concrete, gypsum, or cementaceous coating (see **Figure 10-2**). Fire-resistive construction holds up well under fire conditions, provided the structural protection remains intact. High-rise buildings with steel structural elements encased in concrete are examples of Type I buildings.

While the possibility of collapse during normal fire conditions is remote, fire spread from floor to floor via windows, or through vertical openings that have not been properly firestopped in accordance with the code, is possible. During the First Interstate Bank fire in Los Angeles (1988), the fire extended upward from the 12th to 16th floors. Flames were estimated to have lapped 30 feet up the face of the building. The curtain

walls, including windows, spandrel panels, and mullions, were almost completely destroyed by the fire. Minor fire extension also occurred via poke-through penetrations for electric communications cables, HVAC shafts, and heat conduction through the floor slabs. A minor fire occurred in a storeroom on the 27th floor, ignited by fire products escaping from an HVAC shaft that originated on the 12th floor.[4]

Type II: Noncombustible

In Type II construction, the structural elements are also noncombustible. The difference between Type I Fire Resistive and Type II Noncombustible is in the level of protection of the structural elements. Type I is protected to the highest level. Type II is subdivided into subtypes; one is protected less than Type I and the other has no protection. Type II buildings are noncombustible but are afforded limited or no fire resistance for the structural elements.

> **FIREGROUND FACT**
>
> The typical strip shopping center or big box store with masonry block walls, steel bar joists, unprotected steel columns, and a steel roof deck is a Type IIB (unprotected) building (see **Figure 10-3**). If there is a sprinkler system failure, the roof structure can fail quickly and collapse.

The Sofa Super Store in Charleston, SC, was originally constructed in the 1960s under the *Standard Building Code*'s equivalent of IBC Type IIB construction. Automatic sprinklers were not required by the code at that time. The roof structure failed and collapsed 43 minutes from the time the first hand line was deployed on the fire that started on the exterior loading dock.[5] Steel is a superior building material, with good strength in both tension and compression. However, steel loses almost half of its strength at 900 degrees. In issuing *Preventing Injuries and*

FIGURE 10-2 Type I construction features noncombustible structural elements that are protected from the effects of fire through encasement with concrete, gypsum, or spray appled coatings.

FIGURE 10-3 Type II construction features noncombustible structural elements that may be protected or unprotected depending on the subtype.

Deaths of Fire Fighters due to Truss System Failures in 2005, NIOSH noted:

> Steel has a high thermal conductivity, which means it can transfer heat away from a localized source and act as a heat sink. If an intense fire is evenly distributed along the steel member, the critical temperature may be reached very quickly. Steel also has a high coefficient of expansion that results in the expansion of steel members as they are heated. A 50-foot-long steel beam heated uniformly over its length from 72° to 972° F will expand in length by 3.9 inches. The same beam uniformly heated to 800° F would expand by 3.2 inches; if heated to 1,200° F, the beam would expand by 4.9 inches.[6]

Type III: Ordinary

In Type III construction, the exterior of the building is noncombustible, typically masonry, and may be rated depending on the horizontal distance to exposures. The interior structural elements may be of any approved material, combustible or noncombustible. The building code reference to *any approved material* refers to the referenced standards for building materials such as lumber, steel, concrete, glazing, and plastic. There is a standard of quality for every building material. For example, only graded lumber may be used.

Type III construction is divided into two subtypes: protected and unprotected. The brick, wood-joisted buildings that line our city streets are of Type III construction, or "ordinary construction." The term "ordinary" dates back to the time when most city and town building codes established fire districts where combustible building exteriors were prohibited to reduce conflagration potential (see **Figure 10-4**). It was the least expensive type permitted. Ordinary construction is not very ordinary anymore—it has become too expensive! Buildings with a masonry veneer over combustible framing are not Type III. Most Type III buildings are older and many have been renovated or remodeled. Combustible voids that may allow fire to travel undetected present challenges to the fireground commander.

Type IV: Heavy Timber

Type IV construction dates back to the New England cotton and woolen mills of the mid-1800s. Mill owners banded together and formed mutual fire insurance companies, but only insured the best risks. Type IV Heavy Timber construction is also called *Mill* construction. The exterior walls are noncombustible (masonry), and the interior structural elements are unprotected wood of large cross-sectional dimensions (see **Figure 10-5**). Columns must be at least 8 inches if supporting a floor load, and joists and beams must be a minimum of 6 inches in width and 10 inches in

FIGURE 10-4 Type II construction is referred to as "ordinary" construction because it was the predominate type in American cities and towns through the 1960's.

FIGURE 10-5 Type IV Heavy Timber or Mill construction features large dimension lumber structural members, noncombustible exterior walls and has no void spaces where fire can travel undetected. Its origin dates back to the New England woolen and cotton mills of the mid-1800s.

depth. Type IV is not subdivided. The inherent fire-resistive nature of large-diameter wood members is taken into account. A key strength is that concealed spaces are not permitted within the structure.

Type V: Frame

In Type V construction, the entire structure may be constructed of wood or any other approved material (see **Figure 10-6**). Type V was called "wood frame" in the past, because wood was usually the least expensive material for the job. The high cost of lumber has led to an increase in lightweight combustible composites, and to the use of lightweight steel for studs. How well wood structural members will resist the effects of fire is directly related to their mass. The 6 × 10-inch floor joists in a Type IV building can be expected to support the floor for some time when exposed to fire. Lightweight trusses (see **Figure 10-7**) may fail in minutes. Brick veneer is often added to frame buildings for decorative purposes (usually only to the side facing the street), but the building is still Type V. Type V construction is subdivided into protected and unprotected types, again depending on the protection provided for the structural elements.

PERFORMANCE CODES

Fire walls, fire barriers, and fire partitions are all are designed to resist the spread of fire, but are they are not the same thing. Fire walls are structurally independent and separate buildings. Fire barriers and fire partitions are constructed within a building and separate fire areas, or block the passage of heat and smoke from corridors or other areas.

Modern building codes are for the most part **performance codes.** Rather than specify the exact construction of a component such as a fire wall, the codes identify a performance features. Where a **specification code** might require all fire walls to be of a certain thickness of masonry block, a performance merely

FIGURE 10-6   In Type V construction, any approved material is permitted. A combination of wood, lightweight steel, and combustible composites are common.

FIGURE 10-7   Lightweight trusses have replaced dimension lumber in Type V construction.

code identifies the performance required, such as resisting the passage of heat and some for a given time. The design professional is permitted to use any **listed** assembly that provides that fire-resistance rating.

Fire-resistive assemblies required by the code have performance requirements prescribed. The building designer species the assembly (by design number) to be used to attain the required rating, or uses one of the approved equivalent fire-resistance methods from the code. It's important to remember that rated assemblies such as floor/ceiling assemblies and fire doors are tested and listed together. If any portion of the assembly is left out or altered, the assembly is no longer the listed item, and more importantly, the assembly may not function as designed. Do not confuse performance codes with performance-based design.

Performance-Based Design

Performance-based design is an alternative method for satisfying the fire protection and life safety intent of construction codes. The ICC Performance Code for buildings and facilities defines performance-based design as:

> An engineering approach to design elements of a building based on agreed upon performance goals and objectives, engineering analysis, and quantitative assessment of alternatives against the design goals and objectives using accepted engineering tools, methodologies, and performance criteria.[7]

The IPC requires the owner to retain the services of qualified design professionals and qualified peer reviewers to assess the design. The process involves the use of scientific modeling tools and methods for determining fire growth, exit times, system activation times, smoke movement, and other variables. Extensive computer modeling is involved. The process is very expensive and labor intensive. In all but the largest, tallest, or most unusual buildings, designing to the prescriptive provisions of the IBC is the most cost-effective approach.

In March 2007, the International Forum of Fire Research Directors issued a position paper on performance-based design for fire code applications. The title is slightly confusing. The paper addresses performance-based design for fire safety aspects of building codes. The committee noted the limited potential use for performance based design:

> For the majority of traditional buildings with low hazard occupancies, modern prescriptive building, and fire codes, when enforced, achieve this objective. Nontraditional buildings include many of society's largest and iconic structures, such as opera houses, museums, sports stadiums, transportation centers, super-high-rise structures, and some government buildings. Prescriptive codes cannot anticipate all of the requirements that these nontraditional structures impose; prescriptive codes do not adapt rapidly to changing materials and methods of construction, nor to radical architectural designs; and prescriptive codes based upon historical loss experiences are not designed to deal with very low probability, very high impact events or other threats such as from terrorism.[8]

FUEL LOAD

Fuel load or **fire load** is the total amount of combustible material within a fire area, expressed in terms of pounds per square foot. The fire load concept dates back to early tests of columns, wall and other building components by the Bureau of Standards (now NIST). A correlation between fuel load and fire duration (termed fire severity) was established and became the basis for fire protection ratings that are still in the codes. The tests were based on common combustibles (wood pallets) as very few synthetic materials existed. A more accurate way to determine fire severity is based on the heat release rate (HRR) of the materials. Dr. Richard Bukowski of NIST puts it into plain-speak this way:

> Most plastics have twice the heat of combustion and burn with a higher radiative fraction that accelerates fire development. The development of heat release rate measurements provided the ability to measure energy as an extensible property that explained what was happening. The shortcomings of fuel load became apparent; that the rate at which energy is released can vary greatly for the same fuel load depending on the fuel characteristics (a solid block of wood will burn very differently from a wood crib or a pile of sawdust, each of which may represent the same fuel load).[9]

Cook County Administration Building Fire

In 2003, a fire that started in a storage room of the Cook County, Illinois, administration building resulted in six fatalities and several injuries. The 37-story Type I (fire-resistive) building was constructed under a 1960 edition model code and was not protected with automatic sprinklers (see **Figure 10-8**). NIST scientists were requested to model the fire as part of the overall investigation of the incident.

NIST engineers surveyed the damage (see **Figure 10-9**) and then took undamaged representative samples of office furniture and contents to their lab for testing. Both full scale (see **Figure 10-10**) and cone calorimeter tests of 0.1 meter square (0.33 ft.) samples were tested (see **Table 10-1**).

The difference in the heat released by the paper/cardboard stack and the trash can and letter trays. The test results underscore Dr. Bukowski's comments on the value of heat release rate.

Fuel Load and Fire Behavior

As Dr. Bukowski noted, fuel load is the total amount of available fuel within a fire area, but heat release rate is a better indicator of fire severity. Both factors are critical in developing and implementing a strategy to provide for the safety of the occupants, and to confine and extinguish the fire. A room and contents fire involving common combustible materials in a noncombustible one-story building is entirely different from the exact same fire on the first floor of a multistory frame building. Both started with the same materials, but in the case of the combustible frame building, the building itself is part of the fuel load. The size, shape, form, and arrangement of the combustible structural members will affect the heat release rate and the rapidity at which a fire grows and spreads.

FIRE PROTECTION SYSTEMS

Fire protection systems is a broad term that includes built-in extinguishing systems such as automatic sprinklers, clean agent extinguishing systems, fire detection and alarm devices, smoke control and removal equipment, fixed extinguishing systems for cooking operations and industrial equipment, and even portable devices such as fire extinguishers. Each is designed for a specific hazard and is installed in accordance with a standard to address that hazard.

NOTE

If the hazard changes, the fire protection system must be reevaluated to determine its potential effectiveness given the new condition.

The presence or absence of fire alarm, fire extinguishing, and other fire protection systems will have a direct impact on the strategy and tactics of the incident commander, as will their effectiveness and operational readiness. The lack of a working fire alarm system moves notification of the occupants to the top of the

FIGURE 10-8 Exterior of the fire-resistive Cook County, Illinois Administration Building two days after the fatal fire. (*Source: National Institute of Standards and Technology*)

FIGURE 10-9 View of the 12th floor of the Cook County Administration Building. Six fatalities were discovered in the locked stairwell after the fire was extinguished. (*Source: National Institute of Standards and Technology*)

incident commander's priorities. Lack of a working sprinkler system places the responsibility of confining the fire squarely on the shoulders of the incident commander. In both cases, operational fire protection systems would have shifted the fire department's role to that ensuring that all the occupants were able to self-evacuate or shelter in place, and mopping up the fire. By adequately pre-planning these buildings, fire officers will have a superior advantage to knowing what type and how to access the protection system.

FIRE EXTINGUISHING SYSTEMS

Standards referenced by the model building and fire code reference standards for the installation, testing and maintenance and testing of fire extinguishing systems (see **Table 10-2**). Additional sprinkler requirements are included in NFPA standards that pertain to specific processes and occupancies. NFPA 30, *Flammable and Combustible Liquids Code* and

FIGURE 10-10 Representative samples of furniture and office contents were tested at NIST's Building and Fire Research Laboratory. View of the test enclosure, approximately 300 seconds after ignition. (*Source: National Institute of Standards and Technology*)

TABLE 10-1   Cone Calorimeter Test-Average Peak Heat Release at 35 and 70 kw/m²

Exposure Heat Flux	35 kW/m²	70 kW/m²
Item	Avg. Peak HRR (kW/m²)	Avg. Peak HRR (kW/m²)
Carpeting	260	380
Ceiling Tile (face-down)	10	40
Ceiling Tile (face-up)	70	90
Computer Monitor Case	410	490
Letter Tray	1020	1170
Office Chair	210	350
Paper (stacked flat)	260	340
Paper (stacked on edge)	210	250
Paper (stacked flat – covered with cardboard)	320	460
Plastic Wastebasket	1560	2970
Vinyl Blinds	110	160
Wall Covering	340	460
Workstation—Side Panel	140	230
Workstation—Work Surface	340	590

Source: National Institute of Standards and Technology

NFPA 30B, *Code for the Manufacture and Storage of Aerosol Products* are examples. Each standard provides a significant and different level of protection to address a specific hazard or hazards.

The best designed and installed system will not operate if the valves are closed. Fire officers should make it a habit to observe exterior supply valves and investigate when they appear to be closed (see **Figure 10-11**).

NFPA 13 Sprinkler Systems

NFPA 13 systems are by far the most commonly found installed in buildings in the United States.

TABLE 10-2	Sprinkler Standards
NFPA 13	Installation of Sprinkler Systems
NFPA 13D	Installation of Sprinkler Systems in One- and Two-Family Dwellings and Manufactured Homes
NFPA 13R	Installation of Sprinkler Systems in Residential Occupancies Up to and Including Four Stories in Height
NFPA 25	Inspection, Testing and Maintenance of Water-Based Fire Protection Systems

FIGURE 10-11 Closed valves are a major cause of sprinkler system failures. Fire Officers should make it a habit to observe the condition of exterior sprinkler valves.

NOTE

The notion that all sprinklered buildings are equally protected, regardless of the contents or system design is a dangerous one, embraced unfortunately, by too many in the fire and building regulation communities. Take for instance a building that contains hazardous materials. Depending on the type and quantity of the material will determine what type of system is to be installed. The problem is that some of the buildings that house hazardous materials have either increased their amounts of the material or changed the type of material without proper notification. This is why it is essential for fire inspectors to review hazardous material management plans (HMMP) and hazardous material inventory statements on a frequent basis.

OFFICER ADVICE

All responders and fire officers should understand the NFPA 704 placard system. In essence the system was designed to show three principal categories of markings: health, flammability, and instability. They are color coded to make them easily identifiable.

■ Blue is the health category
■ Red is the flammability category
■ Yellow represents the instability category

The ratings, however, are issued by "competent and experienced" persons who are trained in the interpretation of the material and the hazards that it poses. The problem with these signs is the same as the sprinkler systems. They must be inspected and updated as needed.

NFPA 13, *Standard for the Installation of Sprinklers*, is the direct descendent of the original sprinkler standard developed by NFPA in 1896.

NOTE

The NFPA 13 system is a property protection system that has had remarkable success at protecting people by quickly suppressing and often extinguishing fires in their incipient stages.

Heat from the fire rises to the ceiling where sprinklers are installed and act as combination heat detection and fire suppression devices. Once the operating elements of the sprinkler head (generally a solder link or frangible bulb) reach their operating temperature, they fail and water is discharged onto the fire through the open orifice.

Standard automatic sprinklers are designed to operate at specific temperatures. Ordinary-temperature rated heads that operate at between 135 and 170 degrees are required to be installed throughout buildings unless the area is subject to high heat conditions. Boiler rooms, skylights, attic spaces, and areas above certain machinery and equipment may all require the use of sprinkler heads that have higher operating temperatures.

The standard attempts to minimize the amount of time that it takes for the sprinkler system to discharge water on a fire by requiring that ordinary-temperature heads be installed. This is important if fire spread is to be checked and early notification accomplished. In the design of sprinkler heads, **thermal lag** becomes the enemy.

A sprinkler head that is rated to operate at 165 degrees does not operate the minute that the surrounding air reaches that temperature. The detection device itself has to become heated to that temperature before it will operate.[10] In the case of a fire involving readily combustible material or flammable liquids, the surrounding temperature will have continued to rise and may be significantly higher by the time the head operates and water is discharged.

Thermal lag is the difference between the operating temperature of a fire detection device such as a sprinkler head and the actual air temperature when the device activates. It is minimized by designing sprinkler heads and other detection devices with less mass and greater surface area. This concept played an important role in the design of sprinklers for residential applications and will be discussed with the other sprinkler standards.

NFPA 13 uses an occupancy type system of hazard classification to determine the operational requirements of sprinkler systems, which is not to be confused with the use group classification from the building codes.

NOTE

Sprinkler head spacing and discharge per square foot known as density are dependent upon occupancy classification. Occupancy classes are based on the amount and combustibility of the contents and processes that occur within the building.

NFPA 13 systems provide excellent property protection by requiring that the entire building, including combustible void spaces such as attics be protected with sprinklers.

Residential Sprinklers

NFPA 13R systems are designed to protect residential occupancies up to four stories in height. The systems enhance life safety by using listed residential sprinkler heads, which are designed to minimize thermal lag and respond quickly in the event of fire.

NOTE

By quickly suppressing the fire, flashover is prevented and occupants are able to escape from the building.

NOTE

NFPA 13R systems do not provide the same level of property protection as afforded by NFPA 13 systems.

Combustible voids, as well as closets not exceeding 24 square feet and bathrooms not exceeding 55 square feet, are not required to be sprinklered. The total building area exempted from sprinkler protection has been estimated as high as 67%.[11] The potential property damage resulting from fires that "burn around, over and under" the sprinkler system is significant. The possibility of a "total loss" in a 13R sprinklered

building is very real. Hotels, motels, boarding houses, and apartment buildings are typically protected with NFPA 13R systems.

NFPA 13D Systems are designed to protect one- and two-family dwellings. Like NFPA 13R systems, they are primarily designed to protect the occupants by preventing flashover from occurring and giving the occupants time to escape. They have similar limitations to NFPA 13R systems, in that bathrooms and closets as well as garages, attached porches, carports, and some foyers are not required to be sprinklered. All sprinkler systems are not equal. They are designed for specific hazards and to perform specific functions.

Types of Sprinkler Systems

There are four basic types of sprinkler systems. What differentiates them is the method in which water is supplied to the system. The four types are wet pipe, dry pipe, preaction, and deluge. Each has specific applications and system specific requirements from the sprinkler standards and in a few cases from the building codes themselves.

Wet Pipe Sprinklers

Wet pipe sprinkler systems are by far the most commonly found systems in most jurisdictions. Water enters the sprinkler system from a dedicated supply or fire line, unless the system is a limited area sprinkler system or a system designed to protect one- and two-family dwellings and manufactured homes. The supply valve is left open and water at street pressure is always on the system. For this reason, they are the quickest at getting water on the fire and are the simplest to maintain.

Wet pipe systems are installed where indoor temperatures can be maintained at 40 degrees. Below that temperature, there is the danger of freezing pipes. If the outside temperature is below freezing and the interior temperature is less than 40 degrees, steel sprinkler piping, which rapidly conducts heat and rapidly loses it, will drop below freezing. The frozen area may be isolated and near an opening or noninsulated portion of the building. It may be a small area, but it could be enough to put the whole system out of service.

Antifreeze systems are sometimes used where freezing is expected, but these systems are usually limited to 40 gallons or less. Each year, antifreeze systems must be drained into containers and the antifreeze solution must be replenished to restore the specific gravity required to prevent freezing. This is costly and labor intensive. Larger systems subject to freezing are normally designed as dry pipe systems.

The wet pipe system's advantage of getting water to the fire area quickly is also its weakness in areas where water damage is of great concern. Mechanical damage from machinery such as forklifts or from vandals can lead to the soaking of valuable equipment, artifacts, merchandise or records. The public's perception that all the sprinkler heads within a system discharge at once has caused undo concern on the part of many.

Dry Pipe Systems

Dry pipe sprinkler systems are installed in warehouses, parking garages, factories, and other buildings where there is a danger of freezing. They may also be installed in unheated attics or portions of buildings where providing heat is impractical. A dry pipe valve is installed at the sprinkler riser and keeps water out of the system piping until a fire activates a sprinkler head or heads. Dry pipe valves are designed so that a moderate amount of air pressure in the system above the valve is capable of holding back a much greater water pressure. The ratio of air pressure to water pressure at which the valve will open or trip is called the *differential*. Dry pipe systems are equipped with two pressure gauges. One gauge measures the incoming water pressure below the dry pipe valve, and the other measures the air pressure on the system above the valve.

Preaction Systems

Preaction sprinkler systems are installed in properties where potential water damage from broken piping or sprinkler heads is of particular concern. Preaction systems are dry systems in which the water supply valve is opened on a signal from detection devices such as heat or smoke detectors. The potential for accidental water damage is minimized since no water is in the system unless a detection device has been activated or the valve is manually opened.

The sprinkler heads in a preaction system are traditional closed heads, which must be fused by heat to open, discharging water only over the fire. Early notification also minimizes water damage associated with suppression of the fire since an alarm is transmitted upon activation of the detection device, which precedes the fusing of sprinkler heads and activation of a sprinkler flow switch.

Deluge Systems

Deluge sprinkler systems are installed in extra hazard occupancies where there is the possibility of a flash fire or fire growth so rapid that the response of a standard sprinkler system is too slow. Facilities with large quantities of flammable liquids or materials that pose a deflagration hazard are protected with these systems. The activation of the system is designed to apply water over the entire area covered by the system, rapidly and simultaneously. Deluge systems are dry systems that use open sprinkler heads and a deluge

valve. The deluge valve operates much like the supply valve in a preaction system. When activated by a fire detection device, the deluge valve automatically opens and water is discharged from all of the heads. Manual activation is also provided. Aircraft hangers, flammable liquid tank vehicle loading racks, and industrial facilities that process flammable or explosive materials are among those protected by these systems.

Standpipes

Standpipe systems are required in tall and very large buildings where advancing hand lines the entire distance would be prohibitive. Standpipe systems are classified according to their intended use, by the building occupants, fire department, or both. Standpipes are generally required in high-rise buildings, covered mall buildings, buildings over three stories, large-assembly occupancies, and adjacent to stages in theaters. Generally, standpipe connections are required on each floor at every exit stairway and on each side of a horizontal exit. Buildings under construction are generally required to have at least one standpipe or temporary standpipe, capable of flowing 500 gpm within one floor of the top of the building.

Fire Department Connections

Fire department connections are required for all water-based fire extinguishing systems and standpipe systems with the exception of limited area sprinkler systems of less than 20 heads and NPPA 13D sprinkler systems.

Wet and Dry Chemical Extinguishing Systems

Wet and dry chemical extinguishing systems are used in various applications including range hood and duct fire protection, paint spray booth protection, and even at unattended self-service motor vehicle fueling sites (see **Figure 10-12**). Each has a standard that

FIGURE 10-12 Wet chemical extinguishing systems protect commercial cooking operations.

addresses the design, installation, and maintenance of the system. They use four significant mechanisms of extinguishing fires: smothering, cooling, radiation shielding, and chain breaking.[12]

Halogenated and Clean Agent Fire Extinguishing Systems

Halon and other extinguishing agents that do not leave a residue are the agent of choice for certain high-value commodities and equipment. Halogenated extinguishing agents, or halon, are chemical compounds that contain carbon and one or more elements from the halogen series (fluorine, chlorine, bromine, and iodine). Most of the halogenated agents are toxic and unsuitable as extinguishing agents. Two compounds, Halon 1211 and Halon 1301, are effective extinguishing agents and are considered nontoxic.

Halon 1301 is used in total flooding systems that discharge quantities sufficient to create up to a 10% concentration in the areas protected. Halogenated agents suppress fires by interrupting the chemical chain reaction and will not extinguish deep-seated fire in class A materials. Under the Clean Air Act, the U.S. government banned the production and importation of halons 1211 and 1301 effective January 1, 1994. The ban did not affect the continued use of halon systems, and replacement halon can be purchased for existing systems. In light of the government ban on manufacture and importation, other clean extinguishing agents such as Great Lakes Corporation's FM200, Dupont's FE-13, and Ansul's INERGEN were developed. Fire protection for areas where the prevention of accidental water damage is a concern has also been rethought. The use of preaction sprinkler systems, which nearly eliminate the possibility of accidental discharge, has increased significantly.

Carbon Dioxide Extinguishing Systems

Carbon dioxide is one of the most plentiful compounds on earth. It is a byproduct of combustion and fermentation. We exhale carbon dioxide with every breath. It is cheap, leaves no residue and is a highly effective extinguishing agent for Class B and C fires. The major drawback to carbon dioxide systems is that carbon dioxide in concentrations of over about 9% will render persons unconscious almost immediately,[13] so they are not suited for the total flooding of computer rooms and other occupied areas. NFPA 12, *Standard on Carbon Dioxide Extinguishing Systems* contains installation, maintenance, and testing procedures.

Fixed Foam Systems and Water Spray Systems

Fixed foam and water spray systems are generally installed to protect hazards outside of buildings. Foam systems protect flammable liquid storage facilities and water spray systems are generally for exposure protection. Both have associated NFPA standards and specific acceptance tests that should always be witnessed by code officials, preferably by the end user, the fire department.

The following standards were put into place to govern the maintenance and inspection of fire protection systems:

- NFPA 12, *Standard on Carbon Dioxide Extinguishing Systems*
- NFPA 12A, *Standard on Halon1301 Fire Extinguishing Systems*
- NFPA 13, *Standard for the Installation of Sprinkler Systems*
- NFPA 14, *Standard for the Installation of Standpipes and Hose Systems*
- NFPA 15, *Standard for Water Spray Fixed Systems for Fire Protection*
- NFPA 17, *Standard for Dry Chemical Extinguishing Systems*
- NFPA 17A, *Standard for Wet Chemical Extinguishing Systems*
- NFPA 25, *Standard for Inspection, Testing, and Maintenance of Water-Based Fire Protection Systems*
- NFPA 72, *National Fire Alarm Code*

Fire Alarm Systems

Fire alarm systems are required in buildings so that occupants will receive prompt notification to evacuate, or in the case of institutional occupancies, to take appropriate action. In addition to the requirements for manual fire alarms by use group, fire alarm components are required in specific locations in all buildings. Smoke detectors are required in elevator lobbies to ensure that elevators do not return to the fire floor. Air handling units over 2,000 cfm must automatically shut down upon activation of a duct smoke detector installed upstream from the air handler. Smoke alarms are required in residential occupancies and must be interconnected within individual dwelling units.

The design, installation and maintenance of fire alarm systems are regulated by NFPA 72, the *National Fire Alarm Code*. Unlike most humans, fire alarm systems are quite good at doing several things at once. Fire alarm systems are comprised of a series of devices and circuits normally linked by a control panel. Detection devices sense the presence of heat or smoke, sprinkler flow or manual activation by a building occupant, and signal the control panel. Signaling devices display audio and visual signals to the building occupants that there is a potential or actual fire in the building. Fire alarm systems also perform auxiliary functions such as elevator recall, automatic actuation of smoke removal or stairwell pressurization fans and shut-down of certain air handlers. The system may also notify the fire department or a central station monitoring company.

Fire alarm systems range in size from the single-station battery-powered smoke alarm found in an existing one-story home, to a complex system in a high-rise building. Each has detection devices, signaling devices, and a control center. In the $5 battery-powered unit, they are all housed in a 6-inch plastic box. In a high-rise building, the circuitry may go for miles and cost many thousands of dollars.

Most complex modern systems are computerized. Fire alarm technicians do a significant amount of work during installation within the system's software program. When the flow switches are found to be annunciating in the fire control room as duct smoke detectors, technicians simply go into the program and make the correction with a few key strokes. Before the advent of computerized systems, a technician had to rewire the panel to accomplish such a change—and that is the potential problem.

Schedules for routine testing and maintenance of systems are contained in NFPA 72, as well as requirements for completion documents and for records of all tests. A review of these documents should be a routine part of the inspection or pre-incident of the building. Do not wait until after the fire when the system has malfunctioned to ask if the system was being maintained.

Smoke Control Systems and Smoke and Heat Vents

Systems that control the movement of smoke or provide for the rapid exhaust of smoke are required in atriums, covered malls, high-piled combustible storage facilities, underground structures and large theaters. High rise buildings and malls are required to provide manual controls for all air handling equipment so that fire fighters can use the equipment for ventilation of the structure.

Smoke and heat vents are required in large factory and storage buildings where the length of exit access travel is long, or over stages due to the large quantity of combustible sets. Curtain boards, which extend from the ceiling a minimum of 6 feet (but not within 8 feet of the floor), are installed to retard the lateral movement of smoke and gases. Smoke and heat vents must operate automatically and may be required to be provided with a means for manual activation by the fire department.

TWO FIREFIGHTERS LOST WHEN WOOD TRUSS ROOF COLLAPSES[14]

Two firefighters died when they were trapped by a rapidly spreading fire in an auto parts store and a pre-engineered wood truss roof assembly collapsed on them. The cause of the fire was an electrical short created when a power company truck working in the rear of the building drove away with its boom in an elevated position, accidentally pulling an electrical feed line from the main breaker panel at the rear of the store. Post incident investigations indicate that the electrical fault may have sparked multiple points of fire origin throughout the roof structure of the building, due to improperly grounded wiring.

This is another incident illustrating the rapid failure of lightweight construction systems when key support components are involved in a fire. It points out the importance of pre-emergency planning and accurate size-up by fire companies to determine the risk factors associated with a fire in this type of construction. Lessons regarding importance of initial company actions, constant re-evaluation of action plans, strong command and coordination of units on the fireground, and recognition of signs of impending structural failure were also reinforced.

Summary of Key Issues

- *Staffing.* The first-alarm response provided a small attack force with limited capabilities. The full response brought only 10 personnel.
- *Size-up.* The first-arriving company officer was not able to determine the location and extent of the hidden fire.
- *Pre-emergency plan information.* This complex required a pre-emergency plan due to the complex arrangement, multiple occupancies, mixed construction, lack of fixed protection, limited access and difficult water supply problems. The first-due company did carry

a pre-emergency plan that showed the layout of the shopping center and the floor plan for the auto parts store, but the pre-emergency plan was not referenced by the crew during the fire.

- *Delayed response.* The first-arriving company was on the scene alone for several minutes with only three personnel. The backup companies had long response times. The lack of evidence of a working fire prompted the initial incident commander to return some of the responding units, resulting in even longer response times.
- *Water supply.* The first-in company did not establish a water supply. This required the second engine company to be committed to this task.
- *Incident command.* The battalion chief was faced with a complicated and rapidly changing situation. He was not able to effectively transfer command from the initial officer and direct the operations of widely separated units.
- *Operational risk management.* The officers involved in the initial part of the operation had to make critical risk management decisions with limited information.
- *Accountability.* Accountability for the personnel operating in the hazardous area was not established prior to the structural collapse. As the situation became critical, no one realized that a crew was still inside the building.
- *Rapid intervention crew.* Additional crews did not arrive in time to assist the crew that was in trouble inside the building.
- *Radio communications.* The lack of clear radio channel for fireground communications caused serious problems with command and control of the incident, including the failure to maintain communications with the crew inside and the failure to hear their request for assistance.
- *Lightweight construction.* The roof collapsed quickly and with very little warning. This should be anticipated with a lightweight wood truss roof assembly. This hazard was not recognized by the crews. If the construction of the building had been known or recognized, the early failure of the roof structure should have been anticipated and the interior crew should have been withdrawn. This requires pre-emergency planning to identify high-risk properties and a reliable system to label the building or to inform the responding units of the risk factors of the building. It is usually difficult or impossible to make this determination when the building is burning.

This fire reinforced many operational issues that are essential to safe and effective firefighting operations.

- RISK ASSESSMENT is the primary responsibility of the incident commander. This incident presented a very high risk to the firefighters who were attempting to make an interior attack. However, the risk factors were not recognized and the interior crew was not directed to abandon the building. Risk assessment should be a continual process, particularly when a situation is changing very quickly.
- ACCOUNTABILITY is an essential function of the incident command system. The location and operation of the initial attack crew was not tracked according to the incident command system that was in effect at the time of the fire. The system must keep track of the location, function, status, and assignment of every individual unit or company operating at the scene of an emergency incident. In order to be effective, the accountability process must be routinely initiated at the beginning of every incident, and updated as the incident progresses and units are reassigned to different tasks.
- TACTICAL RADIO CHANNELS are essential for firefighter safety. The fireground operations were conducted on the same radio channel as the routine dispatch and transfer of additional units, hampering the fireground communications during the important early stages of the incident. Designated radio channels should be set aside specifically for communications between the incident commander and the units operating at the scene of an incident. The exchange of information, orders, instructions, warnings, and progress reports is essential to support safe and effective operations. Tactical channels should be assigned early and routinely to avoid the confusion that occurs when units that are already working are directed to switch to a different radio channel.
- FIRE OPERATIONS must be limited to those functions that can be performed safely with the number of personnel that are available at the scene of an incident. The initial response to this incident did not provide enough resources to safely initiate an effective interior attack for the situation that was encountered. The first-arriving company initiated interior operations that could not be adequately performed or supported with the limited number of personnel at the scene or responding. The delayed arrival of backup companies increased the risk exposure of the first-due company. The situation called for a more conservative initial attack plan and/or an early retreat when the magnitude of the fire became evident.
- WATER SUPPLY is a critical component of a safe and successful operation. The failed attempt to establish an adequate and reliable water supply for the interior attack was a critical problem at this incident. This task occupied the second-due engine company, which was needed to provide either a backup hose line to support the interior attack or a rapid intervention crew.
- LIGHTWEIGHT WOOD TRUSS CONSTRUCTION is prone to rapid failure under fire conditions.

DUPLICATION OF OR GAPS IN PROTECTION

NFPA 550 outlines a concept called the Fire Safety Concept Tree. This concept shows the link between fire prevention activities and fire damage control strategies. It provides an overall structure with which to analyze potential impact of various codes and standards on a particular safety problem. Departments can use this model as a way to communicate with building designers to assist in developing the role of fire protection requirements within a structure. Often times this communication does not take place and there can be a duplication of fire protection features with a building. In contrast, failure to communicate effectively can lead to gaps in fire protection.

OFFICER ADVICE

You should understand these systems and how to control them so that you can use them safely and effectively, or secure them, as appropriate.

FIRE BEHAVIOR

As a fire officer, you should also thoroughly understand fire behavior. **Fire behavior** is a term used to define the way fire performs or reacts in given situations. Every fire is unique, but there are some general concepts that help us better understand fire behavior. Understanding fire behavior helps us to be more effective at fire suppression. Understanding fire behavior is also important from a safety perspective, both for the occupants of the structure and for the firefighters.

We need three things to start a fire: fuel, heat, and an oxidizer, represented by the three sides of the fire triangle (see **Figure 10-13**). The process is interactive. The fuel must be heated to its **ignition temperature,** and the continuation of the process is dependent upon the heat being fed back into the process. This heat, in turn, helps warm the fuel. This process continues until all of the fuel is consumed, the oxidizing agent (usually as oxygen in the air) is reduced to a level where it can no longer support combustion, the heat is removed at a rate that prevents the feedback process from continuing, or the process is altered by the addition of some chemical.

All fires eventually go out. Some go out because all of the fuel was consumed; some go out because you altered one of the other conditions or ingredients. All are used as fire extinguishment techniques. The Fire Tetrahedron better explains how fires burn and how fires are extinguished (shown in **Figure 10-14**). By truly understanding the behavior of a fire, a fire officer can assess the fire growth capabilities and employ acceptable strategies and tactics to combat the situation.

Fire Science

The processes involved in fire ignition and growth are essential for fire to survive, and our understanding of these processes is essential for effective fire extinguishment. The following brief summary is not intended to be an exhaustive treatment of the subject of fire behavior; however, a brief review may help in your understanding of the information in this section.

NOTE

Fire science is the body of knowledge concerning the study of fire and related subjects (such as combustion, flame, products of combustion, heat release, heat transfer, fire and explosion chemistry, fire and explosion dynamics, thermodynamics, kinetics, fluid mechanics, fire safety) and their interaction with people, structures, and the environment.

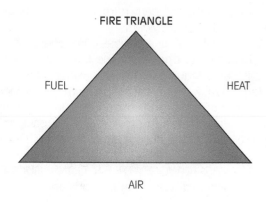

FIGURE 10-13 The fire triangle helps explain how fires start. To have fire, fuel, heat, and air must be present.

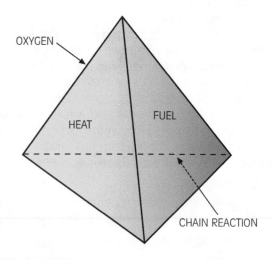

FIGURE 10-14 The fire tetrahedron helps explain how fires burn and how they are extinguished. A fourth side, called a continuous chain reaction, recognizes the presence of an ongoing complex chemical process. Fires can be extinguished by removing one or more of the four sides.

Fire usually develops from a small flame to where there is significant involvement of the fuel. When this process takes place within an enclosed building, it usually passes through three stages.

1. First stage—called the **incipient phase.** A fire may smolder for a period of time that ranges from seconds to hours before there is sufficient heat to produce open flame. During this time, there is little heat buildup, but some smoke is evident.

2. Second stage—called the **free-burning phase,** as the fuel is burning free of nearly any constraint. Flames become visible along with an increased rate of heat generation and fuel consumption. Since the heat expands the air, there is a slight increase in air pressure within the structure. Where there are openings in the structure, smoke and heat are pushed from the building. Hot gases heat the fuel above the fire, accelerating its ignition. The process continues and accelerates. Because of the dynamic forces at work, we see rapid fire extension during this phase.

3. Third stage—called the **smoldering stage.** Assuming that the structure is still intact, the fire will have consumed a portion of the oxygen in the air, and the fire is reduced to where it just smolders again. However, unlike the first phase, the room is filled with hot flammable gases.

The change from a free-burning fire where there is 21 percent oxygen to a smoldering fire with an oxygen-deficient atmosphere requires only a slight reduction in the oxygen concentration. Anyone who has ever adjusted the burners on the gas grill knows that it takes only a slight adjustment to significantly change the characteristic of the flame.

Fire Phenomena

When fire burns in structures, several fire-related phenomena occur. These include thermal stratification, rollover, flashover, and backdraft explosions.

Thermal Stratification

When a fire burns in an open area, the heat rises and is dissipated into the atmosphere. When a fire occurs in an enclosed space, a very different process occurs. Part of this difference is due to the fact that the hot gases cannot escape as fast as they are created. While the feedback process that allows some of these hot gases to be reintroduced back into the fire presents some very complex concepts of fire dynamics, we know from our own experience that most of the heat from a fire rises and stratifies within an enclosed space. This natural process is called **thermal stratification.** The heat is at the top of the room, and the cooler air is near the floor. This is why we tell people to crawl low in smoke.

When firefighters enter a space and start fire suppression operations, they can easily disturb the natural process that sends the heat to the highest areas in the room. When this occurs, firefighters endanger occupants and themselves alike. Because most fire operations deal with ordinary combustibles, water is used as an extinguishing agent. It works well; however, the inappropriate application of water disrupts the natural process of thermal balance.

Rollover

When materials burn, they consume oxygen and give off combustible gases. These combustible gases are hot and, therefore, less dense. These fuel-rich gases rush upward in the column of heat above the fire and fill the area. As additional gases are heated, they also rise and try to displace the gases that are already there.

In an open environment, these heated gases escape (see **Figure 10-15**). However, in an enclosed environment such as found in structures, the gases spread out and travel along the ceiling well in advance of the front of the actual fire (see **Figure 10-16**). In areas

FIGURE 10-15 In an outdoor fire, the heat dissipates into the air.

FIGURE 10-16 In a fire inside a structure, the heat banks down, accelerating the chain reaction and placing civilians and firefighters in grave danger.

where there is limited opportunity for these hot gases to expand, such as in a long hallway, they may precede the main body of fire by as much as 20 feet. When these hot, fuel-rich gases meet fresh air, they ignite and burn. This dynamic process appears to roll along the ceiling and hence the name, **rollover.** This situation has this potential for placing firefighters in considerable danger.

Flashover

Flashover is another fire phenomenon that presents significant risks to firefighters. As a fire burns in an enclosed space, the contents of the room are gradually heated. Everything in the room has an ignition temperature, and as the room temperature approaches the ignition temperature of the furnishings, they will ignite. **Flashover** can occur within a few minutes after ignition. Many factors determine when flashover will occur. Unfortunately, you seldom know all of these factors in a given fire situation.

FIREGOUND FACT

Flashover is another fire phenomenon that presents significant risks to firefighters.

Numerous demonstrations of flashover have been captured on film and many firefighters have been trained in a device called a "flashover trainer." While not always predicable, the process is very real. Results on the other hand are predictable and are very real. Anyone in the room at the moment of flashover has little chance of survival. Firefighters have experienced and survived flashover conditions. Those who survived their experience had and were using all the components of their personal protective clothing. In spite of their protection, most were injured, and all tell of the tremendous heat they experienced.

FIREGOUND FACT

Someone in the room at the moment of flashover has little chance of survival.

Investigation of Training Accident Addresses Dangers of Flashover. A live-fire exercise conducted in a vacant one-story single-family dwelling constructed of cement block, turned tragic for two firefighters.[15] During the exercise, two firefighters entered as a crew without a hoseline to simulate a search and rescue operation. They were to look for a mannequin dressed as a firefighter. To reach the fire room, they had to pass through two doorways and two narrow spaces. A crew with a hoseline was located in the next room by the doorway that led into the fire room. A flashover occurred, and that hoseline crew retreated due to severe heat conditions (at least one crew member received thermal burns).

After the fire was knocked down by a third crew, the two search and rescue team firefighters were found, evacuated to the exterior, medically treated, and transported to a hospital where they were declared dead. The deaths were the "result of smoke inhalation and thermal injuries suffered during the training exercise."

While a universal definition of flashover does not exist, this event is generally associated with rapid transition in fire behavior from localized burning of fuel to involvement of all the combustibles in the enclosure. Experimental work indicates that this transition can occur when upper room temperatures are between 750 and 1112°F (400 and 600°C). The critical importance of flashover is explained in many firefighter training courses, such as the following from *Firefighter's Handbook* (2000): "Firefighter Fact: At best, the survival time of a firefighter in bunker gear and breathing apparatus, fully encapsulated with gloves, hood, and helmet flaps down, is estimated to be less than 10 seconds." Unless the firefighters can immediately exit the room or have a hoseline that can cool the fire down, they are doomed. The key to preventing firefighter deaths and injuries from flashovers is to prevent the flashover, as noted in the NFPA 1403, *Standard on Live Fire Training Evolutions:*

- 4.3.5 The fuel load shall be limited to avoid conditions that could cause an uncontrolled flashover or backdraft.

- 4.3.7* The instructor-in-charge shall assess the selected fire room environment for factors that can affect the growth, development, and spread of fire [*refers to the following annex designated as A.4.3.7].

- A4.3.7 The instructor-in-charge is concerned with the safety of participants and the assessment of conditions that can lead to rapid, uncontrolled burning, commonly referred to as "flashover." Flashover can trap, injure, and kill firefighters. Conditions known to be variables affecting the attainment of flashover are as follows:

  1. The heat release characteristics of materials used as primary fuels
  2. The preheating of combustibles
  3. The combustibility of wall and ceiling materials
  4. The room geometry (e.g., ceiling height, openings to rooms)

In addition, the arrangement of the initial materials to be ignited, particularly the proximity to walls and ceilings, and the ventilation openings are important factors to be considered when assessing the potential fire growth.

Flashover is a very complex phenomenon containing several variables. For example, adequate ventilation can prevent flashover by removing heat and carbon monoxide gases from the room. However, in a training fire, the scenario may be to produce enough smoke to obscure vision, which would directly conflict with the ventilation needed to prevent flashover. In addition, closing the door and preventing any ventilation will also prevent flashover (although this can create a situation that may cause a backdraft). This is a very complex subject with many variables. Since the NFPA 1403 standard speaks only about the effect of fuel in preventing a backdraft or flashover, the following should be included in a mandatory training program for instructors and safety officers:

- Flashover must always be considered.

- Appropriate training on techniques to prevent or reverse a flashover must be provided to lead instructors and lead safety officers who organize and supervise live-fire exercises. Combustible materials that may create a very fast, hot or smoky fire should only be used with a thorough, deliberate evaluation and analysis of the fuel load. For example, the ability to provide emergency ventilation and/or apply water directly onto the main body of fire must be provided. Openings into the fire room can be closed off, using materials that can be quickly removed.

- A safety hoseline crew must be pre-positioned where they have an unobstructed view of the fire, allowing them to apply water directly onto the fire. One technique to accomplish this would be the opening in a wall (preferably an outside wall) of the fire room at the floor level. This would not affect the fire other than to provide additional ventilation. In most cases, except a flashover or a nonflaming fire (oxygen deprived), the visibility at the floor level would allow this hoseline crew to see the fire and any firefighters in the room.

- Portable temperature meters (thermocouples) can also be used to monitor the temperature in the room. Appropriate techniques to include the height above the floor and the maximum temperature that would indicate a possible flashover situation should be formulated. These temperature meters are used in some training center burn buildings to detect unusual and dangerously high temperatures. Consultation with subject matter experts would help facilitate this technique.

- Firefighters participating in a live-fire exercise should be thoroughly briefed on the nature of a flashover and techniques for immediately (15 to 30 seconds maximum) exiting the danger area.

- Firefighters should be equipped with a hoseline that can either be used to immediately apply water onto the fire or provide an unobstructed path out of the room by following the hose. Other personnel in the building should be instructed not to block this exit path, since exiting must be immediate.

- The fire room should have access to one exit (the entry door) and also a second remote means of escape, such as a door or a window large enough for a firefighter with SCBA to immediately exit.

NOTE

The 2007 edition of NFPA 1403 allows utilization of normally closed roof ventilation openings that can be opened in the event of an emergency. This can consist of precut panels or hinged covers.

Summary of Findings to Reduce Risk of Similar Incident[16]

1. Agencies conducting live-fire training must comply with NFPA 1403.

2. Agencies conducting "search and rescue" training should limit their use of live fire. Realistic conditions can be simulated without the danger of live fire.

3. When live fire is used the following safety measures are needed:

   - It is preferable to have any crew being trained be equipped with a charged hoseline.

   - The safety team must be in place to monitor the progress of the crew being trained and the fire. The safety team must have a hoseline of sufficient flow (minimum of 95 gpm) to extinguish a fire involving the entire fire room.

   - Training mannequins should be readily identifiable as such, so as not to be confused with a firefighter.

   - No fire room shall be used where two separate means of egress/escape are not available.

   - Live fire used in training should never be in a position to block the main or planned secondary exit of firefighters.

   - Emergency ventilation must be planned to limit fire spread and to improve tenability in the event such action is necessary. Normal room venting shall not be through the primary or secondary egress point.

4. Training fires are no different than hostile fires, and must be ventilated sufficiently to reduce the chance of unexpected flashover and to maintain visibility at the floor level. Adequate emergency ventilation must be provided for the fire room and any combustible attic space above.

5. Live-fire training in acquired structures must include planning for secondary means of egress or escape in case of unexpected fire condition changes. The use of rooms with limited access should not be considered.

6. Firefighters should be trained to constantly identify hazards and alternative escape routes during interior fire suppression operations inclusive of training exercises. Prior to live fire training drills, firefighters need to identify two ways out of each area.

7. Firefighters need training specific to recognizing the signs of a flashover, backdraft, proper ventilation techniques, and/or water application to reduce the chance of flashover.

8. All participants' protective clothing, PASS, and SCBA must be inspected for proper use and serviceability.

9. Formal communications plans must be in place inclusive of alternate channels, incident command communications versus crew-to-crew, emergency operations, medical operations and scene to dispatch.

10. Formal training for instructors conducting live-fire training is necessary.

Backdraft

Of all the fire-related events described here, **backdraft explosions** are the most forceful. As a fire develops in an enclosed space, it consumes oxygen and produces hot combustible gases. We have already described how this process produces what is known as flashover and rollover. As this process continues, it moves into the third phase of fire where the atmosphere within the room or building reaches a point where there is insufficient oxygen to support the open combustion process.

As this process continues, the entire space becomes filled with hot combustible gases. One side (the oxidizer) of the fire triangle has been removed, and this has interrupted the complete combustion

process. Eventually, the flames diminish and the materials smolder, producing hot flammable gases. When firefighters open a door or window, they allow oxygen to enter the space. When the right combination of heat, fuel (as a combustible gas), and oxygen are present in the space, the triangle is complete once again. Since the fuel is in a gaseous form, and since it is at or above its ignition temperature and spread fairly uniformly throughout the space, a rapid ignition of these gases occurs, sometimes with explosive force.

These events are commonplace and dangerous events that present great danger to every firefighter. There are usually clues that can be used to forecast these events; fire officers should understand all of them and avoid the conditions that kill and injure firefighters.

PUTTING IT ALL TOGETHER

There are two other issues that deal with both building construction and fire behavior.

The first term, **fire extension,** describes the spread of the fire through the fuel load and the structure itself. Remember that in Type III, IV, and V buildings, some or all of the structure itself is made of combustible material. Fire will follow fairly predictable paths to find the fuel needed to sustain its own life. It will generally travel upward, sometimes travel horizontally, and sometimes even downward, moving through the walls and ceilings, through voids and shafts where both fuel and oxygen are available to satisfy the fire's ravenous appetite.

While engaged in structural firefighting, we should check for fire extension as quickly as possible. Some will be obvious, and you will deal with that while extinguishing the original fuel. But in many cases, the fire cannot be seen, as when it is within the wall or concealed space. Firefighters have to open up these spaces to check for signs for fire extension. Whenever openings are made in walls or other spaces of a building that contain fire, or are even suspected of containing fire, firefighters should be immediately ready to deal with fire suppression. This means full turnout gear, SCBA, and a charged hoseline in hand. This also applies to opening doors and windows, cutting holes in roofs, or pulling ceilings to check for hidden fire extension.

The second term we want to look at is **ventilation.** Ventilation can be described as the procedures necessary to effect the planned and systematic direction and removal of smoke, heat, and fire gases from within a structure. Ventilation is a key tactical operation that affects how the fire behaves. Firefighters in many departments do not fully understand, nor effectively use, good ventilation practices in structural

---

FIREGOUND FACT

Of all the fire-related events described here, backdraft explosions are the most forceful.

firefighting. Ventilation is often considered as an afterthought, when the main body of the fire has been knocked down, and we want to dissipate the smoke. Good ventilation is needed to support all fireground activities from rescue to investigation, and it is needed to some degree in every type of building. Good timely ventilation greatly alters the conditions within the structure, allowing for the more rapid removal of heat and fire gases, thus affecting the fire-related phenomena we discussed in the preceding section and virtually eliminating the dangerous conditions that lead to backdraft and flashover.

Ventilation can be accomplished in two ways, generally referred to as natural ventilation and mechanical ventilation. **Natural ventilation** is effective for small fires and may require nothing more than opening a few doors and windows. Natural ventilation has its limitations, however, and, in some cases, can be both time consuming and labor intensive. If not done early, the real benefits are lost.

Mechanical ventilation is an alternative that uses fans or blowers to force the movement of air through the structure. Fire departments have been using fans to pull smoke and gases from buildings for years, and, in the 1980s, some departments started using fans to push fresh air into the building, thus forcing some of the smoke and fire gases out. Those who understood the concept used this approach quite effectively. Those who did not understand were skeptical about the idea of forcing fresh air into a structure during interior fire attack or were just unwilling to learn a new approach to an old problem.

Change comes slowly, but many departments are now routinely using positive pressure fire attack. However accomplished, ventilation activities during firefighting operations, especially in large structures, must be carefully coordinated with other fire suppression activities. If used effectively, ventilation is a key tactical activity that can greatly affect the positive outcome of any fire suppression effort.

OFFICER ADVICE

Whenever openings are made in a building that contains a fire or suspected fire, whether to fight the fire or to check for fire extension, firefighters should have a charged hoseline available to extinguish any fire that is found.

SAFETY

Ventilation alters most of the fire-related phenomena described previously and reduces the probability of backdraft and flashover.

QUESTIONS THAT YOU FACE AT EVERY FIRE

- Where is the fire?
- Where is the fire likely to spread?
- Is there an immediate likelihood of a flashover?
- Is there an immediate likelihood of a backdraft explosion?
- Is there a likelihood of a collapse?
- Is there any building feature that will help slow the fire's advance?
- Are there any installed fire protection features to assist in fire suppression efforts?
- Are there civilians present in the building? How will your activities affect them?

What Makes the Fire Go Out?

Most structural fires are extinguished with water. Water has many desirable properties that make it an effective extinguishing agent. It is cheap, readily available, and it has a great capacity to absorb heat. Many factors affect both the ways and the efficiency with which water absorbs heat from a particular burning combustible. The application rate and droplet size of the water impact on its effectiveness. The form and type of material burning have a significant impact on the amount of heat being generated as well as the ability of the water to cool the burning material and control the fire.

During firefighting, it appears that water cools the burning materials. In reality, effective fire suppression occurs when the heat from the involved materials is transferred to the water being applied. When applied effectively, a large portion of that water is converted to steam. Under ideal conditions, nearly all of the water is converted to steam in this process.

Let us talk science for a minute. The amount of heat required to raise the temperature of one pound of water one degree Fahrenheit is called a **British thermal unit** (BTU). You might think that a British thermal unit is a metric unit, but it is not. We use BTUs for many purposes in life, but we use them to our advantage here, not so much to show that water works well as an extinguishing agent, but rather to estimate how much water is needed.

Another scientific term we need is **specific heat,** a term used to describe the heat-absorbing capacity of a substance. As we have already noted, water is very effective in absorbing heat. In extinguishing fires, we apply water to absorb the heat generated by the fire. What we have is a heat transfer process: We reduce the temperature of the burning fuel and raise the temperature of the water.

If you were to raise the temperature of the water from 60°F to 212°F when applying it to burning materials,

152 BTUs would be absorbed for every pound of water applied. Water weighs about 8.3 pounds per gallon. Thus, each gallon of water has the potential for absorbing 1,266 BTUs when heated from 60° to 212°.

$$212 - 60 = 152 \text{ BTUs/pound}$$

$$152 \text{ BTUs/pound} \times 8.3 \text{ pounds/gallon}$$

$$= 1,266 \text{ BTUs/gallon}$$

If you continue to heat the water, it will turn to steam.

Another advantageous property of water is that in the process of converting to steam, water will absorb many more BTUs. We know from research that every pound of water will absorb 970 additional BTUs. Scientifically this phenomena is called the **latent heat of vaporization**—the quantity of heat absorbed by a substance when it changes from a liquid to a vapor. Again, multiplying by 8.3 pounds per gallon, the latent heat of vaporization for a gallon of water is 8,080 BTUs. Adding 1,266 and 8,080 we get 9,346 BTUs per gallon.

$$970 \text{ BTUs/pound} \times 8.3 \text{ pounds/gallon}$$

$$= 8,080 \text{ BTUs/gallon}$$

Combining the results of the two:

$$1,266 \text{ BTUs/gallon} + 8,080 \text{ BTUs/gallon}$$

$$= 9,346 \text{ BTUs/gallon}$$

One gallon of water will absorb 9,346 BTUs of energy. What does this mean to you as a company officer? You will see the significance of this in a moment. If you continue to heat the steam, it will continue to expand, as evidenced by billowing of steam during fire fighting operations.

We now look at the other side of the equation: How much heat do you have? Except for the kind of fires that fire protection engineers and scientists build for research purposes, you usually do not really know how many BTUs are being generated during a fire. But that research allows you to estimate the heat that is produced in the fires you typically encounter.

We do know that when materials burn, they generate heat. Scientists use the term **heat of combustion** when measuring the quantity of heat or energy released per unit of weight. Looking at some of the types of materials we commonly see involved in fire, we can get some idea of the approximate amount of heat being generated. For example, wood and paper yield about 8,000 BTUs/pound, polystyrene yields 18,000 BTUs/pound, and gasoline yields 19,000 BTUs/pound. In general, flammable liquids and gases burn more readily than solids. Some gases burn very hot. Not surprisingly they are used as fuel gases. These include methane, propane, and natural gas. However, for purposes of illustration, assume that the fuel is wood.

Research has shown that the average load of fuel in typical residential structures is about 4 pounds per square foot. Of course, this factor is much higher in warehouses and in some industrial and commercial facilities. Here is where the knowing this benefits the company officer. Most structure fires are extinguished by using water to absorb the heat created by the fire. From basic training, we recall that reducing the temperatures of the items that are burning below their ignition temperature will cause the fire to go out. In order to control the fire, the quantity of water applied to control the fire must be able to absorb the heat being produced.

FIREGROUND FACT

Research has shown that the average load of fuel in typical residential structures is about 4 pounds per square foot.

If you can estimate the amount of heat being produced, you can estimate the amount of water needed.

In real life, you do not know exactly what is burning, and, even if you did, you would not have time to do the math. What you do want to know is how fast you have to apply the water to make an effective fire attack. The rate of water needed to control the fire is called the **theoretical fire flow.** There are several formulas for determining the fire flow, but one of the easiest ones to use is the one currently taught by the National Fire Academy (NFA). It is based on the area in square feet of the burning structure. The area is determined by multiplying the length by the width. The NFA formula follows:

$$\text{Fire flow (in gpm)} = \frac{\text{Area in square feet}}{3}$$

NOTE

In order to control the fire, the quantity of water applied to control the fire must be able to absorb the heat being produced.

For example, take a room of a building 20 by 30 feet, or 600 square feet. Divide the area, 600, by 3, and you have 200. The answer is in gallons per minute, so the answer is 200 gpm. Remember that we are talking about flow rate, not the total quantity of water required. Knowing the fire flow helps in preplanning and firefighting operations.

OFFICER ADVICE

Knowing the fire flow helps in preplanning and firefighting operations.

This brief explanation of the NFA formula has not addressed how to deal with additional floors for fire involvement, exposures, and other important issues associated with fire flow. We will discuss this topic further in Chapter 13.

How can this help you in preplanning and firefighting? If you estimate the area, fuel, and level of involvement at the time you start fire suppression operations, you can

- Determine number and size of lines to provide the needed fire flow
- Determine the pumping capacity to supply these lines
- Determine if existing water supply resources are adequate
- Determine the number of personnel and companies needed to initiate a safe fire attack

This information will serve you in two ways. It helps you preplan for fire attack. You have to make assumptions about the size of the fire at the time you start fire suppression operations, but you can make some estimate of the needed fire flow while preplanning for the emergency. From this, you can determine how many and what size hoselines are needed, the pumping capacity required, and the water supply and the number of personnel needed.

A second application of the NFA formula facilitates **size-up** of the situation after arrival. Lacking a preplan, you can use the NFA formula to deal with the same questions regarding fire flow and resources that are previously listed. If you estimate the size of the involved area and level of involvement, you can better determine number and size of lines needed, if you have adequate pumping capacity to supply these lines, if the existing water supply resources are adequate, and if there are adequate personnel to initiate a safe fire attack. In most cases, the number of fire fighting personnel present is the critical element. Where a hydrant is available, water supply, pumper capacity,

NOTE

Where a hydrant is available, water supply, pumper capacity, and available hoselines generally exceed the capabilities of the personnel of the first arriving companies.

UNDERSTANDING BUILDING CONSTRUCTION

Christopher Naum, Captain,
Battalion 2, Moyers Corners Fire Department,
Liverpool, New York

The call came in at 2:30 AM—an incident occurring in a garden apartment of woodframe construction. The 200 foot × 100 foot building, which appeared to have been constructed in the 1960s (when building codes may have been more lax), housed 24 units. When Captain Christopher Naum's company got to the scene, the firefighters found fire originating in a below-grade parking garage.

The company members sized up the situation and made some assumptions based on their training—the fire should have stayed in one compartment. But it broke all the rules. Instead, it got into voids in the walls and traveled horizontally, then vertically, breaking out into various apartments. As it did so, the company then had to deal with a rescue component; one firefighter was on search detail when two floors collapsed, and he fell through to the basement, just missing the burning cars. Fortunately, Naum's company could hear his PASS alarm and was able to save him.

What should have been a contained fire ended up being four-alarm, taking 100 firefighters 6 hours to extinguish. For Naum, the lesson learned was that company officers should anticipate construction features in building types—but also be acutely aware of what can go wrong and anticipate what to do if the worst happens. That means understanding that construction, code deficiencies, or materials might not be in your favor; be cautious and aware when looking at size-up factors.

It also provided a lesson to the fire department for the future. There were more than 35 other similar structures in the community, so from now on, the fire department knew to treat those buildings as it would an older balloon-type structure. It would change its strategy and placement of interior personnel to be more aggressive. Based on the performance of this building, the fire department could build a profile to help assess the high risk potential for rapid fire spread in others.

Questions

1. What are the fire-related problems associated with woodframe construction?
2. What can we learn from this incident?
3. Can the lessons learned at one fire be applied to similar structures?
4. Can these lessons be applied in your jurisdiction?
5. We learned from the story at the start of this chapter that in today's world, fires burn hotter and buildings fail

LESSONS LEARNED

We recognize that this chapter was a review of material you may have learned while in recruit school. We make no apology, for as a company officer, you must have thorough understanding of fire behavior and building construction. (If you are an EMT, consider how much human anatomy and physiology you learned to get that certificate.) Think of these topics as the anatomy and physiology courses for firefighters as these areas of knowledge are fundamental to the task of fire suppression. Understanding and being able to control the forces that control the growth and extension of a fire will likely result in less fire loss, and fewer injuries to firefighters. You should keep these goals in mind as you plan, train, and operate at structural fires.

In addition to understanding these topics, it is important that we apply them to specific buildings. This planning activity allows companies to prepare for certain expected operations by better identifying and analyzing the building, the potential for fire, and the hazards associated with firefighting in that particular structure. Planning also allows for better assessment of the fire department's needs during a fire, and identifies limitations that may suggest alternative strategies to interior firefighting operations. We will discuss this further in Chapter 13.

As we close Chapter 10, we urge you to continue your studies in this area. If you are fortunate enough to have a community college with a fire science program in your area, they likely offer courses in building construction and fire suppression. These are in-depth studies of what we have only briefly addressed here. You will also find worthwhile articles on these topics in nearly every issue of the major trade journals. Study them and discuss them with your teammates. The more you know, the better you will be. Improvements in fire suppression skills mean less fire loss and a safer work place, two goals we should all embrace.

KEY TERMS

automatic fire protection sprinkler   self-operating thermosensitive device that releases a spray of water over a designed area to control or extinguish a fire

backdraft explosion   a type of explosion caused by a sudden influx of air into a mixture of burning gases that have been heated to the ignition temperature of at least one of them

British thermal unit   the amount of heat required to raise the temperature of one pound of water one degree Fahrenheit

community consequences   an assessment of the consequences on the community, which includes the people, their property, and the environment

fire behavior   the science of the phenomena and consequences of fire

fire extension   the movement of fire from one area to another

fire resistance   the resistance of a building to collapse or to total involvement in fire or the property of materials and their assemblies, which prevents or retards the passage of excessive heat, hot gases or flames under conditions of use

flashover   a dramatic event in a room fire that rapidly leads to full involvement of all combustible materials present

free-burning phase   second phase of fire growth, has sufficient fuel and oxygen to allow for continued fire growth

fuel load or fire load   the total amount of combustible material within a fire area, expressed in terms of pounds per square foot.

heat of combustion   the amount of heat given off by a particular substance during the combustion process

ignition temperature   the minimum temperature to which a substance must be heated to start combustion after an ignition source is introduced

incipient phase   first stage of fire growth, limited to the material originally ignited

life risk factors   the number of people in danger, the immediacy of their danger, and their ability to provide for their own safety

listed   equipment or materials included in a document prepared by an approved testing agency indicating that the equipment or materials were tested in accordance with an approved test protocol and found suitable for a specific use.

noncombustible construction   in broad terms, constructed of materials that will not burn

performance-based design   an alternative method for satisfying the fire protection and life safety intent of construction codes based on agreed-on performance goals and objectives, engineering analysis, and quantitative assessment of alternatives against the design goals and objectives

performance code a prescriptive code that assigns an objective to be met and establishes criteria for determining compliance

protected construction protected from the effects of fire by encasement; concrete, gypsum, and sprayed on fire-resistive coatings are all used to protect structural elements

rollover ignition of gases that have risen and encountered fresh air, and thus a new supply of oxygen

size-up mental assessment of the situation; gathering and analyzing information that is critical to the outcome of an event

smoldering phase third stage of fire growth; once the oxygen has been reduced, visible fire diminishes

specification code a prescriptive code that specifies a type of construction or materials to be used

specific heat the heat-absorbing capacity of a substance

structural factors an assessment of the age, condition, and structure type of a building, and the proximity of exposures

theoretical fire flow the water flow requirements expressed in gallons per minute needed to control a fire in a given area

thermal lag the difference between the operating temperature of a fire detection device such a sprinkler head as and the actual air temperature when the device activates

thermal stratification rising of hotter gases in an enclosed space

ventilation a systematic process to enhance the removal of smoke and fire by-products and the entry of cooler air to facilitate rescue and fire-fighting operations

REVIEW QUESTIONS

1. What is meant by life risk factors?
2. What is meant by property risk factors?
3. What is meant by community consequences?
4. How do each of the following factors affect fire-fighting operations?
   - physical factors
   - access factors
   - structural factors
   - survival factors
5. Define the five building classifications.
6. Identify a fire-related hazard associated with each type of building.
7. Define the following:
   - rollover
   - flashover
   - backdraft
8. What is the NFA's fire flow formula?
9. How does the NFA fire flow formula help in pre-fire planning and fire suppression operations?
10. Why is an understanding of building construction and fire behavior essential for you, as the company officer?

DISCUSSION QUESTIONS

1. What are the significant risk factors in your community?
2. What are the five principle building types? What is the significant problem from a firefighter safety perspective while firefighting in each type?
3. Which types of buildings are found in your community? What are the significant hazards of these building?
4. List some of the target hazards in your community.
5. How has building construction changed in your community in the last 50 years?
6. How have the contents of buildings changed in the last 50 years?
7. How have these changes affected fire suppression activities?

8. We hear a lot about firefighters going where they have never gone before—that firefighters, protected by better protective clothing and SCBA, can penetrate further into a building to locate and control fires. In light of your answer to question 7, how has firefighting changed in the last 10 years?

9. Explain the use of the fire flow formula. What formula is used in your community?

10. Have you used a fire flow formula during preplanning activities? Have you used it during size-up and fire suppression activities?

ENDNOTES

1. Lloyd Layman. *Fire Fighting Tactics.* Quincy, MA: National Fire Protection Association, 1953, 8. Reprinted with permission from National Fire Protection Association, Quincy, MA 02269.

2. Emanual Fried. *Fireground Tactics.* Chicago: Marvin Ginn Corporation, 1972, p. 357.

3. Francis Brannigan. *Building Construction for the Fire Service.* Boston: National Fire Protection Association, 1971, p. 2.

4. J. Gordon Routly. Interstate Bank Building Fire, Los Angeles, California USFA-TR-022/May 1988.

5. J. Gordon Routley. City of Charleston Post Incident Assessment and Review Team Phase II Report, Sofa Super Store Fire, May 15, 2008.

6. *Preventing Injuries and Deaths of Fire Fighters due to Truss System Failures,* National Institute of Occupational Safety and Health, 2005, NIOSH publication 2005-132, p. 23.

7. 2006 ICC Performance Code for Buildings and Facilities, 2006, International Code Council, p. 10.

8. Paul A. Croce, et al.. *Fire Safety Journal, 43,* 234–236, 2008.

9. Richard W. Bukowski, P.E. *Determining Design Fires for Design-level and Extreme Events.* Gaithersburg, MD: NIST Building and Fire Research Laboratory, 2006, p. 5.

10. Richard E. Hughey, P.E. "Property Protection or Life Safety: Can We Have Both?" *Fire Marshal Quarterly,* September 1996, p. 20.

11. Edward J. Kaminski, P.E. "Practical Aspects of Wet and Dry Chemical Extinguishing System Inspection and Acceptance." *BOCA Magazine,* 4 (July/August 1995), p. 34.

12. *Fire Protection Handbook.* 13th ed. Quincy, MA: National Fire Protection Association, pp. 15-23.

13. For additional information about this incident, go to the USFA web site at www.fema.gov and look for Technical Report #087.

14. This information was extracted from the official investigation report, prepared by the Bureau of Fire Standards and Training, Florida Division of State Fire Marshal. For the complete report, go to www.fsfc.ufl.edu. Go to link for state fire marshal and look for report of July 30, 2002, incident.

15. For additional information regarding the NFA fire flow formula, see the preparation course student manual for the NFA training program entitled *Managing Company Tactical Operations.* Additional information about this course is contained at the end of Chapter 12 of this book.

16. Reproduced with permission from NFPA 1403, *Live Fire Training Evolutions,* Copyright©2007, National Fire Protection Association. This reprinted material is not the complete and official position of the NFPA on the referenced subject, which is represented only by the standard in its entirety.

ADDITIONAL RESOURCES

Brannigan, Francis L. "Effect of Building Construction and Fire Protection Systems." In *Fire Protection Handbook.* 19th ed. Quincy, MA: National Fire Protection Association, 2003.

Custer, Richard L. "Dynamics of Compartment Fire Growth." In *Fire Protection Handbook.* 19th ed. Quincy, MA: National Fire Protection Association, 2003.

Davis, Richard J. "Building Construction." In *Fire Protection Handbook.* 19th ed. Quincy, MA: National Fire Protection Association, 2003.

Drysdale, D. D. "Chemistry and Physics of Fire." In *Fire Protection Handbook.* 19th ed. Quincy, MA: National Fire Protection Association, 2003.

Fried, Emanual. *Fireground Tactics.* Chicago: Marvin Ginn Corporation, 1972.

Friedman, Raymond. "Theory of Fire Extinguishment." In *Fire Protection Handbook.* 19th ed. Quincy, MA: National Fire Protection Association, 2003.

Hicks, Harold, D. "Confinement of Fire in Buildings." In *Fire Protection Handbook.* 19th ed. Quincy, MA: National Fire Protection Association, 2003.

Klinoff, Robert W. *Introduction to Fire Protection.* 2nd ed. Clifton Park, NY: Delmar Publishers, 2003.

Lloyd Layman. *Attacking and Extinguishing Interior Fires.* Quincy, MA: National Fire Protection Association, 1955.

Nelson, Floyd W. *Qualitative Fire Behavior.* Stafford, VA: International Society of Fire Service Instructors, 1991.

Quintiere, James G. *Principles of Fire Behavior.* Albany, NY: Delmar Publishers, 1997.

Watts, John M. "Fundamentals of Fire-Safe Building Design." In *Fire Protection Handbook.* 19th ed. Quincy, MA: National Fire Protection Association, 2003.

11

The Company Officer's Role in Fire Investigation

It was a quiet, spring morning, about 4:00 A.M., when we were jarred awake with the notification of a fire alarm. Our ladder company pulled up just behind the battalion chief and the first engine. Heavy black smoke and flames were visible in the large front window of a darkened diner. We listened to the chief's size-up as we approached, "One story ordinary construction, 30 feet by 100 feet, heavy fire conditions in a restaurant." This was followed quickly by a second message, "Battalion 5 to dispatch, send us an ambulance, a civilian driving by saw a victim inside."

As the lieutenant in charge of the truck company, I sent two firefighters to the roof and took two with me for search and rescue. We quickly vented the front window and forced the front door. As the heavy smoke billowed out, I detected the strong scent of gasoline. Via the radio, I relayed this information to the chief, and he called for arson investigators to respond.

By relaying the information over the radio that this fire was possibly an arson fire, all firefighters were now aware of the arson potential. All the firefighters on scene knew to use extra caution because of the presence of accelerants, which can make the fire much more volatile and harder to control. Another potential danger is that arsonists may sometimes weaken structural components to aid in collapse, adding an additional element of danger for firefighters.

The engine company led a 2½-inch line into the front door and we started our search behind them. The rear door was locked tight and no victim was found during our primary search of the restaurant. We found a stairway to the basement and continued our search down there. The basement was free of fire and had only a moderate amount of smoke. The rear door to the outside was forced open. My first reaction was that this is how the arsonist gained entry. On closer inspection, it was obvious by the damage to the door that it was forced from the inside. This was important evidence because the rear stairwell was in a well-lit area, which faced a large apartment building. This might indicate someone wanted it to look like forcible entry, but did not want to be seen while attempting this cover-up. It was clear to me this was an inside job, arson for profit.

We treated the fire scene like a crime scene and disturbed as little evidence as possible. Another team of firefighters found and preserved a gas can in the restaurant. I documented our actions thoroughly in the company journal and forwarded additional forms to arson investigators and my district chief. I also kept records in my own personal file. This turned out to be critical when, 4 years later, I was a key witness in the arson trial of the restaurant owner. Without good documentation of information, this criminal may have gotten away with arson.

Hundreds of people die in arson fires every year. Sometimes the victims of arson are firefighters. All company officers should understand that arson investigation starts with the first-arriving firefighters. Suspicious looking flame or smoke, low burn patterns, smells of accelerants, the presence of gas cans or timing devices, and forced doors are just some of the examples of arson evidence. We must detect evidence, preserve it, relay the information to arson investigators and chief officers, and ensure the safety of our firefighters in these dangerous situations.

*—Battalion Chief Steve Chikerotis, Battalion Chief/Assistant
Director of Training, Chicago Fire Department, Chicago, Illinois*

LEARNING OBJECTIVES

After completing this chapter the reader should be able to:

11-1 Describe the methods that are most commonly used by individuals that set fires.

11-2 Describe the responsibilities of a fire officer while interacting with other agencies involved in a fire investigation.

11-3 Describe the importance of a fire scene assessment.

11-4 Identify evidence that will assist in determining fire origin and cause.

11-5 Identify procedures employed to determine the origin and cause of a fire

11-6 List the elements or the indicators of the fire growth and development to assist in determining the need for a fire investigator.

11-7 Describe procedures for securing and preserving evidence at a fire scene.

11-8 Define the term chain of custody.

11-9 Describe the importance of maintaining the chain of custody.

11-10 Describe the types of evidence that must be preserved.

11-11 Identify methods used to protect evidence from damage or destruction (spoiliation).

11-12 List the steps necessary to secure an emergency scene.

11-13 List problems caused by evidence removal from a fire scene prior to an investigation occurring.

11-14 Explain the importance of evidence preservation.

11-15 Explain procedures used to document a preliminary origin and cause investigation

*The FO I and II levels, as defined by the NFPA 1021, the *Standard for Fire Officer Professional Qualifications*, are identified in different colors: FO I = black, FO II = red.

INTRODUCTION

A major part of any active fire prevention program should be the investigation of fires. Fire investigations help identify the community's fire risk by determining the frequency and causes of the fire. This information can be used to provide the basis for fire prevention programs and for identifying the resources needed for fire suppression. When a fire may have been intentionally set, a good fire investigation is the first of several steps needed to bring the perpetrator to justice. Fire officers have a significant role in this process.

All Fires Should Be Investigated

Every fire should be investigated to determine its origin and cause. We often equate fire investigation with arson and arson investigation. Although arson investigation is a part of the fire investigator's job, fire investigation has other purposes. A majority of fires (70 to 80 percent overall) are accidental in nature. (Statistics in your particular response area may suggest a different value.)

All fires, including those that are obviously accidental, should be investigated to determine their cause

OFFICER ADVICE

Every fire should be investigated to determine its origin and cause.

so that patterns of human behavior or equipment failure can be discovered and corrected.

Before going much further, we need to clearly define several words we are using in this chapter. First, we use the term **accidental** here to include all types of fires that are not intentionally set. Some would argue that accidental is too kind a term, that while accidents do happen and mechanical devices do fail, and lightning and other natural causes do occur, human carelessness is the cause of all too many fires. In many jurisdictions, all fires are considered to be intentionally set until proven otherwise by a fire investigator.

NOTE

Fire investigators try to eliminate all possible accidental causes before considering arson.

The term intentionally set is probably a more appropriate term for what is often referred to as arson.

In the past, the terms incendiary or suspicious were often used here, but the current National Fire Incident Reporting System (NFIRS) uses the term "intentionally set," and current National Fire Protection Association data on fire loss use the same term. We will follow their lead.

> **NOTE**
>
> The crime of arson requires criminal intent as well as the act itself.

Arson, on the other hand, is a legal term. **Arson** is the result of a deliberately set fire, but not all intentionally set fires are arson: It is not a crime to light a fire in your fireplace. And where authorized, it is not illegal to burn brush or trash. But the intentional setting of a fire where there is criminal intent is arson. Thus, arson has two elements: a criminal intent, and a deliberate or intentional act. When you cannot find reasonable evidence of an accidental cause, it would, indeed, be appropriate to consider the investigator's logic: All fires are considered to be intentionally set until proven otherwise. Fire investigators use this approach in both their fire investigation and in their presentation of evidence in courts. They rule out all accidental causes first and then prove that the fire was intentionally set.

> **NOTE**
>
> All fires, including those that are obviously accidental, should be investigated to determine their cause so that patterns of human behavior or equipment failure can be discovered and corrected.

As company officers, you are often in a unique position for conducting fire investigations. You are likely to be among the first emergency responders at the scene and therefore have an opportunity to see the fire while it is still burning. For smaller events, you may be the senior fire department official at the scene, so you must complete the fire incident report. To complete the report, you have to determine the origin and cause of the fire. Where an intentionally set fire is involved, your observations and actions play a significant role in any legal proceedings that follow. With these thoughts in mind, let us consider the common causes of fire.

What Is Fire?

In Chapter 10, we reviewed fire behavior. We noted that a fire will not occur unless the three essential elements that comprise the fire triangle are present.

HAVE YOU HEARD MS. VINA DRENNAN TALK ABOUT THE COST OF CARELESSNESS?

Vina Drennan understands carelessness when it comes to fires; she is the widow of New York City Fire Department Captain John Drennan who, along with two other firefighters, died in a fire caused by a citizen's careless act. (A bag of trash was left on top of a gas stove.) Captain Drennan survived the fire but died several agonizing weeks later as a result of his injuries.

Shortly after his death, someone said to Ms. Drennan, "I'll bet you would like to get your hands on those who were responsible for the fire." She was surprised at the suggestion. She responded by saying, "Oh no, they didn't mean it, they were just careless. Just careless . . . we tolerate a lot of carelessness in America."

She has traveled all over the country and talked about John Drennan's life. She also talks about fires; she has learned a lot about fires in the last several years. She compares the attitude toward fire and the fire loss in America with other countries, noting that we do not compare very well. She tells those who will listen that Captain Drennan loved his job, and he died in that job because of someone's carelessness.

She challenges all of you in this fire service who hear or heed her message! "Tell the public that carelessness can kill!" If we can reduce the number who suffer injuries from fires, and if we can reduce the number of fatalities from fires, then we can find even greater meaning in John Drennan's life.

One of these is heat. Contemporary fire science textbooks refer to this side of the fire triangle as **energy** (see **Figure 11-1**). Heat, or energy, warms the fuel to cause vapors to form and provides the ignition source in the fire triangle. Heat also promotes fire growth and flame spread. There are many commonly found ignition sources including chemical reactions, electrical energy, and mechanical energy.

The second essential element of fire is fuel. Fuel can be anything that is combustible. Fuel can be present as a solid, a liquid, or a vapor. Fuel for accidental fires includes items that are all around us, ranging from household furnishings to trash.

Oxygen is the third requirement. Oxygen is naturally present in the air. In fact, fire is usually defined in simple scientific terms as the process of **oxidation** of a fuel by atmospheric oxygen. Fire is often defined as a rapid, self-sustaining oxidation process, usually accompanied by the evolution of heat and light in varying intensities.

FIRE TRIANGLE (TRADITIONAL)

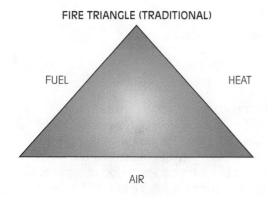

FUEL   HEAT

AIR

FIRE TRIANGLE (MODERN)

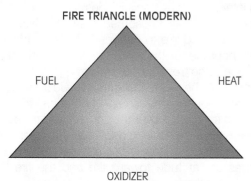

FUEL   HEAT

OXIDIZER

FIGURE 11-1 In Chapter 10, we noted that fuel, heat, and air are needed to start a fire. To better understand the chemistry and physics involved in the combustion processes, air is replaced with the more scientific term oxidizer; likewise, heat is replaced with the term energy.

COMMON CAUSES OF FIRE

Accidental Causes

In Chapter 9, we discussed the common causes of fire. Those same causes are discussed here briefly, with the focus on looking at their origin and cause.

Cooking

Cooking is the leading cause of fires in residential occupancies. Cooking is also by far the leading cause of fire-related injuries. When used properly, cooking equipment is not hazardous. Unsafe practices, inattentiveness, poor housekeeping, inappropriate attire, and lack of maintenance can lead to fires associated with cooking equipment. In most cases, it is not really the equipment that is at fault; it is human error (see **Figure 11-2**).

Heating Equipment

Heating equipment is another leading cause of fires. By design, heating devices get hot. When sufficient quantities of that heat get close enough to combustible materials, you have a fire. Most heating equipment is safe; what causes problems is usually not the equipment, but rather the way the equipment is used. Fires occur when heating equipment is misused, improperly installed or maintained, or placed too close to combustible materials, like bedding and clothing.

FIREGROUND FACT

Most heating equipment is safe; what causes problems is usually not the equipment, but rather the way the equipment is used.

Chimneys are designed to vent the hot and hazardous by-products of the fire to the outside. When the chimneys are not properly maintained, those by-products accumulate in the chimney. Eventually, when they get hot enough, they ignite and cause a chimney fire. When the chimney is not properly maintained,

FIGURE 11-2 Fires in the kitchen can have serious consequences.

the hot gases that are intended to go up the chimney can escape through the walls of the chimney. If they escape into the structure, they may ignite nearby combustible materials.

Hot fireplace ashes are another major concern in many parts of the country. Regardless of the type of fire that produced the ashes, ashes that remain warm can set fire to any nearby combustible material. Many people have carelessly set fires by placing hot fireplace ashes in a plastic can or paper bag in their garages or on their wooden patio decks.

Firefighters everywhere attend fires like these and can tell stories about how fires they have seen were started. But not many homeowners know what you know about fires, and some of those who do know, do not seem to care. So, we keep having the same kinds of fires. Nearly every fire department in the country attends fires involving heating equipment during the winter season.

> **FIREGROUND FACT**
>
> Leading accidental causes of the fire include the following: cooking and cooking equipment heating equipment, the careless use and disposal of smoking materials, and electrical distribution systems.

Smoking Materials

Smoking and the careless discarding of smoking materials are the source of many of the nation's fire fatalities. Such fires often involve a delayed ignition of upholstery materials, such as in a bed, chair, or sofa. The results are a slowly developing fire that eventually produces enough heat and toxic vapors to kill everyone present. When undetected, these fires are often significant in terms of their intensity, and because they frequently occur after bedtime, or when the smoker is intoxicated, the results are fatal.

Electricity

Electricity is all around us. When properly used, it is a safe form of energy. However, that energy has tremendous potential for causing damage. Electrical appliances and electrical distribution systems cause fires. Safeguards are built into appliances and electrical distribution systems to protect us from fire and personal injury. However, the safeguards do not always protect individuals from careless acts.

One of the major protection devices in any electrical system is the electrical panel. It is a good place to check after any fire. You should try to determine if the electrical circuit was energized at the time of the fire. Examine the panel itself as well as the area around the meter to determine if there was any tampering

before the fire or any evidence of arcing or other malfunction during the fire.

When you are looking at wiring inside a structure after a fire, it is often difficult to determine if the fire caused damage to the wire or if the wire caused the fire. The nature of the damage is a function of the cause of the fires, the extent of the exposure of the wiring to heat, the duration of the fire, and the protection on the wire. In a typical residential fire involving a room and its contents, the wiring does not get hot enough to melt. On the other hand, wiring can be damaged by the heat generated by an excessive amount of electrical current. In this case, you may see melting of insulation in areas of the structure where there was no fire damage and evidence that the wire was warm enough to deform due to the heat generated by the electrical current.

> **FIREGROUND FACT**
>
> In a typical residential fire involving a room and its contents, the wiring does not get hot enough to melt.

Arcing, as a result of poor connections, can damage wiring. Arcing usually leaves a distinctive mark on the wire either in the form of beaded ends or one or more cavities in the wire itself. The arcing usually causes sparks, which may ignite nearby combustible materials. Finding one or more clues associated with electrical distribution systems and electrical equipment may help you determine the cause of the fire. Electrical problems are often missed as a cause of fire. Fires are also unfairly blamed on electrical problems.

> **FIREGROUND FACT**
>
> Arcing usually leaves a distinctive mark on the wire either in the form of beaded ends or one or more cavities in the wire itself.

Natural Causes

There are other causes of fires. Many wildland fires are the result of lightning. Lightning can strike anywhere, even in the middle of congested areas. Storms often bring lightning strikes to structures, downed powerlines, and other problems. Fires in the wildland environment endanger any structure that is present. Even in the urban or suburban environment, fires can start as a result of exposure to other fires, or from flaming materials carried downwind from the original fire.

Another natural cause of fires is the result of spontaneous combustion or autoignition. Autoignition is

FIGURE 11-3 This furniture store was completely destroyed by a fire that started in an on-site refinishing facility. A thorough fire investigation revealed the cause of the fire as autoignition of cleaning rags in a trash can in the refinishing shop.

the result of a chemical reaction; it does not require a spark or flame to start the fire. The process is usually associated with rags or other combustible materials containing a residue of flammable liquid. The fuel must be sufficient and arranged to generate enough heat to reach the ignition temperature of the surrounding materials. When cleaning rags are improperly stored or casually discarded, they can be the source of ignition (see **Figure 11-3**).

Arson

The *intentional setting of fires* is a pervasive problem affecting many types of property. Every year, we read of statistics dealing with structure fires, wildland fires, and even vehicular fires. In all three categories, those fires that were intentionally set contribute to the total loss. In the case of structure fires, the loss also includes civilian fatalities and injuries, and to some extent, firefighter deaths and injuries (see **Table 11-1**).

WHAT IS ARSON?

Arson is a legal term. Under common law, arson was defined simply as the malicious burning of someone else's house. Today, the definition of arson varies from state to state, but generally the definition has been extended to include any property including one's own property, and to designate four levels or degrees of arson:

- The burning of dwellings
- The burning of buildings other than dwellings
- The burning of other property
- The attempted burning of buildings or property

FIREGROUND FACT

NFPA reports that there were 32,500 intentionally set fires in the United States in 2007. About half of these fires were in structures, resulting in almost 300 deaths and $733,000,000 in property loss. Most of the remaining intentionally set fires were set in vehicles, resulting in an additional $145,000,000 in property loss.[1]

TABLE 11-1   Leading Causes of Residential Fires[2]

Rank	Civilian Deaths	Civilian Injuries	Dollar Loss	Number of Fires
1	Smoking	Cooking	Arson	Cooking
2	Arson	Arson	Electrical	Arson
3	Cooking	Smoking	Heating	Heating

Source: Fire in the United States, 12th ed. p. 62.

Individuals set fires for many reasons. Many think of arson as a scheme to fraudulently collect money from the insurance company. In this context, an arsonist might be thought of as the hired torch that intentionally set the fire. To some extent, that is still true, but today, we are seeing increasing use of intentionally set fires and explosions by spiteful individuals or groups to hurt and kill people for revenge. These acts range from retaliatory acts against an individual or family to large-scale events such as those in Oklahoma City in 1995 (see **Figure 11-4**) and in New York and elsewhere in 2001. In the last case, the results were the acts of terrorists. Many think of terrorism as something new, but it has been around a long time, and when we look at the number of events worldwide, we should realize that the threat is very real. We will talk more about this in Chapter 13.

NOTE

Motives for arson include spite and revenge, fraud, intimidation, concealment of another crime, vanity, pyromania, emotional dysfunction, civil disorder, and the acts of juveniles.

MOTIVE

The reason a person sets a fire is known as the **motive**. There are several reasons or motives for arson-related fires. Many fires are set to obtain money. The usual scenario is that property is deliberately destroyed by fire in an effort to collect money from an insurance company or others. The motive here is fraud: an unfair or dishonest act to obtain something, in this case, to obtain money.

Another common motive for arson is revenge. Revenge fires are usually associated with rage related to a broken romantic relationship or particularly hostile labor-relations problems in the workplace. Because of the attitude of the arsonist, these fires are frequently started with a flammable substance or even an explosive device with devastating results.

THE VALUE OF FIRE INVESTIGATION

In the state laws that regulate your fire department's operations, there should be a statement regarding fire investigation. The following is typical:

> The local fire official shall make an investigation into the origin and cause of every fire occurring within the limits of the jurisdiction.

Taken literally, that statement suggests that you would call the fire chief to every fire. For most fires, calling

FIGURE 11-4   The Alfred P. Murrah Federal Building. *(Photo courtesy of Michael Regan)*

an investigator, much less the fire chief, is not practical or necessary. One of the reasons we fail to call out investigators is we may have to wait for them to arrive and may have to wait even longer until the investigation is completed. Another reason is that we are often quick to assume the fire was accidental.

OFFICER ADVICE

Every fire officer should become proficient in determining the "origin and cause" of every fire.

In many jurisdictions, fire investigators do not respond to fires that are considered accidental or that result in minor property damage. In such locations, it is important that every fire officer become proficient in determining the "origin and cause" of every fire. The average fire is attended with one or two engine companies. One of the officers is usually assigned the task of gathering the important information and filling out the fire report. Part of the report asks for information about the cause and origin of the fire. Experienced fire officers should be able to determine the origin and cause of nearly all fires they attend. These reports are important in that they help the community, state, and to some extent, the nation, better identify our true fire problem. Therefore, we encourage every fire department and every fire officer to complete the report that supports the gathering of data. Most of the states participate in the National Fire Incident Reporting System (NFIRS). Reporting continues to get easier as new computer technology is introduced.

NOTE

Experienced fire officers should be able to determine the origin and cause of nearly all fires they attend.

Although fire officers should be able to determine the origin and cause of most fires, there will be times when you have to call the investigator. The request for an investigator might be made as early as during the arrival of the first-arriving companies, or it may not be evident until overhaul activity has started. Local policies and protocols suggest exactly how this should take place. If the company officer believes there is a reasonable cause for belief in existence of facts an investigator should be summoned as soon as possible. The investigator's arrival during fire suppression or at least during overhaul will assist in determining the cause of the fire, help build a better case if arson is involved, and may even allow suppression companies to leave sooner than they might have otherwise.

OFFICER ADVICE

If in doubt, ask for the fire investigator and ask as soon as possible.

In any event in which you have concerns about these or other issues, you should talk with a fire investigator by phone or other means before leaving the scene.

Who does the fire investigation? A needs assessment of the U.S. Fire Service provides information that helps answer that question. While company officers and incident commanders are investigating many fires, frequently there is a need for a certified fire investigator. Most large departments have such personnel on staff. In fact, in 90 percent of the jurisdictions of 50,000 or more, the fire department provides the investigator. In jurisdictions of less than 50,000, a variety of resources are used including fire investigators from the fire department, from regional task forces, from the state fire marshal's office, and from the state police.[3]

GUIDELINES FOR CALLING THE FIRE INVESTIGATOR

It is suggested that an investigator should be called in the following situations:
- Any event that involves an obvious incendiary fire
- Any fire that results in death or serious injury
- Any event that produces burn injuries, especially those involving direct flame contact, fireworks, or is the result of an assault
- Any event that involves an exploding or incendiary device
- Any event that involves fire damage beyond the original room of origin
- Any event that involves damage to government property
- Any vehicle fire that is not the result of an accident
- Any event in which the officer in charge is unable to determine the cause

Although we are still assuming that the fire is accidental 80 to 90 percent of the time, we should still accurately determine the cause of the fire all of the time. In cases where the fire is determined to have been accidental, timely and accurate fire cause determination is important for the owner, the insurance company, and the fire department. In addition, timely and accurate fire investigations help identify faulty equipment, unsafe habits, and patterns of activity. For example, a series of similar events may look like an unrelated series of accidents until one notices a pattern of time, addresses, or causes.

OFFICER ADVICE

Timely and careful gathering of evidence will aid in building a case that may bring someone to justice.

Although most fires are accidental, you should be thinking about the possibility of arson on every call. In cases where you have a reason to believe that the fire may have been deliberately set, timely and careful gathering of evidence will aid in building a case that may bring someone to justice. It may take some effort to note the things identified in the following paragraphs while they are occurring, but these observations may be useful during the size-up process, regardless of the cause of the fire, and they may be very useful if it is later determined that the fire was the result of arson. By performing a fire **scene assessment**, the investigator can define the extent of the scene. This assessment is a complete walk-around of the scene to determine the extent of damage. Most investigators prefer to start with the area that has the least damage and finish at the area with the most damage. This should be done on every fire scene to protect the safety of personnel and to protect possible evidence. The investigator should adjust the scene perimeter based on the site examination. In many situations, it is very difficult to put these facts together after the fire is over. With that thought in mind, let us look at the things that the first responders may be able to note.

Observations of First Responders

Firefighters, especially those arriving first at the scene of a working fire, have a unique opportunity to make observations regarding a fire's behavior. Although their primary purpose is to save lives, protect property, and extinguish the fire, a moment of time making observations about the fire and the surrounding conditions may help in both the suppression effort and in the fire investigation that follows. Since even the best of fire investigators is usually not present during these first few minutes, and since you cannot recreate these condition once an investigator is present, it is important that you retain enough of the details so that they can be recalled and shared with the investigator and included in your fire report.

First responders have unique opportunities for making observations while en route to the fire, upon arrival, during the initial stages of suppression, and during overhaul activities. Some information can even be obtained during the first notification of the fire to the fire department.

NOTE

First responders have unique opportunities for making observations while en route to the fire, upon arrival, during the initial stages of suppression, and during overhaul activities.

During the Initial Notification

Most fires are reported today by telephone. The first clue of a fire is usually a citizen's call to the fire department or emergency dispatch center. The caller may be an occupant or someone who just noticed the fire. Multiple calls pertaining to the same event indicate that it is probably significant. The information from the caller, as well as the information previously gathered in conjunction with the company's preplanning efforts, is integrated to form the basis for the response action. The call taker should get the caller's name, phone number, and location; you may want to talk to the caller later.

While en Route

Weather, traffic conditions, time of day, and other factors should be noted and also incorporated into the response action. All of this information is important for the responders. Some of it may become important as you try to determine the cause of the fire.

Upon Arrival

The first emergency personnel to respond to the scene of a fire have a unique opportunity to help determine the origin and cause of the fire. These first responders can observe the amount of smoke and fire showing, the condition of the building, and activities in the immediate vicinity. The first responders may be able to note who is present at the scene upon their arrival, especially someone who is alone or demonstrating unusual behavior.

Upon arrival at the scene of a fire, you would expect people to be exiting the building and showing concern for those who may remain in the building. You should note the appearance and mental state of any occupants of the building. In most cases one or more of the occupants will meet the fire officer. You would expect people to be dressed appropriately for the time of day. Finding the occupant in nightclothes at 4:00 AM is normal; finding the occupant of a residence dressed for travel at that hour should raise some questions.

Fires also generally attract spectators. For all sorts of reasons, people will stop and watch a fire, and many will linger to watch the fire department's suppression activities. This is normal behavior, too. We usually do not have time to observe the crowd of spectators that gather at the scene of a fire, but noting any who may be demonstrating unusual behavior may be very useful. Concern and curiosity are normal behaviors, but any unusual behavior, especially indications of excitement or pleasure should be noted. If there has been a series of fires in a neighborhood and a particular person has been present at several of these events, this person's appearance should be carefully noted and pointed out to the investigator or police.

OFFICER ADVICE

Noting the appearance of the fire itself is one of the most important observations that first responders can make.

We should also be concerned about those who might be leaving. It is difficult to note details of a person's appearance, activity, or even a description of the person's vehicle during this time, but simply noting that the event occurred has its value. Persons who may have been working on the premises should be carefully noted. Construction activity, especially in cold weather, may have been the cause of the fire. Careful observations help you note and later recall the details that may be useful to the fire investigator (see **Figure 11-5**).

Noting the appearance of the fire itself is one of the most important observations that first responders can make. Information regarding the color and location of flames, and the color, quantity, and location of smoke provides valuable information for both fire suppression and for fire investigation. if and where the fire has **vented** itself, the number of rooms and floors involved, whether it was one fire or perhaps several separate fires burning simultaneously, any evidence of collapse or pending collapse, and the presence of explosions or other unusual sounds. Pay particular attention to the presence of sounds from any fire alarm systems.

The appearance of the property itself should be noted. The question "Was the building occupied?" really has two meanings. Clearer understanding might occur if we were to ask, "Was the building in use at the time of the fire?" and "Was anyone inside the building at the time of the fire?"

Unoccupied buildings, that is, those abandoned by their owner, become a frequent shelter for homeless persons. In this case, the answer to the first question (Was the building in use?) might be no, while the answer to the second question (Was anyone inside?) might be yes. These "unoccupied buildings" are often the scene of significant fires, especially during colder weather, when the occupants build fires in the structures to keep warm. A homeless person is still a human being.

A completely separate issue deals with the security of the building. For buildings in which there were people present at the time the fire ignited, you would expect the doors, and possibly even the windows, to be unlocked and maybe even open, depending upon the weather. Conversely, for buildings that are in use, but unoccupied at the time of the fire, you would expect to find these openings secured. Where automatic fire doors are installed, you should expect to find them closed. Deviations from these conditions should definitely be noted and reported both to the incident commander and the fire investigator.

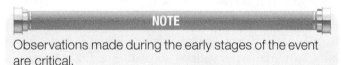

NOTE

Observations made during the early stages of the event are critical.

The observations made during the early stages of the event are critical. Whether it be food on the stove or a fully involved structure as a result of arson, there are always indicators present that help in the investigation.

During the Initial Stages of Suppression

Once entry is gained and fire suppression activity is started, there is often another opportunity for the first responders to note the conditions present. Any unusual fire conditions, including the fire's behavior during fire suppression, should be the cause of concern both during fire suppression and later during the fire investigation. From experience, you learn how fire acts when appropriate fire suppression actions

FIGURE 11-5 Fire suppression personnel should be prepared to describe their observations and activities to the fire investigator.

are started. In all properties, you should listen and note the sound of the fire alarm system. If there was a sprinkler system in the building, you should be able to determine if the sprinkler system was operating and if the sprinklers were able to help control the fire.

During Overhaul

For first responders, overhaul activities may present the best time to determine the cause of the fire. Although the purpose of overhaul is to be sure that the fire is completely extinguished, at the same time you can search for and gather clues to help determine the cause of the fire. It is important for those conducting the fire investigation to maintain a good command organization and to wear proper protective clothing during overhaul operation, but most of the urgency needed for effective fire suppression activities is no longer required. With a more deliberate pace, you can work safely, carefully, and systematically to overhaul and look for the origin and cause of the fire.

During overhaul, it may be necessary to pull down ceilings and wall materials. If present, the investigator may want to look around before this activity takes place. When this occurs, the overall investigation may be completed sooner. Any effort to move and remove the contents of the room to facilitate overhaul impedes the investigator's efforts. If there is time to look at this material before it is moved and discarded, the investigator will have a much better picture of the fire scene and may see evidence that would otherwise be missed. There are many different types of evidence. However, they can often be placed into two distinct categories. The first category includes non-fire evidence (e.g., victims, blood stains, forcible entry marks). The second category is known as fire-related evidence (e.g., unburned ignition sources, fuel containers, ignitable liquids or solids in and around the area). In the final analysis, if you can help the investigator save some time and effort, everyone may be able to go home a little sooner. If circumstances preclude an investigator being at the scene, then the company officer and others present must do some of the work of the investigator. The first thing to look for might be whether any items appear to be missing. In most cases, pictures, personal property, and other items that one would expect to find in the places where you would expect to find them should still be in place. Fire crews may divide spoilage piles in separate location if there is concern that evidence may be lost if overhaul must continue in order to extinguish the fire before the arrival of the investigator. Overhaul crews will need to try to preserve possible evidences from destruction even if items are removed. Company officers should be able to identify the different

types of evidence and have knowledge of how to preserve the evidence if there is a delayed response for the investigator.

The chain of custody of evidence must be preserved and documented should the fire become a crime. The chronological order in which evidence is moved starts with the first responders. Company officers should document the order in which content is removed from the structure, who removed the items, and how the evidence was transferred. Strict control must be maintained until the evidence is transferred to the investigator, and the process in which the evidence is transferred will also need to be documented. Often, the chain of custody of evidence is called into question during criminal trials.

Chain of custody refers to the chronological documentation, and/or *paper trail*, showing the seizure, custody, control, transfer, analysis, and disposition of *evidence*.

> **NOTE**
>
> Cooperation during the investigation and overhaul phases provides better results and may allow everyone to go home sooner.

Sometimes the fire investigator may be several hours away. This situation presents the fire department with some very difficult choices. Although the fire suppression personnel may be most concerned with completing the overhaul of the fire and getting back into service, they should realize that their overhaul activities will likely destroy most of the evidence the investigator needs to determine the cause of the fire. Therefore, it is essential that firefighters and fire officers secure the scene. They should restrict entry into any fire affected structure and they should secure the structure against weather and any non authorized entry by posting personnel at each entry point.

Firefighters should also be able to note the construction features that allowed for fire, heat, and smoke to travel through the structure. Open stairways, open windows, laundry chutes, and concealed voids all present opportunities for fire travel. Where burn patterns are apparent, do they indicate normal fire spread? Normally, these patterns help determine the nature of the fire as well as its path of travel from the point of origin and preferably photograph as much evidence of the fire's behavior as possible. These methodical and careful observations will assist in making a complete investigation.

The information gathered at the time of the alarm, observations made during the response, size-up, and suppression activities and post fire information can be combined to help determine the cause of the fire.

FIRST RESPONDERS' OPPORTUNITIES FOR MAKING OBSERVATIONS

During the initial notification of the event:

- Identification and location of the caller
- Background noises

While en route to the fire:

- Additional information that may be provided by the caller, weather, time of day, and so forth
- Delays due to highway construction, trains, and so forth

Upon arrival:

- Any persons present and what they are doing
- Any vehicles present and whether they are leaving
- The fire conditions: location and intensity of the fire, color of the flames and smoke

During size-up:

- Operation of any alarm or suppression equipment
- Any unusual observations
- Methods of escape of any occupants

During the initial stages of suppression:

- Were furnishings and inventory in place?
- Was the fire alarm sounding?
- Was the sprinkler system operating?

During overhaul activities:

- Recheck items noted earlier
- Determine origin and cause
- Determine path of fire travel

Each has an important part in the overall investigation, and for the most part, only first responders are in a position to gather the information.

Legal Issues for Fire Investigations

As company officers, you should know several important concepts about the legal issues of fire investigation. Remembering these will help keep you out of trouble and, at the same time, reduce the chance that you interfere with the fire investigation. This information may seem to be more appropriate for the fire investigator, but every firefighter and fire officer should realize that actions taken before the investigator arrives are significant in setting the stage for any possible criminal action that may follow.

OFFICER ADVICE

Every firefighter and fire officer should realize that actions taken before the investigator arrives are significant in setting the stage for any possible criminal action that may follow.

Let us briefly review the legal rights of firefighters to enter the property of others. The Fourth Amendment of the U.S. Constitution and other laws hold that all citizens have the right to be secure in persons, houses, papers and effects, against unreasonable searches and seizures. The fire service has some rights too. In general, members of the fire service have the right to enter the property of another to extinguish a fire. The logic of this law is that the fire may spread to other property and that the concerns of the community and its citizens outweigh the rights of any one individual. This same right generally means that the fire department has the right, within reason, to enter an adjacent property as needed to check for extension or facilitate fire extinguishment.

In addition, the fire department also has a responsibility to determine the origin and cause of every fire. As a result, conflicts may arise between the owners' constitutional rights of security in their property, and the community's needs, represented by the legal responsibilities of the fire department to conduct an investigation. These conflicting issues raise significant legal questions:

- How long can the fire department remain on the property after the fire is extinguished?
- What is required for the fire department to reenter private property to start or conclude a fire investigation?

More than 20 years ago, a fire in Michigan provided us with guidelines to help answer these questions. To briefly summarize the events, the fire started about midnight in January. At 2:00 AM, the fire chief arrived and found evidence that the fire had been intentionally set. Efforts to conduct a thorough investigation at that hour were hampered by darkness, cold weather, poor visibility, and other dangerous conditions. At about 4:00 AM, and with suppression activity complete, the fire department departed the scene.

Around 8:00 AM that same morning, the assistant chief, who was the primary fire investigator for the department, went to the property to determine the cause of the fire. Several weeks later, a representative of the state police entered the property to take photographs and seize evidence. Other searches followed. The owner was eventually tried for arson. The owner tried to suppress the evidence, saying that the searches were in violation of the protection provided by the U.S. Constitution. The case eventually made its way to the U.S. Supreme Court.

Three Legal Ways to Search the Property of Another

Just as any firefighter has the right to enter the property of another to suppress fire, the person in charge has the right to enter the property to determine the cause of the fire. But as the Supreme Court noted, there are conditions that must be satisfied.

There are three methods by which the fire department may reenter the property after the fire is extinguished. The first is with the owner's consent. In many cases, simply explaining what needs to be done and asking permission to enter is all that is required to allow you to reenter the property and conduct an investigation. It is recommended that this consent be documented with a "consent to search" form. Most fire departments have a standard form for this purpose.

Another alternative is to obtain an **administrative search warrant** that the stated intent is to determine the origin and cause of the fire, an **administrative process**. So far, there is no reason to presume any criminal action. On the other hand, suppose that the evidence gathered during the department's initial investigation, or at a subsequent investigation, does in fact suggest that there is a possibility of intent.

Now the rules change. In this situation, the fire department will need a **search warrant**. There is reason to suspect that a crime has occurred, and this is now a criminal process. A criminal search warrant requires **probable cause** and spells out the conditions of the search, the area or areas that may be searched, and the material that may be seized. Evidence obtained under other conditions is inadmissible in court.

Hopefully at some point along this path, the fire department has been able to obtain the assistance of a fire investigator or a police officer who is more familiar with this process (see **Figure 11-6**). The point of our brief discussion here is to make you aware of these basic rules so that your actions do not jeopardize the rights of the citizens or the opportunity to obtain legal evidence that might be used in court. As company officers, you should remember that you have a right to complete an investigation and that you should have continuous presence on the property until that investigation is complete or an investigator has released you. Maintaining security of the scene is important throughout these activities (see **Figure 11-7**).

NOTE

As company officers, you should remember that you have a right to complete an investigation and that you should have continuous presence on the property until that investigation is complete or an investigator has released you.

MICHIGAN v. TYLER

In 1978, the U.S. Supreme Court laid down the rules for post fire searches in a landmark case known as *Michigan v. Tyler*. In summary, the Court ruled that the fire department has the right to enter property under emergency conditions and has the right to remain on the property until the emergency is over to extinguish the fire.

The court further ruled that the fire department could remain on the scene for a reasonable time for purposes of determining the origin and cause of the fire. Once the initial stage of the investigation is complete, the fire department should leave the property.

If a member of the fire department wishes to return to the property, it should be under the auspices of an administrative search warrant or a criminal search warrant, as appropriate, or with the freely given consent of the owner. Evidence obtained under other conditions is in admissible.

As we can see, the court said that the fire department has a legal right to enter private property to extinguish the fire and determine its cause. And as we can also see, the court said that citizens, under the Fourth Amendment of the U.S. Constitution, have a right to privacy in their property, and that evidence obtained during warrantless searches will not be admissible in an effort to convict someone of arson. The major function of a warrant is to provide property owners with sufficient information to reassure them that the law enforcement agency's (in this case the fire department) entry onto their property is legal.

In summary, the court held that entry into a burning building for fire suppression is acceptable, and that once inside, firefighters have the right to determine the cause of the fire and to seize evidence that is within plain view. Evidence obtained under these conditions is admissible.

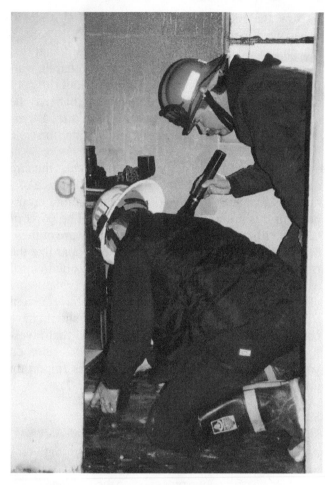

FIGURE 11-6 Fire investigators must work carefully to locate, gather, and preserve evidence.

FIGURE 11-7 Maintaining security of the fire scene is important for the safety of the firefighter and for the preservation of evidence.

Recording the Information

It is important to record the observations made at a fire. We do not expect to be questioned about every fire we attend; however, we might be questioned about an occasional fire. Unfortunately, we do not know in advance which one it will be, so it would be wise to make a few notes about every fire.

Why would you be questioned about the fire or about your actions at a particular fire? The owners might be interested in your actions. The insurance company might be interested in what you saw upon arrival and what you did. Certainly, if there is any evidence of arson, you will be asked about what you saw and what you did. Months may pass before someone starts asking you to recall the events of a particular fire. In the meantime, you may have been to other fires in similar structures. How will you be able to recall the facts regarding a particular event?

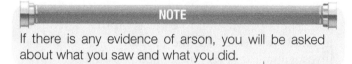

NOTE

If there is any evidence of arson, you will be asked about what you saw and what you did.

First responders should note the address of the fire, the type of building, the location and intensity of the fire, the color and intensity of the smoke, and action taken. Your report should also note if anything was unusual within the structure. How did your firefighters get in, and what did they see after they gained access? These are simple questions that can be easily answered shortly after the call is complete. Trying to recall the events of a particular fire a few weeks or months later may be impossible.

In most fires, there are usual answers to these questions. In those cases where arson is involved, the answers may suggest unusual conditions that warrant closer examination. Some good questions to assist in the process follow:

- Who discovered the fire?
- When was the fire first discovered?
- When was the fire reported?
- Who provided the first report of the fire?
- Who extinguished the fire?
- Who provided scene security?
- Who has pertinent knowledge regarding the fire?
- Who had a motive for setting the fire?
- What happened during the fire?
- What actions were taken by firefighters?
- What damage occurred?
- What do the witnesses know?
- What evidence was found?
- Where did the fire start?
- Where did the fire travel?
- Where were the occupants?

The Fire Report

Careful recording of the observations made during the course of any fire may be useful later, especially if the circumstances led to a trial. The information should include recorded observations regarding the response, size-up, suppression, and overhaul phases of the firefighting activity. Many of the questions will be prompted by the requirements to complete an official report. For most departments, this is on an official form that prompts many of these questions.

However, the most important part, and for some the most difficult part, is the written narrative statement. It should answer any questions not required in the form, explain any unusual circumstances, and otherwise complete any missing information. It is generally acceptable to express these statements using a personal style. "Upon arrival, I saw . . . Firefighter Smith and I forced the front door. We advanced a 1 3/4-inch hoseline. . . ." Usually, it is easier to write and easier to read this sort of narrative if it is written in the first person and if the facts are reported in the order in which the events occurred.

When writing your report, you should avoid using terminology that might not be fully understood by people who are not in the fire service. Do not make assumptions and avoid opinions. If you do not know it is factual do not put it in your report or verify it before you do. One error in your report can cause the entire report to be suspect. Also, be aware that not only can your reports be called into court as evidence, but all of your notess related to the incident can as well. So only write down accurate, factual information.

After you have written the first draft, proofread your work, and clean up the grammar and spelling errors. Drawings are also useful and should be attached to the report. Complete the fire incident report as fully and as accurately as possible (see **Figure 11-8**). When you are writing this report, you have no knowledge of its potential use. It is wise to assume that every report could be reviewed by the department or used in court.

Documenting the Fire Scene

The fire scene can be documented by notes, diagrams, and photographs. Notes should be consistent, impersonal, accurate, time and date stamped, and numbered. We have already discussed the fire report. Notes are also important. Written notes can provide as much detail as necessary to help recall the facts at a later time. Some investigators use a tape recorder to facilitate the note-taking process, transcribing the recorded information to written documentation as soon as possible.

FIGURE 11-8   Writing a complete and accurate fire report is an important part of the company officer's job.

Photography provides a visual image of the fire scene. Photographs can be used to assist the investigator in recalling what was seen and may be used as evidence in court. If possible, the photographs should include scenes taken during fire suppression operations and bystanders at the scene. As soon as the fire is over and conditions permit, an organized series of pictures should be taken to document the entire scene.

> **OFFICER ADVICE**
>
> An organized series of pictures should be taken to document the entire scene.

Just as the investigator follows a path toward the point of origin, so should the photographer follow a path that starts with external views, followed by internal views. The pictures should tell a story and work in a logical way toward the point of origin. This can also be done with a video recording. Regardless of the cause of the fire, or the quality of the written report, a picture may well be worth a thousand words if questions later arise regarding the department's actions, where items were located, the extent of the damage, evidence of other crimes, and so forth. Diagrams are also useful in documenting the fire scene. Here, the entire fire scene is captured on one piece of paper. Diagrams can be useful when used in conjunction with photographs, especially if the location of each photograph is noted on the diagram as the picture is being taken.

Regardless of the method by which the documentation is obtained, it is important that the person be competent in using the equipment involved. Competency does not imply that one is an expert; rather that one knows how to operate the equipment and has a reasonable expectation of getting the desired results. Photographing a fire scene can be quite difficult. Challenges include the lack of good lighting, poor visibility, and the tendency of everything involved in the fire to look black. All of these can be easily overcome with training and experience.

Fire investigators take care of these activities when they are on the scene, but in many situations, you may not have the luxury of an investigator. The fire may be considered accidental or too small to justify an official investigation, or an investigator may not be readily available. In such cases, the origin and cause of the fire must be investigated and documented by you, the company officer.

OFFICER ADVICE

The fire may be considered accidental or too small to justify an official investigation, or an investigator may not be readily available. In such cases, the origin and cause of the fire must be investigated and documented by you, the company officer.

Courtroom Demeanor

In cases involving arson, it is not unusual for fire officers and even firefighters of the first-arriving company to be asked to report their observations in the courtroom. Since the trial might take place many months after the actual fire, it is important to review the facts prior to appearing in court. In this case, the photographs, diagrams, and notes that you made during and shortly after the fire prove their value. Review questions with the prosecuting attorney and anticipate defense questions prior to the court date.

The choice of wearing a uniform or civilian attire is often decided by the custom of the local court. Find out what is appropriate. Regardless of the attire, it is important that you appear professional. This means that you are well groomed in every regard from having a neat haircut to having well-shined shoes. The following are also suggested:

- Sit upright with your feet flat on the floor.
- Keep hands in a natural position and try to avoid gesturing.
- Speak to the jury if you can, rather than to the attorney.
- Refer to the defendant courteously as "the defendant" or Mr. or Ms. _____.
- Answer questions with confidence gained by being familiar with the facts.
- Take time to be sure you understand the question. If unsure of the questions, ask to have it repeated or rephrased. After all, you are not on trial.
- If the opposing attorney objects to a question, wait until the judge rules on the question before answering.
- Tell the truth at all times.
- If you do not know the answer, say so.
- Do not volunteer information.

FIRE INVESTIGATION

George Lucia, Sr., Fire Marshall,
City of Vista Fire Department, Vista,
California

It was a cold January evening, about 5 PM, when the call came in about a passenger vehicle burning behind a local shopping center. The late model BMW was abandoned; the company officer on duty noticed that the seat was burned and that there was a plastic container on it. Being a bit suspicious, he called in a trained company officer and fire investigator, George Lucia, to the scene within 5 minutes.

When Lucia arrived, he secured the scene. He walked around with the company officer and a police detective at his side. Witnesses in the area were interviewed. Involving suppression crews allows them to understand his needs as an investigator.

The area of origin was determined to be the right front passenger seat. A container of ignitable liquid was found on the front floor. All areas were photographed and samples were taken, documented, and secured for testing. He also got the firefighters involved, showed them the evidence and told them how he could look at the evidence and start trying to determine how the fire started. A search of the car to ensure no fire extension found containers of auto coolant in the trunk, some open and used. The car displayed collision damage to the front grill. This damage appeared old and had some type of hair embedded in the grill. This hair was also removed for testing. Auto window glass was found under the front seat area indicating a prior damage incident. The vehicle was removed to the impound yard for secure storage.

Initially, it seemed to Lucia that maybe this was a case of kids stealing a car and taking it for a joy ride, but he had learned not to take anything as a slam dunk case. And some of the factors—for example, that this happened in daylight on an icy day when a joy ride would not have been wise—made him think twice. As a company officer, he had learned the importance of a broad evaluation.

At 6 PM, the owner reported the vehicle stolen from a shopping center 5 miles away from where it was found. The owner was asked to come in for an interview with fire investigators and local police. He stated that he was at the mall and came out to find his car gone. He also stated that there was no damage to his car when he parked it and that he knew of no reason why someone would take his property.

Further investigation found that the owner had left his employment early the day of the incident to take care of some "personal business." His insurance carrier forwarded records of previous claims for theft of a car stereo system, and the financial institution forwarded records of a delinquent payment history for this car and owner. Lab results found the ignitable liquid to be gasoline and a match to the interior seat samples. Lab results found clear fingerprints on the ignitable liquid container and the coolant containers in the trunk. The lab also found the hair samples from the grill area to be from a deer.

The owner was called back for a follow-up interview, in which he was apprised of the evidence and lab results. Overwhelmed with the facts, the owner admitted that he had crashed his car a week ago and had impacted a deer on the road. The damage had caused a cooling system leak, causing overheating and the need to add additional fluid to the cooling system. The owner stated that because of hard financial times, being behind in his payments, and the new damage to the car, he planned the theft and arson to gain the insurance payment. A plea agreement was reached for charges of arson, and the insurance company levied an administrative penalty to the policyholder for insurance fraud. Lucia credits the success of the case to good cooperation between the suppression companies, the fire marshal's office, the police department, and the insurance company. Lucia adds an important note—if the fire officer had not taken note of the situation and called the investigator, the owner would have gotten away with his scam.

Questions

1. What type of fire is this? What was the motive of the owner?

2. What evidence did the suppression officer have to arouse his suspicions?

3. What evidence did the fire investigator have to arouse his suspicions?

4. What can firefighters learn from participating in such investigations?

5. How would this situation be handled in your jurisdiction?

LESSONS LEARNED

As company officers, you should be aware of the causes of fire and the impact of arson in your community. You should be able to determine the cause of most fires and recognize the signs of arson. You should be able to conduct a basic investigation to properly determine and document the origin and cause of the fire. You should understand both the rights of the property owner and the requirements for conducting a lawful investigation.

This chapter has only scratched the surface of the topic of fire investigation. It is an area that deserves the attention of every fire professional. You are encouraged to read additional information on this important topic and to take training and educational programs whenever the opportunity presents itself.

Training for firefighters and fire officers should include courses in determining the origin and cause of fires and arson recognition. (NFPA standards for both firefighter and fire officer certification include this requirement.) The fire science program at most community colleges includes a course in fire investigation. Many states offer training for firefighters and officers and certification programs for fire investigators.

The National Fire Academy, recognizing the need for additional training in these topics, provides a series of training programs to help you learn about fire investigation and arson recognition. A listing of these programs, starting with their basic awareness-level training program, follows:

Course Name	Course Length
Arson Detection for First Responders	2 days
Fire Cause Determination for Company Officers	6 days
Fire/Arson Investigation	2 weeks
Management for Arson Prevention and Control	2 weeks

Contact your training officer or state fire training director for additional information about these training programs.

Publications Available from the U.S. Fire Administration

Arson in the United States (FA 174)
Rural Arson Control (FA 87)

These and many other relevant reports are listed on USFA's Web site. Go to www.FEMA.gov. All of these reports can be ordered free of charges and many can be downloaded.

KEY TERMS

accidental   refers to those fires that are the result of unplanned or unintentional events

administrative process   a body of law that creates public regulatory agencies and defines their powers and duties

administrative search warrant   a written order issued by a court specifying the place to be searched and the reason for the search

arson   the intentional setting of a fire where there is criminal intent

energy   the capacity to do work; in the fire triangle, heat represents energy

motive   the goal or object of one's actions; the reason one sets a fire

chain of custody   refers to the chronological documentation, and/or *paper trail*, showing the seizure, custody, control, transfer, analysis, and disposition of *evidence*

oxidation   a chemical reaction in which oxygen combines with other substances causing fire, explosions, and rust

probable cause   a reasonable cause for belief in the existence of facts

search warrant   legal writ issued by a judge, magistrate, or other legal officer that directs certain law enforcement officers to conduct a search of certain property for certain things or persons, and if found, to bring them to court

vented   opened to the atmosphere by the fire burning through windows or walls through which heat and fire byproducts are released, and through which fresh air may enter

REVIEW QUESTIONS

1. What are the most common causes of fire?
2. What is arson?
3. What is meant by the term "motive"?
4. What are some common motives for arson fires?
5. What information should be noted by the first-arriving personnel at the scene of any fire?
6. What information should be noted by fire personnel during fire suppression operations?
7. What information should be noted during overhaul?
8. What are the ways you can enter private property to conduct a fire investigation
9. Why is it important to complete a fire incident report for every event?
10. What is your role as the company officer in fire cause determination?
11. How is evidence maintained and documented?

DISCUSSION QUESTIONS

1. Why should firefighters be concerned about intentionally set fires?
2. What causes people to set fires?
3. What is the cost of these intentionally set fires in this country?
4. What is the extent of these intentionally set fires in your community?
5. What can be done to reduce the incidence of such fires in your community?
6. Explain how a fire should be documented.
7. What legal considerations must be kept in mind while searching for evidence of the origin and cause of a fire?
8. Discuss the impact of the *Michigan v. Tyler* case as it applies to the actions that should be taken by the incident commander during and following a fire?
9. Under what circumstances would you call a fire investigator?
10. As a company officer, what can you do to reduce the number of fires in your response area?
11. What does chain of custody mean?

ENDNOTES

1. *Fire Losses in the United States 2007*. Reprinted with permission from the NFPA Copyright © 2008, National Fire Protection Association.
2. *Fire in the United States*. 14th ed. Emmitsburg, MD: U.S. Fire Administration, 2007, p. 3.
3. *A Needs Assessment of the U.S. Fire Service*. Emmitsburg, MD: U.S. Fire Administration, 2002, 53.

ADDITIONAL RESOURCES

Brannigan, Francis L. "Effect of Building Construction and Fire Protection Systems." In *Fire Protection Handbook*. 19th ed. Edited by Arthur E. Cote. Quincy, MA: National Fire Protection Association, 2003.

Codding, George and John Boyn. "Incident Commander's Guide to Preserving Evidence." *Fire Engineering*, August 2006.

Coleman, John. "SCBA Use During Overhaul." *Fire Engineering*, September 2007.

Custer, Richard P. "Fire Loss Investigation." In *Fire Protection Handbook*. 19th ed. Edited by Arthur E. Cote. Quincy, MA: National Fire Protection Association, 2003.

Dehann, John D. *Kirk's Fire Investigation*. 5th ed. Saddle Brook, NJ: Prentice Hall, 2002.

Donahue, Michael. "Fire Investigations Anonymous." *Firehouse*, April 2007.

Miller, Geoff. "Fire Cause Determination." In *The Firefighter's Handbook*. Clifton Park, NY: Delmar Publishers, 1999.

NFPA 921. *Guide to Fire and Explosion Investigation*. Current ed. Quincy, MA: National Fire Protection Association.

Watts, John. M. Jr. "Fundamentals of Fire-Safe Building Design." In *Fire Protection Handbook*. 19th ed. Edited by Arthur E. Cote. Quincy, MA: National Fire Protection Association, 2003.

Zerbe, Jess. "Determining the Fire's Point of Origin. *Fire Engineering*, September 2007.

12

The Company Officer's Role in Planning and Readiness

Although I had always associated preplanning with structural fire responses to dwellings or commercial structures, I learned a valuable lesson about preplanning and its tremendous value on September 11, 2001.

In the mid-summer of 2001, the Washington metropolitan area was facing the possibility of large drains on emergency service resources because of the World Trade Organization (WTO) protests in Washington, DC, scheduled for mid-September. There were several meetings and planning initiatives that took place over that summer to preplan this event so that jurisdictions would not be overwhelmed, safety would be paramount, and mass chaos would not ensue.

The results of these planning efforts were not finalized, but there were high-level discussions and decisions about concentric forward staging areas throughout the region, responses based on task forces and functional groupings, and simultaneous activations of Emergency Operating Centers (EOCs).

In addition, August of 2001 saw parts of the metro region attending to a large mass gathering that turned into a mass casualty incident. This also required preplanning efforts on the part of many of the jurisdictions and prompted debriefings that attempted to solve perceived flaws in the acquisition of logistics and interagency cooperation and communication.

Then, out of the blue, the events of September 11, 2001, threw the Washington metropolitan region a more outrageous challenge than could ever have been expected. The Pentagon incident involved all the surrounding jurisdictions and required a massive deployment of resources and coordinated planning efforts.

The framework for a major multijurisdictional response had been laid out because of the WTO planning effort; therefore, the jurisdictions latched on to these preliminary plans to facilitate the Pentagon response. The forward staging areas were used in Loudoun, Fairfax, and Arlington Counties, the task force concept was put in place, and EOCs across the region were activated early to manage the mobilization of resources from the numerous jurisdictions. I am certain that the lessons we learned in my county from the mass casualty incident also assisted us in responding that day as well.

I think every member of the fire and emergency services had lessons that they learned from September 11, 2001. I discovered the great importance of preplanning and the much more far-reaching abilities of preplanning beyond water supply locations at commercial buildings or subdivision street layouts. In this case, a little planning went a long way towards the operational capabilities of our region. The preplanning processes that took place across the jurisdictions facilitated what I believe to be the largest coordinated response of resources in the Washington metropolitan area.

—David Short, Senior Training Analyst, MPRI, and Assistant Chief, Sterling
Volunteer Fire Department, Sterling, Virginia

LEARNING OBJECTIVES

After completing this chapter the reader should be able to:

12-1 Explain the value of fire inspection and fire incident reports regarding the development of pre-incident plans or surveys.

12-2 Identify the policies and procedures regarding the preparation and/or modification of pre-incident plans.

12-3 Describe the information that must be collected for a pre-incident plan.

12-4 List common human failures that may be identified during the pre-planning process.

12-5 Describe how fire detection alarm and suppression systems affect a pre-incident plan.

12-6 Identify local water supply sources.

12-7 Explain how codes, ordinances and standards contribute to effective pre-incident plans.

12-8 Describe the types of services or equipment that can be supplied by the agencies.

12-9 Describe potential benefits from cooperative ventures with selected community agencies.

12-10 List allied agencies and potential contributions to an incident or issue.

12-11 Identify the proper approval process for the final pre-incident plan.

12-12 Describe the elements of standard operating procedures concerning emergency operations.

12-13 Describe the elements of standard operating procedures concerning dispatch of emergency units.

12-14 Describe the elements of standard operating procedures concerning tactics and emergency operations.

12-15 Describe the elements of standard operating procedures concerning customer service.

12-16 Select the instructional methods necessary to meet the needs of the unit members.

12-17 Describe the procedure for issuing multiple tasks to unit members for a training exercise.

12-18 Identify the company training needs to select the proper evolution to be completed.

12-19 Describe the planning and preparation procedures so that they comply with departmental and applicable NFPA standards.

12-20 Identify the training policies and procedures for the department.

*The FO I and II levels, as defined by the NFPA 1021, the *Standard for Fire Officer Professional Qualifications*, are identified in different colors: FO I = black, FO II = red.

INTRODUCTION

Managing emergencies is part of the fire department's business. Up to now in this book, we have been looking at the administrative activities that make up your day as a company officer. We now move to managing emergencies.

Emergency management is a combination of anticipation, preparation, and action. For emergency service providers, anticipation is represented by pre-planning; preparation is, in part, based in training. Adequate preplanning and training are important activities for fire departments and the companies that make up fire departments. As this book focuses on the role of the company officer, the emphasis of this chapter is on your role in preplanning and training at the company level. We will save the action part for Chapter 13.

Today, the emphasis in emergency management is on all-risk preparation. Certainly the events of September 11, 2001, brought national attention to the role of emergency responders. It also brought to light the need for fire departments and other emergency responders to be better prepared for large-scale events. Communities expect their fire and police departments to be prepared to handle not only the normal daily events but also to be prepared to deal with hazardous materials incidents, floods, tornados, hurricanes, earthquakes, and building collapse. In recent years, a new emergency situation has been added—response to terrorist activities.

MANAGING EMERGENCIES

Consider these words of warning: "As a result of recent events, significant threats over the past few years, and the increased ability and proliferation of nuclear, biological, and chemical materials, there is an increasing concern for the potential of terrorist incidents occurring within the United States." This sounds like something that might have been said last week. Actually, these words were written nearly 10 years ago by Kay Goss, director for preparedness, training, and exercises at FEMA. Her remarks were based on the then-recent events that included the bombings of the New York World Trade Center (1993) and the Alfred P. Murrah Federal Building in Oklahoma

City (1995). But other events had taken their toll as well. In fact, during the 10 years that preceded the attack on the World Trade Center in 1993, there were over 18,000 explosive and incendiary bombing incidents within the United States. These attacks resulted in 11,178 actual explosions, resulting in 236 deaths, 3,215 injuries, and over $500 million in damages.

If the message had not been heard before the events of September 11, certainly that day was a wake-up call for many individuals and organizations. Many suddenly realized that what others were saying about the threat of terrorism was quite real. Suddenly, terrorism was a threat to all of us, not something that happened in other places. Fire stations and other public places became a little less accessible. Security was increased at airports and seaports, and at other vital installations. A new government agency, the Department of Homeland Security, was established to better coordinate our nation's security efforts.

All of this may seem as if it is a long way from your role as the company officer in preparation and training. Although there are national and state activities that deal with both preparation and response, everyone recognizes that the first responders on the scene of any emergency include representatives of the local fire and police departments. The local fire department will, in all probability, be initially represented by you—the company officer.

Regardless of the nature of the emergency, there is a need for preplanning and preparation at all levels. The National Fire Academy and the Emergency Management Institute, both part of FEMA and both located at Emmitsburg, Maryland, have led the way in helping train representatives of federal, state, and local governments. Much of their training is based on the concept that comprehensive emergency management is comprised of four activities. These activities are common to all types of events, ranging from weather-related disasters to terrorist attacks. These four activities can be summarized as follows:

1. *Mitigation.* Mitigation refers to the activities that actually eliminate or reduce the chance of the occurrence of an event or the effects of the event if it does occur. For example, building dikes to keep out flood water would reduce the chance of occurrence, while eliminating the use of the flood plain for residential or commercial purposes would reduce the effect of the flood should one occur.

2. *Preparedness.* Preparedness includes the preplanning activity on how to respond and the effort to increase resources available to respond effectively. To respond effectively, local jurisdictions must have an emergency plan, must have trained personnel, and must have the necessary equipment to respond to anticipated emergencies. The purpose of this activity is to help save lives and minimize damage by preparing emergency responders for their role.

PRE-INCIDENT PLANNING DOES WORK

A review of the emergency response to an Amtrak derailment in Iowa showed the benefits of the all-hazards approach to incident planning.[1] In this case, the local responders had not prepared specifically for a train derailment, but because they had planned for other types of large-scale events, they were much better prepared than most for the derailment of a passenger train in their community.

The accident occurred at night in a remote area during cold weather. Many were injured. All aboard were suddenly plunged into a cold, dark, and unfamiliar condition. The responders overcame these challenges and rescued 240 people from the train, nearly 120 of them with injuries. Because of the planning and preparation, many things about the rescue went very well.

In spite of their planning, the rescuers ran into several challenges that frequently confront responders at any large-scale event:

■ The volume of emergency communications traffic quickly overwhelmed both the department's radios and the local 911 telephone system, and cellular phones were ineffective. Because of the vital role in coordinating emergency response activities, every possible effort must be made to assure that there will be adequate effective communications channels for emergency response personnel.

■ Because of the number of injured, the rescuers soon ran out of ambulances. School buses and private vehicles were pressed into service to move some of the passengers to a safe shelter. Spouses of the members of the local fire department were organized to assist with the injured at the local hospital and in providing food and shelter at the fire station.

There is a lot to be learned from this incident. In this case, a small community fire department had prepared for large-scale events. Although its focus was understandably more on tornado preparedness than on train derailments, it was ready and responded well. The rescue operation was quick, effective, and successful.

3. *Response.* Response activities are designed to provide emergency assistance to victims. In addition, response activities usually include stabilizing the event and, thus, reduce the likelihood of secondary damage. In many cases, the fire department is in the lead role here, providing both leadership and resources.

4. *Recovery.* Recovery includes both public and private efforts to restore the situation to normal. Since they are already on-scene, emergency service responders will likely be involved in the initial phases of the recovery operation. But the recovery operation may go on for months or even years. In such cases, the role of emergency responders will be gradually reduced to be taken over by other government or private organizations.

Everyday Implications

Good preplanning is essential for the large-scale events, but it is important in our everyday activities as well. Good preplanning is essential for emergency scene management. When dealing with fire situations, the plan should include information about layout, access, construction features, occupancy, and other important information needed to effectively and safely manage the event.

OFFICER ADVICE

Good pre-incident planning is essential for safe emergency scene management.

OFFICER ADVICE

Pre-incident planning should address the very topics that may likely become an issue during any significant emergency in the building.

The advantages of pre-incident planning are often overlooked. We continue to read of firefighters who are injured and killed in structural fire situations where there was no pre-incident plan, or if there was a preplan, the plan was not followed. Pre-incident planning should address the very topics that may likely become an issue during any significant emergency in the building. These include layout, contents, construction, and type and location of installed fire protection equipment. This is why inspection and incident reports are vital to the pre-incident planning development process. By evaluating these reports and adding the information to the pre-incident plan, the fire officer will have a better understanding of the building layout and contents before arrival on a scene.

The information and plan have a variety of uses. The primary purpose is to provide information during the response and initial fire fighting operations in a particular building, or other situation, but you can also use the same information for training purposes. For example, when the preplan calls for special evolutions that are not undertaken on a regular basis, this should suggest that these evolutions become a regular part of your training program. The preplan itself should be a part of the training program. A regular review of the plan by every member of every company that will respond to a working event at that address will help everyone to better remember the features and hazards of a particular address.

NOTE

Most of the material in this chapter deals with fire-related situations, in part because this book was written to address the requirements listed in NFPA 1021. You should be undertaking the same planning and training efforts for dealing with all of the emergency situations that you are likely to encounter.

Fortunately, many of the tools used in planning for fire emergencies can be used to address the other problem areas as well. As a minimum, your planning and training programs should certainly address the medical emergencies, hazardous materials events, weather events, and rescue situations that you are likely to encounter in your particular community.

PRE-INCIDENT PLANNING

Pre-incident planning and preplanning are all terms that mean essentially the same thing. **Pre-incident planning** is a process of preparing a plan for emergency operations at a given occupancy or hazard. Pre-incident planning enhances effective and safer operations, helps you save lives, and protect property.

WHAT IS PRE-INCIDENT PLANNING?

The *Fire Protection Handbook* defines pre-incident planning as follows[2]:

> A written document resulting from the gathering of general and detailed information/data to be used by public emergency response agencies and private industry for determining the response to reasonably anticipated incidents. In simple terms, pre-incident planning is ensuring that responding emergency personnel know as much as they can about a facility's construction, occupancy, and fire protection systems before an incident occurs.

NOTE

Pre-incident planning often starts with a survey of the property.

PRE-INCIDENT PLANNING ACTIVITIES

Pre-incident planning consists of these interrelated activities:

- The pre-incident survey
- The development of information resources that would be useful during the event
- The development of procedures that would be used during an emergency

Pre-incident planning includes the gathering and storing of information to facilitate the strategic dispatch and placement of personnel and equipment to deal with fire, rescue, and hazardous materials emergencies.

OFFICER ADVICE

Look at the fire scene as a problem that needs to be solved. The first two steps in the pre-incident planning process are to identify the problems and gather the facts needed to find a solution.

Pre-incident planning often starts with a survey of the property. This activity is called a **pre-incident survey.** Pre-incident surveys are essential for developing pre-incident plans. Pre-incident plans, in turn, are essential for developing strategies for safe and effective fireground operations.

In Chapter 5, we described the problem-solving process. We said that the first step in the problem-solving process is to recognize that we have a problem. Succeeding steps were to define the problem, collect data, analyze that information, develop alternatives, select the best alternative, implement, monitor, and correct as needed.

Look at the fire scene as a problem that needs to be solved. The first two steps in the pre-incident planning process are to identify the problems and gather the facts needed to find a solution.

Pre-incident planning is a problem-solving process. You want to know as much as you can about the facilities and the activities that are within your area so that you are ready to deal with any emergencies that do occur. You are in a unique position of being able to define the problems, gather information, and develop plans for dealing with the emergency situations that might be encountered.

Pre-incident planning information should be recorded and stored so that it is readily accessible to you in emergencies. The pre-incident plan or simply, "preplan," allows you to make informed decisions while en route to and at the scene of an emergency. Preplanning benefits everyone involved with the event and contributes to more effective and safe incident management.

Always review your plan during training activities or at a real emergency. You will find that real incidents do not always turn out exactly like your plan, but the fact that you have a plan gives you a far better starting point than if you had nothing at all.

After you have a training exercise or an event in which the plan is used, you should review the plan and ask others for feedback. This is similar to the controlling or evaluation phase of the management cycle. From the feedback, you should be able to make some improvements to your plan. That entire process is what this chapter is all about.

USING FIRST-DUE COMPANIES FOR PRE-INCIDENT PLANNING ACTIVITY

Most fire departments have standard policies for the pre-incident planning activity, including how to conduct a pre-incident survey. Your department can gather this information in several ways, but the most logical way is to assign the company or emergency response unit that is most likely to be the first to arrive at the address during an emergency to make the survey. This response unit, usually referred to as the first-due company, is likely to be there first during any emergency and will probably have the greatest impact on mitigating any event that does occur. It is logical that it should be tasked with planning the actions that should be taken.

This procedure has several other advantages. First, it allows all of the firefighters from the company to visit the site and become familiar with the layout of the building under favorable conditions. A second benefit is that it permits all of the firefighters to have a chance to look at the process and storage hazards that may be present and how these would impact on any fire or other emergency that might occur. Pre-incident surveys are also an opportunity to let the building owner, occupants, and even the public knows that the fire department is interested in protecting their property, just as it says in the fire department's mission statement. Pre-incident surveys are valuable in many ways.

OFFICER ADVICE

Using first-due companies for pre-incident planning activities allows personnel to become familiar with the property and allows the public to see the firefighters' interest in protecting the property.

OFFICER ADVICE

Given that you cannot plan for every situation, you should plan for those that present the greatest risk.

Where to Start

If your fire department has been involved with this activity for a number of years, your major task is simply to maintain the system, keep the existing information up to date, and pick up new structures as they are built and occupied. On the other hand, if your department has not taken advantage of preplanning, you may be asking, "Where do I start?"

Look first at target hazards. It would be nice to plan for all of the emergencies that might occur at all of the facilities within your jurisdiction, but it is unlikely that this will happen. Given that you cannot plan for every situation, you should plan for those that present the greatest risk. We will call these facilities target hazards. **Target hazards** are those occupancies that present high risk to life safety and property.

Target hazards usually include the following:

- Occupancies such as health-care facilities, including convalescent homes

- Large public assembly facilities, including airports, theaters, libraries
- Multiple residential occupancies, including apartment buildings, motels, and dormitories
- Schools and educational centers
- Occupancies such as jails and prisons
- Occupancies that present difficult challenges—high-rise buildings, and large buildings of any classification
- Occupancies where access by fire department vehicles precludes the normal method of operations
- Occupancies where there may be high property value
- Occupancies where there is a high likelihood of a fire
- Occupancies where there may be water supply issues
- Large mercantile and business occupancies
- Industrial facilities and hazardous storage facilities
- Large unoccupied buildings
- Buildings under construction or demolition

Even though residential occupancies (single-family homes, individual apartments, and so forth) collectively represent the locations for the majority of fire calls, specific single-family homes are generally exempt from the type of formal pre-incident planning discussed here. However, all departments should have standard procedures for dealing with

HOSPITALS—A CONCERN IN EVERY COMMUNITY

Hospitals should be identified as target hazards for almost all fire departments. Hospitals are large buildings with continuous occupancy. Many of the occupants are nonambulatory or unable to rescue themselves due to physical illness, mental illness, age, and/or security measures and may require ongoing medical care while being moved (i.e., ventilatory assistance). Hospitals and all health care facilities can present significant logistical issues when it comes to evacuation, especially down stairways of multistory buildings. As a result, most health care facilities plan to limit patient movement and rely on fire doors, fire-resistant corridor walls, smoke and fire barriers, and fire detection and suppression systems.

In addition to these challenges, consider the following:

- Significant occupant load consists of a mix of patients, staff, and visitors, most of whom are not familiar with the facility.
- Most hospitals include a series of interconnected buildings constructed over a period of time when floors are added on top of existing buildings.
- Limited access points may require you walk a great distance just to investigate the situation.
- Parking around hospitals is usually at a premium.
- Fire department access can be a challenge.
- Fire apparatus can easily choke exits and assembly points.
- Patients and staff gathering at predesignated assembly points can block fire department access.

We hope that we never get a fire call at the hospital, but we must always be ready in case we do.

fires and other emergencies that would likely occur in residences in the area they serve.

NOTE

The pre-incident survey is the fact-finding portion of the pre-incident planning process.

Gathering Information

Pre-incident surveys are an essential part of developing pre-incident plan-of-action procedures to mitigate fire, hazardous materials incidents, and medical service calls in advance of an emergency situation. The pre-incident survey is the fact-finding portion of pre-incident planning. Pre-incident surveys provide valuable information to facilitate an effective response assignment and provide responding personnel with timely information regarding the features and hazards of the facility. Features can include items such as fire alarm and sprinkler systems, location of water sources, and location of the electrical room. Hazards can include anything from hazardous materials storage to the types of building materials used to construct the building. In addition to ensuring the effective use of resources during an emergency, such surveys provide documents to help train new personnel and help all personnel become aware of the hazards that may result in an accident or other unexpected event during emergency operations.

OFFICER ADVICE

While conducting these surveys, remember that you are a guest on someone's property and that good manners and good appearances are important while you are on the premises.

The first step in the survey process is to make an appointment to visit the facility at a mutually convenient time. Try to reduce the interference your intrusion may bring. Shopping facilities should be avoided during peak shopping days. Restaurants should be avoided during their busiest rush hours.

Once on the site, you should look for the facility owner or manager. Once that person is found, you should introduce everyone (including all of the members of your company) and clearly state the purpose of your visit. Establish a pleasant but businesslike relationship. While conducting these surveys, remember that you are a guest on someone's property and that good manners and good appearances are important.

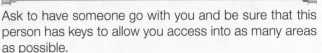

OFFICER ADVICE

Ask to have someone go with you and be sure that this person has keys to allow you access into as many areas as possible.

Ask to have someone go with you and be sure that this person has keys to allow you access into as many areas as possible. This does several things. First and most important, for liability reasons, you do not want to be walking around the facility unescorted. Second, you may need to have access to areas that are generally kept locked, or you may want to ask questions about a particular area or operation. Having someone along with you facilitates the entire process.

The next step is to develop a systematic plan for seeing all of the facility. The complexity and size of the facility determines how long it will take to make the survey.

It is important that you get a good feel for the layout of the building. Sometimes a picture, or in this case, a drawing is worth a thousand words. Actually, two drawings are useful. One of these is called a **plot plan,** an overhead view of the entire property showing the structure's placement, access, utilities, water supply, and other major external features (see **Figure 12-1**). A second drawing is the **floor plan** (see **Figure 12-2**). This drawing shows the major internal structural features of the building and special hazards. In buildings of more than one story, it may be necessary to have floor plans for each floor, unless all of them are alike.

If possible, try to obtain a copy of these drawings. This will save you considerable time in taking and recording dimensions. During the survey, note any variations between the drawing and the building as it appears. Be sure to return the documents if you are asked to do so.

Allow enough time to complete the survey properly. **Figure 12-3** shows some examples of forms to facilitate pre-incident planning. In most cases, your company will be in service and available for calls. However, many jurisdictions have established procedures whereby companies that are involved with pre-incident surveys are not called unless they are really needed.

Even when a preplan already exists, it is advisable to visit the site once a year to make sure that the plan is still up to date, to see if there have been any structural or other alterations that might affect the plan, because it may be more hazardous to operate with an outdated plan than with no plan at all. Unknowingly using an outdated plan can cause you to make incorrect and perhaps deadly assumptions about the layout

PLOT PLAN

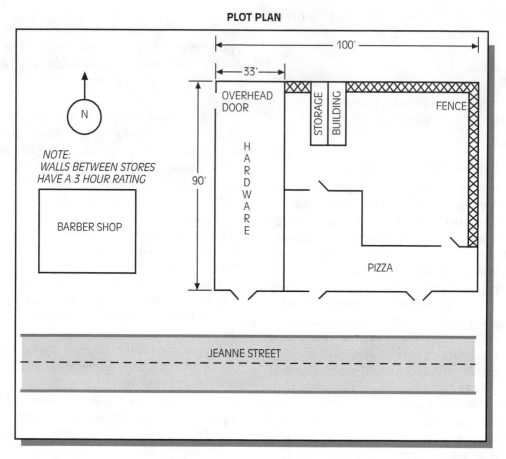

FIGURE 12-1 Sample plot plan.

FLOOR PLAN – MAIN LEVEL

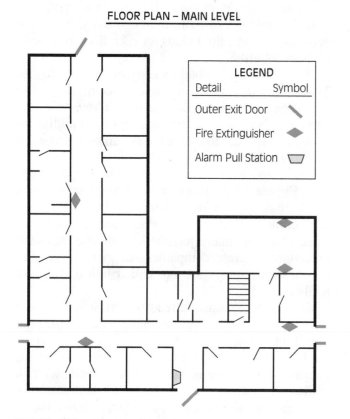

FIGURE 12-2 Sample floor plan.

FIGURE 12-3 Sample pre-incident planning form.

Skyland Fire Department
Fire Suppression Systems/Alarm & Smoke Management

Occupancy Name | Revision Date:
Revised By:

Alarm System

Manual ☐   Smoke Detectors ☐   Guard On Duty ☐   FACP (& Remote Annunciator Panel(s))
Automatic ☐   Heat Detectors ☐   Hours: _____   Location(s): _____
Combination ☐   Carbon Monoxide ☐   _____

Area of Coverage: _____   Occupant Notification: _____

Notes: _____

Automatic Sprinkler System

Type of System: _____   Area of Coverage: _____   Risers: _____

Control Valves: _____   Notes: _____

Standpipe System

Type of System: _____   Location of Oulets _____   Risers: _____

Control Valves: _____   Notes: _____

Fire Pump	Add. Suppression System	Smoke Mgmt. System
Location: _____	Type: _____	Type: _____
Type of Driver: _____		
	Area Served: _____	Area Served: _____
Capacity: _____		
Water Supply Source: _____	Control Valves: _____	Control Valves: _____
Area Served: _____		
Notes: _____	Notes: _____	Notes: _____

(b)

Skyland Fire Department
Occupants, Facility Site & Construction, & Notes

Occupancy Name | Revision Date:
Revised By:

General Occupant Information

Maximum No. of Occupants: _____

Expected Variations: _____

Special Occupant Hazards: _____

Egress Considerations: _____

Emergency Action Plan: _____

Notes: _____

Construction

Construction Type: _____
No. of Floors: _____ at Grade: _____
Building Height: _____ Stories: _____
Square Footage: _____
Fire Flow at 100%: _____ Flow at 50%: _____
Roof Construction: _____

Floor Construction: _____

Interior Finishes: _____

Structural Barriers: _____

Concealed Spaces: _____

Notes: _____

Notes

_____

Site Considerations

Restrictions: _____

Exposures: _____

Notes: _____

(c)

Skyland Fire Department
Hazards/Exposures/Processes (HEP)

Occupancy Name | Revision Date:
Revised By:

HEP$_1$	HEP$_2$	HEP$_3$
Type: _____	Type: _____	Type: _____
Location: _____	Location: _____	Location: _____
Notes: _____	Notes: _____	Notes: _____

HEP$_4$	HEP$_5$	HEP$_6$
Type: _____	Type: _____	Type: _____
Location: _____	Location: _____	Location: _____
Notes: _____	Notes: _____	Notes: _____

HEP$_7$	HEP$_8$	HEP$_9$
Type: _____	Type: _____	Type: _____
Location: _____	Location: _____	Location: _____
Notes: _____	Notes: _____	Notes: _____

(d)

G	Gas Cut-off	PIV	Post Indicator Valve
CNG	Compressed Nat. Gas Cut-off	RV	Riser Valve
NG	Natural Gas Cut-off	ZV	Sprinkler Zone Valve
LPG	LP Gas Shut-off	HC	Hose Cabinet/Connection
W	Domestic Water Shut-off	WH	Wall Hydrant
E	Electric Shut-off	TH	Test Header
SV	Smoke Vent	TC	Inspector's Test Connection
SL	Skylight	FDC	Fire Department Connection
AP	Fire Alarm Control Panel	FH	Fire Hydrant
RP	Fire Alarm Reset Panel	DS	Drafting Site
SP	Smoke Control & Pressurization	WT	Water Tank
WB	Water Flow Bell	AC	Air-Conditioning Eqpt Room
FD	Fire Department Access Point	EE	Elevator Equipment Room
RA	Roof Access	EG	Emergency Generator Room
KB	Knox Box	FP	Fire Pump Room
HL	Halon System	TE	Telephone Equipment Room
DC	Dry Chemical System	BR	Boiler Room
CO	CO$_2$ System	ET	Electrical/Transformer Room
WC	Wet Chemical System		
FO	Foam System		
CA	Clean Agent System		

(e)

FIGURE 12-3   Continued.

FIGURE 12-4 A photograph of the building also helps. This is the same building shown in Figure 10-2, but a year later, after it was completed and occupied.

of the building and its contents. The visit also allows new members of your company to see the building for the first time (see **Figure 12-4**).

> **NOTE**
> The survey information and the pre-incident plan should give first consideration to life safety.

What to Look for during the Survey

Reflecting on your organization's mission statement, your first obligation is to save lives. The survey information and the pre-incident plan should follow the same principles and the incident command system, giving first consideration to **life safety,** following the order of the fire service priorities: life safety, incident stabilization, property conservation. We usually do this by making sure that everyone is safely evacuated from the building and accounted for.

> **NOTE**
> You should be aware of the number of people who would be present and their physical ability to exit the building.

Your second priority should be to attempt to save property loss by stopping the fire, usually referred to as **incident stabilization.** Your final concern should be to consider what you can do to reduce the loss that will result from the fire and the fire suppression efforts. Information regarding this aspect of the operation, usually referred to as **property conservation,** should also be addressed during the prefire planning.

Just as your first priority during an emergency focuses on life safety, your first priority during the survey should be to look for hazards that may impact on life safety. This means you should look for the location of means of egress for everyone in the building, making sure they are clear of obstacles and properly lighted.

Life Safety Concerns

You should be aware of the number of people who would be present and their physical ability to exit the building. Dealing with ambulatory children in the middle of the day is quite different from dealing with nonambulatory adults in the middle of the night. Age, mental and physical conditions, and time of day are all important life safety considerations. They should be addressed in your preplanning activities.

There are several alternatives available to protect the civilians who may be in the building. Although the most common approach is evacuation, there are several situations where evacuation is not the appropriate answer. An alternative may be to relocate the occupants to another part of the building, to an **area of refuge.** This is a likely situation in a large high-rise building or where you do not have sufficient resources to perform rescue/evacuation and fire control.

Another alternative may be to have the occupants remain in place or **shelter in place,** as would likely occur in a hospital. Obviously, in either of these situations where the occupants are in the building, life safety considerations must be incorporated into the building's design and into your pre-incident planning. In most of these situations there are physical barriers like fire doors and ventilations systems that prevent the occupants from being exposed to the fire or its products of combustion. It is not acceptable to shelter occupants in place if they are exposed or have potential to be exposed to either.

> **SAFETY**
> The building type, age, and condition are all important for the occupants' and firefighters' safety, as well as the way you fight the fire.

Building Type and Condition

You should also take interest in the structure itself. We addressed this topic at length in Chapter 11, but certainly the building type, age, and condition are all important for the occupants' and firefighters' safety, as well as the way you fight the fire. The design and condition of the roof are important when planning strategies for safe fire fighting. Think about how the fire will likely spread and how you stop that fire. These should also be addressed in good pre-incident planning efforts (see **Figure 12-5**).

Building Contents

Ask the owner or the manager about high-value items. Ask about where the files are stored and where other items that would prevent the occupant from resuming operations are located. Make note of these items. You will want to give priority to these items during the salvage operations that would follow any fire.

OFFICER ADVICE

Your pre-incident plan should address the community's overall risk. Sometimes, it may be more prudent to let a fire burn and focus efforts on protecting exposures and the environment.

Contents Hazards

Some buildings contain chemicals and other hazards that when heated and mixed together, present significant safety, health, or environmental problems. Again, preplanning should allow you to identify these hazards and how effective your initial control efforts will be. If a significant fire should occur at one of these buildings, it may be better to allow the fire to burn and focus personnel energies on protecting the exposures and the environment. More than one fire suppression effort has resulted in creating a lake of highly contaminated water. If this contaminated water reaches the community's drinking water, there could be long-term problems associated with the incident. The overall mission of the fire department is to protect the community. Saving the community's water supply for the next generation may have far greater value than saving part of a building that will likely be torn down after the fire is over. Along these lines, consideration should also be given to Environmental Protection Agency (EPA) guidelines. If contaminated water is allowed to run off in rivers streams and/or groundwater, the implications could be costly and long lasting and potentially shut down a community.

The more you know about these factors, the better you will be able to deal with the emergency problems that occur. While the resource and time constraints may remain, many of the unknown factors can be eliminated. Elimination of unknown conditions will make the working conditions safer and more effective.

Building Systems

You should also note the location and function of elevators, material handling equipment, and other mechanical devices that may become hazards during fire conditions. These hazards may present problems

FIGURE 12-5   Preplanning includes a visit to the site on a regular basis.

DEADLY 1999 WORCESTER, MASSACHUSETTS, FIRE PROVIDES VALUABLE LESSONS ON ABANDONED BUILDINGS

In December of 1999, a Worcester, Massachusetts, fire in an abandoned building claimed the lives of six firefighters. The fatal fire started when homeless individuals overturned a candle inside the warehouse. Eventually going to five alarms, the fire took more than 20 hours to extinguish. The six-story warehouse building was in the heart of the town's former warehousing and cold storage district. It had been abandoned for a decade before the fire but was frequented by homeless individuals.

Following the fire, the U.S. Fire Administration (USFA) investigated the actions taken.[3] That review emphasized that abandoned buildings are a serious threat to firefighters and fire departments must make a concerted effort to maintain pre-incident data on abandoned buildings in their response districts. The review also notes that delayed reporting contributed to the warehouse fire's spread and that fire services should initiate rapid intervention teams earlier in a structure fire response and use a strict system of personnel tracking on the scene.

Other lessons learned cited in the report include the following:

■ Fire prevention efforts should target abandoned and even temporarily vacated buildings to avoid fires.
■ Proper permitting and ongoing building inspections for construction changes within businesses can help reduce noncompliant interior finishes that contribute to combustion.
■ Large buildings such as warehouses and high-rises require special search techniques and tools, including additional air tanks.
■ Better techniques must be developed to better track the movements of firefighters within a structure.
■ Alternative radio channels should be explored as radio channels can be overloaded at multiple alarm fires.
■ Thermal imaging cameras, while expensive, are invaluable equipment for all fire departments.

"The Worcester fire dealt a serious blow to the nation's fire service," said U.S. Fire Administrator David Paulison. "It was one of the largest firefighter death tolls for a single event before the World Trade Center tragedy in 2001. It merited our study and resulted in findings relevant to every department in the nation."

---

for safe evacuation, as well as means by which the fire may rapidly spread. Give attention to storage and process operations involving flammable materials.

Check the ventilation system. Determine if it will help spread the smoke, heat, and fire. Determine if the ventilation system would be used to control these same factors. Look for and make note of openings in the building that would facilitate ventilation. Most modern buildings have skylights and other fixtures that allow the roof to be opened without cutting. Timely ventilation will significantly help in the evacuation of the occupants, permit safer and more effective firefighting, and greatly reduce property loss due to the spread of heat and smoke throughout the building. While you are not expected to know how to run the buildings' ventilation system, you should note how to contact the appropriate personnel who do.

OFFICER ADVICE

You should always be interested in any installed fire protection equipment.

You should always be interested in any installed fire protection equipment. Clearly understand the scope and function of any installed alarm systems. Your survey should capture appropriate information on how suppression personnel can safely reset these

systems after false activations. However, do not automatically reset an alarm system. If upon your initial investigation, you find nothing is showing and further investigation is required, silence the system rather than resetting it. This will not only keep from disturbing the occupants; it will also allow you to better communicate and allow the alarm company to identify the problem. If the system is always reset, not only will you not be able to locate the problem, but the alarm company will not fix the problem and you will likely return. And if sprinkler systems and standpipes are included, be sure that you understand the pipe arrangement and how to augment the existing water supply. When any of these features are present, they should be noted in the preplan. Tactical planning considerations should work in harmony with the installed fire protection equipment, not compromise its intended efforts (see **Figure 12-6**).

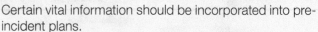

OFFICER ADVICE

Certain vital information should be incorporated into pre-incident plans.

Water Supply

Take careful note of the water supply that may be available, both on-site and within the immediate

FIGURE 12-6 During site visits, be sure to note the fire alarm and protection systems.

vicinity of the building. It seems elementary to mention that water supply is a critical item, but we point out that the best fire departments with the best of pre-incident information and plans will need water to put out a fire. Where a hydrant system is available, the hydrant's capabilities should be verified (see

INFORMATION NEEDS FOR PRE-INCIDENT PLANNING

Pre-incident planning is a process of preparing a plan for emergency operations at a given building or hazard. Most pre-incident planning is directed to a specific location or occupancy. When the target of the planning process is a fixed location, the following information should be incorporated into the plan information:

- Building number and street address
- Emergency contact information
- Other names commonly used to describe the occupancy
- Occupancy load during the day and night
- Building description (number of floors, dimensions, or square feet of floor space)
- Hazards to personnel
- Installed fire detection equipment
- Installed fire suppression equipment
- Water supply
- Access to utility cutoffs
- Priority salvage area
- Special considerations

FIGURE 12-7 If in doubt, have the hydrant flow tested to be sure that it can provide adequate water during a time of need.

Figure 12-7). Where water may come from other sources, plans should be made and resources identified for moving the water to the scene of the fire.

Exposures

There may be as many as six exposures in a fire. Each should be considered in pre-incident planning. An exposure can present greater risks than the fire building itself. In such cases, the planning effort should identify these risks. It may be that with limited resources, the exposures deserve the full attention of the fire department, rather than the target building. These factors should be carefully addressed in any pre-incident planning documents.

Recording the Information

Use a checklist or a copy of the form itself to identify and record important information. Start by getting the correct address and the name of the building. Get the name of the owner, manager, and other key representatives, and their telephone numbers both during working hours and nonworking hours; it may be important to contact one of these individuals when the facility is closed.

OFFICER ADVICE

Use a checklist to identify and record important information.

A program for conducting pre-incident surveys does not replace the need for conducting regular fire prevention inspections. Both should be ongoing activities, done for different reasons. Pre-incident surveys are best done by suppression personnel within their own response area.

Fire prevention inspections may be done by suppression personnel or others. In any case, fire code violations found during a pre-incident facility survey shall not be overlooked. Severe violations should be corrected at once. When immediate correction is not possible, immediately advise your fire prevention division of the situation. In all cases, hazards should be reported using the appropriate reporting forms. Copies of the report should go to the person responsible for the building as well as to the fire department's fire prevention division. Review the plan with the fire prevention division personnel and see if they have found any additional issues with regard to the occupancy. In addition, find out if there are any local ordinances or additional standards that should be recorded on the plan. This is important because sometimes codes and ordinances allow for exceptions based on the type of occupancy.

Converting Your Information into a Plan

Turning all of the information gathered during the survey into a pre-incident plan is the second step in pre-incident planning. Pre-incident planning gathers information from a variety of sources. We have the survey, of course. The information gathered during the survey, information about the department's resources, and the department's standard operating procedures suggest appropriate actions for planning the initial strategies that would be needed during a real emergency. The topics listed on the previous pages will provide a guide for these planning efforts.

NOTE

Turning all of the information gathered during the survey into a pre-incident plan is the second step in the pre-incident planning process.

You may also have information from the fire prevention division of the fire department. It may be inspecting the facility on a regular basis due to the nature of its occupancy. It may have issued a permit to store certain material or conduct certain operations, either of which you should know about.

Upon completion of the survey, the company should work together to develop a plan that will address the most likely scenarios that would occur at that address (see **Figure 12-8**). While we often think of the threat of fire, we should also consider EMS and hazardous materials control, and rescue situations as well.

The planning effort should address the same factors that were addressed during the survey. In addition, the plan should address the resources needed to deal with the emergency and how these resources will be initially organized. For example, the plan should include placement of first-arriving companies, identify water supply sources, and provide a starting point for fireground strategy. No building is worth risking a firefighter's life. Some buildings, because of their age, fire load, type of construction, extent of fire involvement, or physical condition, may necessitate operating in defensive mode from the start. When these buildings are identified, the initial mode of operations should clearly indicate that all firefighting

FIGURE 12-8 Get as much participation as possible in the preplanning activity.

activity will be from outside the structure. This is one of the most critical tasks an officer will perform, making the risk versus benefit analysis and determining the operational mode. If there is any doubt about which mode to assume, always error on the side of caution and make sure your crew goes home at the end of their shift.

NOTE

The planning effort should address the same factors that were addressed during the survey.

PRE-INCIDENT PLANS SHOULD ADDRESS

- Life hazards problems, for occupants and firefighters
- Availability of egress for the occupants
- Availability of access for firefighters
- Places where a fire would most likely start
- Factors that would influence the spread of the fire
- Factors that would influence the intensity of the fire
- Initial strategies for effective fire containment and control
- Factors that would limit the fire department operations
- Resource needs including those of other agencies
- Apparatus placement
- Location of the command post

Resource Needs to Deal with the Emergency

Part of the preplanning process includes forecasting the resources that may be needed to mitigate the situation. Matching resources to the needs of routine operations is easily accomplished. Room-and-contents fires, vehicle accidents, and standard EMS calls make up the bulk of most fire department operations. The events can usually be controlled with first alarm resources. But for large events, you will need additional resources. Your pre-incident planning efforts should address the potential for needing additional resources (i.e., rescue companies, aerial apparatus, and hazardous materials). If additional resources are likely to be needed, add that information to the pre-incident planning information. Many jurisdictions provide special dispatch assignments, based on the pre-incident planning information.

NOTE

Part of preplanning includes forecasting the resources that may be needed to mitigate the situation.

For example, when planning for operations at target hazards, you should always be thinking of the potential of an event that will require resources beyond the normal first-alarm assignment. In shopping centers and high-rise buildings, the mere transportation of firefighters and their equipment to the fire requires a tremendous support team. Large or multistory occupancies may necessitate the use of multiple or specialized ventilation equipment. Hazardous materials incidents also require considerable support. Pre-incident planning should look at these needs. Even during routine firefighting, we should have relief crews on scene, ready to step in when needed.

Once you have identified your resource needs, you should consider how these needs will be met and how rapidly these resources can be on-scene. Even in large cities, getting second- and third-alarm companies to the scene can take some time. In smaller communities, especially those communities where volunteer firefighters provide fire protection, there may be appreciable time delays as the firefighters and their equipment travel considerable distances to the scene.

OFFICER ADVICE

Once you have identified what your resource needs are, you should consider how these resource needs will be met and how rapidly these resources can be on-scene.

In addition to fire department resources, you should consider the resources of other agencies. For example, will the police department be able to help with traffic and crowd control? What resources may be available on-site and from other industrial or commercial facilities in the immediate vicinity? What about resources from federal and state agencies? While preplanning activities may identify a need for these resources should the situation be significant enough, officers should always use the normal chain of command processes to request these resources. It should be noted on the pre-incident plan the types of services and the equipment that can be supplied by these resources. While it is difficult to list all of the possible scenarios for a given occupancy, you can list the possible resources that will be needed.

OFFICER ADVICE

In addition to fire department resources, you should consider the resources of other agencies.

Interagency Cooperation

Identifying those resources is only the first step. Resources from agencies outside of your own should have an opportunity to review your plans and understand their role. The time to develop a clear understanding of the lines of communication and authority are during the planning stages, not during the time of an actual emergency. There should also be a clear understanding of who is authorized to make the request for assistance, what will be sent, and a resolution of questions regarding liability issues that may arise during an actual emergency. These are normally identified in Memorandums of Understanding (MOUs) or Mutual Aid Agreements. These are typically legal binding documents that require administrative approval above the level of the company officer.

NOTE

Interagency cooperation must be carefully planned to reduce conflict, confrontation, and confusion during an emergency.

Interagency cooperation must be carefully planned to reduce conflict, confrontation, and confusion during an emergency. Effective cooperation includes a study of each agency's capabilities, regular sharing of information, and joint training exercises so that everyone understands the capabilities and limitations of all of the other agencies. In short, you need to establish effective working relationships before the emergency occurs. By doing so, you enter an emergency incident with a level of trust and understanding that permits you and your respective agency to immediately begin dealing with the issues at hand.

Allied Agency Cooperation. It is important to plan your department's interagency support. Some of the agencies that might be useful include:

- *Police or sheriff departments.* They can help with traffic issues, hostile patients or citizens, and assistance with evacuation procedures.

- *Adjoining fire departments.*Most fire departments have areas within their districts that have inadequate or insufficient fire protection. By having these departments as allied agencies, the department can ensure that their citizens' needs will be met.

- *Emergency management.* They can assist with equipment issues on large incidents. These agencies often have the "key to the state" when it comes to needed equipment.

- *EMS agencies.* While most departments have either their own EMS system or a city or county system, it is recommended that the department have additional support in the event of a large scale emergency.

- *Department of Transportation.* These agencies can be of extreme benefit for highway issues or even hazardous materials incidents.

While this list is small, the reader can grasp the importance of adjoining themselves with other agencies. Most departments cannot do it by themselves; therefore, it becomes necessary to join forces to protect the citizens.

PRE-INCIDENT PLAN INFORMATION SHOULD INCLUDE

- Address
- Owner
- Occupancy
- Means of access and entry
- Personnel hazards
- Fire behavior predictions
- Locations of stairs and elevators
- Ventilation system
- Installed fire protection systems
- Exposures
- Special hazards
- Resource needs
- Company assignments
- Estimated fire flow
- Water supply
- Predicted strategies

The National Fire Protection Association (NFPA) publishes several documents that aid the pre-incident planning process. NFPA 170, *Standard for Fire Safety and Emergency Symbols,* provides a good listing of standard symbols used in preplanning and other fire-related applications.

Common Human Failures in the Pre-incident Planning Process

- Failure to become familiar with the location and surrounding areas of buildings within the response area
- Failure to adequately identify the type of building construction and its hazards
- Failure to adequately identify the hazards associated with the contents
- Failure to keep the pre-incident planning information up to date
- Failure to keep personnel up to date on the changes
- Failure to train and practice for the event
- Failure to use the information during a real emergency

While this section has focused on what information to collect, how to conduct, and how to prepare a preplan, officers should always know what departmental policies and procedures may exist concerning preplans. Many departments have specific guidelines on what information to collect and how to process that information. There may also be some sort of approval or submittal process that your preplan is required to go through. Many times these policies are in place to ensure consistency and uniformity among all preplans. It may also be a requirement to ensure the preplan is compatible with the department's computer applications. Once approved, it may be the company officer's responsibility to distribute the preplan to appropriate personnel. This may mean making paper copies and inserting them into manuals, or it may be a simple as scanning it into the designated computer file. Regardless, if it is not done correctly, then all the effort up to this point will have been wasted. So be sure that you follow internal procedures related to conducting, submitting, and approval processes for the preplan.

PRE-INCIDENT PLANNING FOR OTHER EVENTS AND LOCATIONS

Over the last several pages, we have discussed the need for pre-incident planning at target hazards. We usually think of target hazards as fixed sites with specific addresses. You should also be thinking about the typical incidents that would benefit from the same pre-incident planning process. In this case, we are thinking more in terms of typical or recurring events rather than specific hazards.

NOTE

In addition to events that occur at structures, departments should be prepared to deal with other situations that are likely to occur within their jurisdiction.

During some of your pre-incident planning efforts, you may lack a specific address, but the general characteristics of the occupancy allow you to develop a pre-incident plan for the type of hazard, rather than for a specific address. Examples include various types of residential structures common to the first-due assignment area, high-rise buildings, shopping centers, service stations, and schools. This allows you to communicate your expectations to your crew members before the incident occurs. If you expect a member of the crew to carry in a pressurized extinguisher when responding to a systems alarm in a multistory

building, it should be communicated to him or her during these preplanning visits. It makes for a much more efficient, smooth, and safe operation when you do this.

In addition to incidents that occur at structures, departments should be prepared to deal with other situations that are likely to occur within their jurisdiction. Nearly every fire department has roads and highways, railroads, pipelines, and waterways. Each of these modes of transportation presents opportunities for fire and rescue situations that will involve the fire department. You should drive them and discuss how to best access them given a wide variety of weather and enduring conditions.

The preplanning for these events often leads to a department wide **standard operating procedure,** or SOP. The SOP is usually a several-page document that follows an established format and lists the various hazards associated with the type of hazard, and establishes standard procedures for first arriving companies. Unlike the pre-incident plan information for site-specific hazards in which you will have time to look at information while en route to the emergency scene, many fire departments expect officers to be able to recall from memory the general procedures to follow at these selected target hazards.

CAN YOUR DEPARTMENT RESPOND TO A LARGE-SCALE INCIDENT?

A Needs Assessment of the U.S. Fire Service (U.S. Fire Administration) provides an indication of the ability of the nation's fire departments to deal with selected large-scale events with locally trained personnel.

- Only 11 percent of the departments indicated that they could handle a technical rescue with EMS at a structural collapse of a building with fifty or more occupants.
- Only 13 percent of the fire departments indicated that they could handle a hazmat and EMS event involving chemical or biological agents and ten injuries.
- Only 26 percent of the fire departments indicated that they could handle a wildland/urban interface fire affecting 500 acres with locally trained personnel.
- Only 12 percent of the fire departments indicated that they could handle mitigation of a developing flood with locally trained personnel.

Storing and Retrieving the Information

The National Fire Academy suggests a very simple form for recording the information for the pre-incident plan called a **Quick Access Prefire Plan**

Quick Access Prefire Plan

Building Address: *1020 Jeanne Street*

Building Description: *1-story, ordinary construction; 2 occupancies in building firewall between occupancies*

Roof Construction: *Wooden 2" x 10" rafters, plywood, composition roof covering*

Floor Construction: *Concrete slab*

Occupancy Type: *Commercial*	Initial Resources Required: *2 Eng., 1 Truck or Squad, 1 Amb., BC*

Hazards to Personnel: *Pesticides, flammable/combustible liquids*

Location of Water Supply: *Hydrant across Jeanne Street*	Available Flow: *1200 GPM*

	Estimated Fire Flow			
Level of Involvement	25%	50%	75%	100%
Estimated Fire Flow	375	750	1125	1500

Fire flow of largest fire area–hardware store and 2 exposures

Fire Behavior Prediction: *Rapid horizontal spread within one occupancy*

Predicted Strategies: *Confinement, ventilation, extinguishment*

Problems Anticipated: *Poor rear access, limited horizontal ventilation*

☐ Standpipe: *None*	☐ Sprinklers: *None*	☐ Fire Detection: *Smoke Detectors*

FIGURE 12-9 The National Fire Academy's Quick Access Prefire Plan.

(see **Figure 12-9**). Usually a follow-on page of typed information, a drawing of the plot plan, and a drawing of the floor plan accompany this basic information.

There are a variety of ways of storing pre-incident planning information. Regardless of the method selected, the main purpose of having the information is so that it is available when and where it is needed. If an emergency arises at a particular address, the preplan should be reviewed by companies while they are en route to the emergency. The information is also needed by companies operating at the scene. This information could be saved on a computer or on paper in a notebook or folder.

Each company should have access to pre-incident plans for its first-due response area in its apparatus. An additional set of the pre-incident plans should be kept at the station near the watch desk for use by companies that may be reassigned to fill in at the station. These companies would likely be called up if additional alarms are needed. This same copy can be used for training and drill purposes.

OFFICER ADVICE

Each company should have access to pre-incident plans for their first-due response area in their apparatus.

A computer provides a compact way of storing a great deal of information and retrieving it rapidly. The information may be stored at the dispatch center and transmitted to a company when it is needed, or the information may be stored on a portable computer located in the vehicle. Although there are many other modern ways of storing and using this information, many fire departments still rely on the reliable three-ring notebook.

Good information in a readily available format permits timely access and allows for better analysis and the quicker start of solutions to mitigate the problem.

When there is a lot of stored information about a particular facility, some departments keep the information on the facility site in a lock box. The lock box should be located in an accessible location on the property and can contain pre-incident planning information, material safety data sheets, the names and phone numbers of key personnel, and, most importantly, keys that allow emergency responders to enter the facility. Obviously, the fire department must have access to the lock box. The lock box has many advantages, especially in that it allows for items such as keys to be left at the site in a secure place. Having the detailed pre-incident planning information there may also be advantageous, especially when such information requires storage space beyond the capabilities of first-arriving units. If a lock box is used, the basic information (address, and so on) should be available to companies while en route (see **Figure 12-10**).

Training and Practice

Training is a vital part of pre-incident planning. If personnel do not know of the plan or do not know how to implement it during a time of need, the effort to gather the information and to plan for the emergency has been largely wasted. Training and practicing for dealing with emergency situations should be a continuing process.

NOTE

Pre-incident plans should be exercised through regular training and practice.

LOOK TO THE NFPA STANDARDS FOR MORE INFORMATION

The Standard for Providing Emergency Services to the Public, NFPA 1201 is intended for the use and guidance of those charged with providing fire protection (safety) services to protect lives, property, and the environment from the effects of fire and, in many cases, other perils. With regard to planning and emergency management, it includes the following:

■ This standard is intended for the use and guidance of persons charged with providing emergency services to protect lives, property, critical infrastructure and the environment from the effects of hazards (e.g., fire, medical emergency, hazardous materials, natural disaster, critical infrastructure disruption).

■ It is the requirement of the authority having jurisdiction to develop and utilize a policy or guideline for organizations directives for the operation of the fire department (standard operating procedures).

■ A master plan shall be created to coordinate the vision, mission, values and goals of the Emergency Service Organization (ESO). Research and planning within each ESO shall include maintaining ongoing relationships with other agencies involved in planning for the service area.

■ The emergency service system shall include a continuing program of research and planning that encompasses any or all aspects of the emergency services system, both generally and specifically.

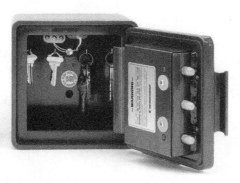

FIGURE 12-10 A lock box provides information as well as a means for entry.
(Photo courtesy of The Knox Company)

Developing SOPs from Pre-Incident Plans

Gathering of information through a good pre-incident plan process will often find information that may require an update to or a whole new SOP. The following information outlines some of the SOPs that have a direct link to pre-incident plans.

- *Emergency Operations*—Based on the information obtained in the pre-incident plan process, the following may need to be evaluated:
  - ☐ Does the department have SOPs in place so that all department members understand their role on the emergency scene?
  - ☐ Does the department have the resources needed to mitigate the situation?
  - ☐ Does the department have contracts in place to request equipment from outside agencies?
- *Dispatch of Emergency Units*—Information from the pre-incident plan may show a need to update an SOP for the dispatch of emergency units. Departments often have an SOP that map out the types of units that are to respond to different types of emergencies. For an example, it might be that the department has said that 2 engines, a ladder company, and a battalion chief are to respond to all reports of fire in business occupancies. While this might work for stand-alone business, what if the business is in a strip mall? Or what if the business is 400,000 square feet?
- *Strategies and Tactics*—SOPs written for strategies and tactics can be widely misinterpreted. They are often subjective to the views of the incident commander. However, the pre-incident plan can add some objectivity by pointing out items that are not opinioned based. The pre-incident plan shows items such as whether the building has a suppression system and where the location of the water source is located and pinpoints the building occupancy and classification. Knowing these objective items will have a direct impact on the strategies and tactics for a given situation.
- *Customer Service*—Pre-incident plans and SOPs are written for a lot of reasons. They are written for the safety of the department members, to reduce property loss, and to stabilize incidents. However, they are also written for the customers that the department serves. By properly conducting pre-incident plans the department will know the life safety concerns of the buildings within the jurisdiction, they will know where the hazards are, and they will have a written plan on how to mitigate those hazards.

TRAINING NEEDS

Many firefighters are not well prepared to safely and effectively deal with the emergencies they are likely to encounter. The U.S. Fire Administration recently reported the following:[4]

- An estimated 233,000 firefighters, most of them volunteers serving in communities with less than 2,500 population, are involved in structural firefighting but lack formal training in those duties.
- An estimated 153,000 firefighters, most of them volunteers serving in communities with less than 2,500 population, are involved in structural firefighting but lack certification in those duties.
- An estimated 27 percent of fire department personnel involved in delivering emergency medical services (EMS) lack formal training in those duties, most of them serving in communities with less than 10,000 population. The majority of fire departments do not have all their personnel involved in emergency medical services (EMS) certified to the level of Basic Life Support, and almost no departments have all those personnel certified to the level of Advanced Life Support.
- An estimated 40 percent of fire department personnel involved in hazardous material response lack formal training in those duties, most of them serving in smaller communities. More than four out of five departments do not have all their personnel involved in hazardous material response certified to the Operational Level, and almost no departments have all those personnel certified to the Technician Level.
- An estimated 41 percent of fire department personnel involved in wildland firefighting lack formal training in those duties, with substantial needs in all sizes of communities.
- An estimated 53 percent of fire department personnel involved in technical rescue service lack formal training in those duties. In every population group of communities with less than 500,000 population, at least 40 percent of fire department personnel involved in technical rescue service lack formal training in those duties.
- An estimated 792,000 firefighters serve in fire departments with no program to maintain basic firefighter fitness and health, most of them volunteers serving communities with less than 5,000 population.

COMPANY TRAINING

The remainder of this chapter discusses fire department training at the company level. Training is the key to the firefighter's safety as well as the department's effectiveness. No factor has more effect on a fire department's operations than its training program. Because the fire department's main function is to save lives and protect property, members of the department must be qualified to successfully and efficiently perform a wide range of tasks in carrying out these important functions. No one has more impact on a firefighter's attitude toward training than the company officer. Once the alarm goes off, it has to be right the first time. You need to drive to the right address, select the right size hoseline, and select the right tactic the first time. There is seldom, if ever, a chance for a redo.

The realistic financial constraints of the world in which we work combined with the mission of the fire service require the greatest efficiency from every agency and from every member of these agencies. With decreased fire activity, most firefighters are not called upon to save lives and protect property on a daily basis. However, all firefighters are expected to maintain a high level of readiness, should that need arise. Efficient use of training time provides significant opportunities to increase the readiness of the fire department and its personnel.

FIREGROUND FACT

The realistic financial constraints of the world in which we work combined with the mission of the fire service require the greatest efficiency from every agency and from every member of these agencies.

During the recent past, fire service training has changed immensely. For many years, teaching a few basic manipulative skills was considered to have met the requirements of an adequate fire service training program. But the role of the fire service is constantly changing and becoming more complex. New tasks are being undertaken, and new challenges to existing procedures are being faced every day. The training process starts when an individual joins the fire service and should continue throughout that individual's work years. The fire service training program must prepare the member to meet the known challenges of today as well as to be ready to meet the future.

Educators tell us that training precedes learning. Learning is the process of acquiring new knowledge and skills. To be effective, a training program must provide a process that encourages firefighters to learn information and develop skills that can be retained and used. Effective fire department training programs are essential for the safe and efficient operation of any fire service organization and for its members.

All company officers in a fire department should fully support and augment the fire department's training program and should take personal interest in making sure that the department's training activities are enthusiastically carried out within their area of responsibility.

OFFICER ADVICE

To be effective, a training program must provide a process that will encourage firefighters to learn information and develop skills that can be retained and used.

In this regard, you are responsible for training members assigned to you. Typically, most of the department's training program is conducted at the company level under your direction. Such training activity strengthens the bonds among personnel while developing proficiency in individual and company evolutions. You should periodically evaluate the abilities of your personnel to both determine the effectiveness of the training and to provide a valid basis for individual performance evaluations.

You should observe the performance of your personnel during emergencies to see that they are using the same techniques that were presented in training activities. Critiques of company operations at fires and other emergencies can help identify the company's performance as a team. You are probably the best qualified individuals in the department to assist the department's overall training efforts by identifying topics, procedures, and conditions where training is required.

The overall relevancy of the department's training program should also be continually evaluated by the department's senior staff. The department and company training program, as well as the results of that training, should be reviewed on a regular basis. The results of that review should suggest topics for future training activities.

New Personnel

New personnel should receive comprehensive training during their recruit school. During the period following the successful completion of that school, they are usually required to serve as probationary firefighters. This period is important because the opportunity and motivation for individual learning is at its greatest during these first few months. It is also a time to teach good habits and reinforce the training provided by the training academy. The training accomplished during this period will become the basis for that which follows throughout an individual's work years. New personnel should be trained consistent with the performance objectives outlined in NFPA 1001, *The Standard for Fire Fighter Professional Qualifications*.

Individual and Company Skills

NFPA 1001 provides a list of the requirements for basic firefighter skills. These are skills that should be trained on and tested frequently. All firefighters should be able to pass any part of that test on any given day. To maintain this level of proficiency, company-level training activities should also include a regular review of the individual tasks required of all firefighters during their probationary period, because the basic knowledge and skills are often quickly lost without regular use (see **Figure 12-11**).

Regular training on all the skills required of the firefighter may well exceed the time available, so some selectivity in training topics is usually required. As company officers, you should be able to recognize the needs of your company and emphasize particular areas of training. These training needs can usually be recognized by observing the performance of individuals and companies during training evolutions and at actual emergency situations. You should also encourage your firefighters to individually study during any free time.

Multiple-Company Evolutions

Multiple-company training evolutions provide an opportunity to put it all together, to test the system as it will operate during a real situation. The pace of activity is sometimes slowed to permit monitoring the progress and directing the training activities, but the events should all occur and should be conducted under realistic conditions whenever possible.

OFFICER ADVICE

As company officers, you should note the performance of your personnel during training and actual emergency situations and should provide training in those areas in which skills do not meet expectations.

Multiple-company training evolutions permit firefighters to use their skills in realistic activities under conditions that can approach those that can be expected during a real emergency (see **Figure 12-12**). By assigning multiple tasks to unit members, you can evaluate them on how the tasks were carried out. Were all of the tasks carried out? Were all of the tasks carried out safely? Were all of the tasks completed in a timely manner? Were communication and accountability maintained during the assignment? In such evolutions, you—the company officers—take your place as leaders in the operation. Your leadership ability as well as your prior efforts in providing training for your personnel will be quickly apparent. Although these activities provide a test for the readiness of all individuals and their units, remember that these activities represent a training exercise: This is a time for learning, a time to identify and correct problems that may exist.

FIGURE 12-11 Company personnel should maintain basic skills while learning new ones.

FIGURE 12-12 Multicompany training evolutions provide an opportunity to exercise the incident management system.

Multiple-company training evolutions should also provide an opportunity to train and test the fire department's incident command and communication systems. Such training activities provide an ideal opportunity for officers to move into positions of increased responsibility in managing emergency incidents and in controlling increased numbers of resources. These activities also provide a realistic opportunity for using and evaluating the department's communications resources.

Multiple-company training evolutions provide a unique opportunity to work with companies that would probably be encountered on a mutual aid response. These training sessions provide an opportunity for the mutual aid company to become familiar with the neighboring department's procedures, response area, and high-risk targets. This is a good time to test those preplans and discover incompatibilities in equipment and communications capabilities. As a company officer, you do not have to wait until a more senior officer suggests the need for such evolutions; you should be able to make it happen on your own initiative.

Suggested training scenarios for individuals, company, and multiple-company training are included in Appendix A. These are just suggestions: You should have additional ideas for your company.

OFFICER ADVICE

This is a good time to test those preplans and discover incompatibilities in equipment and communications capabilities.

Live-Fire Training

Live-fire training provides unique opportunities for developing skills and self-confidence at all levels. Live-fire training permits those who have completed their basic training to operate under realistic conditions (see **Figure 12-13**). This training provides an ongoing opportunity for maintaining and improving acquired skills for all personnel. Live-fire training provides a means where the firefighter can display many combinations of skills, and at the same time,

NFPA 1410, *STANDARD ON TRAINING FOR INITIAL EMERGENCY SCENE OPERATIONS*

This standard contains the minimum requirements for evaluating training for initial fire suppression and rescue procedures used by fire department personnel engaged in emergency scene operations. NFPA 1410 specifies basic evolutions that can be adapted to local conditions and serves as a standard mechanism for the evaluation of minimum acceptable performance during training for initial fire suppression and rescue activities. NFPA 1410 can provide fire departments with an objective method of measuring performance for initial fire suppression and rescue procedures using available personnel and equipment.

It is recognized that most successful emergency scene operation efforts involve a coordinated engine, ladder, and rescue company operation. When performing the evolutions included in this standard for the purpose of training, departments should use the number of personnel normally assigned to perform the initial operations at the scene of an emergency incident. The evolutions should be planned according to department policies and procedures. These procedures should outline steps for the planning of a drill. They often include bulleted lists for the type of equipment needed for a drill, the location of the drill, and who is required to attend.

Two aspects of initial fire attack are covered in NFPA 1410: (1) engine company operations, including handline operations, supply and operation of master streams, and automatic fire sprinkler system support; and (2) truck company operations, including ladder evolutions, the use of hoisting tools and appliances, the use of SCBA, and ventilation and illumination of an incident.

Individual firefighting evolutions involving the placement and connection of hoselines, the operation of hose streams and apparatus, the setting of ground ladders, the use of hoisting tools and appliances, the use of SCBA, and ventilation and illumination of an incident are the essentials of good fire department procedures. This standard provides the fire chief and other department officers with a method of measuring the effectiveness of evolutions that involve fire suppression and related tasks based on their normal first-alarm engine and truck company response.

With the exception of very small communities and isolated rural areas, the standard response to an emergency incident on the initial alarm is generally a minimum of two engine companies and a truck company. This practice is for several reasons. First, one engine company ordinarily cannot be expected both to operate the proper streams promptly for fast attack and to provide the necessary backup stream(s). Experience frequently has shown that small streams often prove to be inadequate. Second, fires commonly necessitate prompt application of hose streams from at least two positions. Finally, the possibility that an accident or mechanical failure will delay the arrival of one company is always present.

FIGURE 12-13 Live-fire training allows personnel to maintain both skills and confidence. When possible such training should be conducted at appropriate training structures, rather than acquired structures.

develop an appreciation of the necessity for personal safety during actual emergency evolutions. In order to ensure safety for all of the students and trainers during live-fire training exercises, all participants should have achieved a basic level of training by having completed the requirements for Firefighter I certification.

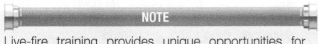

NOTE

Live-fire training provides unique opportunities for developing skills and self-confidence at all levels.

When and where possible, training-facility burn buildings should be used instead of acquired structures for all live-fire training events. When acquired structures are used, the building must be carefully inspected and prepared for the training evolution. Consideration shall be given to the presence of hazards, including utilities, pressure vessels, closed tanks, and interior finishes that may contribute to the rapid spread of fire. These hazards should be removed before the training starts. Consideration shall also be given to the age and condition of the structure itself, the presence of live and dead loads, and the probable effect that the fire will have on the integrity of the structure.

SAFETY

When acquired structures are used, the building must be carefully inspected and prepared for the training evolution.

In all cases, live-fire training evolutions should be conducted in accordance with the current edition of NFPA 1403, *Standard on Live Fire Training Evolutions.*

In addition to the safety considerations for their own personnel during live-fire training, fire departments should carefully consider the safety, health, and environmental consequences of their live-fire training activities on the community, both during and after the training activity.

Training for Officers

To the extent possible, fire departments should encourage the continuing education and training of their personnel, not only to meet the minimum qualifications for promotion to positions of greater responsibility but also to provide these officers with the knowledge, skills, and abilities needed to be good managers and leaders.

Such education and training is available through many community colleges, state and regional fire schools, various training conferences and seminars, and, of course, the National Fire Academy at Emmitsburg, Maryland. When such training programs are offered in-house, the subject matter of officer training courses should be designed to permit individuals to enhance their abilities and at the same time, meet the performance objectives outlined in NFPA 1021, *Standard for Fire Officer Professional Qualifications.*

Other Training

Specialized and advanced training should take place throughout a firefighter's work years. Such training includes courses in emergency medical procedures, recognition and control of hazardous materials, specialized rescue procedures, driving of emergency vehicles, equipment operation, fire investigation, fire prevention, fire safety education programs, and customer service. Fire departments should be alert to the changing needs of the fire service and to the needs of their personnel. As we said at the start of this section, training is the key to a firefighter's safety and the fire department's efficiency.

To ensure consistency in the delivery of training among various components of the fire department and over a span of time, all training should be conducted according to standardized lesson plans. Such lesson plans are intended only to assist instructors

NOTE

Training activities allow you to prepare your personnel and evaluate your equipment in what might be considered a practice session for the real event. No sports team or show company would dare perform without practicing. Why should you?

in providing effective, consistent, and standardized training and should be looked upon as providing the minimum amount of material to be presented. Lesson plans should never be considered as restricting new approaches for presenting material, improving teaching effectiveness, or maintaining student interest.

Training activities allow you to prepare your personnel and evaluate your equipment in what might be considered a practice session for the real event. No sports team or show company would dare perform without practicing. Why should you?

Company-level training includes classroom instructions, practice of evolutions, drills, street and building familiarizations, inspections, and pre-incident planning, to name just a few. Several hours of each workday should be devoted to such activity. Most departments have training policies and procedures that outline the required training activities. These policies often include what types of training are to be included in the training schedule, how and when they should be conducted, the re-certification steps, and how the drills will be evaluated.

Physical Training Is Important, Too

In Chapter 8, we noted that the requirements of firefighting are physically demanding. Part of having your company ready for dealing with emergency situations includes that all company personnel be physically and mentally ready as well. Good physical condition increases all firefighters' ability to perform their duties and reduces the likelihood of injuries while performing those duties. Once again, we see an important role in this process for the company officer. As the company officer, you should demonstrate good leadership and provide, by personal example, the benefits of a good fitness program. The readiness, success, and safety of your company depend upon it.

THE COMPANY OFFICER'S RESPONSIBILITY FOR COMPANY READINESS

Throughout this book, we have talked about your role as the company officer. Nowhere is that more evident than in the performance of company personnel during emergency response. The ability of the company to respond in a timely manner, to operate effectively to resolve the emergency, and to return safely is clearly within your area of responsibility. The pre-planning process discussed earlier in this chapter is but one part of the process. Let us take a look at the larger picture.

The entire life cycle of fire can be described by a series of steps starting with ignition and ending with extinguishment. In between these ten steps are nine intervals of time.

Step	Interval
1. Ignition point	Free-burning time
2. Recognition point	Permitted burn time
3. Detection point	Transmission time
4. Notification point	Alarm handling time
5. Alert point	Turnout time
6. Response point	Travel time
7. Arrival point	Setup time
8. Attack point	Control time
9. Fire-contained point	Overhaul time
10. Extinguishment point	

Starting with step 1, our fire ignites and grows. What happens over the next few seconds, or minutes, will depend upon many factors. This is where good fire prevention efforts pay off. Prevention activities like good housekeeping practices will minimize the fuel available for the fire. With regard to detecting and suppressing incipient fires, we also see the advantages of automatic equipment here, ranging from residential smoke detectors to elaborate integrated fire detection and suppression systems.

Time is of the essence throughout this process. Unchecked, the fire will grow exponentially. As it grows, it presents increasing dangers to civilians and firefighters alike. Given the nature of construction and furnishing today, flashover and collapse can be anticipated in many occupancies. As the fire extends beyond the room of origin, the risk to human life and property loss goes up dramatically.

At step 4, the fire department, or the emergency communications center, is notified of the fire. The fire department may or may not operate the emergency communications center, but it certainly should be involved with its operation. Ideally, calls are quickly processed and passed to the fire department for response.

At step 5, the company or companies assigned to respond to the situation are alerted. From this moment on, the timeliness and efficiency of the response is your responsibility, as the company officer. Turnout time, the next interval, is a reflection of the readiness of the company. For paid departments, the ability of firefighters to get to the vehicle, don protective clothing, and take their place within the vehicle should be a matter of a well-practiced routine. Running is not appropriate, but certainly every member of the crew must be focused and ready to respond at all times. NFPA 1710 allows 1 minute for this activity. (See "Why Time I So Important" in the adjacent box.)

For volunteer departments where no one is in the station, there will be additional time while members respond to the station. For the safety of all concerned, volunteer departments should have and enforce policies about safe driving of personnel vehicles and about how many and who must be on the apparatus before it can leave the station. However, once a volunteer crew is assembled, the events from here on should look the same in both volunteer and paid organizations.

Once the first vehicle leaves the station (step 6), we enter the interval known as travel or response time. Driver proficiency, knowledge of the response area and safety of operation are critical. All are your responsibility as the company officer. Safe arrival on-scene is marked by step 7, arrival time. NFPA calls for all of this to occur within 7 minutes or less. What happens over the next few seconds is also your responsibility and, to a large extent, your responsibility alone. Events here include a good size-up, giving a good size-up report, and giving directions to the personnel on-scene.

If preplanning was done, this is where it pays off. Firefighting always presents unknowns, but prior knowledge of the building, its contents, operations, and hazards will give the first-arriving officer the advantage of information and saved time. The proficiency of the company will be revealed here as well. Once given their assignments, well-trained personnel will be able to quickly go to work. This includes finding and establishing a reliable water supply, advancing hoselines, making entry, and locating the fire (see **Figure 12-14**). Compare the fire propagation curve in Figure 12-15 to the time of arrival and the attack point. This clearly demonstrates the need for first arriving crews to be efficient and timely in their actions. Time wasted here can not be made up.

The next benchmark is the attack point, step 8. What happens next is the control time. Again, proficiency in fire suppression operations is essential for the activity here to be effective. Fire officers and their companies should be constantly preparing for this moment. Because actual firefighting is down in many communities (good news), some firefighters do not get the firefighting experience of earlier generations (bad news). This deficit must be overcome at the

company level by training, practice, and discussion. (Every trade journal has an article every month on firefighting. Make sure your teammates read these articles.)

By now, you should see the importance of time in this series of events. And by now, you should see your importance as a company officer managing your resources to reduce these intervals of time.

ONE FINAL COMMENT

We started this chapter by drawing a parallel between the pre-incident planning process and the problem-solving process. You have identified the problem, gathered information, considered alternative solutions, and implemented a solution. Remember the last steps in the process; after training activities or after a real event, take time to go back and review the pre-incident plan to see how it worked. When appropriate, make the necessary changes. This process will enable you to have plans that really work and that people will readily accept.

FIGURE 12-14 Companies should practice basic skills on a regular basis.

WHY TIME IS SO IMPORTANT

NFPA 1710, *Standard for the Organization and Deployment of Fire Suppression Operations, Emergency Medical Operations, and Special Operations to the Public by Career Fire Departments*, provides the following guidelines for response times by career departments:

The fire department shall establish the following time objectives:

1. One minute (60 seconds) for turnout time

2. Four minutes (240 seconds) or less for the arrival of the first-arriving engine company at a fire suppression incident and/or 8 minutes (480 seconds) or less for the deployment of a full first-alarm assignment at a fire suppression incident

3. Four minutes (240 seconds) or less for the arrival of a unit with first responder or higher level capability at an emergency medical incident

4. Eight minutes (480 seconds) or less for the arrival of an advanced life support unit at an emergency medical incident, when this service is provided by the fire department without compromising safety. The fire department shall establish a performance objective of not less than 90 percent for the achievement of each response time objective specified above.[5]

NFPA 1710 spells out the specific parameters of services to be provided that allow the fire department to plan, staff, equip, train, and deploy members to perform these duties. This service delivery requirement is intended to have a fire department plan and situate its resources to consistently meet a 4-minute initial company fire suppression response and an 8-minute full first-alarm fire response assignment.

The logic for these requirements is based on the knowledge that an early aggressive and offensive primary interior attack on a working fire, where feasible, is usually the most effective strategy to reduce loss of lives and property damage. The graph shown in **Figure 12-15** represents a rate of fire propagation, which combines temperature rise and time. It roughly corresponds to the percentage of property destruction. At approximately 10 minutes into the fire sequence, the hypothetical room of origin flashes over. Extension outside the room begins at this point.

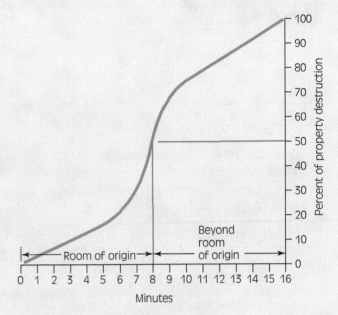

FIGURE 12-15   Fire propagation curve.

Consequently, given that the progression of a structural fire to the point of flashover (that is, the very rapid spreading of the fire due to superheating of room contents and other combustibles) generally occurs in less than 10 minutes, two of the most important elements in limiting fire spread are the quick arrival of sufficient numbers of personnel and equipment to attack and extinguish the fire as close to the point of its origin as possible.

The ability of adequate fire suppression forces to greatly influence the outcome of a structural fire is undeniable and predictable. Data generated by NFPA provides empirical data that rapid and aggressive interior attack can substantially reduce the human and property losses associated with structural fires (see **Table 12-1**).

Although NFPA 1710 is for career departments, we must remember that fires burn pretty much the same as in communities served by volunteer departments. Accordingly, response is just as important for volunteer organizations as it is in career departments. NFPA 1720, *Standard for the Organization and Deployment of Fire Suppression Operations, Emergency Medical Operations, and Special Operations to the Public by Volunteer Fire Departments*, provides similar performance standards, without the above standards for response time, for volunteer departments.

TABLE 12-1   Fire Extension in Home Structure Fires: 2002–2005

| | Rate per 1000 Fires | | |
Extension	Civilian Deaths	Civilian Injuries	Average Loss Per Fire
Confined fires (identified by incident type)*	0.08	9.52	309
Confined to room of origin	4.91	47.35	8,989
Confined to room of origin, including confined fires by incident type	*2.14*	*25.64*	*4,008*
Beyond the room, but confined to floor of origin	17.54	80.83	34,024
Beyond the floor of origin	27.77	60.33	59,317

* NFIRS 5.0 has six categories of confined structure fires, including cooking fires confined to the cooking vessel, confined chimney or flue fires, confined incinerator fire, confined fuel burner or boiler fire or delayed ignition, confined commercial compactor fire, and trash or rubbish fires in a structure with no flame damage to the structure or its contents. Although causal information is not required for these incidents, it is provided in some cases. In this analysis, all confined fires were assumed to be confined to the room of origin.

Note: Homes include one-and two-family dwellings, apartments, townhouses row houses and manufactured housing. These are national estimates of fires reported to U.S. municipal fire departments and so exclude fires reported only to Federal or state agencies or industrial fire brigades. National estimates are projections. Casualty and loss projections can be heavily influenced by the inclusion or exclusion of one unusually serious fire. Property damage has not been adjusted for inflation.

Source: Marty Ahrens, Fire Extension in Home and Residential Structure Fires, NFPA, January 2008.

PLANNING AND READINESS

Aaron Heller, Captain, Hamilton Township Fire District #9, Mercer County, New Jersey

The after-hours alarm came in: a fire in a 100,000-square-foot warehouse. Captain Aaron Heller's company members had done a preplan reconnaissance of the building and had a blueprint of it—marked with sprinklers, entrances, materials, and so forth. Because of that, they knew the building housed cosmetic chemical compounds, none of which was harmful and should not have instigated the alarm, even if chemicals were involved in the fire. They also had the number of a contact for the building, whom they immediately called.

The problem was they had been getting calls to go to the same building over and over when something little would trigger the alarm. That allowed the company members to let their guard down. On this night, it was a water flow alarm. When they came inside, they saw water running; Heller's company and a company that mutually aids it split up to find the cause.

It turned out that the thermostat on a processing oven had failed, and the oven was overbaking its contents. The oven was smoldering, and the heat had activated the sprinkler. Once the fire was extinguished, the company members turned a corner to leave . . . and walked straight into an unexpected cloud. Heller and his firefighters immediately reached for their SCBA masks. They radioed command, who called the hazmat team and ordered them to evacuate. Their contact for the building, who had come out in his pajamas, confirmed that the material would not stimulate combustion, but it WAS semitoxic and could have been dangerous if enough was inhaled.

For Heller, it was a dual lesson: It reminded him of the importance of preplanning, but it also emphasized that he should never allow his firefighters to get lackadaisical on their calls. You can go to the same site every day for weeks on a nuisance call, but the day you pull up and there is smoke coming out, you are already behind the eight ball.

Questions

1. What are the potential hazards faced in a large building such as that described in the story?

2. What are the lessons reinforced by this story?

3. What can fire departments do to reduce the frequency of repeat "nuisance" calls?

4. How can officers help avoid complacency in such situations?

5. What would you do in a situation like this?

LESSONS LEARNED

As company officers, you should develop a personal plan that will allow you to effectively lead your company to accomplish all of the tasks we have considered in this chapter. This will help the company become proficient in its tasks and ready to safely respond to help the citizens of the community it serves. The ability of the company to provide these essential services lies with your knowledge and dedication. Understanding the problems and the solutions increases readiness and reduces the challenges. Readiness also involves being ready to perform. Constant planning, training, and physical fitness are essential for company readiness.

KEY TERMS

area of refuge   portion of a structure that is relatively safe from fire and the products of combustion

fire control   activities associated with confining and extinguishing a fire

floor plan   a bird's-eye view of the structure with the roof removed showing walls, doors, stairs, and so on

life safety   the first priority during all emergency operations that addresses the safety of occupants and emergency responders

lock box   a locked container on the premises that can be opened by the fire department containing pre-planning information, material safety data sheets, names and phone numbers of key personnel, and keys that allow emergency responders access to the property

plot plan   a bird's-eye view of a property showing existing structures for the purpose of pre-incident planning, such as primary access points, barriers to access, utilities, water supply, and so on

pre-incident planning   preparing for operations at the scene of a given hazard or occupance

pre-incident survey   the fact-finding part of the pre-incident planning process in which the facility is visited to gather information regarding the building and its contents property conservation the efforts to reduce primary and secondary (as a result of firefighting operations) damage

Quick Access Prefire Plan   a document that provides emergency responders with vital information pertaining to a particular occupancy

standard operating procedure (SOP)   an organized directive that establishes a standard course of action

shelter in place   selecting a small, interior room, with no or few windows, and taking refuge there.

target hazards   locations where there are unusual hazards, or where an incident would likely overload the department's resources, or where there is a need for interagency cooperation to mitigate the hazard

REVIEW QUESTIONS

1. **a.** What is a pre-incident survey?
   b. What is a pre-incident plan?
2. Why is it beneficial for fire suppression personnel to develop pre-plans?
3. What information should be contained in a pre-incident plan?
4. What features of the building should be noted in the pre-incident survey?
5. What types of occupancies should be preplanned?
6. How is conducting a pre-incident survey or developing a pre-plan related to serving the best interest of the community?
7. How can the pre-incident planning information be made available during a time of need?
8. Should all members of the company be familiar with pre-incident plans for occupancies in their area? Why?
9. Why is it important to have a continuing program of company level training?
10. What are the benefits of multiple-company training activities?
11. What is your role as the company officer in company readiness?

DISCUSSION QUESTIONS

1. Identify several target hazards in your community. What makes these "target hazards"?

2. Take one of the occupancies you identified in the previous question and describe the process of gathering the information and developing a preplan for that situation.

3. How well do you preplan target hazards in your first-due response area?

4. What use do you make of these preplans?

5. What processes are in place for developing submitting and approving a preplan?

6. How can the preplanning process be extended to emergencies other than fire (for example, EMS, hazardous materials release, or technical rescue)?

7. How does your company training program measure up in preparing firefighters for emergency response?

8. Why is time so critical in response to a fire or EMS event?

9. FEMA's programs are focused on mitigation, preparedness, response, and recovery. Given the type of natural disaster most likely in your area, how should local, state, and federal resources prepare for and respond to such a disaster?

10. How well prepared is your fire department to deal with:

    a. A structural collapse with 50 or more injured occupants

    b. A HAZMAT event involving a toxic chemical and 10 or more affected citizens

    c. A response to an event such as you proposed in question 8

11. How does your company's training program measure up in preparing firefighters for emergency response?

ENDNOTES

1. For more information about this event, go to www.usfa.fema.gov/applications/publications/tr143.shtm.

2. Reproduced with permission from NFPA's, *Fire Protection Handbook*®, Copyright © 2008, National Fire Protection Association. This reprinted material is not the complete and official position of the NFPA on the referenced subject, which is represented only by the standard in its entirety.

3. A copy of the full report can be ordered by going to http:/www.usfa.fema.gov/fire-service/techreports/tr134.shtm.

4. *A Needs Assessment of the U.S. Fire Service.* Emmitsburg, MD: U.S. Fire Administration, 2003.

5. Reproduced with permission from NFPA 1710, *Organization and Deployment of Fire Suppression Operations, Emergency Medical Operations, and Special Operations to the Public by Career Fire Departments*, Copyright © 2007, National Fire Protection Association. This reprinted material is not the complete and official position of the NFPA on the referenced subject, which is represented only by the standard in its entirety.

ADDITIONAL RESOURCES

Arnold, Brian. "Training to Increase Base Knowledge." *Fire Engineering,* April 2008.

Bachman, Eric. "Preparing Facility Emergency Plans." *Fire Engineering,* January 2007.

Brannigan, Francis L. "Effect of Building Construction and Fire Protection Systems on Firefighter Safety." In *Fire Protection Handbook.* 19th ed. Quincy, MA: National Fire Protection Association, 2003.

Foresman, Douglas P. "Training Fire and Emergency Services." In *Fire Protection Handbook.* 19th ed. Quincy, MA: National Fire Protection Association, 2003.

Halton, Bobby. "Training Every Day is the Answer." *Fire Engineering,* March 2007.

Hewitt, Joel and Paul Landreville. "The Reluctant Company Officer." *Fire Engineering,* November, 2006.

Hildebrand, Michael S., and Gregory G. Noll. "Public Fire Protection and Hazmat Management." In *Fire Protection Handbook.* 19th ed. Quincy, MA: National Fire Protection Association, 2003.

Managing Company Tactical Operations, published by the National Fire Academy, Emmitsburg, Maryland. This three-part training course focuses on preparation, preplanning, and managing typical structural firefighting activities. The first two parts, called "Preparation," and "Decisionmaking," are particularly relevant. Contact your training officer or your state fire training director for additional information.

McAuliff, Pat. "Hot Drills." *Fire Chief*, April 2007.

Mills, Steve. "The *Firehouse* Meal: Where the Learning Continues." *Fire Engineering*, April 2008.

Oliver, Donald. "Multi-Use Mapping." *Fire Chief*, March 2008.

Prziborowski, Steve. "The Qualities of an Effective Engineer-Driver." *Fire Engineering*, February 2008.

Salka, John. Basic Skills and Tactics that Firefighters and Officers Must Perform. *Firehouse*, April 2008.

Schultz, Gerald R. "Determining Water Supply Adequacy." In Fire Protection Handbook. 19th ed. Quincy, MA: National Fire Protection Association, 2003.

Serapiglia, Michael. "Pre-incident Planning for Industrial and Commercial Facilities." In Fire Protection Handbook. 19th ed. Quincy, MA: National Fire Protection Association, 2003.

Shupe, Jeff. "25 Pointers for Your Engine Company." *Fire Engineering*, October 2006.

Smith, James P. "Vacant Buildings." Firehouse (December 2004)

Wallace, Mark. "Fire Department Strategic Planning 101." *Fire Engineering*, February 2009.

Wilson, Dean. "Fire Alarm Systems." In Fire Protection Handbook. 18th ed. Quincy, MA: National Fire Protection Association, 1997.

Wood, Thomas R. "Formats for Fire Hazard Inspecting, Surveying and Mapping." In Fire Protection Handbook. 19th ed. Quincy, MA: National Fire Protection Association, 2003.

Wright, Richard. "Organizing Rescue Operations." In Fire Protection Handbook. 19th ed. Quincy, MA: National Fire Protection Association, 2003.

13

The Company Officer's Role in Incident Management

As an incident commander, your job is to ensure the safety of personnel on the scene. It is very hard to stand back and not get involved in the nuts and bolts of the operation, especially if you are a new company officer. More often than not, however, the personnel operating on the scene are better off with you standing back with the "big picture," ordering needed resources and watching their backs.

A great example of this type of situation happened one night when my engine was dispatched to a motor vehicle accident. We were only the second engine to arrive on the scene. The first engine's company officer had given the size-up and passed command. There were two patients, one critical, the other serious, and the first engine crew was going to start extricating the critical patient.

I directed my crew to the serious patient and took command of the incident. The medic unit had been dispatched, and I asked for an additional engine because we needed another Hurst tool. As the police arrived I asked them to take care of traffic. It was at that time that my crew signaled me to help with our patient who had just lost consciousness.

As I was assisting my crew with the now critical patient, other cars started driving slowly through the scene; the medic unit had arrived and was assisting the other engine crew, and the additional engine I had requested arrived and the crew was asking for instructions—a request I did not even hear because I was so busy assisting my own crew.

Finally the company officer from the third engine came and asked me if I was still in command, and if so, what could he and his crew do to help. It was then that I realized that even though I had taken command of the incident, I was not literally in command, and the incident was rapidly getting out of control.

As a new company officer myself, I had fallen into the trap of being a worker and not an incident commander. I was comfortable working the line, and had taken up that task instead of calling for more resources and controlling the scene. Fortunately, I was able to regroup. I gave better directions to the police to stop all the traffic, and I split the medics so that both of the patients received advanced life support and rapid transport. Both patients did survive.

Had I been doing my job as an incident commander, I would have noticed a few things earlier—the battalion chief was not dispatched to the call because of a dispatch error, and the time from the first engine on scene to the third was about 14 minutes. Luckily, both patients made it to the Level 1 trauma center within the hour and survived. I was lucky no one got hurt, and I definitely learned a lesson.

—*Brian Dodge, Captain, Kent Fire Department, Kent, Washington*

LEARNING OBJECTIVES

After completing this chapter the reader should be able to:

13-1   Describe the supervisor's responsibilities regarding safety at the incident scene.

13-2   Explain how to initiate and continually apply, throughout the incident, the personnel accountability system.

13-3   Describe the elements of size-up.

13-4   Describe the function of a personnel accountability system.

13-5   Identify the principles of effective communications used to make clear assignments to subordinates.

13-6   List the types of emergency and non-emergency service units available for use at the incident.

13-7   List the elements of the communication model.

13-8   Describe order confirmation procedures during an emergency.

13-9   Describe the purpose of a standard operating procedure for emergency operations.

13-10   Explain how to initiate an evacuation order.

13-11   List the elements of a verbal command.

13-12   Describe the standard operating procedures govern response strategy, tactics, and operations, and customer service.

13-13   Identify departmental operating procedures that describe the application of an incident management system.

13-14   Evaluate the need for various components as they apply to the emergency incident.

13-15   Describe the purpose of performing a post-incident analysis.

13-16   Outline the components typically covered in a post-incident analysis.

13-17   Describe the parts of a unit critique.

13-18   Explain how to conduct a unit critique.

13-19   Explain the collection of data and how to distribute the results of the critique

13-20   List potential informational resources available for emergency response.

13-21   List the main components of an incident management system.

*The FO I and II levels, as defined by the NFPA 1021, the *Standard for Fire Officer Professional Qualifications*, are identified in different colors: FO I = black, FO II = red.

INTRODUCTION

As you pull up in front of Ms. Carmen Garcia's house, you see enough smoke coming out of the back of the house to know that this is no false alarm, nor is it someone's backyard barbecue. Ms. Garcia is in her front yard, her hands clutching a wet towel. She is understandably upset. Her kitchen is on fire!

For you or any first-arriving officer, this is show time, time to put all of your training, all of your experience, and all of your equipment to work. And for you, and for all of those overqualified, hard-charging firefighters you brought with you, this is why you joined the fire service. Will things go well? Will you stop the fire with no additional damage to Ms. Garcia's house? Will you and your firefighters operate safely?

Strong leadership is needed at events like these. We know that every group has a leader, and in this case we expect the officer to be the leader, to be in charge, to be making decisions and give directions. You—the company officer—should be managing the event using a proactive management style.

NOTE

Good scene management focuses on the priorities of the incident, provides for the safety of all concerned, and seeks to reduce further loss.

Good scene management focuses on the priorities of the incident, provides for the safety of all concerned, and seeks to reduce further loss. The old philosophy "grab a line and put the wet stuff on the red stuff" has long been outdated. Today, as the company officer, you are expected, within a very short timeframe, to take command, provide leadership, identify problems, establish priorities, allocate resources, coordinate activities, and protect the safety of firefighters. When these things occur, you are in a proactive mode and are properly fulfilling the leadership role of the position you hold.

This chapter only introduces these concepts. We cannot cover all the aspects of incident management in one chapter, or even in one book. Clearly, considerable learning and experience must be integrated to become an effective incident commander. But, as company officer, you should be able to command first-alarm resources. For small events, you should be able to manage the event from start to finish. For larger events, you should be able to set the foundation for what will follow, pending the arrival of a more senior officer. In either case, action taken during the first 5 minutes sets the tone for the entire alarm and has a significant impact on the overall outcome of the event.

> **NOTE**
>
> As the company officer, you are expected to take command, provide leadership, identify problems, establish priorities, allocate resources, coordinate activities, and protect the safety of firefighters.

At an emergency event, the incident commander is responsible for the overall management of the incident and for the safety of everyone involved. At such events, the incident commander must assume command, evaluate the situation, initiate, maintain, and control communications, formulate strategy, and develop an organization to fit the needs of the event (see **Figure 13-1**).

> **FIREGROUND FACT**
>
> Action taken during the first 5 minutes sets the tone for the entire alarm and has a significant impact on the overall outcome of the event.

INCIDENT PRIORITIES

The Plain City fire department's mission statement includes a phrase about saving lives and property. Your decision-making process at the scene of any emergency should focus on this same statement: Save lives and property. The order of the fire service priorities; life safety, incident stabilization and property conservation, should always serve as the guiding principles for prioritizing activities at an emergency incident.

> **NOTE**
>
> The order of the fire service priorities; life safety, incident stabilization and property conservation, should always serve as the guiding principles for prioritizing activities at an emergency incident.

Your first priority is life safety for the civilians who may be at risk and for the firefighters under your command. Life safety must always be the first consideration of emergency responders. Your first responsibility is to safeguard human safety; to protect the safety of those under your direct supervision; and to take all reasonable risks in protecting the lives of the public you are sworn to protect. We stressed safety in Chapters 8 and 12, but it cannot be stressed enough that you, as company

FIGURE 13-1   Action taken during the first 5 minutes sets the tone for the entire alarm and has a significant impact on the overall outcome of the event. (*Photo courtesy of Andy Thomas*)

officer, are responsible for maintaining the accountability of the firefighters assigned to you, even if your department has some other type of scene accountability system. You need to know where they are and that they are operating safely. It is your job to make sure they go home at the end of their shift (see **Figure 13-2**).

In structural firefighting, we use the term rescue to describe the activities associated with the civilians who may be at risk. Rescue includes protecting the occupants, removing those who are at risk, and searching for those who may not be able to help themselves. In some cases, based on the number of victims, the location of the fire and available resources, stretching a hose line may be the most appropriate action to save lives and affect a rescue.

The second priority is incident stabilization. You cannot be held accountable for what happened on scene before you arrived, but you are clearly accountable for what happens after you arrive. You must be sure that your information, your planning, and your actions all

FIGURE 13-2 The incident commander's first priority should be life safety, including the safety of the firefighters involved in the operation. (*Photo courtesy of Michael Regan*)

support an effective, coordinated, and safe operation. In firefighting operations, we call this fire control. In a hazardous material spill, we attempt to contain or mitigate the product. In both cases, we want to keep from doing any additional damage.

The third priority is property conservation. Firefighters often think of property conservation in terms of salvage operations, often done after the fire is controlled or even extinguished. Property conservation should be addressed in pre-incident planning efforts and started upon arrival of fire suppression forces. Property conservation involves using the least destructive means for entering, using early and aggressive ventilation, using a coordinated fire attack with the proper selection of hose and nozzles, effective fire stream management, and throwing salvage covers early in the event before water has a chance to do any damage. Property conservation does not mean putting out the fire at the expense of losing the building. Ms. Garcia will be pleased if you put her fire out. She will be delighted if you do it without destroying her house.

OFFICER ADVICE

Positive property conservation measures can have a significant positive impact on the property owner and the community.

Positive property conservation measures can have a significant positive impact on the property owner and the community. Irreplaceable possessions can be saved, damage reduced, and the building reoccupied within a minimum period of time. If the building is a commercial occupancy, the rapid resumption of operations will mean that workers are back at work earning money and spending that money in the community.

Effective property conservation was demonstrated by a department when a they responded to an alarm at a local computer manufacture. A plumber caught some insulation on fire within the walls of the facilities test lab. The fire crew during a preplan inspection noted that the outside of the lab walls had copper sheeting behind the sheetrock to limit electrical interference. Therefore, they accessed the walls from the opposite side, which not only made access easier, it reduced the cost of repairing the wall.

Some departments have **benchmarks** to note the accomplishment of these three priorities. These benchmarks are transmitted to the incident commander who records the time. In some departments, they are also transmitted to the communications center. This transmission officially records the time of the event and provides a concise progress report for senior fire department staff officers who may be monitoring the event by radio.

For fire-related activities, these conditions are as follows:

Incident Priority	Transmission
Life safety search, rescue	"all clear"
Incident stabilization	"under control"
Property conservation	"loss stopped"

Things are not always tidy during emergencies, and there are times when the incident commander must overlap or mix activities to make these priorities happen. For example, during interior fire operations, it may be necessary to control the fire while searching and rescuing trapped civilians. In the auto accident scenario you may have to stabilize the vehicle before you start to treat or remove victims (see **Figure 13-3**). Both activities protect your own safety as well as the safety of the victims. In both examples, life safety (the first priority) is the focus of all of the activity.

At the scene of incidents, you have to quickly sort out a lot of information and give directions. This process is complicated, but it can be simplified by using the standard decision-making process at all emergency events. Using a standardized approach helps you focus on the incident priorities just mentioned.

There are other benefits of using a standardized approach. When incident commanders follow a consistent pattern, they reduce the tendency to overlook important activities. For others at the scene, your plans and your directions will more likely be understood when using a standardized approach. A good approach is to use the command sequence.

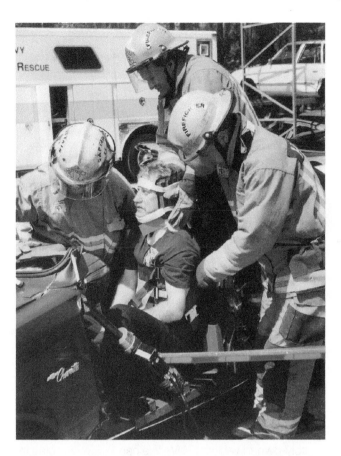

FIGURE 13-3 Incident stabilization focuses on reducing injuries and loss after operations are started. (*Photo courtesy of Michael Regan*)

THE COMMAND SEQUENCE

It is critical you know, understand, and follow these command priorities, as they are fundamental to all incident command operations and the basis of the application of the Incident Management System (IMS). If you do not follow these simple rules, then it is likely that chaos and unsafe operations will follow. The steps involve size-up, developing an action plan, and implementing the action plan. This system provides a tool to assist you in establishing the necessary objectives in order to achieve the goals of life safety, incident stabilization, and property conservation. The steps involve size-up, developing an action plan, and implementing the action plan. Here the incident commander gathers information, analyzes the information, determines and ranks the problems, defines solutions, and selects tactics. Finally, the incident commander issues orders to implement the tactics selected.

NOTE

The command sequence is size-up, developing an action plan, and implementing the action plan.

Understanding Size-Up

Before any action is undertaken, you should take a moment to analyze the problem and think about the action that you are going to take. That statement is true in nearly all of your endeavors and certainly true for those where you are leading personnel at the scene of emergency activities. The first part of this process is called size-up. Size-up is process of gathering and analyzing information that is critical to the outcome of the event.

"WHAT'S SHOWING?"

For fires, there are three possible situations: nothing showing, smoke showing, and fire showing. Each situation should cue the first-arriving officer on the appropriate action to take, and for the instructions to be given to other units. Your pre-incident planning, training, and department SOPs should address these three commonly encountered situations.

Most of us think of size-up as what is seen through the windshield when you pull up in front of a place such as Ms. Garcia's residence. It is all of that and a whole lot more (see **Figure 13-4**). Size-up includes all of the information available. It includes the information gathered during your pre-incident survey. It incorporates your preplanning efforts, based on the information gathered from those surveys. During size-up, you should also consider your department's SOPs and the resources available to deal with the situation.

Hopefully, you have good pre-incident plans for your community. Given the time of day and the nature of the occupancy, the first-arriving officer should be able to make some estimate of the nature of the life hazard encountered. Even without a formal or written

FIGURE 13-4 Size-up involves more than what you see upon arrival.

SAFE OPERATION ON THE HIGHWAY
Ron Moore

As our line-of-duty-death statistics continue to show with each passing year, responders working in or near moving traffic are placed at significant risk of injury or death. Regardless of whether you volunteer your services or are paid for what you do, when you are at a crash or fire scene and are working in or near moving traffic, you are considered a "highway worker" and fall under federal Department of Transportation (DOT) standards and regulations. The content of these national regulations is just now becoming known to the fire service and the impact of these federal highway standards is beginning to have an effect on our incident scene operations.

Laws and regulations did not stop the alcohol-impaired drivers that killed several of our firefighters last year in struck-by incidents. Even DOT standards will not prevent the speeding 18-wheeler from crashing into your emergency scene. Proper highway response training, improved highway safety, PPE, special techniques for advance warning to approaching motorists, and the skills necessary to create a physical barrier between you and the moving traffic will.

There is an art to properly and effectively blocking with an emergency vehicle at a highway incident scene. The driver must have an uncanny feel for the size of their vehicle regardless of whether it is a sedan or a 40-ton, tandem-axle ladder truck. The process of blocking is done by the apparatus driver just as they come to a stop at the incident scene. With the intent being to physically block the shoulder of the road and the closest lane of traffic or to block off several lanes of traffic, the emergency vehicle slows and before coming to a complete stop, makes a sharp turn to the right or left. This slants their vehicle at an angle across the lanes of traffic.

You may call it the Expressway; your partner may call it the Interstate. It may be known as the Thruway, the Toll Way or by other names in your local area. What it is to you is a firefighter killer. A multi-lane, divided highway, having a high posted speed limit, lots of traffic day and night, few access points to get on and off, and few if any intersections is technically a limited access highway, the leading type of highway incident location known to kill emergency responders.

More firefighters and EMS personnel have been struck and injured or killed on limited access highways than any other type of roadway system in the U.S. Contributing factors such as higher speeds, heavy volumes of traffic plus a greater possibility of heavy trucks approaching the scene make these roadways extremely dangerous. Two major reasons for responder deaths on these highways have emerged. Lack of proper advance warning to approaching traffic is one of the major causes. Attempting

A fire department vehicle provides a visible barrier between the workers and oncoming traffic.

to cross the multiple lanes of the highway to get to the other side on foot was the last thing many dead firefighters were trying to do as they were struck and killed.

There are ways to improve our safety on these big roads. By department policy, we can forbid responding directly to a limited access highway scene in a privately-owned vehicle (POV). We can forbid stopping in lanes traveling in one direction and crossing the highway median or barrier to access a crash or fire scene in lanes of traffic traveling the opposite direction. That's a sure way to get killed! We can also send a second major apparatus to establish an upstream block that increases the advance warning at the incident scene.

The Second Company to Block

When a call is received for an incident on a limited access highway, an additional apparatus should be dispatched along with the first-due companies. A tandem-axle ladder truck is preferred due to its long length and heavy weight. In lieu of that, a tanker (tender) is a good vehicle to send. If no ladder truck or tanker/tender is available in your department, the recommendation is made to add a second engine company to your initial assignment.

The primary function of this second vehicle is NOT to work at the crash or vehicle fire scene. Their principal function is to establish a second upstream block at least 1/2-mile from the main activity area. They block whatever lanes of the highway must be shutdown and any additional shoulder areas with their large vehicle. They assume their vital blocking position and set up so that their apparatus and its warning devices make approaching motorists aware that there is an emergency scene ahead. Traffic cones can be extended downstream of the blocking apparatus towards the activity area to keep vehicles out of the shadow area. A DOT color-coded retro-reflective sign can also be deployed upstream to further expand the advance warning area.

All slowing of approaching traffic, lane changes, and merging of traffic should happen upstream of this blocking company. A PD unit should be working upstream of this blocking company to assist with upstream traffic control. When done properly, by the time traffic actually passes the main activity area, vehicles have slowed to a manageable speed, are moving smoothly in the open lanes, and are following each other in a controlled and predictable manner.

Remember, the ambulance, first-due engine and possibly your heavy rescue unit, if that is what you send to crash scenes, should not be used as this upstream second blocker. Those units need to be close to the patients and the wrecked vehicles. The first-due engine should already be in a blocking position upstream of the crashed vehicles and the ambulance and heavy rescue should be in the downstream shadow.[1]

pre-incident survey, officers should know enough about their first-due response area to know the type of construction, access points, water supply, probable fire behavior, and safety concerns that are likely to be encountered. You should not wait until an alarm comes in to learn about the structure that is located just down the street from your fire station!

Your knowledge of the first-due area should also include information about the best response route, the natural barriers (rivers and terrain), the human-made barriers (railroad tracks and highways), and all of the significant hazards that may be encountered. Your mind should be like a little computer, constantly storing information about the area, and analyzing the information to deal with the problems that might be encountered.

Along with an assessment of the problem, you should have a good assessment of the resources that can be expected to help deal with the problem. Again, the time of day may be important, not just in terms of the conditions that are likely to be encountered but also in terms of the available resources.

OFFICER ADVICE

Remember that size-up is the first step in the action plan. With good size-up information, you are better able to develop an action plan.

Depending upon the situation, there may be some non-fire department resources on-scene as well. Many communities notify the police when the fire department is called out, and they often arrive ahead of the firefighters. In some cases, they may undertake some of the size-up activities for you. A good working relationship needs to be established before this is done. Help the police by offering them training on providing the information you need and on what to do and what not do before your arrival (i.e., where to park and not park their police cars; do not kick in the doors; do not enter the structure without proper PPE, training, and resources, etc.). Private ambulance companies often respond to structural fires, for the same reasons that

a fire department emergency medical services (EMS) unit is assigned on working fires in many larger cities. Regional or state hazardous materials teams may be called when the local department is not prepared to deal with hazmat problems. You should know about these resources and how they can help at the scene of any emergency. If you do not know what resources are available in your community, how can you call for them?

Along with all of this information, you will make your own observations while en route and upon arrival at the incident. When you see Ms. Garcia's house, you can focus on the specific problems you face (one excited woman; one kitchen on fire). You now have a moment to make a rapid mental evaluation of the situation and of the relevant factors that may be critical to this unique incident.

OFFICER ADVICE

As the first-arriving officer, you must assess the conditions calmly and be proactive rather than reactive to the situation.

For many officers, the first few moments at the scene create a lot of stress. For a new officer, every call is a significant event. With experience and maturity, most officers routinely deal with these as ordinary events. However, even seasoned veterans sometimes encounter situations beyond their previous experience. Hopefully, they will be able to draw on their previous experiences and remain calm, even though the event seems overwhelming. In any case, as the first-arriving officer, you must take a moment to assess the conditions calmly and be proactive rather than reactive to the situation. Many refer to this as accessing the slide tray of pictures in your head, comparing the incident before you to other similar incidents you have dealt with and noting what actions were and were not successful at those incidents. This forms the basis for what actions you are likely to take to mitigate the current emergency. Pre-incident plans and standard operating procedures need to be considered and used

LESSONS FROM A DEADLY HOUSE FIRE IN STOCKTON, CALIFORNIA

On February 6, 1997, in Stockton, California, a fire began early in the morning in a second-floor master bedroom occupied by an 82-year-old resident. The small lot size, a crowded driveway, and limited access from the sides and rear complicated the fire response. A second-story addition not visible to firefighters entering the building from the front subsequently collapsed, killing two firefighters. A third firefighter was seriously injured. The resident also died from carbon monoxide poisoning.

The U.S. Fire Administration (USFA) conducted a review and noted that large, open areas and nonstandard construction should have been indicators of potential collapse in a burning residence and that the existence of these hazards must be conveyed to interior fire crews whose vantage point may prevent them from being aware of the potential danger.[2]

The review also pointed out that a standard accountability of personnel system is a critical part of an incident response, noting that when something does happen, the system should be in place, and the crews should have the unit discipline and self-control to quickly account for all personnel and report their status to the incident commander. Therefore, it is essential that a standardized accountability system be initiated on every scene. The system should remain in place until the conclusion of the incident.

Other lessons learned cited in the report include the following:

- An incident commander must be responsible for overall operations and should establish and maintain command throughout an operation as self-initiated and self-directed efforts of firefighting crews lack coordination and control.
- An operator can be assigned to handle radio traffic, but the incident commander must be aware of all important messages, direct all assignments, and control the communications process.
- Rapid intervention teams outfitted with appropriate equipment should be provided to rescue trapped or missing firefighters.
- An insufficient number of radios for firefighting personnel impacts the success of a fire response.
- Smoke detectors in homes and second exits from rooms for evacuation of occupants enhance the likelihood inhabitants of a burning residence can escape.

"This was a terrible fire that caused the death of two firefighters and an elderly resident," said U.S. Fire Administrator David Paulison. "Lessons learned from this review are relevant to every department in the nation and underscore the importance of communication, training and established (accountability) systems."

---

if possible. Hazards need to be identified and incorporated into the planning process (see **Figure 13-5**). When initiating a size-up, it may be helpful to answer these three questions:[3]

- What have I got?
- Where is it going?
- How can I stop it?

FIGURE 13-5 The first arriving officer should take note of any hazards that may be present.

During fire-related events, you should be concerned about the following.

Building Construction

The important thing to remember here is that each type of building presents us with a different set of problems in dealing with fire spread and fire control. See Chapter 10.

Location and Access

Finding the building is part of your job. In addition, you need to know how to get past any fences and doors you may encounter. You should also have obtained information about exposures, water supply, terrain problems, and other features that impact on your operations (see **Figure 13-6**).

Occupancy and Contents

For Ms. Garcia's house and its 2,000 square feet of living space, the first-arriving officer can anticipate the size and contents of the structure just from knowing the neighborhood. Your pre-incident survey and planning efforts really pay off for larger structures.

FIGURE 13-6 Gaining access to the building can be a challenge.

FIGURE 13-7 Certain structures have obvious hazards. What hazards do you see here?

Location of the Event within the Structure

You also need to know where the situation is located within the structure. Again, your prior knowledge will be useful here. If your problem is a fire, knowing its location helps you answer all three of the questions: "What have I got, where is it going, and how do I stop it?" Lacking this information will adversely affect your ability to answer any of these questions.

THE IMPORTANCE OF PREPLANNING

Imagine the difference of pulling up to a well-involved fire in a structure of 20,000 square feet of floor space in a building that you have carefully surveyed and planned. Contrast that with your situation while standing in front of a similar building that you have never seen before (see **Figure 13-7**). You may have trouble finding the front door.

Time of Day and Weather Conditions

Of course, you will know the time of day and the weather situation, but you should always be thinking about how these conditions will affect both the situation and how you can mitigate the situation. For example, a fire at night in any residential structure presents us with significant life safety concerns. A fire in a commercial building at night may present reduced life safety concerns but will likely have a delayed discovery. Your ability to get to that fire may also be delayed while you attempt to gain access. Again, you see some advantages to your pre-incident planning activity.

You will also need to consider the weather and other **environmental factors** that impact on your operations. Extremely hot or extremely cold weather affects firefighting operations and firefighters. Cold weather

FIGURE 13-8 Severe weather can hamper operations.

can make life difficult, especially when the cold is mixed with precipitation (see **Figure 13-8**). Cold weather will probably slow your response and your operations once you are on the scene. Cold weather can also have an impact on any citizens affected by the event. Imagine evacuating all the occupants of an apartment house or office building while your firefighting activities are taking place. In nice weather, they will have a chance to watch your operations and visit with one another. In cold weather, they may be at greater risk from the elements than they were from the fire. The same can be said for those involved in some sort of transportation accident. We should not lose sight of the first priority: human life.

Several acronyms can be used for keeping track of all of these factors. One of the traditional ones is COAL WAS WEALTH:

- **C**onstruction
- **O**ccupancy
- **A**pparatus and personnel
- **L**ife safety
- **W**ater supply
- **A**uxiliary appliances
- **S**treet conditions
- **W**eather
- **E**xposures
- **A**rea
- **L**ocation
- **T**ime
- **H**eight

Keeping track of all this information can be quite a challenge. Many departments use a tactical worksheet for larger events, but there are benefits of having some sort of checklist for all events. Keep in mind that the main reasons for using an incident command system and some form of checklist are for firefighter safety and to maintain an accountability of the firefighters at the emergency scene. Many officers keep a copy of some sort of worksheet folded up in the pocket of their turnout coat. A sample tactical work sheet is shown in **Figure 13-9**. This sheet provides a checklist for size-up factors, a place to record assignments, and a space for a simple diagram.

RISK ANALYSIS

Although there are dangers associated with all activities, firefighting certainly ranks high as a risky activity. One of the considerations for the first-arriving officer is a risk/benefit analysis that determines the strategy to be initially used for mitigating the event. Where lives are at risk, any reasonable risk that may result in saving those lives is warranted. However, where lives are not at risk or cannot be saved, you should never place firefighters at risk. We have been reading about this for years, yet we continue to read of firefighters taking unnecessary risks to save buildings that are scheduled for demolition. If you encounter residential fire where the structure is almost fully involved and a dark hot smoke is emitting from every

crack and crevice in the structure, it is very unlikely that anyone in the structure is still alive or savable. This is one of the toughest decisions that officers will face, but you have to be realistic about what you can and cannot accomplish. Is it worth risking a firefighter's life to save a corpse?

To sum it all up, in the simplest of terms this means after you look at what is showing, consider the type of building construction, consider the location and access, examine occupancy and contents, consider the location of the event within the structure, and the environmental conditions you look at what you can save based on what you risk; you make a go or no go decision. This is all done in your head, in seconds. This may all sound overwhelming and for an inexperienced officer it is. That is why training is so important. Failure to examine any of these factors could lead to an ill fated decision.

SAFETY

Where the risk justifies entering the building, you should realize that just because you can safely enter the building does not mean that you can safely stay in the building.

Where the risk justifies entering the building, you should realize that just because you can safely enter the building does not mean that you can safely stay in the building. For many types of buildings, modern construction techniques are not designed to enhance structural integrity during fire conditions, and some modern buildings will collapse, in some cases, sooner than one might expect. Normally the arriving firefighters do not know how long the fire has been burning before they arrive, nor are they likely to know how the fire has impacted on the structural integrity of the building. The length of time firefighters can be expected to sustain an interior attack is usually very limited at best.

If there are sufficient resources on hand to mount an aggressive interior fire attack, the save might be worth the risk. But in many cases, especially where resources may be limited or where the building's condition is doubtful, it may be best to focus resources on protecting the exposures. Decisions regarding interior operations are among the hardest you can expect to face. Remember that the fire is not your fault and that you may not be able to save all of the burning buildings you are likely to encounter. Property that is already lost is not worth losing a firefighter (see **Figure 13-10**).

Multi Agency Tactical Worksheet

Address: _____

Incident # _____

Time of Alarm: _____

OFFENSIVE MODE   TIME: _____
DEFENSIVE MODE   TIME: _____
OFF/DEF MODE   TIME: _____

S=STAGED   R=REHAB

Unit	S	Assigned	Reassign	R	Bldg. Fire	10-48 PIN	Command	Personnel Accountability	TIME	PAR
					Fire attack	Patient care	Safety	OFF/DEF		
					RIC	Safety	PIO	Major Event		
					S & R	Extrication	Staging	30 Minutes		
					Water Supply	1 3/4 line	Investigation	Code 4		
					Ventilation	Landing Zone		Primary AC		
					Back-up line	Basic		Secondary AC		
					Utilities	Major				

WEATHER

	TIME
WIND SP	
WIND DIR	
TEMP	
HUMIDITY	

MEDICAL SUPPORT

Command
Ems unit
Ems unit
Ems unit
Ems unit

VICTIMS

Green
Yellow
Red
Blue
Black

INCIDENT SUPPORT

KG&E CODES
A=Gas meter/elec. at transformer
B=Gas/Electric meters
X=Gas:CO Leak
Y=Elec: Public Hazard
Z=Elec: Non-Public Hazard
TRAFFIC CONTROL
RED CROSS
SALVATION ARMY
HEALTH DEPARTMENT

Mutual Aid

COMMAND ASSIGNMENTS	FIRE ATTACK	RIC	SEARCH & RESCUE	WATER SUPPLY	VENTILATION
OFFICER AND/OR UNIT					
LOCATION					
CHANNEL					

UNITS ASSIGNED

FIGURE 13-9 A tactical worksheet helps keep track of the information.

FIGURE 13-10 Property that is already lost does not justify risking firefighters.

NOTE

Property that is already lost is not worth losing a firefighter.

Selecting the Correct Mode

There are three modes of operation at a fire: offensive, defensive, and transitional.[4] When in the **offensive mode**, operations are conducted in an aggressive interior attack mode, taking the attack to the fire. You must determine that the interior attack is worth the risk. You must also determine if you have sufficient resources to safely mount and sustain a coordinated attack. The offensive mode is a good choice: Many fires are put out this way and lots of lives and property are saved with this mode of attack.

FIREGROUND FACT

There are three modes of operation at a fire: offensive, defensive, and transitional.

When fire conditions have advanced to the point where there is little chance of saving lives or property, or when there are insufficient resources on hand to safely mount and sustain an interior fire attack, you must resort to defensive operations (see **Figure 13-11**). In the **defensive mode**, operations are conducted from a safe distance outside of the structure and may focus more on containing the fire rather than on extinguishing it.

The third mode is called transitional. During the **transitional mode**, operations are changing from either an offensive to a defensive mode, or from a defensive to an offensive mode. Obviously, the transitional mode is not something you would start with, but rather a phase you pass through as

FIGURE 13-11 At times, the mode will be obvious.

operations are shifted from one mode to another, Operations may have started in the offensive mode, but firefighters were unable to make any headway, so the officers prudently withdrew their forces from within the building. Conversely, the event may have started as a defensive operation, with firefighters knocking down some of the fire or providing a holding operation while sufficient companies arrived and prepared for a coordinated interior operation. In either case, the process of shifting from one mode to another presents its own risks and has to be carefully managed. Effective fireground communications become very important during these precarious moments.

Command Decisions

Another critical decision that has to be made is in regard to your personal role in the initial operations. There are three modes of command: command, attack, or a combination of both. Once again, you have a choice to make: You can take command, elect to start combating the situation, or seek a hybrid solution of trying to do both.

FIREGROUND FACT

There are three modes of command: command, attack, or a combination of both.

In selecting the **command role**, the first-arriving officer elects to maintain control of the incident and coordinates the activities of the first-arriving companies. In essence, the company officer becomes the first incident commander. As the incident commander, you have overall control of the event, doing the size-up, directing personnel on scene, and giving instructions to arriving companies by radio. For

RISK MANAGEMENT DURING EMERGENCY OPERATIONS

The IMS starts with the arrival of the first fire department company. The first company to arrive integrates risk management into the routine functions of incident command. As indicated in NFPA 1500, *Standard on Fire Department Occupational Safety and Health Program*, the concept of risk management shall be utilized on the basis of the following principles:

1. Activities that present a significant risk to the safety of members shall be limited to situations where there is a potential to save endangered lives.

2. Activities that are routinely employed to protect property shall be recognized as inherent risks to the safety of members, and actions shall be taken to reduce or avoid these risks.

3. No risk to the safety of members shall be acceptable when there is no possibility to save lives or property.

4. In situations where risks to fire department members are excessive, activities shall be limited to defensive operations.[5]

As indicated in item 2, "actions shall be taken to reduce or avoid these risks." Identifying potential safety concerns to members and taking actions to reduce risks to firefighters are without a doubt two of the most important things that can be accomplished.

The following are just some of the ways to reduce the overall risks to members operating at the scene of emergency incidents and that reinforce the purpose and function of the personal accountability system:

1. Written guidelines shall be established and used that provide for the tracking and inventory of all members operating in an emergency incident.

2. All members operating in an emergency are responsible to actively participate in the department's accountability system.

3. The incident commander shall be responsible for the overall personnel accountability for the incident. The incident commander shall initiate an accountability worksheet at the beginning of the incident and maintain the system throughout the operation.

4. The incident commander shall maintain an awareness of the location and function of all companies assigned to an incident.

5. The incident commander shall implement branch directors, division/group supervisors or team leaders when needed to reduce the span of control for the incident commander.

6. Branch directors, division/group supervisors or team leaders shall directly supervise and account for companies operating under their command.

7. Company commanders are accountable for all company members, and company members are responsible to remain under the supervision of their assigned company commander. Members shall be responsible for following the personnel accountability system procedures, which shall be used at all incidents.

8. The IMS shall provide for additional accountability personnel based on the size, complexity, or needs of an incident. The implementation of division/group supervisors or team leaders can assist the incident commander in this area by reducing the span of control.

9. The incident commander shall provide for control of access to the incident scene.

10. A department shall adopt and routinely use a standard personnel identification system to maintain accountability for each member assigned to an incident. There are several accountability systems used during structural firefighting.

11. The personnel accountability system shall provide an accounting of those members actually responding to the scene in each company or apparatus.

12. The IMS shall include standard operating guidelines that use "emergency traffic" communications to evacuate personnel from an area where an imminent hazard is found to exist and to account for their safety.

smaller events, this is standard procedure. You may decide that one or two companies are sufficient to handle the event and release the remaining companies from the assignment. In such cases, you, as the first-arriving company officer, will likely remain in charge of the event, complete the report, and perform all of the other duties of the incident commander.

The second option that you have is to personally go into action. This option is referred to as the **attack role**. In this case, you decide that the urgency of the situation warrants your immediate personal action and that such action will likely have a significant positive influence on the outcome of the event. This option may be justified where an immediate rescue is required and can be safely undertaken or where a quick fire attack may knock the fire down and prevent significant extension. Once again, you are asked to make an on-the-spot risk/benefit analysis and then take immediate action with limited resources based on that analysis. Times like these can be difficult.

When you select the option of going into action, you should pass command to another officer who is or who soon will be on-scene. There has been some misunderstanding about this process, especially in departments where it is not often practiced. When you select to commit to the attack mode, you communicate this over the radio to the next-arriving company; however, you essentially remain in command until another officer is on scene and officially assumes command. You are also effectively the scene safety officer until that function is assigned to another officer. If the second-arriving company is another engine company, their primary function is usually to establish a reliable water supply. The officer of that company may be in a good position to become the incident commander. Obviously, too much of a good thing can be harmful, and most departments have a rule about passing command again. The general rule is that the next time that command is passed, it must be to a chief officer. This policy precludes the constant turnover of command as companies arrive.

In some situations, you may not really have the luxury of a choice, especially in smaller departments where the initial response might be limited to one or two engines and a handful of firefighters. Additional resources could be about a day away. In this case, you are likely to be involved with a combination of both command and hands-on firefighting activities. This arrangement is obviously

COMMUNICATIONS ARE THE GLUE THAT HOLDS THE COMMAND SEQUENCE TOGETHER

Effective management of emergency situations depends on proper use of command and communications. An effective command system defines how the incident is controlled and how the management of the incident is established and progresses. Communications tie the command system together and provide an opportunity for everyone to pass along and understand essential information. When you use the communications system, your messages should be brief, impersonal, and to the point.

not good as it certainly places you in a position where the best you can hope for is to try to perform one task without serious risk of compromising the other. In such cases, another officer should attempt to get on-scene and relieve you of the responsibilities of command.

OFFICER ADVICE

The initial report should address four questions; What have I got, what am I doing, what do I need, and who is in charge.

Brief Initial Report

Part of the initial size-up is providing a good report of the situation you see. The **initial report** should paint a concise but vivid oral picture of the conditions you observe as well as a quick summary of your intentions and needs. Specifically, the initial report should address four questions: "What have I got, what am I doing, what do I need, and who is in charge?" Advise everyone else what you are doing, your location, and your immediate plans.

The first part is obvious and very important. In a sentence or two, you should describe what is happening in front of you. If a building is on fire, describe the size and occupancy of the building and the evidence of fire conditions you can see. If you are arriving at the scene of a vehicle accident where you have more victims than you can likely handle with the resources on hand, the same rule applies: Describe what you see.

Advise everyone else what you are doing, your location, and your immediate plans. Let everyone know if you are entering the building and if you are planning to operate in the offensive or defensive mode.

Next, give assignments to incoming units and ask for additional resources if needed. If you think that additional resources will be needed, now is the time to ask for them. Better to get them started and find that you really do not need them than to discover 5 or 10 minutes later that you need some help and have to wait another 5 or 10 minutes until help arrives. Your department's prefire plans should list both emergency and nonemergency units that may be needed for use. Emergency units include anything from police units to assist with traffic control to additional ambulances to handle patient care. Nonemergency units can also encompass a wide range of available resources. Resources such as the American Red Cross, wrecker services, and hazardous material clean-up companies are all considered nonemergency units; however, their services cannot be overlooked.

Some departments use the acronym OSCAR (Occupancy, Size, Conditions, Actions, and Resources) to assist in remembering all of the components of an initial size-up report. Occupancy can be described as single-family, multi-family, or commercial structures. The description does not have to be elaborate. Size can be communicated as small, medium, or large. It can also be communicated in terms of square footage or in numbers of floors. When communicating conditions, you could say "heavy black smoke coming from the back door on side C" or "small fire confined to the exterior of side B." The important thing is to clearly and briefly verbalize what you are seeing because you are the eyes for the rest of the responders. Actions should communicate what you and your crew are preparing to do (i.e., "E-1 is initiating fire attack with a 1¾-inch hose line through the front door."). Resources are only communicated if needed. If a water supply or second alarm is needed, you should communicate this at that time. When you put it all together, it should sound something like, "E-1 we are on scene of a single story, single family residence with heavy black smoke coming from the back door on side C. E-1 is passing command and initiating fire attack through the front door. The next arriving engine needs to bring me a water supply."

You may not have specific assignments for all members and that is OK. Let them know that you want them to continue in or go to a staging area. They can wait there, ready to go to work. This gives you a moment to assess the situation. On the other hand, if you have one of those "nothing showing" situations, you may want them to slow down the response. Most departments have standard procedures for doing this.[6]

Who is in charge? Here the command statement enters into the report. If the first-arriving officer is assuming command, that should be clearly stated. The officer or the dispatch center identifies the command, usually by street or building name. If there is

any question as to your exact location, you should clearly indicate your location so the arriving units will know where to find the command post.

A good initial size-up report provides incoming units with the information needed to mentally and physically prepare for the call. The majority of alarms do not involve working fires. As a result, we fall into a pattern of "here we go again" for a report of a fire alarm or sprinkler activation. A good size-up report will jolt firefighters from the usual "here we go again" mindset into the reality that this is a real fire and we have some real work to do. It also alerts the command center and senior officers who may be listening that firefighters are about to go into battle. Here is another example:

"I'm on-scene of a two-story town house residence with heavy smoke and fire showing on the Alpha and Bravo sides of the north unit. Engine 2: supply my line on Dawn Street and then report here for a backup line off Engine 1. I will assume Eight Street Command. Send me a second alarm and police units for traffic control."

Most fire officers do a pretty good job on size-up reports at fire situations, but some do not do a very good job at EMS and hazmat situations. Although both EMS and hazmat events present a host of unknowns, personnel should be able to describe what they see. Let us take an EMS call for example:

"Engine 1 is on-scene of a two-car accident at East Highway and Route 123. Both vehicles are extensively damaged. One car is on its side in the ditch and appears to have a small fire under the hood. Approximately six people are involved, and all are still in their vehicles. Engine 1 is assuming East Highway command and is going to check the fire and assess the patients in the overturned car. Rescue 2; assess the patients in the second car. Send two police units for traffic control and two additional ambulances."

Ongoing Size-Up

At this point, the action starts. However, size-up does not stop when the action starts. Size-up must be ongoing throughout the event. Many responses start without all the information, or firefighters find that the information is incorrect. As additional information is obtained, be sure that it is shared with those who need

> **NOTE**
>
> A good initial size-up report provides incoming units with the information needed to mentally and physically prepare for the call. Some departments use the acronym OSCAR (Occupancy, Size, Conditions, Actions, and Resources) to assist in remembering all of the components of an initial size-up report.

> **NOTE**
>
> Size-up must be ongoing throughout the event.

FIGURE 13-12 Good fireground commanders keep everyone informed of progress and plans.

the information to change tactics, actions, or mode of operation (see **Figure 13-12**).

Two things are certain to change. The first is the situation. Hopefully, your actions will have some positive impact on the situation. Report those changes on a regular basis. Many departments have a system in place that prompts a status report every 10 or 15 minutes. These automatic prompts also provide an opportunity to be sure that the incident commander has everyone accounted for. After 10 or 15 minutes of interior firefighting operations, you should see significant progress. If that progress is not apparent, it may be time to withdraw and shift to a defensive mode. Sustained interior operations invite disaster!

Along with a status of the event, the incident commander should keep track of the resources available for the operation. Are additional units needed? Are additional firefighters needed to relieve those now engaged? Are support units needed to sustain the personnel and apparatus that are on-scene? Will you need lights, food, shelter, toilets, compressed air to recharge the (SCBA) units, and so forth? Only those at the scene can anticipate these needs. A good incident commander thinks about these things (or has someone else doing it) and has them on-scene before they are needed. If it starts getting dark before you think to call for scene lighting, then you are being reactive to the situation and not proactive. As incident commander, you need to think ahead!

OFFICER ADVICE

Along with a status of the event, the incident commander should keep track of the resources available for the operation.

Developing an Action Plan

An **action plan** is an organized course of action that addresses all phases of incident control within a specified time frame. Note the keywords in this sentence. First, it is an organized course of action. All of the operations should be conducted in an organized manner. This does not infer that it is necessary to have a meeting and vote on the actions you are taking; it just means that someone is in charge and that someone has analyzed the situation and issued orders that will permit safe and effective operations. This has to be done quickly. You can not take a lot of time to do this, or you have crew members standing around while the situation progresses. You are already behind from the time the call comes in. You can not afford to waste any more. Seldom will you have all the facts to start with, so you have to start operations based on what you do know, collect information as you go and modify your actions, and that of your crew, based on that additional information. Conceptionally this may sound like a "seat of your pants" management style, but this differs in that your actions are deliberate and not flippant or without consideration to the outcome of those actions.

Second, the action plan should address all phases of the emergency. Action planning does not stop until the last unit leaves the scene. Action planning, like the ongoing size-up, should be a continuing process.

NOTE

From Sections 4.6.1 and 4.6.2 of the 2009 edition of NFPA 1021 Fire Officer 1, the candidate shall be able to

- ". . . Develop an initial action plan, given size-up information for an incident and assigned emergency response resources, so that resources are deployed to control the emergency."
- "Implement an action plan at an emergency operation, given assigned resources, type of incident, and a preliminary plan, so that resources are deployed to mitigate the situation."[7]

The third key element in the definition of action plan is within a specified time frame. With a little experience, you should be able to estimate how long certain actions will take. Your action plan should allow for the fact that it takes several minutes from the time you, as the incident commander, give an order, until the action starts. It may take additional time for the results to be seen. Meanwhile, the fire is still burning, or the victim is still trapped. You may need to think about your actions in slices of time, especially when coordinating the efforts of several companies.

LESSONS OF STAFFING AND INCIDENT MANAGEMENT REINFORCED AT IOWA FIRE

At approximately 8:24 AM on Wednesday, December 22, 1999, a fire was reported in a multifamily dwelling in Keokuk, Iowa.[8] Several neighbors phoned the Keokuk 911 center to report smoke coming from a residence, and that a woman was outside screaming that there were children trapped inside. At the time the fire was reported, the on-duty force from the Keokuk Fire Department (an assistant chief, a lieutenant, and three firefighters) were completing operations at a motor vehicle accident at a major intersection, two miles northwest of the fire scene. The dispatcher notified the units of the fire and the report of people trapped. Both units at the accident (Rescue 3 and Aerial 2) responded to the fire.

During the response, additional calls were made to the 911 center reporting heavy smoke coming from the house. Upon arrival at 8:28 AM, the units found heavy smoke showing from a two-story multifamily dwelling. A water supply was established from a hydrant one block southwest of the scene.

The assistant chief requested a recall of six off-duty firefighters back as he arrived at the house in Aerial 2. As the two truck operators set up the apparatus, the assistant chief reportedly spoke to the female resident of the burning apartment. She reported that three of her children were still inside the apartment. The assistant chief completed donning his protective clothing, including SCBA, and entered the right side apartment door.

The chief arrived not long after the assistant chief entered the building. The chief ordered the two apparatus operators into the building to assist the assistant chief with the search for the children. In the meantime, a firefighter that arrived with the fire chief stretched an 1½-inch hoseline to the front door of the fire apartment and returned to the apparatus to don her SCBA. When the hoseline was charged, she noticed that the hoseline had burned through at the entrance to the apartment. The burned length of hose was removed, and the nozzle reconnected to the line as it was charged again. The firefighter played a hose stream into the burning apartment. She was only able to advance 6–8 ft into the apartment before being driven back by the intense heat.

Efforts continued to contact the three firefighters that were in the fire apartment. As additional call-back firefighters arrived they were ordered to begin to search for the missing firefighters in the original fire apartment. As the fire was knocked back and a search could begin, firefighters quickly found one firefighter in the first floor room to the right of the main entrance corridor. The assistant chief's body was then found at the top of the stairs, not far from the body of the remaining child, a 7-year-old girl. The third firefighter was found in the master bedroom to the right of the top of the stairs. All had perished.

On the basis of the fire investigation and analysis, the NFPA has determined that the following significant factors may have contributed to the deaths of the three firefighters:

- Lack of a proper building/incident size-up (risk vs. benefit analysis)
- Lack of an established IMS
- Lack of an accountability system
- Insufficient resources (such as personnel and equipment) to mount interior fire suppression and rescue activities
- Absence of an established rapid intervention team (RIT) and a lack of a standard operating procedure requiring an RIT

On the basis of the fire investigation and analysis, the NFPA has determined that the following significant factor may have contributed to the deaths of the three children:

- Lack of functioning smoke detectors within the apartment to provide early warning of a fire

REALITY CHECK

For some readers, this discussion about planning and applications of management science at the fireground may look like a lot of modern textbook stuff with no practical application. However, these ideas have been around for a long time and they do have lots of practical application. Consider that Chief Layman's books were both published in the 1950s and Chief Emanual Fried's classic *Fireground Tactics* was published in 1972.

Charles Walsh and Leonard Marks, both deputy chiefs of large fire departments, wrote a book in the 1970s entitled *Firefighting Strategy and Leadership*. Chapter 2 of their book is entitled "Action Plan." They compare developing an action plan at the scene of a fire or other emergencies to the same problem-solving process we discussed earlier.

What's the point? These ideas have been around for a long time. Many aspects of firefighting have changed remarkably in that period of time, and some aspects are still as they were described years ago. As a good officer, it is important that you not only be up-to-date but that you also remember the basic and timeless concepts of effective firefighting and fireground management.

Strategy

Strategy is the overall plan that is used to gain control of an incident.

Chief Lloyd Layman, author of *Fire Fighting Tactics*, gave us many ideas about firefighting that are still with us. Chief Layman listed seven basic strategies of firefighting as follows[9]:

	rescue	
	exposures	
[size-up]	confinement	ventilation
	extinguishment	salvage
	overhaul	

Notice that even in his time (the late 1940s), Chief Layman felt that the first step was to conduct a proper size-up. In the second column, you see Chief Layman's list of basic strategies related to firefighting in a logical sequence. Rescue comes first, then protect the exposures, if needed. Next, focus on confining and extinguishing the fire. Ventilation and salvage are also important strategies but do not lend themselves to a fixed location in this sequence, and so Chief Layman wisely listed them separately. They should be used whenever needed.

Defining the strategy has many advantages. One of the advantages is that when the strategy is well defined, all personnel understand the tasks at hand and can focus their efforts on making it happen. Without a strategic vision, everyone will contribute, but in a less organized manner. This leads to freelancing and confusion.

NOTE

When the strategy is well defined, all personnel understand the tasks at hand and can focus their efforts on making it happen.

If strategy is the broad picture, then tactics will help you narrow your focus a bit and look at how to reach your strategic goals. As the strategic goals were related to time, so are the tactics. They must also be focused on a specific location and measurable so that you can see if the desired results are happening.

Strategy and tactics lead us to an action plan. After thinking about the event (size-up), and planning (strategy and tactics), you finally get to make it happen. While it may initially seem like a waste of time, the first two steps will save time in the long run by reducing redundancy and amount of communications to get the job done and improving the overall understanding of incident strategies. We have simply said that where

the firefighters' activities are preceded by the incident commander taking a moment to identify the problems and considering the various alternatives, we are more likely to have a satisfactory outcome. The action plan puts the planning and thinking phases into motion.

When assigning resources to carry out the action plan, the incident commander should have a realistic assessment of what can be accomplished with the resources on hand. This applies to the resources, collectively and individually. If the incident commander is blessed with three well-staffed engine companies and two truck companies, a lot can be accomplished. But even in this example, there are limitations. The crew from one of these companies can only be expected to deal with one problem at a time and be in one place at a time. If there are not enough resources, now is the time to ask for more.

NOTE

The action plan puts the planning and thinking phases into motion.

Putting Your Action Plan to Work

To get things started, the incident commander communicates orders. It is critical that the national communications order model be followed. Essentially, that means that the receiver provide a brief repeat of the original order given, allowing the sender to evaluate if the order was received, understood, and resulted in the correct action being taken. Understand that a brief repeat is more than simply saying "10-4." You are to restate what you heard.

Assigning tactics to individual companies is an effective way to communicate what you want done. The incident commander can direct a company to confine the fire or conduct a primary search. By assigning tactics, the incident commander allows the companies to determine the tasks that best accomplish the required action. The directions should be brief, clear, direct, and concise. Assigning tactics usually reduces radio traffic. When giving a verbal order, you should describe what job the receiver needs to perform, not how he or she should perform the job. Some commanders tell crews, "Engine 1, the fire appears to be in the back of the structure. I want you and your crew to put on your gear and airpacks and bring a saw with you to the rear of the structure and vent the roof on the back side near the top." While this communicates what the commander wants done, it is not really stating the objective. Does the commander want to vent the fire, vent the smoke out of the attic, or create an access hole to get to the fire? Additionally, it takes up a lot of air time. If the commander had stated "Engine 1, perform vertical ventilation over the fire," it would not only have accomplished the same thing in a much shorter period of time, but the officer performing the work would then be the one determining what tools are needed and where to make the cut. This should provide for a more efficient and safe operation, because the company officer should know best what PPE, what type of cutting tool, and what length of ladder are needed. The company officer, after determining the location of the fire and obtaining a visual of the roof, should determine if it is first safe to get on the roof and second where in the roof the vent hole should be cut. A lot of commanders resort to this type of communication for several reasons. First, they were or are very comfortable performing these actions. This is what they did before they were elevated to the command function and under pressure they resort to what they are comfortable with. Another reason is that they may lack confidence in their company officers. This again points out why training is so important. Training is the place to build confidence and work out these issues—not the fireground!

Once the task is under way, the incident commander should get some feedback about progress. If the news is good, the incident commander wants to hear about that. And if the news is bad, the incident commander needs to hear about that too. The incident commander depends upon timely reports to make decisions. A progress report should be given anytime you experience a sudden change in conditions (good or bad), change levels (up or down), have completed your assigned task, or need additional resources to complete your assignment. Just like the command assignments the progress reports should be brief, clear, and concise. You should know what you want to say before you depress the mic button. Also be aware that you cannot typically breathe and talk at the same time. That is why it is so important to practice communicating with an airpack on—so you learn where to hold the radio mic and at what rate you are best understood.

A commander may assign **tasks** with very specific directions. This option will probably increase the amount of radio traffic and will require a little more of the incident commander's time and attention, but there are times when this approach is appropriate, for example, when the task is critical to the overall operation, or the incident commander wants the task accomplished in a way that departs from normal procedures, or when critical safety considerations are paramount. It also is appropriate to use this approach when the company involved is inexperienced. The best example of this might be when a **mutual aid** company is assigned the task.

Good Training and SOPs Pay Off

Many departments use standard operating procedures (SOPs) that outline what certain companies are to do upon arrival. These departmental policies provide a predetermined course of action for a given situation. With SOPs the companies can get to work and perform their assignments with little direction from the incident commander (see **Figure 13-13**). SOPs can work well, but you should be aware that some companies will race ahead with their assigned task, without the incident commander having had an opportunity to complete the size-up or give some coordinating instructions. Incident commanders can avoid this problem by making assignments to companies as soon as possible or by telling companies to wait for instructions. It may also be appropriate for the incident commander to tell arriving units that because of some unusual circumstances, the SOPs are not appropriate and will not be used.

NOTE

With SOPs, the companies can get to work and perform their assignments with little direction from the incident commander.

Basic communications skills are important at times like these. There is a lot going on, and things need to be resolved as rapidly as possibly. Everyone is looking to the incident commander to give a few orders to put the operation into motion. Yelling, raising your voice, and speaking rapidly will certainly reveal to

FIGURE 13-13 Good training and SOPs allow companies to get to work with a minimum of direction from the incident commander.

everyone listening that you are excited. At this time, you want a calm, reassuring voice. As much as anything else, the incident commander's calm voice tells us that someone understands what is happening and has the solution well in hand. That is exactly what the incident commander is supposed to be doing. No one wants to hear an excited incident commander on the radio or see an excited incident commander running around the fireground.

One of the most hectic moments on a fireground occurs when some unexpected or catastrophic event occurs and the decision is made to shift from offensive to defensive operations. This is usually done when either the incident commander or the safety officer has determined that the risk outweighs the benefits and as a result it is unsafe to be inside. Consequently, there is an urgency to get everyone out as soon as possible. If the commander is yelling into the radio and the order is not understood, it may actually delay the crews from hearing and understanding the order. This is no time to vary from the SOPs. A preestablished, understood, and practiced evacuation procedure should be implemented. Any variance from this could result in confusion and delay in getting the crews out.

Just as the incident commander gets the big picture and breaks it down into strategy, tactics, and tasks for companies, the company commander should do the same thing for the individuals who comprise the company. All persons need to have an idea of the big picture and the role they will play in the overall outcome of the event. All must also clearly understand their individual assignment. Only with understanding and coordination can you operate safely and efficiently.

Fire Attack

The effectiveness of the initial attack often determines the outcome of the event. The effectiveness of the initial attack's leader, company officer, chief, or whomever, is dependent on both that individual's knowledge of fire behavior and decisions with regard to the size, number, and placement of hoselines during that initial attack.

The most common fires encountered are considered free burning; that is, they are in the second of the three stages of fire. When possible, such fire should be fought by a direct, offensive attack. Direct attack is the surest and quickest way of controlling the fire and reducing the loss. You should make every possible effort to attack the fire from the unburned side and to push the fire back to those areas where it has already caused damage. This is a fundamental concept of good firefighting. It reduces property loss and usually enhances the safety of the operation (see **Figure 13-14**).

FIGURE 13-14 Most firefighting is done with hoses and water. This activity is best accomplished by a direct attack from the unburned side.

You may not be able to attack the fire from the unburned side in certain situations. These might include a commercial occupancy where access is limited to one end of the building, in multiunit residences (apartments, town houses, and so on) where you do not have access to the unburned side, and where such action might endanger occupants or firefighters.

Fire Control, Confinement, and Extinguishment

These three words are used quite a bit in connection with firefighting operations. Let us take a moment to agree on what they mean. Fire control includes both confinement and extinguishment of the fire.

Confinement means to cut off the fire from further advance, to keep it from spreading and doing any further damage.

Fire extinguishment means to extinguish all visible fire. Not all of the fire may be extinguished at this point; that will occur during **overhaul**. To be effective in fire control, one must understand fire behavior, building construction, and fire flow requirements.

Most firefighting is done with fire hoses and firefighters. In spite of all of the advances we have made in the last few years, it still takes time to get a water supply established and a hoseline directing water onto the fire in operation. As a fire officer, you have to realize this and allow for this delay and factor it into your thinking. If the fire has already passed you before your firefighters can open the nozzle, then you were not realistic in your projections to get this accomplished. Do not be overly optimistic in your calculations to get everything in place. It is easier to advance forward to the fire than back out and play catch-up.

During your planning, you should consider the number of firefighters available and how many hoselines they can put into operation. From this information, you will have a good idea of how much water you can put on the fire. And you have to decide if this flow rate, or fire flow, is adequate to confine and extinguish the fire. Finally, you need to coordinate the advance of the hoselines with other activities such as forcible entry, ventilation, search and rescue, and salvage.

Let us get back to Ms. Carmen Garcia's house for a moment. You have learned that she was in the house alone and was cooking. She went into the next room to answer the telephone and while she was talking, "something happened in the kitchen." The next thing she knew the smoke detector was going off, and the kitchen was filling with smoke. You may have seen this before, but Ms. Garcia has not. She was smart enough to call 911 right away. Then, she tried to get to the stove to turn things off. When she did that, she instinctively grabbed a pan handle that was very hot.

What do you do? You get one of your medics to take care of Ms. Garcia. Meanwhile, you calmly evaluate the fire situation, select to go for an offensive operation with an interior attack, mask up, take an attack line in through the front door, find your way back to the kitchen (not really hard to do because the fire is free burning and the interior is not filled with smoke which allows you to put the fire out with a few gallons of water (see **Figure 13-15**).

One of your colleagues was smart enough to go around to the back door and carefully open it, allowing some of the smoke and steam to vent outside through the doorway. You look around and find smoke throughout the house and considerable damage in the kitchen. Although you will check further, it looks like the fire did not get into the structure itself. You have been on-scene less than 5 minutes.

FIGURE 13-15   Firefighter in Ms. Garcia's kitchen.

About this time, you hear the battalion chief say over the radio: "Engine 1 from Battalion 1 I'm on-scene. What have you got?" (see **Figure 13-16**).

Interagency cooperation was discussed in Chapter 12; however, it is worth mentioning again that an effective company officer should know what resources are available. During an alarm is not the time to figure out what resources are available, where they are located, and how to request them. This means that prior to the alarm, you visit with your chief officer, visit surrounding stations, and review established mutual aid agreements, memorandums of understanding, and other related documents to garner this information. If you do not know a resource exists, you cannot request it. And just because you know a resource does exist does not mean that you can get it when you call for it. You need to know and follow the proper request channels. This may sound like a lot

FIGURE 13-16   "What have we got?"
(Photo courtesy of Andy Thomas)

TAKING CUSTOMER SERVICE TO MS. GARCIA'S HOUSE

We have already discussed what we are doing for Ms. Garcia and her fire. Let us look at what else we might do for her. Suppose that as you make your way to the kitchen, you notice several family pictures sitting on a table in the hall. These will certainly be knocked over and likely broken as you and your firefighters advance a hoseline or start overhaul. And you also notice a very nice antique chair that will certainly get in someone's way.

So what, you ask; we have work to do in the kitchen. Yes, but property conservation includes little things that can mean a lot. Pick up those pictures and put them out of the way. You might even slide them into a drawer if you can find one close at hand. And pick up the antique chair that is in the hallway and set it in the dining room where it will be out of your way and out of harm's way. After the fire is out, control the ventilation. Open windows or close doors to reduce the effects of the smoke that will smell up everything in the house for a month. Cover her hall carpet with a runner when you get a chance. Lots of little things that cost little but will mean a lot after you have gone (see **Figure 13-17** and **Figure 13-18**). Remember, we want her to be satisfied. But would it be nice if she was delighted with your work?

FIGURE 13-17 Fire departments can do much to help their citizens recover from the consequences of a fire. The first step is providing materials to assist during salvage operations. Boxes, bags, and plastic sheeting can help save a lot of property. (*Photo courtesy of Phoenix, Arizona, Fire Department*)

FIGURE 13-18 Firefighters can also help citizens recover from the consequences of fire by moving their personal effects to a safe area. (*Photo courtesy of Phoenix, Arizona, Fire Department*)

NOTE

On a typical fire or other emergency, command is transferred from the first-arriving officer to a chief officer upon the latter's arrival. The officer being relieved should be prepared to provide the chief officer with an assessment of the general situation and tactical priorities, the assignments of companies, the identification of companies available, and the need for additional resources.

of bureaucracy, but in reality it saves time if proper channels are followed.

This very situation occurred one night in a rural community in the Midwest, when at the height of a severe storm, a car was washed off a rural road, trapping several occupants in the middle of a raging stream. The local fire department did not have the training or equipment to deal with the situation; however, they attempted to rescue the three trapped occupants. After one of the firefighters drowned, assistance was requested from a swift water rescue team in a larger, neighboring department. Because the rural chief did not use the pre-established mutual aid request process, it took much longer than it should have for the

metro department to get administrative approval to send resources outside their normal response district. Fortunately, the occupants were saved, but not knowing how to access available resources or follow established mutual aid request procedures put the occupants in danger for a much longer period of time than needed.

As previously stated, other emergency and nonemergency agencies can also be used as resources. For example, law enforcement agencies may be able to assist in establishing traffic control, securing a fire scene and/or a crime scene. A public works agency such as the

PHOENIX FIRE DEPARTMENT STANDARD OPERATING PROCEDURES

Phoenix Fire Department

Occupant Services Sector

The purpose of this procedure is to establish the role and responsibilities of the Occupant Services Sector.

The Occupant Services Sector shall be established by the Incident Commander at all working structure fires, and as early in the incident as is practical. The Occupant Services Sector should also be established at any incident where the need is identified: Fire, EMS, Special Operations, etc.

The Occupant Sector is a critical extension of our service delivery, and serves as the liaison between the Fire Department and those citizens (responsible parties) directly, or perhaps indirectly, involved in or affected by the incident.

If necessary, Command will request additional resources in order to establish the Occupant Services Sector. An additional engine, ladder, or battalion chief is acceptable. If necessary, at prolonged incidents, in order to return fire companies and personnel to service, Command may assign staff personnel to this function. The Occupation Services Sector responsibilities may extend beyond the termination of the incident.

Responsibilities

The Occupant Services Sector should consider offering the following services to the occupant/responsible parties. It should be noted that other occupant service needs may be identified and should be addressed as part of the Department's customer service goals.

- Carry out responsibilities under supervision of loss control officer.
- Explain what happened, what we are doing and why, how long we expect to take until the incident is under control.
- Obtain from occupant/responsible party, any significant information regarding the structure and/or its contents that might assist Command tactically with the operation. Inform Command of this information.
- Provide cellular telephone access.
- Communicate the location to which evacuees have been sent. (Notify the Investigations Sector of this location also when passing on this information.)
- Identify any mental health needs of occupants/responsible parties, as well as any spectators or evacuees (i.e., effects of shootings, mass casualty, highly visible critical rescue, etc).

- Notify Red Cross, Salvation Army, or other relief agencies.
- Notify other necessary agencies and/or individuals.
- Provide coordination of salvage efforts with the loss control officer.
- Where safe to do so, and after approval from Investigations Sector, coordinate a "walk-through" of the structure with the responsible party.
- Determine the location of valuables in the structure and notify command/loss control officer.
- Work with loss control and proper utility services to restore power, gas and water, as quickly as possible to reduce additional losses through a loss of business to affected occupants.
- Provide use of service vans as necessary.
- Coordinate site security.
  - □ Fire watch
  - □ Private security company
  - □ Necessary insurance services
  - □ Any services identified as necessary and possible
- Handout and explain the "After the Fire" brochure.
- Assist the occupant in notifying insurance agents, security services, restoration company, etc.
- Provide blankets, and a shelter, where practical to do so (i.e. an apparatus cab, neighbor's house, etc.) to get occupants out of the weather and at a single location.
- Provide ongoing service and support until the customer indicates our services are no longer needed.

The Occupant Services Sector shall report to Command unless a loss control branch/section is assigned, at which time he/she shall report to the loss control officer.

Mental Health Needs

Occasionally, the public is witness to a critical life threatening event that can have substantial psychological impacts on them. These persons may be survivors of a critical event or a witness to a mass casualty, or a parent of a severely injured child, or a witness to the death of a family member, etc.

Additionally, witnesses may have misunderstandings of fire department operations that cause a delay in removal of the patient (i.e. trench collapse, an electrocution rescue that is delayed due to energized contact, etc.). Addressing these issues early, on-site, or as soon as possible following the event, can minimize these misunderstandings, and reduce psychological effects, and produce improved relations with the public.

The Occupant Services Sector should consider additional help for these needs. Assistance and advice on availability of mental health services can be obtained through the department's Critical Incident Debriefing Team, the Employee Assistance Program Contractor, the

American Red Cross, and in some cases, through the victim's personal medical insurance. Support from the fire department chaplain or local clergy may also be available.

American Red Cross Services

For residential fires where the occupant has suffered a loss of living quarters and clothing, the American Red Cross may be used to provide support. The American Red Cross can provide some clothing, food, toiletries,

and arrange for temporary shelter/housing for the occupants. When contacting the Red Cross, provide the following information:

- Address of the incident
- Address where victims can be contacted
- Phone number of contact location
- Number of displaced persons with information on age, sex, etc.
- Fire department incident number

state's Department of Transportation may be able to assist in removing debris or setting up road blocks.

Other resources include informational resources. An effective officer knows how to access information to assist in mitigating an alarm. For example, you should know how to use a DOT guidebook. Do you know what the different color sections of the book tell you? Informational resources may also be located on the computer in the engine or the chief's car. Information may be available in the lock box or in a pre-plan. Most dispatch systems have access to additional information through look-up systems or 800 call centers such as Chemtrec and the Poison Control Center. But if you are not familiar with these or do not know they exist, then they are of little value to you during an emergency.

Within the last several pages we have attempted to provide you with direction on how to manage an emergency scene until you are relieved by a chief officer. The first 5 minutes of operation at an emergency scene are the most critical. They set the stage and tempo for how the event will play out. As a good company officer, you should be able to get on-scene, conduct a size-up, give a report, and get the show started without having to wait for a more senior officer to arrive.

In situations like those at Ms. Garcia's house, a good company officer should be able to coordinate the entire operation. The chief should show up and relieve the officer if needed, but if all is going well, should let the company officer continue to run the show. That is how company officers get trained to become good fireground commanders. On the other hand, if the situation is escalating, getting out of hand, or is going to be a long-term event, it may be time for the chief to take over. In any case, an IMS should be used.

Jim Crawford, a well-known fire official, writer, and past president of the International Fire Marshals Association, suggests that limiting damage once the event starts should also be part of our risk-management process. He suggests that emergency response

is the fourth E in the fire prevention triad; engineering, enforcement, and education discussed in Chapter 9. Emergency response is the traditional role of every fire department. But in the overall scheme of risk management, firefighting must be part of a continuum of effort that seeks to reduce life and property loss by using education, engineering, enforcement, and emergency response.

THE INCIDENT MANAGEMENT SYSTEM

Background

Prior to the 1970s, the concept of incident command was pretty much a local function. Since then, several significant events have helped the nation's fire service move to a standardized program. During the 1970s, the FIRESCOPE incident command system and the Phoenix Fireground Command system became popular.

FIRESCOPE stands for FIre REsources of Southern California Organized for Potential Emergencies. Serious wildland fires in southern California in those years challenged the firefighting resources of southern California. As a result, federal, state, and local agencies combined their talents to form a more organized response to the disastrous fires that seemed to be occurring every year. FIRESCOPE focused on command procedures, resource management, terminology, and communications.

Chief Alan Brunacini of the Phoenix Fire Department realized that the organization and management concerns that faced firefighters in wildland firefighting operations were also faced by firefighters dealing with structural firefighting back home. He took some of the ideas from FIRESCOPE and applied them to his department. Others followed suit. This led to the **incident command system** (ICS). While originally intended to serve as a command system for large structural fires, ICS became a tool for dealing with emergencies of all kinds.

Incident management system (IMS) implies something more than just incident command; it implies that every aspect of controlling an incident is accomplished in an orderly and logical way. IMS looks upon emergency management as an extension of the type of management you should be doing on a daily basis. IMS is a management tool—it allows a department and its personnel to plan, organize, command, coordinate, and control emergency incidents (see **Figure 13-19**). The focus of IMS is on managing people and resources. You cannot do much to manage fire and floods: Those events have already happened before you arrive. However, IMS allows you to manage response and mitigation efforts at these events.

As the incident grows in size or complexity IMS allows you to smoothly integrate with other responders and grow the management system with it. At the largest incidents you will be operating under the **National Response Framework (NRF)**, formerly known as the National Response Plan. The National Response Framework defines the principles, roles, and structures that organize how we respond as a nation. The National Response Framework describes how communities, tribes, states, the federal government, private sectors, and nongovernment partners work together to coordinate national response; describes specific authorities and best practices for managing incidents; and builds upon the National Incident Management System (NIMS), which provides a consistent template for managing incidents.

The command function must be clearly established with the first-arriving unit. This implies that the first-arriving unit is qualified and has the authority (and the ability) to take command and that it is clear to all responding personnel who is in charge. IMS also incorporates the rules of organization and management discussed in Chapters 3 and 4. When applied to the management of resources at an **emergency incident**, they improve safety, accountability, efficiency, and the overall operation while reducing confusion.

While IMS defines how the management of an incident starts and progresses during the incident, the success of the operation is usually directly related to how well the initial command function is established and how effective the communications are handled during the event. Effective communications are vital to incident management.

"NFPA 1500, *Standard on Fire Department Occupational Safety and Health Program,* states that fire departments shall have written operating procedures for managing emergency incidents. The standard also requires training in incident management for all personnel involved in emergency operations. NFPA 1561, *Standard on Emergency Services Incident Management System,* provides details regarding such a system. Incident Management Systems can provide better management of firefighters during emergency operations. Better management will enhance firefighter safety."

> **NOTE**
>
> The focus of IMS is on effective management of people and resources.

As we know it today, the IMS is an all-risk all-situation emergency management concept. The focus of IMS is on effective management of people and resources. To implement IMS you must first accept IMS as a workable concept in your organization. Next, you must plan, prepare, and train for the events that you are likely to face. You should remember the basic rules of effective management and use them when you are managing emergency activities.

IMS can be used for nonemergency activities too. Many use IMS to manage large-scale training activities. There are other applications as well. Large-scale operations of any kind, from preparing for the Super Bowl size event in your city to having a fire department funeral, can benefit from the organizational concepts of IMS. IMS works across political and organizational lines. It can be used to manage any large-scale event.

For managing resources during emergency situations, IMS is designed to work from the time of alarm until the incident is concluded. The title "incident commander" may apply to an engine company officer or a chief of the department. And depending upon the situation, the commander may be an individual associated with another emergency organization. A county sheriff was in command of the Columbine school shooting. The structure of the IMS can be established and expanded depending upon the needs of the event.

FIGURE 13-19 The incident commander is responsible for the overall management of the event.

National Incident Management System

On March 1, 2004, the Department of Homeland Security published the *National Incident Management System* (NIMS). NIMS provides a consistent nationwide template to enable federal, State, tribal, and local governments, the private sector, and nongovernment organizations to work together to prepare for, prevent, respond to, recover from, and mitigate the effects of incidents, regardless of cause, size, location, or complexity, in order to reduce the loss of life, property, and harm to the environment. This consistency provides the foundation for utilization of NIMS for all incidents, ranging from daily occurrences to incidents requiring coordinated Federal response. NIMS was updated in 2007 based on input from stakeholders at every level within the nation's response community and lessons learned during recent incidents.

NIMS incorporates incident management's best practices and represents a core set of doctrine, concepts, principles, terminology, and organizational processes that enables effective, efficient, and collaborative incident management across all emergency management and incident response organizations and disciplines. The president of the United States of America has directed federal agencies to adopt NIMS and encouraged adoption of NIMS by all stakeholders—federal, state, territorial, tribal, substate regional, and local governments, private sector organizations, critical infrastructure owners and operators, and nongovernment organizations involved in emergency management and/or incident response. As initially laid out in Homeland Security Presidential Directive (HSPD), *Management of Domestic Incidents*, which established NIMS, adoption and implementation of NIMS by state, tribal, and local organizations are one of the conditions for receiving federal preparedness assistance (through grants, contracts, and other activities).[10]

NIMS Courses

Currently, there are six required courses for command and general staff, select department heads with multi-agency coordination system responsibilities, area commanders, emergency managers to be NIMS compliant.

- ICS-100: Introduction to ICS
- ICS-200: Basic ICS
- ICS-300: Intermediate ICS
- ICS-400: Advanced ICS
- FEMA IS-700: NIMS, An Introduction
- FEMA IS-800: National Response Plan (NRP), An Introduction

ICS-300 and ICS-400 courses are courses conducted in a classroom. The Emergency Management Institute and the National Fire Academy both sponsor NIMS compliant ICS-300 and 400 training.

Additional NIMS-related recommended online courses are:

- IS-701 NIMS Multi-Agency Coordination System
- IS-702 NIMS Public Information System
- IS-703 NIMS Resource Management

See **Figure 13-20** to determine what level of training an individual needs according to their level of responsibility during a multi-jurisdiction, multi-agency, multi-discipline incident. This is also available online at: www.fema.gov/pdf/emergency/nims/TrainingGdlMatrix.pdf

For a complete listing of all online and live NIMS courses visit: http://training.fema.gov/IS/NIMS.asp

Key Features of NIMS

- *Incident Command System (ICS).* NIMS establishes ICS as a standard incident management organization with five functional areas—command, operations, planning, logistics, and finance/administration—for management of all major incidents. To ensure further coordination, and during incidents involving multiple jurisdictions or agencies, the principle of unified command has been universally incorporated into NIMS. This unified command not only coordinates the efforts of many jurisdictions but also provides for and assures joint decisions on objectives, strategies, plans, priorities, and public communications.

- *Communications and Information Management.* Standardized communications during an incident are essential, and NIMS prescribes interoperable communications systems for both incident and information management. Responders and managers across all agencies and jurisdictions must have a common operating picture for a more efficient and effective incident response. Interoperable can mean a number of different things. It does necessarily mean everyone has be able to talk on the same radio channel, but you do need to have predetermined procedures for how you are going to communicate and there should be redundancy built into the procedures so if one system fails you can immediately begin to use another.

- *Preparedness.* Preparedness is essential for effective response. According to the National

FY07NIMS Training Guidelines

Audience	Required Training
Federal/State/Local/Tribal/Private Sector & Non-governmental personnel to include: *Entry level first responders & disaster workers* • Emergency Medical Service personnel • Firefighters • Hospital staff • Law Enforcement personnel • Public Health personnel • Public Works/Utility personnel • Skilled Support Personnel • Other emergency management response, support, volunteer personnel at all levels	• ICS-100: Introduction to ICS or equivalent • FEMA IS-700: NIMS, An Introduction
Federal/State/Local/Tribal/Private Sector & Non-governmental personnel to include: *First line supervisors*, single resource leaders, field supervisors, and other emergency management/response personnel that require a higher level of ICS/NIMS Training.	• ICS-100: Introduction to ICS or equivalent • ICS-200: Basic ICS or equivalent • FEMA IS-700: NIMS, An Introduction
Federal/State/Local/Tribal/Private Sector & Nongovernmental personnel to include: *Required:* Mid-level management including strike team leaders, task force leaders, unit leaders, division/group supervisors, branch directors, and; *Recommended:* Emergency operations center staff.	• ICS-100: Introduction to ICS or equivalent • ICS-200: Basic ICS or equivalent • ICS-300: Intermediate ICS or equivalent • FEMA IS-700: NIMS, An Introduction • FEMA IS-800.A: National Response Plan (NRP), An Introduction*
Federal/State/Local/Tribal/Private Sector & Nongovernmental personnel to include: *Required:* Command and general staff, select department heads with multi-agency coordination system responsibilities, area commanders, emergency managers, and; *Recommended:* Emergency operations center managers.	• ICS-100: Introduction to ICS or equivalent • ICS-200: Basic ICS or equivalent • ICS-300: Intermediate ICS or equivalent • ICS-400: Advanced ICS or equivalent • FEMA IS-700: NIMS, An Introduction • FEMA IS-800.A: National Response Plan (NRP), An Introduction*

* NOTE: Not all persons required to take ICS-300 and ICS-400 will need to take IS-800.A. Emergency managers or personnel whose primary responsibility is emergency management must complete this training.

FIGURE 13-20   The FY07 NIMS Training Matrix. (*Source*: FEMA, www.fema.gov/pdf/emergency/nims/TrainingGdlMatrix.pdf)

Response Framework, there are six essential activities for responding to an incident. Those include: plan, organize, train, equip, exercise, and evaluate and improve. We will expand on these functions later. All of these serve to ensure that pre-incident actions are standardized and consistent with mutually agreed doctrine. NIMS further places emphasis on mitigation activities to enhance preparedness. Mitigation includes public education and outreach, structural modifications to lessen the loss of life or destruction of property, code enforcement in support of zoning rules, land management, and building codes, and flood insurance and property buy-out for frequently flooded areas.

- *Joint Information System (JIS).* NIMS organizational measures enhance the public communication effort. The Joint Information System provides the public with timely and accurate incident information and unified public messages. This system employs Joint Information Centers (JIC) and brings incident communicators together during an incident to develop, coordinate, and deliver a unified message. This will ensure that federal, state, and local levels of government are releasing the same information during an incident.

- *NIMS Integration Center (NIC).* To ensure that NIMS remains an accurate and effective management tool, the NIMS NIC will be established by the Department of Homeland Security to assess proposed changes to NIMS, capture, and evaluate lessons learned, and employ best practices. The NIC will provide strategic direction and oversight of the NIMS, supporting both routine maintenance and continuous refinement of the system and its components over the long term. The NIC will develop and facilitate national standards for NIMS education and training, first responder communications and equipment, typing of resources, qualification and credentialing of incident management and responder personnel, and standardization of equipment maintenance and resources. The NIC will continue to use the collaborative process of federal, state, tribal, local, multidiscipline and private authorities to assess prospective changes and assure continuity and accuracy. All issues related to training, compliance, certification and other related articles and tools can be found on the National Integration Center (NIC) Incident Management Systems Division's Web site: http://www.fema.gov/emergency/nims/

Organizational and Management Theory in the Incident Management System

Back in Chapter 3, different organizations were introduces along with a few concepts that enhance effectiveness in organizations. One of those concepts was that organizations should have an organizational structure that embraces unity of command, span of control, division of labor, assignment of specific responsibilities, and clear lines of authority, responsibility, and communication. Later chapters explained management and leadership concepts that bring organizations to life.

In Chapter 13, we look at the IMS that underlies incident command, and we see that it embraces all of the principles of organizational theory. It is a predeveloped system of management, with a structure that can be molded for just about any situation. We usually think of the IMS for controlling resources at large-scale disasters, and NIMS was clearly developed with that intent. However, the same approach can be used for managing any event, from a large-scale drill to organizing a large function or exercise.

Several management concepts are so fundamental to success that they are defined in the NIMS example, span of control (the number of individuals a supervisor is responsible for) and unity of command (the concept that each person reports to one person) are defined in the document itself. Clearly, the writers of this document understood organizational theory and the practical limitations of human beings working together in an organized effort.

The IMS is a perfect example of a practical application of organizational and management theory. It reinforces the value of our earlier learning and shows that NIMS is based on sound ideas and principles.

Transfer of Command

The IMS should be applied to every incident from the arrival of the first individual until termination. At small-scale incidents, the assumption of command can be informal, but the principle of one individual in overall command of the incident always should apply. Routine application of the system is intended to increase familiarity with the concepts and procedures, even where the need to apply a formal command structure is not obvious. The officer in charge of the first-arriving company or the first-arriving individual of the emergency services organization, regardless of rank or function, should be the incident commander until relieved by more qualified personnel. All personnel should be sufficiently familiar with basic responsibilities and communications protocols in order to assume the role of initial-arriving incident commander, if only until a more qualified individual arrives.

The emergency services organization should establish a protocol of command authority based on rank structure, assignments, and qualifications to define a hierarchy for transferring command. The qualifications required to perform as incident commander should increase with the size and complexity of the incident. Standard operating procedures should define the circumstances under which an officer at a higher level should respond to an incident and whether the transfer of command to a more qualified officer is mandatory or discretionary.

In certain cases, an individual with a higher level of command authority arriving at the scene can direct the current incident commander to continue in this role. Technically, just by virtue of rank, the higher-level officer is responsible for the command of the incident but acts as an observer or advisor to allow the incident commander to benefit from the experience. This should only be done if the situation is under control or not getting any worse, and at the discretion of the higher-level officer.

Post-Incident Analysis

The incident commander should analyze every incident to look for ways to improve personal, unit, and system performance. Certainly after any large incident, a **post incident analysis** should be conducted to examine the strengths and weaknesses of the response and lessons learned at the event. To simply say "The fire went out and no one got hurt" is NOT an effective post incident analysis. Those types of comments reinforce complacency and unsatisfactory performance and should not be acceptable. We should be striving for perfection with the understanding that we will likely never achieve it, but nevertheless, it should remain our benchmark.

A formal post incident analysis should be conducted in the following circumstances:

- A civilian fatality

- A firefighter injury or fatality

- A hazardous materials incident that results in chemical exposure to either civilians or emergency response personnel

- Any multicasualty event, such as a bus, train, or multivehicle accident

- Any firefighting operation presenting unusual situations

- Any event for which the information obtained during the analysis would benefit others

Each of these situations has the potential for providing valuable information to the organization for a variety of personnel, including off-duty personnel and personnel from other jurisdictions. Many departments use this process as a valuable training tool.

The incident commanders are often the best persons to gather the information needed for the analysis. In many cases, the decisions were theirs, and only they may know the reasoning for those decisions. Certainly, the post incident review should include as much information as can be learned about the structure, the strategic decisions made, the operational issues, and anything else that would be beneficial to others.

As company officers, you should routinely do this for your own unit, even if formal critique is not warranted. This is commonly called a tailboard critique and is a very simple verbal review of how you and your crew did, analyzing what went right and what you can improve on next time. Avoid only doing a critique when things go wrong. If this becomes the norm in your organization everyone will begin associating a negative connotation to the occurrence of a critique. Lessons can also be learned from things that went right.

Most organizations have some sort of outline that provides a standardized approach to the formal post incident analysis that benefits both the writer and the reader. In its simplest form the post incident analysis should review the basic functions that were performed at the emergency (i.e., command, water supply, fire attack, rescue, extrication, patient packaging, demobilization, etc.).

The following is one example and a starting point that may be expanded, or condensed as necessary.

Post-Incident Analysis of a Structural Fire

I. Introduction

   a. Provide a general overview of the incident including an area diagram of the building, exposures, water supply, time of day, weather conditions, and so on.

   b. Indicate any unique circumstances/problems, and so on.

II. Building Structure/Site Layout

   a. Review type of structure.

   b. What construction or design features contributed to the fire spread or prevented fire spread?

c. Did the topography and/or type of fuel affect fire control efforts?

d. Did fire alarm and/or suppression devices work properly?

e. Did personnel or apparatus encounter any problems in gaining access?

f. What is needed to correct these problems?

III. Fire Code History

a. Review relevant fire code requirements and history.

b. Was the structure in compliance?

IV. Communications

a. Did dispatcher provide all information available at the time of dispatch?

b. Was the fireground channel adequate?

c. Were proper communications procedures followed?

d. Were there problems communicating with mutual aid companies?

e. Was the communications network controlled to reduce confusion?

f. Was radio discipline effective?

g. Did the incident commander provide timely updates?

V. Pre-emergency Planning

a. Was a prefire plan needed on the scene?

b. Was it available?

c. Was it useful?

d. Should it be updated?

VI. On-Scene Operations

a. What was the structural integrity of the building upon arrival?

b. Was command identified and maintained throughout the incident?

c. Was a command post established and readily identifiable?

d. Was a proper size-up made?

e. Were additional apparatus requested in a timely manner?

f. Was proper strategy developed?

g. Was there an action plan?

h. Did personnel, units, and teams execute tactics effectively?

i. Were standard operating procedures used?

j. How was risk analysis applied to the incident?

k. Were the sectors used appropriate to the incident's type and complexity?

l. Was a standby team established? If not, why?

m. Was a safety officer designated? How was this officer used?

n. Were actions taken to ensure accurate personnel accountability?

o. Was a rehab sector established?

Once the critique or post incident analysis is complete, a report should be developed noting the highlights of the critique, especially as it relates to specific functions and the ability to satisfactorily or unsatisfactorily complete these functions. The report should be disseminated to everyone who participated in the critique and made available to all. Again, the intent is not to place blame or make anyone look bad but rather to learn from past mistakes as well those things that were done well. A library of these reports should be kept for future reference.

Emergency Management

A disaster or emergency can strike anywhere and at any time. Disasters can be acts of nature such as a tornado, flood, earthquake, or winter storm or can be human-made such as a hazardous materials spill, utility failure, or an act of terrorism. A disaster can build up slowly, taking days or weeks, or impact suddenly without warning. Originally, the emergency management function was known as civil defense. The main objective was protection of the civilian population in the event of enemy attack. In the last two decades, this approach has been redefined to encompass the protection of the civilian population and property damage from the destructive forces of natural and human-made disasters. The approach then changed to a comprehensive emergency management, or an "all hazards approach."

Comprehensive emergency management refers to any government's responsibilities and capability for managing all types of emergencies and disasters by coordinating the actions of many organizations and agencies. Comprehensive emergency management includes the four phases of disaster activity—preparedness, mitigation, response, recovery. Many of these organizations are now the lead agencies for the development and implementation of local disaster plans. If an emergency meets the criteria of a county-wide or statewide disaster, the disaster declaration starts with the local county emergency manager. Once this system is activated, it provides a conduit to access state and/or federal resources.

Preparedness

Preparedness is the measures taken in preparation of a disaster or emergency and essential for effective response. According to the National Response Framework, there are six essential activities for responding to an incident. Those include: plan, organize, train, equip, exercise, and evaluate and improve

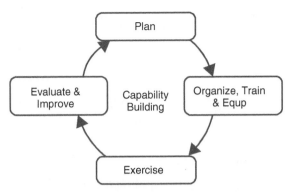

The Preparedness Cycel Builds Capabilities

FIGURE 13-21 The National Response Framework Preparedness Cycle. (*Source: January 2008—National Response Framework.*)

(see **Figure 13-21**). Planning includes the collection and analysis of information, development of policies, procedures, mutual aid agreements, strategies, and other arrangements to perform critical missions and tasks. Organizing includes activities such as developing an overall organizational structure, strengthening leadership, and assembling, equipping, and training teams of paid and/or volunteer staff to perform essential tasks. Exercises provide the opportunity to test plans and improve proficiencies. Evaluation and improvement activities should be an ongoing process that should evaluate performance against relevant capability objectives, identify deficits, and institute corrective action plans.

In addition to organizational preparedness, all citizens should learn what to do in case of an emergency. Preparing in advance by learning what hazards may affect your community and by learning about how to deal with these hazards is an important part of emergency preparedness. This includes how to take the necessary precautions to protect yourself, your family, and your property. This is especially important for ire and EMS personnel, because, if you and or family are not personally prepared to deal with emergencies, it is likely that you, as a first responders to a disaster can focus on your job, if you are worried about your family and/or your home. If you have a preestablished supply kit, escape plan, communications plan, evacuation route, etc., then it provides you a peace of mind, allowing you to concentrate on the tasks at hand.

For more information related to personal preparedness go the FEMA website: http://www.fema.gov/plan/

Response

Response is related to the activities that are taken during and immediately following a disaster. In most jurisdictions, the fire and police departments are the lead agencies for handling day-to-day emergencies. In the event of an actual or threatened large-scale emergency situation, the designated emergency management coordinator is assigned emergency duties in addition to this person's primary day-to-day functions. These functions range from assessing damage to assisting with sheltering operations to maintaining detailed records of disaster-related expenses.

Mitigation

Mitigation refers to activities that actually eliminate or reduce the chance of occurrence or the effects of a disaster. The government has the responsibility for developing specific mitigation measures to reduce the effects of natural or human-made hazards and to identify and develop mitigation measures for other hazards that may occur. These measures include, but are not limited to, the development of zoning laws and land use ordinances, state building code provisions, regulations and licensing for handling and storage of hazardous materials, and the inspection and enforcement of such ordinances, codes, and regulations.

Recovery

Recovery is the activity or phase that involves restoring all systems to normal. Short-term recovery involves returning vital life support systems to minimum operating standards. Long-term recovery may take years and may involve complete redevelopment of the area affected by the disaster. Local and state governments share the responsibility for protecting their citizens from disasters and for helping them to recover when a disaster strikes. In some cases, a disaster is beyond the capabilities of the state and local government to respond.

The Robert T. Stafford Disaster Relief and Emergency Assistance Act, Public Law 93-288, as amended (the Stafford Act), was enacted to support state and local governments and their citizens when disasters overwhelm them. Federal disaster assistance is authorized under the provisions of the Stafford Act when the governor requests and the president declares an emergency or a major disaster to exist in the state. The Stafford Act authorizes two types of assistance: individual assistance and public assistance.

While the Stafford Act is the most familiar mechanism by which the federal government may provide support to state, tribal, and local governments, it is not the only one. Often, federal assistance does not require coordination by DHS and can be provided without a presidential major disaster or emergency declaration. In these instances, federal departments and agencies provide assistance to states, as well as directly to tribes and local jurisdictions, consistent with their own authorities. For example, under the Comprehensive

Environmental Response, Compensation, and Liability Act, local and tribal governments can request assistance directly from the Environmental Protection Agency and/or the U.S. Coast Guard. This support is typically coordinated by the federal agency with primary jurisdiction rather than DHS. The Secretary of Homeland Security may monitor such incidents and may, as requested, activate framework mechanisms to support federal departments and agencies without assuming overall leadership for the incident.[11]

NFPA 1561, *Standard on Emergency Services Incident Management System*[12]

Since 1986, OSHA has required the use of an incident command system to manage the response and coordination of responders at hazardous materials incidents. These regulations also require that personnel operating at hazardous materials incidents be appropriately trained. In addition, OSHA requires that those in supervisory positions have additional training in incident command.

Meanwhile, other incident command systems were emerging including those developed by the National Fire Academy and the Emergency Management Institute. These programs established the beginning of what we know today as a standardized IMS that includes common terminology, modular expansion, and implementation at the incident scene.

Concurrently with these activities, NFPA was addressing many firefighter safety issues. NFPA published the first edition of NFPA 1500 in 1987, and the first edition of NFPA 1561 followed in 1990. You will note that the IMS standards in the 1500 series pertain to firefighter safety and health. Incident management is a critical factor in firefighter safety.

NFPA 1561 was a logical follow-on for the then brand-new NFPA 1500, *Standard on Fire Department Occupational Safety and Health Program.* One of the areas addressed in NFPA 1500 was the need for an effective incident command. There is a very direct and obvious correlation between effective command and firefighter safety.

The most recent changes to NFPA 1561 are the name of the document (from Fire Department to Emergency Services) and extension of its contents to include all emergency service providers, recognizing that many agencies are often present at the scene of emergencies, and they may not be as familiar with IMS as are fire departments. Regardless of the agency, the critical factors of IMS follow:

- Establishment of command
- Routine evaluation of the risks involved

- Well-defined strategic options
- Standard operating procedures
- Effective training
- Appropriate protective clothing and equipment
- Effective incident management
- Effective incident communications
- Safety officers and procedures
- Personnel accountability
- Backup crews for rapid intervention
- Adequate resources
- Provisions for rest and rehabilitation of personnel
- Regular reevaluation of the conditions
- Pessimistic evaluations of the situation

The key to success is knowledge, knowledge gained in knowing, understanding, and using the system. For those who lack experience, training must fill in the void, providing them with the information they will need to operate within the system, play an effective part in solving the problem, and survive.

Incident management, accountability, rapid intervention, risk management, safety factors, two-in-two out, rehabilitation, safety officers. We hear these terms in our training programs and see them in the literature. Where do they come from and how do they fit into the business of firefighting? All of these terms deal with safety during emergency operations (see **Figure 13-22**). All are addressed in NFPA 1500 and were covered briefly in Chapter 8 of this book. NFPA 1500, NFPA 1561, NFPA 1710, NFPA 1720, and other documents provide vital information pertaining to enhancing safety during emergency operations. As company officers, you should be intimately familiar with these concepts

FIGURE 13-22 Firefighter safety is always important.

DESIGN REQUIREMENTS FOR AN INCIDENT MANAGEMENT SYSTEM

- Can be used with a single jurisdiction or agency or can include multiple jurisdictions and/or agencies
- Structure can be adapted to fit any situation; The system is easy to use and can be used at all types of events
- Must have common elements of organization, terminology, and procedures
- Allows for expansion, transfer, and termination of operations

to ensure the safety of your personnel and the ability to operate as an effective unit during large-scale events.

What Makes IMS Work

Several characteristics make IMS work. Two of these have to do with communications. The first is the use of plain language; the second is the use of common terminology. To be effective, the terms and definitions used in IMS must be understood and accepted, and the words must have a single definition over a period of time and the space of distances. If we use the same terms and understand their definitions, we can better communicate our thoughts.

Another characteristic of IMS is that it has a modular organization, meaning that you only apply the parts of the system that are needed, based on the size and complexity of the incident. In simple incidents, the IMS structure is a simple one. There are few problems, the event is resolved quickly, the resources are limited and adequate, and the incident commander (IC) can personally deal with all of the management functions. As the incident becomes more complicated, so does the IMS structure. You will need additional resources, and the event is likely to last longer. The IC may not be able to personally manage all of the activities. In such cases, the IMS structure must be expanded to meet the needs of the situation.

When the IMS expands to fit the needs of a larger situation, the IC has to delegate some of the management functions to others. The IMS allows for an orderly and standardized way of doing this and allows for delegation of specific responsibilities (see **Figure 13-23**). These responsibilities can be along functional lines (fire attack) or at specific locations (third floor.)

IMS is a performance-oriented system rather than a rank-oriented system meaning the function is more important than the rank of the person filling the position. The qualifications of the individual assigned should be the only concern. Let us outline the nine basic components of the incident management system.

NIMS Training

In addition to the previously listed classes three are a number of additional related classes that address a number of different professions from the firefighter to the healthcare worker to the emergency manager (see **Table 13-1** and **13-2**).

INCIDENT MANAGEMENT SYSTEM NATIONAL INTEGRATION CENTER

While most emergency situations are handled locally, when there's a major incident help may be needed from other jurisdictions, the state and the federal government. NIMS was developed so responders from different jurisdictions and disciplines can work together better to respond to natural disasters and emergencies, including acts of terrorism. NIMS benefits include a unified approach to incident management; standard command and management structures; and emphasis on preparedness, mutual aid and resource management.

The **National Integration Center (NIC) Incident Management Systems Integration Division** was established by the Secretary of Homeland Security to provide "strategic direction for and oversight of the National Incident Management System (NIMS) . . . supporting both routine maintenance and the continuous refinement of the system and its components over the long term." The Center oversees all aspects of NIMS including the development of compliance criteria and implementation activities at federal, state and local levels. It provides guidance and support to jurisdictions and incident management and responder organizations as they adopt the system.

The Center is a multidisciplinary entity made up of federal stakeholders and over time, it will include representatives of state, local and tribal incident management and responder organizations. It is situated within the Department of Homeland Security's Federal Emergency Management Agency.[13]

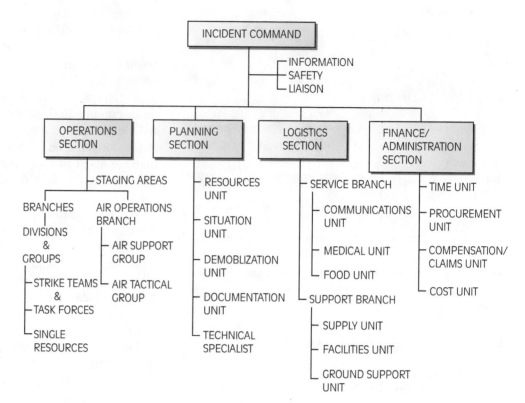

FIGURE   13-23 Incident command system organizational flowchart.

TABLE 13-1 Core Curriculum for the NIMS

Course Grouping	Course ID	Course ID
Overview	IS-700	National Incident Management System (NIMS) an Introduction
	IS-800	National Response Framework (NRF), an Introduction
ICS Courses	ICS-100	Introduction to the Incident Command System
	ICS-200	ICS for Single Resources and Initial Action Incidents
	ICS-300	Intermediate ICS
	ICS-400	Advanced ICS
NIMS Components and Subcomponents	IS-701	NIMS Multiagency Coordination System
	IS-702	NIMS Public Information Systems
	IS-703	NIMS Resource Management
	IS-704	NIMS Communication and Information Management
	IS-705	NIMS Preparedness
	IS-706	NIMS Intrastate Mutual Aid, An Introduction
	IS-707	NIMS Resource Typing

TABLE 13-1 Continued

Course Grouping	Course ID	Course ID
ICS Position-Specific Courses	P-400	All-Hazards Incident Commander
	P-430	All-Hazards Operations Section Chief
	P-440	All-Hazards Planning Section Chief
	P-450	All-Hazards Logistics Section Chief
	P-460	All-Hazards Finance Section Chief
	P-480	All-Hazards Intelligence/Investigations Function
	P-402	All-Hazards Liaison Officer
	P-403	All-Hazards Public Information Officer
	P-404	All-Hazards Safety Officer

COMMUNICATIONS MANAGEMENT = GOOD INCIDENT MANAGEMENT

Communications management is integral to effective incident management. Communications management involves making effective use of communications systems to provide for the timely and useful flow of information. We should have communications procedures that are easily used and widely understood. This implies the use of plain language and common terms that mean the same thing to all personnel.

Fire officers should also establish a procedure for providing initial reports and progress reports on a regular basis. Initial reports provide other responding personnel with necessary information so that they can mentally and physically prepare for action. Initial reports also provide information for fire department managers regarding the commitment of resources and the need to expand the incident command system to meet the particular needs of the situation. Progress reports provide vital information regarding the progress, prognoses, and expected duration of the event. As stated before, a progress report should be given anytime you experience a sudden change in conditions (good or bad), change levels (up or down), have completed your assigned task, or need additional resources to complete your assignment.

Whether you are speaking face-to-face or over a radio or telephone, good communications are enhanced by remembering to use the six Cs of communications: conciseness, clarity, confidence, control, capability, and confirmation. Remember: A moment of thought before speaking usually makes the message shorter and clearer.

- *Conciseness* is important during emergency operations. Usually there is much to think about and to do, and there is usually a lot of radio traffic.
- *Clearly worded* and concise messages speed the entire evolution. By the same token, clear and understandable messages increase the likelihood of comprehension and the probability that the requested action will be accomplished.
- *Confidence* is exhibited with a calm voice and a normal speaking rate. Confidence on the part of the incident commander is communicated with the message and shared by all those involved in the incident.
- *Control of communications* is also important to the success of an operation. Communications control is established by the first-arriving unit and maintained by the incident commander and the dispatch operator throughout the incident. Communications control requires that certain discipline be maintained and that certain rules be followed. Units must identify themselves when calling, and units being called must acknowledge the message. This ensures the message was received and understood.
- *Capability* deals with ability and the willingness to communicate at every level. Communications are a two-way process. Capability implies that the incident commander is willing to listen to others as well as to transmit orders.
- Finally, *confirmation* means that you get an acknowledgement of orders. It is the feedback in the communications loop: "I got your message." One of the main points of discussion in the after action reports of some of the most significant incidents in this country; Katrina; California Fire Storm, World Trade Center, is communications. Once the breakdown in communications starts, it just seems to snowball. From the start, you should learn to communicate right and communicate smart.

TABLE 13-2 NIMS Core Curriculum Aligned with NIMS Components and Level of Training

			Levels of Training		
			Awareness	Advanced	Practicum
Components of NIMS	Preparedness		IS 800 IS-700		
	Communications & Info Management		IS 705 IS 704		
	Resource Management	IS-700	IS 703		
	Command & Management	ICS	ICS 100	ICS 300	Position-specific courses
		MACS	ICS 701		
		Public Info	ICS 702		
	Ongoing Management & Maintenance				

Key ICS Positions

The **incident commander (IC)** is the one position of the IMS that must be staffed at every alarm. Someone has to be in charge, responsible, and accountable for the safe and effective mitigation of the problem. The IC must understand the roles of the various positions within the IMS structure. Initially, the IC is responsible for determining the strategy, selecting tactics, setting the action plan in motion, and developing the IMS organization.

Depending upon the size of the incident, the IC may also be responsible for managing resources, coordinating the activities of these resources, providing for scene safety, releasing information about the incident, and coordinating efforts with other agencies. At smaller incidents, the IC will likely personally retain these tasks. At larger incidents, these functions will likely be delegated. In all cases, however, the IC is responsible for the overall outcome of the event. There is no requirement that the fire chief automatically becomes the IC. The IC should be the most experienced, capable and qualified person, based on the experience level of the officer as well as the dynamics and complexity of the situation. In a hostage situation this may be a law enforcement officer. In a multicasualty event it may be a medical representative. It can also change as the situation changes. An incident may start out with a firefighter as an IC, and once the fire is under control, the position may be transferred to a law enforcement officer because an explosive ordinance was discovered and it is now a crime scene where they are preparing to insert an explosives unit and initiate an investigation.

Volumes could be added about the roles and responsibilities of the IC, but the main thing to take away from this section is to look at things from a strategic perspective and focus on the overall situation. This means the IC has to operate from a fixed vantage point in order to see as much of the entire event as possible. If the commander becomes "hands-on" with tactical operations, then it is likely that he/she will lose that overall perspective. The IC needs to establish and remain at a location that will become known as the command post. The command post may be the hood of someone's car or the inside of a custom vehicle designed expressly as a command vehicle. It must be a visible, fixed location that is capable of supporting the mission

Operations. The first general staff position that should be delegated is the **operations section**. The operations section chief is responsible supervising, managing, and assigning tactical resources associated with the primary mission. The operations chief is responsible for developing and implementing strategies to achieve the objectives developed by command. The operations section may have just one officer, usually one of the first officers to arrive on the scene. The operations chief reports to the incident commander (see **Figure 13-24**).

The terms **divisions** and **groups** are tactical-level management groups that command individual companies, task force, or strike teams. The term *division* indicates a geographical assignment, whereas the term *group* indicates a functional assignment. A **task force** is combination of mixed resources with

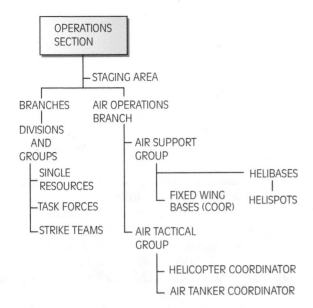

FIGURE 13-24   Operations section.

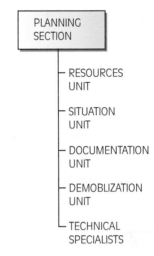

FIGURE 13-25   Planning section.

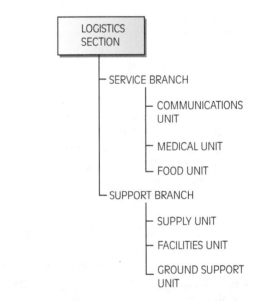

FIGURE 13-26   Logistics section.

a common communications and a single unit leader. A **strike team** is a set number of like resources that operate with common communications and a single unit leader. While not all agencies use these terms, they are standardized NIMS terms that are applicable in all areas of the nation. These positions report directly to the IC unless the operations section is staffed and then they report directly to the operations chief. **Branches** are used when the number of divisions or groups exceeds the span of control. These functions are typically only filled at larger or more complex alarms or where a department has staff personnel available to fill these positions. As a company officer, you should be aware of the functions of each of these positions, and be prepared to occupy any one of these positions if necessary.

Planning. The **planning section** is responsible for collecting, evaluating, and disseminating appropriate information. The planning chief keeps track of the current situation, predicts the probable course of events, and prepares optional strategies and tactics. The planning section should be staffed with a senior officer who has considerable command experience (see **Figure 13-25**). Planning can be expanded to include four units: **resource unit, situation unit, documentation unit**, and **demobilization unit**.

Logistics. The **logistics section** is responsible for obtaining the resources needed at the event. Logistics comes into play at long-duration events (see **Figure 13-26**). Logistics can be developed into two different branches: service branch and support branch. If warranted, the service branch can be expanded into **communications unit, medical unit**, and the **food unit**.

The support branch can be expanded to include the **supply unit, facilities unit**, and **ground unit**.

Finance. Finance is the last of the major components of IMS. The **finance section** chief is responsible for accounting for the costs associated with the event. This is not usually a concern, but for hazardous materials incidents or long-term operations, especially where multiple agencies are involved, the finance section becomes extremely important (see **Figure 13-27**). The cost of a large alarm could potentially bankrupt a department unless costs are documented and recouped through the reimbursement methods listed above.

Staff Assistants. Depending upon the nature of the incident, the incident commander may fill command **staff positions**. These include a safety officer,

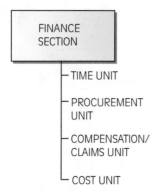

FIGURE 13-27 Finance section.

an information officer, and a liaison officer (see **Figure 13-28**).

The **incident safety officer (ISO)** is one of the most important positions in the IMS organization. This position should be filled as soon as possible. The ISO is responsible for monitoring working conditions and recommending appropriate action. Most organizations have policies that allow the ISO to take direct action when warranted to protect firefighters from immediate danger. In such cases, the ISO should immediately advise the IC of these direct actions. In addition to spotting critical situations, the ISO acts as an advisor to the IC. The ISO has the advantage of being able to move about and see all sides of the action. Ideally, the ISO should be an experienced company officer.

The **public information officer (PIO)** is responsible for the development and release of timely information regarding the event and for serving as a point of contact for the media. PIOs should have access to the incident commander so that they have timely and correct information about the present and projected status of the event. However, their job is to feed that information to the media, thus allowing the incident commander to focus on the job at hand. Anyone can serve as a PIO, but they should have good communications skills and enough understanding of the situation and the department's operations so that they can answer questions.

Finally, we come to the **liaison officer** who works to liaison with other agencies. In some cases, these

THE MEDIA CAN PROVIDE INFORMATION AND ASSISTANCE

We often overlook the value of the media at events, thinking of them as a liability rather than an asset. The media can provide valuable information and services. If an evacuation is needed, the media are an excellent away to reach the public with timely information. Likewise, when people are missing, getting the word to radio and television stations will bring many helpers. The media can also be used for effective life safety public education messages. While you are on television discussing the fire at Ms. Garcia's house, you have a good opportunity to point out that Ms. Garcia's first notice of her problem was her smoke detector. (Obviously, it was working.) Remind viewers about the value of smoke detectors and how important it is to check them once a month.

TAKING CARE OF MS. GARCIA, OUR CUSTOMER

Some departments have introduced a new position called the **occupant services sector**. The basic function of this sector is to treat the uninjured survivors of the event. Leaving Ms. Garcia standing on the sidewalk while you attack her house and its fire will not be a good experience for her. Watching a dozen firefighters drag hoses and tools through her super neat house will not make her feel any better. Maybe you should find a nice place for Ms. Garcia to sit down. Someone should check her over; make sure she is okay. An ambulance would be a good choice.

But lacking an ambulance, a staff car, the cab of a fire truck, almost anything will do. You want to move her out of your way and get her to a safe and comfortable place, protect her from the weather, and keep track of her whereabouts. Someone should stay with her, treat her injuries, explain what is going on, and calm her. You might offer her the use of your cell phone to call a friend or relative.

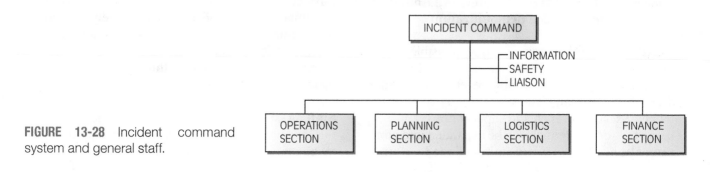

FIGURE 13-28 Incident command system and general staff.

NOTE

An easy way to keep from confusing command staff with general staff is that command staff is made up of officers—safety officer, public information officer, and liaison officer—and the general staff is composed of chiefs—operations section chief, planning chief, logistics chief, and finance chief.

supporting agencies are not represented elsewhere in the command structure, so to ensure their effective use and to prevent conflicts, you need to communicate with them. The liaison officer works with the representatives of the other agencies and forwards any pertinent information to the IC. Any pertinent information that the IC needs to be relay to the agencies is communicated though the liaison officer

LEGAL ISSUES FOR FIRE OFFICERS

The concept of standard of care is well known the firefighters who have some form of training in emergency medical care. Most textbooks in that field deal with this important topic in the first chapter. In this book, we are discussing legal issues last, in hopes that you will remember what you are reading here.

Unlike the emergency medical care business, firefighters and fire officers have not had to concern themselves with the legal ramifications of their actions. For years, firefighters and fire departments have escaped legal scrutiny. That is changing, and although we would be the last to say that you should run your fire department in ways tantamount to the legal profession, we should consider the laws and the risks associated with breaking those laws. In addition, we should be concerned about the concept of "standard of care," that well-known phrase that says that you should do what another reasonable person would do with the same circumstances, training, and resources.

This standard of care is a dynamic process. Laws change, and so do our standards. Training, equipment,

and procedures also change. We cannot pretend we do not know or do not want to follow current procedure. Not too many years ago, firefighters did not have SCBAs and PASS devices. Imagine sending a firefighter into a structure today without this equipment. A few years ago, we had open cabs and allowed firefighters to ride on the back step. When someone got hurt, we just considered it part of the job. Imagine that same situation today. In our litigious climate, a lawyer very possibly would try to persuade the hurt individual to sue.

Citizens and even other firefighters have certain expectations when it comes to the activities of their fire department and its personnel. Fire officers are occasionally thrust into situations for which they are unprepared. Training and experience help, but even well-trained, experienced officers can find themselves in charge of a response where the circumstances are unlike anything seen before.

Most of the nation's fire departments are a part of local government. Local governments are expected to provide for the public's safety, usually through the services of a police department and a fire department. As a result, fire departments have generally been shielded from liability under a concept known as "government immunity." Government immunity comes from an old concept called "sovereign immunity"; the sovereign could do no wrong. Under this principle, the fire department could not be sued in court, no matter how negligent it may have been.

The concept left no recourse to citizens when the fire department or its personnel were negligent. In recent years, the federal government and the legislatures of most states have enacted some type of reform that has allowed governments to be sued for their actions, just as an individual or private company might be sued. These laws usually cap the liability at a certain level. Maryland, for example, sets the limit at $200,000 per claimant, and $500,000 per occurrence.

Faced with these limits, claimants are naming individuals as defendants in liability cases. Thus, as a company officer, you could well wind up as a defendant, along with your department and your municipality.

INCIDENT MANAGEMENT

Grant Mishoe, Captain, North Charleston Fire
Department, North Charleston,
South Carolina

Captain Grant Mishoe's company arrived at the same time as two other engines on the report of a structure fire. When they arrived, they did a size-up: Heavy smoke was coming out of the rear of the building, a multilevel electrical supply warehouse located between a one-story supermarket and laundramat on one side and a two-story Masonic lodge on the other. At the back of the building, there was a loading dock and two doors. The team forced the first door. The smoke behind the door was cold. But after forcing the second, the team members were faced with black, hot, pushing smoke, which signaled there was fire behind it.

Mishoe was given the role of interior command, in charge of two attack teams that would go inside to fight that fire. No one knew how deep seated it was. Once inside, he radioed to command about the conditions: The room was 30 feet by 325 feet with cinder block on the bottom floor, metal on the top and wooden floors. There were wall-to-wall refrigerators on either side of the 15-foot door and other various old electronics and plastics that Mishoe suspected made for a dangerous fire load. The visible fire was in the corner of the room on the bottom floor, emanating a heat that made Mishoe uncomfortable—it meant it was burning fast. By sight and voice, Mishoe maintained location of his crew.

After working on the fire on the bottom floor, he looked up and saw a glow on the second floor, radioing

"Should we advance the hoseline?" Command gave him the affirmative. As he got halfway up the steps, the building made a peculiar groan—one he knew a concrete and steel structure should not make—and on instinct, he radioed that conditions were worsening; structural failure was imminent; and he was having his crew evacuate. He did not wait for an answer. His team members were hand over foot out the door and were just in time: Within 30 seconds, the ceiling opened up, and refrigerators and fire collapsed into the bottom floor. For Mishoe, it was a split second decision, and one he does not regret. He knew it was a matter of choosing between his men or the building, and the building was not a priority.

At that point, the fire was 30 feet through the roof, and after conducting a personal accountability for his firefighters, Mishoe went along with command's decision to surround and drown. The building turned into a six-alarm fire, and after firefighters had been on the job for 8½ hours, it had burned to the ground. The structure was lost, but the lives of the firefighters had been saved.

Questions

1. What were the visible clues at the scene of this fire?
2. What were the size-up factors?
3. What were the clues that suggested imminent structural failure?
4. What risks should be taken by firefighters to protect such buildings?
5. What important lessons can be learned from this story?

LESSONS LEARNED

This chapter provided a discussion of the role of the company officer during emergency operations as well as an introduction to the Incident Management System. Clearly, one chapter does not do justice to this important topic. As company officers, you should be able to analyze emergency scene conditions, conduct size-up, develop and implement an initial action plan, and deploy resources to safely and effectively control an emergency.

As you already know, the role of the fire service has expanded far beyond fire suppression. The name "fire department" does not begin to accurately describe the real mission: public safety. As important as that is, public safety is more than a mission statement. It is a core value that underlies all of our prevention and response services.

Fire prevention and life safety education programs have already been discussed. Similar prevention programs are needed to help citizens prevent and prepare for such diverse threats as medical emergencies, floods, and wildland fires.

Citizens depend upon their fire department to deal with nearly every conceivable emergency. Recent fire department responses to help citizens dealing with severe weather, floods, earthquakes, and other natural disasters quickly come to mind. Add to that a list of human-made disasters—fires, explosions, structural collapse, and technical rescues of every kind.

Recent events in Oklahoma City, New York City, and Arlington, Virginia, have added a new dimension to the concept of public safety. As first responders on any emergency call, fire service personnel must prepare themselves for new threats: weapons of mass destruction. Tragically, in some cases, the target is the emergency responder.

The challenges are significant. For the leaders of the fire service, the challenge is to define a competent, effective, competitive organization that is both prepared for today and ready for tomorrow. These are not small issues. And we cannot pick and choose from a list of options those activities that are personal favorites: We must be ready to face all of the threats and meet the need of our communities and the nation.

While the focus of this chapter and the entire book is to meet the performance standards of NFPA 1021, as company officers, you should be able to cope with EMS and hazardous materials events as well. Most firefighters today are certified as EMTs and hazardous materials responders at the operations level. Training for both of these programs should include managing incidents in these specialty areas. If you can integrate the IMS information on the preceding pages with what you have learned in those courses, you should be ready to manage routine events involving medical emergencies, hazardous materials incidents, and even a kitchen fire at Ms. Garcia's house.

Good luck.

KEY TERMS

action plan   an organized course of action that addresses all phases of incident control within a specified period of time

attack role   a situation in which the first-arriving officer elects to take immediate action and to pass command on to another officer

benchmarks   significant points in the emergency event usually marking the accomplishment of one of the three incident priorities: life safety, incident stabilization, or property conservation

branches   level of supervision between divisions and groups and operations that is used when number of divisions or groups exceeds acceptable span of control

command role   a situation in which the first-arriving officer takes command until relieved by a senior officer

communications unit   prepares communication plan, distributes and maintains communication plan

confinement   an activity required to prevent fire from extending to an uninvolved area or another structure

defensive mode   actions intended to control a fire by limiting its spread to a defined area

demobilization unit   ensures resources a released from the incident in an orderly, safe, and cost-effective manner

documentation unit   maintains and archives all incident related documentation

division   tactical level management group that commands individual companies, task force, or strike teams; a geographical assignment versus a functional assignment

emergency incident   any situation to which a fire department or other emergency response organization responds to deliver emergency services

environmental factors factors like weather that impact on firefighting operations

facilities unit sets up, maintains facilities, provides facility security

finance section that part of the incident management system that is responsible for facilitating the procurement of resources and for tracking the costs associated with such procurement .

fire extinguishment activities associated with putting out all visible fires

food unit supplies food and potable water, operates food services unit

ground unit arranges for transportation of personnel, supplies, food, and equipment

group tactical level management group that commands individual companies, task force, or strike teams; a functional assignment versus a geographical assignment

incident command system a tool for dealing with emergencies of all kinds

incident commander (IC) the person in overall command of an incident

incident management system an organized system of roles, responsibilities, and standard operating procedures used to manage an emergency operation

incident safety officer (ISO) as part of the IMS, that person responsible for monitoring and assessing safety hazards and ensuring personnel safety

public information officer (PIO) as part of the IMS, that person who acts as the contact between the IC and the news media, responsible for gathering and releasing incident information

initial report a vivid but brief description of the on-scene conditions relevant to the emergency

liaison officer as part of the incident management system, the contact between the incident commander and agencies not represented in the incident command structure

logistics section that part of the incident management system that provides equipment, services, material, and other resources to support the response to an incident

medical unit prepares medical plan, provides first aid and light medical treatment; mutual aid assistance provided by another fire department or agency

National Integration Center (NIC) Incident Management Systems Integration Division established by the Secretary of Homeland Security to provide "strategic direction for and oversight of the National Incident Management System (NIMS)"

National Response Framework (NRF) defines the principles, roles, and structures that organize how we respond as a nation

occupant services sector that part of the incident management system that focuses on the needs of the citizens who are directly or indirectly affected by an incident

offensive mode firefighting operations that make a direct attack on a fire for purposes of control and extinguishment

operations section division in the incident management system that oversees the functions directly involved in rescue, fire suppression, or other activities within the mission of an organization

overhaul searching the fire scene for possible hidden fires or sparks that may rekindle

planning section that part of the incident management system that focuses on the collection, evaluation, dissemination, and use of information (information management) to support the incident command structure

resource unit conducts check-in activities and maintains status of all resources

situation unit collects and analyzes information on the current situation

staff positions designated to aid the IC in fulfilling the IC's responsibilities, typically, an incident safety officer, an information officer, and a liaison officer

strategy sets broad goals and outlines the overall plan to control the incident

strike team set number of like resources that operate with common communications and a single unit leader

supply unit orders, receives, and distributes supplies

tactics various maneuvers that can be used to achieve a strategy while fighting a fire or dealing with a similar emergency

task force mixed number of resources that operate with common communications and a single unit leader

tasks the duties and activities performed by individuals, companies, or teams that lead to successful accomplishments of assigned tactics

transitional mode the critical process of shifting from the offensive mode or from the defensive to the offensive

REVIEW QUESTIONS

1. What are incident priorities?
2. What is meant by command sequence?
3. What is meant by size-up?
4. What are the parts of size-up?
5. What are the three modes of firefighting?
6. What information should be included in the initial report?
7. What is an action plan?
8. How is the action plan effectively communicated to all units on the scene?
9. When should an action plan be revised and why?
10. How does the initial action plan provide direction to assign unit resources to accomplish specific tasks?
11. What is meant by *strategy*? *tactics*?
12. What is IMS? When should it be used?
13. What is your role as company officer in incident management during the early phases of fires and other emergency activity?
14. What are the main components of an IMS?
15. What are the functions of command staff, division, or group officer within ICS?
16. What is the role of the fire officer in personnel accountability system at an emergency incident?
17. What are the elements of the communication model and how can they be used to make clear assignments to subordinates?
18. What are the elements of a verbal command and order confirmation process during an emergency?

DISCUSSION QUESTIONS

The first five questions pertain to the kitchen fire in Ms. Garcia's home.

1. What were the size-up factors at Ms. Garcia's house?
2. As the first-arriving officer at this fire, what would be your size-up report?
3. Considering Chief Layman's list of strategies, which would be relevant at this fire?
4. What could you do for her?
5. Based on ICS principles, what pertinent assignments and actions should be included in an initial report for arriving units?
6. How would you use IMS at the scene of a traffic accident on an interstate highway?
7. Does your department have operating procedures that describe the application of an IMS?
8. Are there any unoccupied buildings in your jurisdiction? If so, what is your department doing in anticipation of a response to a fire or other event in one of these buildings? What preparation is being taken at the company level, and what action would be appropriate if you were called to a structural fire in one of the buildings today?
9. How do you evaluate the effectiveness of the implemented IMS?
10. When and how should you revise the IMS to meet the changing needs of the incident?
11. How do you evaluate and determine the need for various IMS components as they apply to the emergency incident?
12. What factors determine how you should assign personnel at an incident?
13. What factors determine when an incident safety officer(s) should be assigned at an incident?
14. What is the evacuation procedure for your department and how do you initiate it?
15. What interagency agreements exist within your jurisdiction and what are the policies and procedures for implementing them?
16. What resources are available from other local emergency and nonemergency agencies and when should you request them?
17. What potential informational resources are available to you during an emergency response?
18. What interagency cooperation might be required for public safety?
19. When and how do you conduct a unit critique?
20. How does your department conduct a postincident analysis?

ENDNOTES

1. Used with permission. For further information, see Ron Moore's series entitled "University of Extrication" each month in *Firehouse Magazine*. His series on operating safely on the highway ran in several installments starting in the October 2003 issue.

2. The USFA develops reports on selected major fires, usually involving multiple deaths or a large loss of property. The objective reviews are intended to uncover significant "lessons learned" or new knowledge about firefighting or to underscore ongoing issues in fire service. A copy of the full report can be ordered by going to http://www.usfa.fema.gov/fireservice/ techreports/tr102.shtm.

3. These questions seem to focus on firefighting, but remember that we are discussing a thought process that can be applied to all emergency situations.

4. Some suggest that there is really a fourth mode, the no-attack mode. There are two situations where no-attack might be the correct solution. Obviously, if there is no fire, you do not need to mount an attack. The other possibility is where the fire is so involved that saving the building is impossible, or where fighting the fire would present unacceptable risks to the firefighters or the environment, so you elect to let it burn.

5. Reprinted with permission from NFPA 1500, *Standard on Fire Department Occupational Safety and Health Program*, Copyright © 2007, NFPA. This material is not the complete and official position of the NFPA, which is represented by the standard in its entirety.

6. SOPs, training, and good fireground discipline are important at times like these. The incident commander has quite enough to think about without having to decide where each apparatus should be placed. But the incident commander should remember that placement is important. Once pumpers and ladder trucks "go to work," it is hard to relocate them.

7. Copied from 2009 version of NFPA 1021, p. 9.

8. Reprinted with permission of NFPA. For more information, visit the NFPA Web site at www.nfpa.org.

9. Chief Layman's book is entitled *Fire Fighting Tactics*, and his original list was called Basic Divisions of Fire-fighting Tactics. Chief Layman made some remarkable contributions to the fire service and was way ahead of his time in many areas. His choice of words (*tactics* instead of *strategy*) is understandable. Reprinted with permission from National Fire Protection Association, Quincy, MA 02269.

10. Copied from the National Incident Management System (NIMS): Five-Year NIMS Training Plan. http://www.fema.gov/pdf/nims/5year_nims_training_08.pdf.

11. Copied from the National Response Framework. Overview. http://www.fema.gov/pdf/emergency/nrf/nrf-overview.pdf

12. The entire document is available from NFPA. For more information, visit the NFPA Web site at www.nfpa.org.

13. Copied from the National Integration Center (NIC) Web site at http://www.fema.gov/emergency/nims/

ADDITIONAL RESOURCES

Brunacini, Alan. *Fire Command*. Quincy, MA: National Fire Protection Association, 1985.

Brunacini, Alan. "Rules of Engagement." *Fire Engineering*, a monthly column.

Castro, Joseph. "One Nation, Under NIMS." *FireRescue*, April 2008.

Coleman, John. "Actions of the First-In Officer." *Fire Engineering*, March 2007.

Crawford, Jim. "New Term, Old Concept." *FireRescue Magazine* (April 2009)

Emory, Mark. "Introducing the Contemporary Fireground." *Firehouse*, February 2009.

Emory, Mark. "Ten The Commandments of Intelligent and Safe Fireground Operations." *Firehouse*, multi-part series starting in February 2007.

Ennis, Rick. "4 for All." *Fire Chief*, October 2008.

Fried, Emanual. *Fireground Tactics*. Chicago: Marvin Ginn, 1972.

Layman, Lloyd. *Attacking and Extinguishing Interior Fires*. Quincy, MA: National Fire Protection Association, 1955.

Layman, Lloyd. *Fire Fighting Tactics*. Quincy, MA: National Fire Protection Association, 1953.

Managing Company Tactical Operations, developed by the National Fire Academy, Emmitsburg, Maryland, is a three-part training course that focuses on preparation, preplanning, and managing typical structural firefighting activities. The second course, "Decision-making," is particularly relevant to the material covered in this chapter.

Mason, James. "Setting the Stage with the First-Due Radio Report." *Fire Engineering*, January 2007.

The National Fire Academy also presents an excellent resident course entitled "Command and Control of Fire Department Operations at Target Hazards." Contact your training officer or state fire service training director for more information.

NFPA 1500. *Standard on Fire Department Occupational Safety and Health Program*. Quincy, MA: National Fire Protection Association, 2002.

NFPA 1561. *Standard for Emergency Services Incident Management System*. Quincy, MA: National Fire Protection Association.

Walsh, Charles V., and Leonard G. Marks. *Firefighting Strategy and Leadership*. 2nd ed. New York: McGraw-Hill, 1977.

National Integration Center (NIC) Incident Management Systems Integration Division. http://www.fema.gov/emergency/nims/index.shtm.

National Response Framework. http://www.dhs.gov/xprepresp/committees/editorial_0566.shtm.

Homeland Security Presidential Directive (HSPD)-5: Management of Domestic Incidents. Washington, DC: White House, February 2003.

Salka, John. "Follow the Rules!" *Firehouse*, September 2007.

VanDoren, John. "The Debriefing: Another Tool for Your Box." *Fire Engineering*, July 2007.

PROFESSIONAL STATUS: THE FUTURE OF FIRE SERVICE TRAINING AND EDUCATION

Dr. Denis Onieal, Superintendent,
National Fire Academy

Professional status is a term that has been bandied about in the Fire and Emergency Services for years. What constitutes "professional" status is in the eye of the beholder? Were we to look at a "professional" independent of the fire service, to some, it means the performance of a series of skills in a manner that is far above average. To others, a professional is associated with performing skills "full-time," that is to say, for a living. Many feel that the distinguishing characteristics of a professional are years of formal education, approval of an accrediting board, and continuing education requirements. More than likely, it is the last statement with which most would agree.

Definitions aside, it is the walk down the main street in any city or town in America that demonstrates who in the community is professional. The physicians and nurses, the architects and engineers, the attorneys and the accountants are among the top professionals in any community. What makes them so?

Each has a unique set of knowledge and skills that are independent of a particular organization or place; they are "portable" skills and held in equal regard no matter where the person practices. In the process of becoming a professional, there is an accredited and independent testing process that assures competency to the public. Professionals are associated with others in their profession through some formal organization; they typically put service to others as more important than profit; and they assume responsibility for their professional acts. Typically, their profession has some continuing education requirements, and the work is client centered.

Then why aren't we given the professional status of physicians and nurses, architects and engineers, and attorneys and accountants? Well, those professions have some things that the Fire and Emergency Services do not yet have; there are a few more steps.

It is interesting to see where our current "professions" were one hundred years ago. Most people probably don't realize that medical education was haphazard in this country until 1910. In the late 1700s, most physicians apprenticed, and a few attended medical schools in Europe. Many people also don't realize that although Abraham Lincoln was a lawyer, he never went to law school; he apprenticed.

Thirty years ago, the foundations for professional status for the Fire and Emergency Services were laid. Performance standards were established. Colleges and universities recognized the need for formal education and began degree programs. Fire departments began to require certifications, and many began to require degrees or advanced degrees for hiring or promotion. Uncommon thirty years ago, but quite common today, is the hiring of people with professional training and education from outside the organization (instead of through the ranks) to come in to run it. That's the evidence that we're ready to make the next move up the ladder of professions.

One of the principal challenges we have is that the number of independent systems of training and education staggers aspiring fire service professionals. There is no "one way" for the student to determine which is the most appropriate training and/or education. There's no "one way" to become the chief. The problem is exacerbated by the reality that there is little chance that one system will recognize that student's performance in another system. Moving from fire department to fire department (or even more difficult—from a fire department in one state to a fire department in another), training or education already received may not be recognized.

The Fire and Emergency Services today are assuredly further along the path to professional status than those in medicine and law were one hundred years ago. We have a body of knowledge, we have standards and we have processes to assure competency (available through the International Fire Service Accreditation Congress [IFSAC] and the National Board on Fire Service Professional Qualifications [NBFSPQ or ProBoard]). We have places to acquire professional knowledge, but right now, they are locally based—they aren't a part of a system that everyone recognizes. The missing link is a nationally recognized, reciprocal system of training and education. The good news is that we have all the parts; nothing has to be invented or established—they just need to be integrated:

- Training systems (available through local, state and the National Fire Academy);
- Education systems (available through 2-year, 4-year, graduate and National Fire Academy);
- Independent assessment of skills (IFSAC and ProBoard);
- Reciprocity among systems of training and education.

The current roles of local, state and national emergency services training generally establish the boundaries for each to prevent costly duplication. Locally, larger departments are capable of training their own people to certain levels of competency. Smaller departments will either seek training from a larger organization, work with other small departments to combine training resources, or seek training from another government agency—either the county or the state training system. Depending on the size of the organization and its needs, local training tends more towards recruit, refresher and "hands-on" training.

State training organizations generally attempt to provide training that is not available locally—ranging from basic recruit training to courses for chief fire officers, from hazardous materials awareness to firefighting strategies at petroleum facilities, and from farm rescue to wildland firefighting. State training organizations vary in their size and capacity, from a few people to a complex, university based system.

At the national level, each State Fire Training System works with the United States Fire Administration's National Fire Academy (USFA/NFA) to deliver USFA/NFA curriculum. The USFA/NFA develops and delivers the kinds of training that aren't available at the local or state level. Community Risk Reduction, Public Education, Codes and Standards, Detection and Suppression Systems, Executive Development, Terrorism, Command and Control of Incidents, Strategic Planning, Information Systems and Budgeting are among the USFA/NFA's curriculum areas. This system isn't something that is planned for the future—this is the system as it exists today.

At the national level, most of you would probably be surprised to learn that the USFA/NFA does the least amount of training on its Emmitsburg campus—about 8,000 students per year. Most of NFA's training occurs off-campus through the cooperative efforts of State and Metropolitan sized fire training organizations. In 2002, the USFA/NFA trained over 87,000 Fire and Emergency Services personnel in off-campus course deliveries, self-study courses, CD based simulation training and other alternative deliveries through its virtual campus. (For more information see http://www.training.fema.gov).

Locally, many colleges and universities provide two- and four-year degree programs in fire science and/or administration. Over the past several years, a few Master's degree programs have emerged. For those who, for reasons of proximity or time, are unable to attend a local college, the USFA/NFA works with seven schools throughout the country to provide four-year degrees via the Degrees at a Distance Program; there are no resident course requirements for these courses.

Those who have attended two- and four-year programs (currently there are 222 two-year and 26 four-year programs in the US)[1] are usually people who are "in-service," that is to say, are going to school part-time and working a full-time job. These individuals may be career or volunteer, but most don't enjoy the luxury of full-time academia—it is a considerable sacrifice to them and their families.

The titles of "fire" degrees vary—from Fire Technology all the way to Public Administration with a concentration in Fire Administration. Some degrees are called Fire Science, Fire Administration or Fire Department Management, but the disparity creates misunderstanding among employers and other schools of higher education. Everyone understands what a medical, law or nursing degree means. Few understand what a "fire" degree means. This makes it difficult for other schools and employers to assess the education or skill of prospective students or employees. Hence,

[1]Sturtevant, Thomas B. "A Study of Undergraduate Fire Service Degree Programs in the United States—Fall 2000," Doctoral Dissertation, University of Tennessee, Knoxville, May, 2001.

transfers of credits between schools (and true professional salaries) are elusive.

So, let me ask you, have you or someone you know:

- taken fire science courses at a two-year college;
- taken courses at state and local fire training academies and through the National Fire Academy (NFA);
- achieved various levels of certification;
- and all combined, these achievements are "all over the map," meaning none of them evolved in a coherent and planned way?

Most firefighters and officers have earned college credits and training certificates since their first day in the first service. However, this professional development is usually uncoordinated and fragmented, resulting in duplications of effort and inefficiencies for students. Lack of coordination between fire-related training, higher education, and certification contributes to this problem.

Collaboration and coordination are needed between all service providers responsible for Fire and Emergency Services' professional development. Each has a major role to play. This report presents the recommendations that have evolved over the past four annual Fire and Emergency Services Higher Education (FESHE) conferences. Combined, these products and outcomes represent a new strategic approach to professional development. They will help move the Fire and Emergency Services from a technical occupation to a full-fledged profession similar to physicians, nurses, lawyers, and architects, who, unlike fire service personnel, have common course requirements within their respective degree programs.

There are several major tenets on which a "profession" is built, including reciprocity for practicing in different states (with an exam), universally accepted standards of practice, and a professional development model, among others. The work accomplished during the FESHE conferences addresses one tenet—professional development.

The Role of FESHE Conferences

The U.S. Fire Administration (USFA) hosts the annual FESHE conference on its campus in Emmitsburg, Maryland. These conferences are a combination of presentations, problem solving, and consensus-building sessions that result in higher education-related products or recommendations for national adoption.

At the 2000 conference, two panel discussions were conducted. The first panel included fire service leaders representing national fire service organizations, and the second was comprised of state directors of fire service training. Both panels raised issues that formed the basis for these national recommendations, including the need for:

- degree programs that teach critical thinking skills by requiring significant numbers of general education, rather than mostly fire science, courses;
- appropriate recognition of certification for academic credit and vice versa;
- associate degree programs that are transferable to baccalaureate programs;
- a model fire science curriculum at the associate level that universally standardizes what students learn and facilitates the application of these courses towards certification goals; and
- collaboration between fire service certification and training agencies and academic fire programs.

Fire and Emergency Services Professional Development Model

The professional development model is one product finalized at the 2002 FESHE IV conference. It is not a promotion model addressing credentials; rather, it is an experience-based model that recommends an efficient path for fire service professional development supported by collaboration between fire-related training, higher education, and certification providers. The model recommends what these providers' respective roles should be and how they should coordinate their programs (see **Figure A-1**).

FIGURE A-1 Fire service development model.

Another result of the 2000 FESHE conference was the model fire science associate degree curriculum. The FESHE attendees identified six core associate-level courses in the model curriculum, including:

- Building Construction for Fire Protection
- Fire Behavior and Combustion
- Fire Prevention
- Fire Protection Hydraulics and Water Supply
- Fire Protection Systems
- Principles of Emergency Services

In 2001, the National Fire Science Curriculum Committee (NFSCC) was formed to develop standard titles, descriptions, outcomes, and outlines for each of the six core courses. In 2002, the FESHE IV conference attendees approved the model courses and outlines. The major publishers of fire-related textbooks are committed to writing texts for some, or all, of these courses.

It was recommended that all fire science associate degree programs require these courses as the "theoretical core" on which their major is based. The course outlines address the need for a uniformity of curriculum and content among the fire science courses within the United States' two-year programs. Many schools already offer these courses in their programs, while others are in the process of adopting them. Once adopted, these model courses address the need for problem-free student transfers between schools. Likewise, they promote crosswalks for those who apply their academic coursework toward satisfaction of the national qualification standards necessary for firefighter certifications and degrees.

The committee also developed similar outlines for other courses that are commonly offered in fire science programs. If a school offers any of these "non-core" courses, it is suggested these outlines be adopted, as well.

In 2003, at FESHE IV, NFA announced it would release its 13-course upper-level Degrees at a Distance Program (DDP) curriculum to accredited baccalaureate degree programs which have signed agreements with their state's fire service training agency. DDP will remain as NFA's delivery system for the 13 courses; however, release to other schools enables the formation of model curriculum at this level.

With model lower-level (associate) curriculum outlines developed and established upper-level (baccalaureate) courses available, the major components are in place to move towards a national system for fire-related higher education. Most core and non-core courses line up with baccalaureate courses of similar content, thus preparing associate degree graduates for their bachelor degree studies. This national system for fire-related higher education is important because, as with other professions, a theoretical core of academic courses should be a prerequisite for entering these fields. As more schools adopt these curricula, the Fire and Emergency Services move toward becoming a full-fledged profession.

Notes

Dr. Denis Onieal was appointed superintendent of the National Fire Academy at Emmitsburg, Maryland, in July 1995. A native of Jersey City, New Jersey, he has been a professional firefighter in that city since 1971, rising through the ranks to become deputy chief in 1991 and eventually acting chief of the department. His entire professional life has been in operations.

Dr. Onieal earned a doctorate in education from New York University, a masters in public administration from Farleigh Dickinson University, and a bachelors in fire administration from Jersey City State College.

Dr. Onieal's essay about the need for professionalism in the fire service was condensed here and used with permission. You may read his remarks in their entirety in issues of *The Voice*, published by the International Society of Fire Service Instructors, *Firehouse* magazine, and *Fire/Rescue Magazine*. The series was published in 2003.

Additional information about Fire and Emergency Service Conferences and the recommendations of those conferences is available at www.usfa.gov.

Appendix B

BURLESON FIRE DEPARTMENT
Standard Operating Guidelines
Section 105.12 – Emergency and Mayday Radio Traffic

PURPOSE

The purpose of this policy is to provide a clear, systematic set of procedures that will be used by the Burleson Fire Department in the event of a declared emergency.

POLICY

Emergency Radio Procedures:

The Burleson Fire Department has established three levels of emergency radio traffic.

1. MAYDAY!!! MAYDAY!!! MAYDAY!!!
2. Emergency Traffic
3. Urgent

Mayday Procedures:

The call of MAYDAY will be used to signify that a person or crew is in imminent danger and in need of immediate assistance. In order to be effective, the MAYDAY must be organized. This SOP is intended to provide a framework for an effective and efficient MAYDAY response. The MAYDAY will consist of the call, command response, and pre-assigned Mayday contingency and communications plans. If an "Operations" position is staffed at the Incident, the Operations Commander will answer the MAYDAY.

Calling Mayday

MAYDAY + Unit Number and Seat/Position + To Command + MAYDAY

Example: MAYDAY!!! MAYDAY!!! MAYDAY!!!, E1C to Command MAYDAY

Command (or Operations) Response to Mayday

Unit with a MAYDAY go…

Steps:

1. Initiate MAYDAY tone
2. Close Channel
3. Assign Rescue Chief – If staffing is available
4. Move non-essential traffic/personnel to additional channel
5. Initiate Dispatch MAYDAY contingency plan

Mayday Rescue Chief

If staffing permits, Command or Operations may elect to assign a Rescue Chief. The Rescue Chief will immediately assume control over RIT team deployments and the rescue of the distressed personnel.

Dispatch MAYDAY contingency plan

When a MAYDAY is called, the IC or Operations Chief will advise dispatch to initiate the MAYDAY contingency plan. Dispatch will complete the following automatically.

1. Dispatch additional alarm from Ft. Worth Fire Department
2. Dispatch Ambulance to scene
3. Place Air Ambulance on Standby
4. Advise P.D. channel of MAYDAY and maintain priority to fire channel

Emergency Traffic:

The term "Emergency Traffic" will be used in the event that an extremely urgent message is being given that may have life safety information.

Example: "Command to all units EMERGENCY TRAFFIC, evacuate the building, eminent roof collapse."

Urgent:

The term "Urgent" will be used in the event that a priority message must be given over the regular radio traffic.

Example: "Command to all units URGENT, tornado reported headed toward Burleson 2 minute ETA."

Mayday Algorithm:

The following section contains the MAYDAY algorithm that will serve as a guide to Burleson Fire Department Mayday Procedures. Due to the unique nature of a MAYDAY situation, it may become necessary for the IC or Operations Chief to deviate from this algorithm. Therefore, this is provided as a guide only.

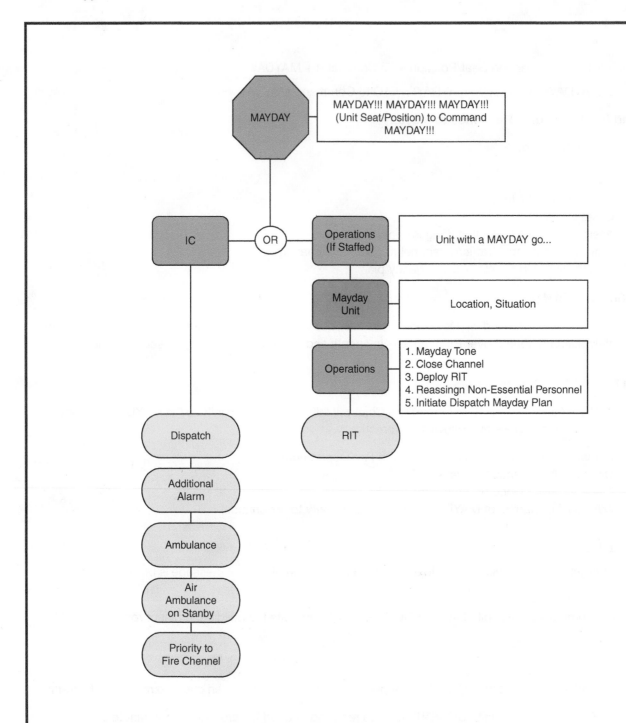

MAYDAY

MAYDAY!!! MAYDAY!!! MAYDAY!!!
(Unit Seat/Position) to Command
MAYDAY!!!

IC — OR — Operations (If Staffed)

Unit with a MAYDAY go...

Mayday Unit

Location, Situation

Operations

1. Mayday Tone
2. Close Channel
3. Deploy RIT
4. Reassingn Non-Essential Personnel
5. Initiate Dispatch Mayday Plan

Dispatch

Additional Alarm

Ambulance

Air Ambulance on Stanby

Priority to Fire Chennel

RIT

ACCOUNTABILITY

Supervisors of the Department are responsible for insuring that this and all guidelines of the Department are followed. Deviations from this or any other guideline are permitted within the scope of authority granted all members of the Department, however, the deviation must be reported to the Chief, in writing, within 24 hours of the deviation.

STANDARD OPERATING PROCEDURE EXAMPLE: NON-EMERGENCY

BURLESON FIRE DEPARTMENT
Standard Operating Procedures
106.02

Rev. July, 2007

UNIFORMS

Purpose:

It is the purpose of this SOP to ensure members reflect a neat and professional appearance. This appearance helps to establish the visibility and authority of Department personnel by fostering community recognition and acceptance. Although it is apparent that no single uniform is appropriate at all times for each of the varied jobs performed by Department personnel, a required conformance to specified standards of dress does help realize these general goals.

Policy:

A. All personnel shall at all times maintain an appearance that is neat, clean and in compliance with the spirit and intent of this policy.

B. The uniform shall be kept clean, pressed and adorned with a full compliment of designated buttons, insignias, badges (as needed), and nametags.

C. No Fire Fighter shall wear or attach any paraphernalia to the designated uniform unless directed to or otherwise authorized.

D. The uniform designated for use by Fire Department personnel is comprised of items specified by the City. All uniform items issued to department personnel shall remain the property of the City.

E. Personnel shall refrain from using the uniform or badge to attempt to influence any person for any cause.

F. Uniforms that have become unfit for service due to poor condition shall be disposed of properly. They shall not be donated or sold to any group.

G. Those uniforms which are listed within this policy shall be the only officially recognized uniforms of the Burleson Fire Department.

H. A minimum of a tee shirt and shorts shall be worn except when employees are in separate restroom, locker, or bedroom areas.

I. The Department baseball cap, tee-shirt, shorts, sweatpants, sweatshirt, and the cold weather jacket are the only parts of the uniform which are authorized to be worn in an off-duty status. The off-duty use of these items must be done in "good taste" and not be worn under any circumstances where the reputation of the Department may be adversely affected. The items are not to be worn by personnel while publicly consuming alcohol or while in places of business where the serving of alcohol is the primary source on income.

J. When wearing a uniform shirt with buttons, the shirt shall be fully buttoned except for the top collar button.

Time Considerations:

A. All personnel shall be in the uniform of the day or dressed for the scheduled activity promptly by 9:30.

B. From 1700 hours and until relieved, uniform regulations while crews are in quarters may be relaxed. Bunker Pants must be worn on all calls during this period, if personnel are in shorts.

(Continues)

Exceptions:

A. Between May 1st and September 31st, personnel will be allowed to wear Fire Department Class "B" shorts and a tee shirt. Providing all personnel on that company are wearing the same.

B. Between May 1st and September 31st, personnel may choose not to wear the Nomex shirt with class "B" uniform when not at a public function. Providing all personnel on that company are wearing the same.

C. The Officer may make the decision as to what uniform is appropriate.

Uniforms:

Fire Department personnel will be issued three types of uniforms.

- Class A – Full "Dress Blues"
- Class B – Nomex work uniform
- Class C – Dress down "work out" uniform

Class A:

The Class "A" Dress Uniform is issued by the Fire Department to fire suppression personnel upon successful completion of their six-month probation. The Class "A" uniform shall be worn during ceremonies, funerals, or when it is appropriate, i.e., public appearance, promotional interviews, or such times as the Fire Chief may direct. The following items comprise and will define the Class "A" Dress Uniform:

1. Dress Cap – Dress cap shall be the standard fire department service cap; navy blue with silver band for Engineers and Firefighters, navy blue with a gold band for Lieutenants and Captains, and white with gold band for Chief Officers. The buttons on the side of the cap shall match the rank color designation. The cap buttons shall be issued determined by rank as follows:

- Chief - five crossed bugles (Gold)
- Assistant Chief – four crossed bugles (Gold)
- Battalion Chief – three crossed bugles (Gold)
- Captain – two parallel bugles (Gold)
- Lieutenant – one bugle (Gold)
- Engineer – "Scramble" (Silver)
- Firefighter – "Scramble" (Silver)

2. Dress Shirt – "Flying Cross" brand. 65% Dacron Polyester, 35% Cotton. Two breast pockets, badge tab, long sleeve, Men's Style #45W6600, and Women's Style #102R660. Name plate, will be worn 1/4" centered below the top seam of the right pocket. Silver for Firefighter and Engineer, Gold for Lieutenant and above. The nameplate will have two lines. Line 1 = Name, Line 2 = Rank. Arial font, block style. Commendation bars will be worn ¼" centered above the top seam of the right pocket. One or Two matching pins (gold/silver) may be worn in the left pocket. Line Personnel will wear blue shirts. Administrative personnel will wear white dress shirts. Shoulder patches will consist of the official Burleson Fire Department patch (silver or gold) on the left side and the appropriate Texas Department of Health patch on the right side. The patch will be sewn on one inch below the shoulder seam and centered over the pressed seam. Dress shirt collar brass shall be 15/16" size and worn with the bugles pointed to the tip of the collar, ¼" from edge of collar. The collar brass will be issued based on the following:

- Chief – five crossed bugles, enameled and "raised" Red/Gold
- Assistant Chief – four crossed bugles, enameled and "raised" Red/Gold
- Battalion Chief – three crossed bugles, enameled and "raised" Red/Gold

- Captain – two parallel bugles, enameled and "raised" Red/Gold
- Lieutenant – one bugle, enameled and "raised" Red/Gold
- Engineer – Silver/Red "scramble", enameled and "raised"
- Firefighter - Silver/Red "scramble", enameled and "raised"

3. Uniform Dress Jacket – navy blue, eight buttons stamped "FD", Double breasted with badge tab and no breast pockets. Buttons shall match rank designation (gold/silver). Shoulder patch will consist of the official Burleson Fire Department patch (silver or gold) on the left side. The patch will be sewn on one inch below the shoulder seam and centered over the pressed seam. The sleeves of the jacket will have a sewn on band designating rank. The bands will be sewn as follows:

- Chief – five bands (Gold)
- Assistant Chief – four bands (Gold)
- Battalion Chief – three bands (Gold)
- Captain – two bands (Gold)
- Lieutenant – one band (Gold)
- Engineer – two bands (Gold)
- Firefighter – one band (Gold)

Dress Jacket collar brass shall be placed in the same manner as the shirt collar brass. The collar brass should be 15/16" and be gold or silver finish to match rank. The Jacket collar brass will be issued by the same rank standards as found in the shirt collar brass above. Nameplates will not be worn with Jackets. "Maltese" service cross(es) shall be worn on the left sleeve only and placed above the rank strip. Members will receive one cross per five years of service to the Burleson Fire Department.

4. Tie – plain, navy blue uniform clip-on tie. Optional tie tack may be any fire service related tie tack with good taste and shall be placed in the center of the tie when worn. The tie tack should match the rank colors gold or silver.

5. Badge – One badge is issued by the Burleson Fire Department. Badges are to be worn on the outside of the jacket if the jacket is being worn, or on the shirt if no jacket is being worn. Badges will be "Rhondium"(Silver) in color for Firefighters and Engineers and Gold for ranks above engineer.

6. Pants – navy blue, matching make of issued jacket.

7. Shoes – black, patent leather, plain toe, low cut, Uniform style, lace up shoe.

8. Socks – Black

9. Belt – Black leather uniform belt. Buckle is to match rank colors of gold or silver.

Class B:

The Class "B" Uniform is issued by the Fire Department to fire suppression personnel. The following items comprise and will define the Class "B" Dress Uniform:

1. Dark Navy Nomex Shirt – Gold Lettering/Patch for Lieutenant and Above, Silver Lettering/Patch for Firefighter and Engineer.

2. Dark Navy Nomex Pants

3. Departmental gray undershirt

4. Departmental approved socks

5. Black boots or roper-type boots. Black shoes if in shorts – non-marking and no color designs.

6. Dark Navy Shorts (designated time)

7. Black Belt

The full Class "B" uniform shall be worn whenever the crew is out of quarters except as specifically allowed to the contrary within this policy. Departmental sweatshirt may be worn in place of Nomex shirt.

Class C:

The Class "C" Uniform is issued by the Fire Department to fire suppression personnel. The following items comprise and will define the Class "C" Uniform:

1. Grey T-Shirt with departmental logo

2. Grey departmental shorts

3. Grey departmental sweat pants

4. Grey departmental sweatshirt – May be worn with Class B

Miscellaneous:

1. Departmental Jacket – May be worn with class B or C

2. Departmental Cap – May be worn with class B or C

Awards Placement:

Please see policy 106.03 regarding the proper placement of awards on uniforms.

Replacement of Lost, Stolen, or Damaged Equipment:

Fire Department personnel shall be individually responsible for the proper care and safe keeping of uniform clothing. In the event of loss, damage or theft of these items, a full report of all appropriate facts and circumstances shall be promptly reported to the Fire Fighter's immediate supervisor. The report shall be reviewed by the Chief and, if appropriates, replacement items shall be authorized.

Accountability:

Supervisors of the Department are responsible for insuring that this and all policies of the Department are followed. Deviations from this or any other policy are permitted within the scope of authority granted all members of the Department, however, the deviation must be reported to the Chief, in writing, within 24 hours of the deviation.

Collar Brass Insignia

Rank	Shirt Insignia	Jacket Insignia
Firefighter		
Engineer		
Lieutenant		
Captain		
Battalion Chief		
Assistant Chief		
Chief		

Appendix C

ELEMENTS OF EFFECTIVE ON-SCENE COMMUNICATIONS[1]

Ensuring that a message is received and understood by the intended parties is obviously a crucial part of communications. Besides the hardware, it is necessary to consider encoding, noise, feedback, and discipline.

Encoding—Fireground verbal communications often have been encoded to increase efficiency by reducing the number of words transmitted through the use of an agreed-upon code. Codes also increase privacy by making it harder for eavesdroppers to understand the message. However, codes can confuse the intended receivers, too. The ten codes may be used incorrectly or be misunderstood. Difficulties in using them include the commitment of time required to learn them and the problem of reliably remembering the less-used codes. Most national organizations now recommend the use of plain text.

Reducing Noise—Anything that can interfere with a sender's message before it arrives at the person receiving it is considered noise. Noise may be induced by the equipment or environment used to communicate the message. Since emergency scenes are, by definition, uncontrolled environments, sources of potential noise must be anticipated and controlled before an incident if possible. That means identifying equipment that can operate efficiently and effectively in a wide range of environments. Furthermore, it means adopting and using an incident management system that controls the sender's and receiver's environment. Though not noise in a strict sense, limiting the number of persons sending and receiving messages in any communication system is essential to ensuring that each sender and receiver gets an adequate opportunity to be heard and understood. The incident command system also is an ideal vehicle for ensuring this.

Improving Feedback—For effective fireground communications, not only must the receiver understand the message transmitted by the sender but also the sender has to know that the message was received and understood. Feedback is a way of communicating that understanding from the receiver back to the sender.

Procedures for letting the sender know that a message was received and understood are a part of communications that needs to be planned in advance. Like the message itself, feedback need not be verbal. In fact, the message and the feedback may not be in a common format. As an example, consider an order to allow smoke and hot gases to escape a building with a deep-seated, smoky fire. Although the truck company officer directing firefighters on the roof may acknowledge the superior's instructions, the true test that the message was understood will come when changes occur in the avenue and character of smoke escaping the building. On the other hand, ordered actions do not always produce expected results, so feedback on the receipt and understanding of orders can be essential.

Fireground Communications Discipline—Officers and firefighters should be aware of the importance of interpersonal and radio communications discipline. The communications network is extremely active at times, and a failure to exercise proper communications discipline may affect operations negatively. In critical situations, improper radio procedures could affect the safety of the operating force. Officers and firefighters using the radio must be alert to avoid improper or untimely transmissions that interfere with effective or critical communications.

Particular care should be exercised to avoid the following breaches of radio discipline:

- *Units with nonemergency transmissions breaking into radio traffic.* Units should not transmit routine traffic until the air is clear.

- *Units transmitting unnecessarily long and detailed messages.* Lengthy messages should be transmitted by telephone or face-to-face.

- *Units calling the dispatcher to determine if they are assigned or should respond.* If a unit has kept the dispatchers apprised of its status and availability, it is not necessary to prompt them or encourage them to dispatch the unit. Dispatchers will assign available units, including those that are available, by radio.

- *Units failing to monitor the radio and acknowledge transmissions as promptly as radio traffic allows.* Chief officers should be alert to radio traffic, improper use, and delayed communications responses. They should assure that unsatisfactory transmissions are investigated and are the subject of appropriate corrective action.

- *Radio users speaking prior to or simultaneously with depressing the microphone button.* This practice causes the first part of the message to be clipped. Users should wait a moment after the microphone has been keyed before speaking.

At the World Trade Center fire and bombing incident in 1993, several hundred firefighters were crowded onto one of the three channels used, causing a major problem in on-air discipline. Similar problems are reported at many large-scale incidents.

Fireground Decisionmaking—Understanding the role of communications in incident management requires an awareness of decision-making methods and the flow of information on the fireground. Thinking about the nature of fireground decision making is useful as one plans the specifics.

Use of Mental Model—Previous personal experience appears to be a critical factor in the decision-making process of fire department incident commanders. The basis for their time-critical decisions often is the ability to recognize a situation as a paradigm and match it with a plausible solution based on experience. They call this "recognition-primed decision making." This finding contradicts traditional decision-making models that assume decisions are based on a functional analysis of available alternatives, followed by the selection of a best-fit solution based on its effectiveness or efficiency. Recognition-primed decision making highlights the importance of experience. Rapid communication of the facts of the situation to the incident commander is crucial in forming the "paradigm" and selecting the "solution."

Situational Awareness—The ability to develop a sense of a situation and formulate a paradigm to aid decision making was based on the ability to develop what is referred to as "situational awareness." Such awareness is difficult to establish in the real-time, large-scale, high-pressure world of an emergency. Incident commanders must develop an ability to filter and focus their attention on the critical details about an incident.

This is precisely what the ICS does. By maintaining the cohesiveness of the company, task force, and battalion units of the fire department, incident commanders can deal more effectively with a large number of resources by maintaining span of control. The ICS simply gives incident commanders the ability to do this without sacrificing unity of command and their control over the management of the incident.

Fireground Communications Philosophy

The importance of communications to achieve situational awareness and to implement command and control makes it essential that fire service professionals understand what needs to be communicated, who at an emergency scene communicates with whom, and which way information must flow.

What Is Communicated?—The so-called navigation analogy is a particularly apt way of describing what must be communicated to achieve effective incident control. To discover how to get some place, we first must know where we are, and then where we wish to be. Knowing how to get somewhere, regardless of how elegant and concise the directions may be, is irrelevant without some knowledge of where we begin and where we want to be. All of your knowledge about the streets and buildings of your city and years of experience fighting fires are useless until you find out where the fire is.

Like the navigation example, communication at fires and emergencies is task-oriented. We start with an emergency, and we orient all actions toward controlling that incident as quickly as possible, with the least cost to people and property. Therefore, after the initial dispatch, emergency scene communications focus on how to do something—for example, fight a fire or control a hazardous materials release—and the progress made to accomplish that objective. To accomplish this, the incident commander must have a clear understanding of what the current situation is (situational awareness) and must have formulated a clear and concise strategy for dealing with it. Establishing control in the midst of the turmoil and confusion of a fire or emergency scene is a primary function of communications.

These types of information then are the focus of most emergency scene communications: (1) orders or instructions for accomplishing the control or mitigation of the incident, (2) reports on progress, and (3) requests for resources to achieve control or mitigation of the incident. Orders and instructions in the task-oriented world of fire and emergency scene operations are messages that command or instruct responders to perform tasks or implement tactics consistent with the incident control strategy selected by the incident commander. Effective orders and instructions are clear, concise, and direct.

However, issuing orders and instructions is not enough to ensure effective communications. Orders must be issued to subordinates capable of responding to them. Therefore, to ensure that fire and emergency scene communication is effective, a feedback mechanism must exist. Subordinate commanders must have a means available to report their progress and request resources to accomplish their tasks to ensure effective communications, and in turn, effective incident control. By keeping this communication back-channel open, the incident commander is not only informed about the nature of the incident and the needs of subordinates but is also equipped with the capability to anticipate future developments and adapt the incident control strategy accordingly.

Who Communicates with Whom?—The principal communicators at any fire are developing strategies and implementing tactics to control the scene. The proliferation of communications hardware makes communications among all emergency scene personnel more effective, and some fire departments are now issuing communications hardware such as mobile or portable radios or telephones to more members.

However, one of the principal means of controlling who communicates verbally with whom at an emergency scene has been to control the assignment of portable radios and other communications devices. The mere presence of radios in the hands of emergency responders encourages increased communications, since these devices enable people to overcome the noise in the communications environment more readily than the unaided human voice. The communications roles of each person on the fireground need to be spelled out and practiced in training to keep communications from turning into noise.

Direction of Communications—As important as what or who is communicating is the place of these persons in the structure of the incident management system and the nature of their communications. Communications between decision makers about the tactics and strategies used to control an incident create an environment where most formal communications about a fire or emergency are vertical, that is, up and down the chain-of-command between commanders and subordinates. However, the larger and more complex an incident becomes, the more horizontal communications become. This results in part from the necessity to limit inputs—filter information—coming to the incident commander.

Decisions that do not affect overall incident strategy should be handled at the lowest possible level, and only major elements that require integration among major incident command functions should be channeled back to the incident commander. Ideally, communications among emergency responders at the task implementation level should be primarily horizontal and with their immediate superior.

Incident Command and Control—The term C^3I is used to describe the four elements of a military command and control system—command, control, communications, and intelligence. The concept also applies to the fireground incident command system. Considering the roles of command, control, and intelligence at fires is another useful model to help understand communications requirements at emergency scenes.

Command is the authority to direct the operation. The incident commander (IC) in the incident command system is the person responsible for making decisions and issuing orders about how to control the incident. Lines of authority must be understood for communications to be effective, and vice versa. The IC decides what resources are needed and how they will be deployed. The IC determines who will oversee subordinate or support functions, and delegates responsibility for these functions. Nonetheless, the IC remains the individual vested with primary responsibility for the outcome of the incident. Although the incident command system makes no assumptions about the rank of the person in charge of the incident, most fire departments see that only experienced officers assume incident command responsibilities.

To effectively control the fireground requires that all personnel responding to, operating with, and communicating with fireground personnel know, agree on, and use only one standard set of rules for communication. The rules must be established prior to the response, be written down, and have someone responsible for enforcing them. These rules and procedures outline what the priority of transmissions is, what transmissions are prohibited, the length and substance of transmission, and any other factors that will enhance, or detract from, effective communications.

Intelligence is the gathering or distribution of information about an incident or that information itself. Prior to the occurrence of an incident, intelligence gathering usually takes the form of prefire plans. However, many fire departments are finding other sources of information helpful in assessing incidents. With the advent of computers, mobile data terminals, mobile facsimile machines, cellular telephones, and sophisticated management information systems, information about a given property or scenario can take many forms. Past event histories, inspection records, the arson information management system (AIMS) data, property tax records, and building permit records are just a few of the data sources available to fire departments. Once an incident is underway, the gathering and distribution of "intelligence" becomes the critical factor in maintaining the situational awareness of the incident commander.

Incident Operations

Perhaps the best way to illustrate the preceding concepts of the communications philosophy is to consider emergency scene operations in chronological order from the time an emergency call is dispatched to the time the incident is terminated.

Notification and Initiation of Response—Earlier chapters dealt extensively with the operation of communications centers and dispatching. This discussion of incident operations starts with the dispatch—the order to emergency forces to respond. It must be received and acknowledged. At a minimum, the dispatch instructions must contain the location of the incident. Beyond that, additional information may provide helpful, even crucial details but still remains subordinate to the location of the incident itself.

Communication En Route—It has been said that size-up begins the moment an alarm is received, as a firefighter's or fire officer's mind becomes focused on the incident once the dispatch is received. Preparation for action actually begins even before an incident is dispatched, in the form of pre-incident or prefire plans.

A good example of how this preparation affects incident management occurred at the Sherwin Williams Paint Warehouse fire in Dayton, Ohio, May 27, 1987, which resulted in a multimillion dollar loss. The fire chief reported that the principal reason for not aggressively attacking the fire was a concern about contaminating the city's primary supply of drinking water, a large aquifer directly beneath the warehouse. In consultation with the city's water department director, the chief had assessed the environmental exposure from firefighting water runoff at the warehouse long before he received notice that the warehouse was burning. Initial reports of the advanced nature of the fire, and progress reports en route confirming that the fire had fully involved the 180,000-square-foot building, simply reinforced his decision not to apply water to the burning volatile organic compounds inside the facility.

Pre-incident plans also exist in the minds of the officer and crew. Armed with these plans, incident commanders and responders are better prepared to assess the risks posed by a fire or emergency. For example, firefighters arriving at a shopping mall during the holiday shopping season with smoke showing from the roof of an anchor store should know from a pre-incident plan whether the building is well isolated from adjacent buildings; is of noncombustible construction; has occupants representing a wide cross-section of the population; may have, at certain times of the year, a larger-than-normal quantity of merchandise on hand, and an increased number of shoppers and staff; and has (or does not have) automatic sprinklers, standpipes, smoke detection, and smoke control systems. Based on this information, one of the IC's first priorities may be to ensure that the fire alarm system is sounding and that occupants are heeding it. Ordering the support of automatic sprinkler and standpipe systems and assessing the operation of any smoke control system also may be planned as the responding apparatus are en route.

In addition to verbal information relayed from the dispatcher, observation en route may provide important clues about the nature of an incident. For example, traffic congestion en route to an automobile accident call usually indicates that the call is legitimate and may be obstructing the thoroughfare. Likewise, a large smoke plume or "loom-up" on the horizon usually indicates that the fire call you are responding to is the real thing. When the observed conditions en route translate into an action plan, or will significantly affect the decisions of the first-arriving companies, this information must be relayed to all responding units.

Communications upon Arrival—Up to this point, nearly all communications have flowed from the dispatch center toward the responding units and the initial incident commander. From this point on, communications from the incident commander back to the dispatcher and out to the other responders will become the focus. The arrival of the first unit at the emergency scene serves as the benchmark for this communications transition. The dispatcher must be notified by the first unit as it arrives on the scene. From then on, the incident commander is the focal point for communications. All communications from that time forward should be directed through the incident commander, unless the IC delegates otherwise.

After the first company arrives, an assessment of conditions (size-up) must be made immediately. This assessment integrates the information already in hand with the observed conditions upon arrival to formulate the initial incident control strategy. The incident commander must quickly take note of the same elements found in the pre-incident plan: the environment, occupants, operating features, and fire protection features. This information then must be relayed to subsequently arriving units and the dispatcher via the initial report.

The initial report is critical because it sets the stage for the actions of each subsequently arriving company and prepares the dispatcher for requests for additional services or the redeployment of resources. A good initial report raises the situational awareness of all responders and the dispatcher. It should include the following five elements:

■ Fire progress (for example, nothing showing, smoke showing, fire showing, "working fire")

■ Fire location (for example, basement, rear of the building, second floor windows, roof, quadrant D-side 2)

■ Building information (for example, six-story brick and steel mercantile building, single-family wood-frame residence, three-story ordinary construction garden apartment)

■ Occupant behavior (for example, occupants are evacuating, people at the windows)

■ Command post location and officer in charge (for example, front of the building or side 1, Lt. Velex in charge)

Some departments include one additional piece of information in the initial report—the capability assessment—in which the IC informs the responding companies and the dispatcher of an overall confidence that the incident can be controlled with the resources and personnel assigned. If conditions upon arrival indicate that control is unlikely or impossible with the assigned complement of personnel and resources, additional resources should be requested in sufficient quantity and capability to assure operational success.

Establishing a Command Post—The initial incident commander must establish a command post. The location chosen for the command post may dramatically affect incident command and control by the way it aids or impairs communications. Once established, the command post becomes the hub of the communications network at the emergency scene. A good command post location must facilitate effective communications by giving the incident commander access to the incident environment without overwhelming the IC's situational awareness. Generally speaking, the larger and more complex the incident, the further the command post should be from the actual incident site. At a small or routine incident, the IC should have good visual access to the scene to receive feedback on the progress or effectiveness of the incident management strategies. A good rule of thumb for ensuring this type of visual contact is for the first-arriving unit to pull past the scene as it approaches. The initial command post is usually established where this first company sets up for initial operations. This approach ensures that the first-arriving officer has seen what is happening on three sides of the incident scene.

However, this approach may be imprudent or impractical at some types of emergencies. Sometimes the terrain, street network, weather, or the nature of the incident itself leave the IC with no choice about how to approach the incident, and therefore, little choice about where to set up the initial command post. Subsequently arriving officers may elect or be required to assume command of the incident. Upon their arrival, the command post may be relocated to a site that remains accessible to the scene but ensures adequate insulation from it. The new command post should be clearly established and identified. Signs, flags, lights, and special vehicles are a few of the ways a command post can be distinguished at the incident scene. The new location of the command post must be clearly announced to all companies operating at or assigned to the incident. This ensures that nonverbal communications remain effective. Furthermore, this ensures that those who must conduct face-to-face communication can locate the IC without delay.

Tactical Communications—Much of the communication at emergency scenes is task oriented. Directing a fire attack will primarily involve communications between the incident commander and subordinate officers, and the companies and crews fighting the fire. Most messages will consist of orders and instructions to perform tasks for controlling the fire; reports about the conditions and progress encountered; and requests for additional resources to fulfill assigned tasks. These communications must be clear, concise, and timely to ensure that communications fully support the incident command and control function. Many of the communications will involve intelligence about the nature of the incident, which increases the situational awareness of the IC and the firefighters alike. The IC's situational awareness is critical to effective management of the incident. The firefighters' situational awareness, keeping abreast of conditions beyond their immediate environment, is critical to their safety.

Even when an incident becomes a defensive operation, a significant degree of communication is required to ensure effective coordination. Defensive operations ordinarily require a significant commitment of personnel and resources. The more people that are involved in an operation, the more difficult communications become. For instance, when an advanced fire threatens to produce a structural collapse and firefighters are ordered out of a building, efforts must be made to quickly account for all personnel who were in, on, or around the involved structure. Once assembled and accounted for, these personnel will require new instructions for their effective deployment. The incident commander must recognize that the functions performed by all personnel prior to the evacuation will be substantially affected by changes in the overall incident control strategy and prepare to communicate new instructions during the time it takes to regroup the forces.

In addition to tactical communications, there often are many other types of communications needed during an incident. Locating, rescuing, and protecting endangered people are usually among the first priorities at any fire or emergency. Often the victims of fires and other emergencies are extremely agitated, anxious, or excited. The stress of the emergency may make civilians behave irrationally, or they may be unresponsive, combative, or uncooperative. The emergency responders need to calm and reassure the victim(s) in order to gain their confidence and cooperation to complete their rescue or secure their protection. Such communications must be authoritative, confident, and direct but should avoid melodrama, hyperbole, or vulgarity. Telling people what is really happening in the most concise and straightforward fashion usually works best, whether you are trying to effect an evacuation or reassure an accident victim. Incident analysis has demonstrated that people usually respond best when confronted with the facts about a dangerous situation. The more complete and accurate the information they receive, the better their response can be.

Managing exposures during an incident often involves some degree of communication with people other than the emergency responders themselves. The location of vital business records or people requiring evacuation assistance, for example, may be identified in pre-incident plans or by contact with the person reporting the incident, or from people at the emergency scene.

Salvage operations often involve establishing a sense of the relative value of exposed property and the degree of danger confronting it. This requires communicating with property owners or occupants as to what is of highest priority to save or if any areas need special protection. For example, the value of paper records themselves may be negligible, but when those records are the accounts receivable or inventory records of a firm, their protection may be considered priceless in the eyes of the owner.

Emergency Traffic—The first service is trained to accept emergency scene operations as "part of the job." In fact, keeping a cool head in hot situations is what firefighting is all about. However, some situations represent emergencies even to seasoned firefighters and emergency responders. These include missing or injured firefighters, deteriorating or extremely hazardous structural conditions, weather changes that will exacerbate the fire situation or further endanger lives, and dramatic changes in smoke conditions. All of these situations require prompt and coordinated action to avert an operational disaster. Effective communications are the key to assuring that appropriate action is implemented quickly. Therefore, procedures for distinguishing and conveying such emergency messages must be adopted.

The use of signal words, such as "Mayday," "emergency traffic," or "urgent," or the broadcast of a predefined signal like a radio alert tone, apparatus siren, or air horn blast must indicate to everyone operating at the incident that important, indeed vital, information follows. To maintain their effectiveness, these signals must be coordinated before an incident occurs and reserved for exclusive use in life-threatening or extremely hazardous situations. Once broadcast, emergency signals should cause the interruption of all other communications until the message is received and acknowledged. Procedures for dealing with such emergencies should be defined before an incident and implemented immediately. For instance, if an alert tone is received from a personal alert safety system (PASS) device in a trapped firefighter's radio, the IC should immediately begin efforts to identify, locate, and rescue the missing firefighter. (Radio equipment that integrates the features of PASS devices also may give an immediate identification of the radio assigned to the firefighter in need of assistance. If radio assignment data are accurate, the identity of both the radio operator and the assignment can then be readily determined.) Similarly, if sector reports indicate that a structural collapse is imminent and defensive operations should be implemented, a signal should be sounded both outside on the fireground (for example, three long blasts on an airhorn) and on the radio (for example, three long alert tones activated by the dispatcher or field communications personnel) to direct firefighters to evacuate the building, secure defensive tactical positions, and await further instructions.

Status Reports—Within a fixed time interval following the preliminary or on-scene initial report, the IC should give a progress report to the CC. A progress report is a verbal picture of what is occurring at the scene and should include (1) a description of the fire building or emergency zone, (2) location of the fire, (3) description of the exposures, (4) description of the strategy, tactics, and resources being employed to control the situation, (5) an appraisal of the expected outcome of the current actions, and (6) any other facts that will be important to superior officers or surrounding companies listening to the report.

Some departments feel that progress reports should be transmitted at the discretion of the IC and should only be transmitted when there is a change in operations or when the operation is complete (such as an escalation or de-escalation). Another option is to require the IC to formulate and transmit progress reports at regular time intervals. Those who favor nonprogrammed progress reports want to give the IC an opportunity to manage the operation without higher levels of command interceding and to limit the radio traffic to essential dispatch traffic. Those who practice programmed progress reports believe the practice of regularly communicating a verbal picture of on-scene conditions helps the IC to better understand and appreciate the problems; keeps the higher level of command better informed, thus allowing them to make judgments on the effectiveness of operations and the need to respond; creates a permanent record of the actions and reported conditions at regular intervals; and helps communications center personnel to have a clearer understanding of the incident and its development. A suggested time interval for programmed progress reports is to give the first progress report approximately 10 minutes after the preliminary report and, for the next hour of the operation, to give one every 10 to 15 minutes and whenever a significant escalation of operations occurs. The transmission of additional alarms would be defined as escalation and would require the IC to inform the communications center of the reasons for this additional response.

Where fire departments use multiple radio channels, such as a primary dispatch or operations channel and a command or tactical channel, the operations channel is the routine communications link from the incident to the communications center. The dispatcher should rebroadcast all progress reports of major operations on both the primary channel and the command or tactical channel. The procedure for rebroadcasting reports should be predetermined and based upon the number of units working at the scene. Preliminary reports and progress reports for fires or emergencies smaller than the predetermined size need not be rebroadcast.

Status reports help personnel maintain their situational awareness and prepare them for operational changes and new tasks. The reports help ensure safety by keeping personnel abreast of changing conditions. And the reports prepare the dispatcher and other potential responders for the deployment of additional resources or redeployment of those still uncommitted.

Requests for Additional Resources—Requests for additional resources during an incident should include the nature and number of the required resources. Most departments group their companies in alarm assignments based on the task force concept. With this concept, companies are grouped by capability. Typically, each task force consists of one truck, aerial, ladder or service company assigned with two engine or pumper companies. For each group of two task forces, an officer of battalion or district chief rank is assigned to command. (While these groups of companies often are referred to as alarm assignments, an alarm assignment also may refer to all of the companies from a particular station or fire district.) Occasionally, special service companies like breathing apparatus supply units, canteen trucks, light towers or utility trucks, manpower squads, or medical units are assigned to round out the complement of capabilities of an assignment.

When additional companies or alarms are requested to control an incident, instructions also must be given to ensure their orderly integration into the incident. Units needed at the scene should be directed to their assignment and instructed as to whom they are to report. Units not immediately required at the scene, but whose assistance is needed for tools, equipment, or personnel, should be assigned to a staging area. The staging area should be established at a location with good access to the incident but far enough away to prevent uncommitted resources from interfering with emergency scene operations. Once established, the staging area represents a resource pool for the IC. At protracted incidents, a relief area and logistics sector may be established adjacent to the staging area. This way, as personnel and resources become available for reassignment, they can easily be attached to the staging sector.

Requests for Special Services—Often at large incidents, the assistance of other agencies is required to mitigate a situation. This assistance may include Red Cross disaster relief teams, social welfare agencies, environmental protection officials, utility companies, emergency management officials, or National Guard personnel. Ordinarily, these services remain under the direction and supervision of their own agency supervisors or commanders when assigned. To ensure effective communications with these outside agencies, someone from each agency should be included in the incident command hierarchy according to the agency's role in the incident. For example, organizations furnishing relief supplies like cots, bedding, food, water, or shelter should be assigned to the logistics division.

Communications with outside agencies will probably remain difficult even if integrated into the incident command system. Hardware and operating procedures often vary considerably between fire service and nonfire service agencies. Outside agencies with radios often cannot communicate with the fire service on a common frequency. One way to overcome such difficulties is to request each outside agency to supply the incident commander or division commander with one of its radios and an identifier. However, even with a radio, the fire service IC or division commander may have communications difficulties with the outside agencies, because they often have less formal communications protocols and standard operating procedures than those of the fire or EMS department. Face-to-face communications may be the most reliable method in some cases. Once again, planning and practicing integrated emergency management operations can help overcome many of these difficulties.

Avoiding Unnecessary Traffic—Most communications at the scene of a fire or emergency will be oral. Radios have become the primary means of communicating information among emergency responders. However, some communications, particularly at the operational or attack level, are most effective when conducted face-to-face. Bogging down radio communications with traffic about where to direct a nozzle or how bad smoke conditions are can choke a communications network.

Some departments have been issuing radios to each firefighter on duty to improve firefighter safety, the rationale being that each member should be able to immediately report extremely hazardous conditions or emergencies such as a missing or injured firefighter or a structural collapse. Procedures and controls must be adopted to ensure that all radio operators are maintaining radio discipline, and sending and receiving the right information to the right people at the right time. It does no good for all personnel at an incident to have a radio if only half of them are listening to the right channel for instructions or if many are speaking at the same time.

Terminating an Incident

As an incident is brought under control, the need for personnel and resources diminishes. However, incident commands must be terminated in an orderly fashion so that control of the incident does not end prematurely and so that resources are only placed back in service when fully prepared to engage in new assignments. The life cycle of any incident readily facilitates scaling back in an orderly fashion.

All Clear—The first significant benchmark in the control of an incident is usually the completion of a primary search or the removal of all trapped or injured victims from danger. Announcing the all clear indicates to all personnel operating at the incident that all known or immediately apparent victims have been removed, and the danger to civilians has been mitigated within reasonable certainty. In other words, there may still be some chance that lives are in jeopardy, but their status could not be determined and must wait for the secondary search, which will be conducted when the incident is controlled or conditions improve. This incident management priority holds for automobile accidents with trapped victims and hazardous materials incidents with contaminated or exposed victims as well as wildland and structural fires. The actions of firefighters and other emergency service personnel usually are dominated by actions to save lives during the early stages of any incident. Announcing the all clear tells all assigned personnel that the focus has shifted to fire control or property protection.

Under Control (Situation Contained)—Once an incident is no longer growing in magnitude or severity, or is at least doing so in a predictable, controlled fashion, an incident may be referred to as contained or under control. By announcing that the incident is contained or controlled, the incident commander is declaring that the incident can be managed with the resources assigned. Furthermore, this declaration prepares the dispatcher and assigned companies for an orderly return to normal operations. If companies are still responding to the incident scene or staging area, they should be considered assigned and continue their response, since they may be required for relief or other duties, unless released by the incident commander.

Available/In Service/Ready for Service—As companies are released by the incident commander through their sector officers, they should report their status to the dispatcher or communications center. Companies that are fully staffed, equipped, and prepared for another assignment should report themselves as "available" or "in service," or "ready" for service (depending on local jargon). Companies with reduced capabilities that do not impair their readiness or response status should indicate their capabilities in their message to the dispatcher. For example, a four-person engine company that sends a firefighter/EMT with a medic unit to the hospital might radio the dispatcher, "Engine 23 in service with three. One member assisting Medic 12." Occasionally, a company must return to quarters but still be out of service. Although the unit is unavailable for reassignment, the dispatcher still must be notified that the company is no longer assigned to the incident. When the last company clears an incident, the command should be formally terminated. This indicates that no further resources are committed to the incident and that the scene is no longer under the control of the fire service. This is a particularly important distinction if the incident is a suspicious fire or is associated with other criminal activity. When an incident scene is left under the control or supervision of another agency, the responsible agency should be reported to the dispatcher as a matter of record.

Postincident Operations

After an incident is over, there may be an investigation, critique, and a variety of dealings with the media.

Investigations—Every fire and emergency incident should be investigated to determine its origin and causes as well as the contributing and mitigating factors. The actions and assessments of emergency responders aid immeasurably in the reconstruction of emergency scenes. Communications between emergency responders and investigators are vital to reconstructing the events surrounding most fires and emergencies. Therefore, emergency response personnel should become skilled in observing, recording, and recollecting details about the incidents to which they responded. This information will aid them as well as the investigators.

Critiques, Postmortems, and After-Action Assessments—Even the best-trained and equipped fire departments and emergency response organizations can learn from their experiences. Every incident is an opportunity to test responders' knowledge while gaining new insights through experience. Postincident critiques, postmortems, and after-action

assessments are an opportunity to share these insights among peers. Sharing the lessons learned from every fire and emergency operation from those with diverse perspectives helps iron out operational deficiencies, build teamwork, establish mutual respect among team members, and build a common core of experience that will serve the responders well at other incidents. Postincident critiques should be critical to the extent that they examine every detail that contributes to or detracts from effective incident management.

Press Releases—One of the most difficult and challenging aspects of communications for any significant incident is dealing with the media. Often it is vital that communications begin with the media while an incident is in progress. In fact, the sooner the demand for information about an incident can be met the better, provided the information is accurate. After an incident, all the loose ends about an incident should be cleared up. A complete press release should be issued following every major incident. Many departments also issue daily press released or briefings through the communications center. A good press release should answer the five Ws of journalism: who, what, when, where, and why (how). Keep press releases clear, complete, and to the point, and the media usually will be satisfied. And, when details are sketchy or incomplete, remember that accuracy is more important than timeliness.

Note

1. Adapted from *Fire Department Communications Manual* (Emitsburg, MD: U.S. Fire Administration, 1995).

Glossary

acceptable risk the level of hazard potential that the public is willing to live with, that constantly changes based on public perception of the news and current events

accidental refers to those fires that are the result of unplanned or unintentional events

accountability being responsible for one's personal activities; in the organizational context, accountability includes being responsible for the actions of one's subordinates

action plan an organized course of action that addresses all phases of incident control within a specified period of time

active listening deliberate and apparent process by which one focuses attention on the communications of another

administrative law body of law created by administrative agencies in the form of rules, regulations, orders, and court decisions

administrative process a body of law that creates public regulatory agencies and defines their powers and duties

administrative search warrant a written order issued by a court specifying the place to be searched and the reason for the search

adoption to formally accept through legislative action and put into effect

adoption by reference a method of code adoption in which the specific edition of a model code is referred to within the adopting ordinance

adoption by transcription a method of code adoption in which the entire text of the code is published within the adopting ordinance

agency shop an arrangement whereby employees are not required to join a union, but are required to pay a service charge for representation

arbitration the process by which contract disputes are resolved with a decision by a third party

area of refuge portion of a structure that is relatively safe from fire and the products of combustion

arson the intentional setting of a fire where there is criminal intent

attack role a situation in which the first-arriving officer elects to take immediate action and to pass command on to another officer

authority the right and power to command

automatic fire protection sprinkler self-operating thermosensitive device that releases a spray of water over a designed area to control or extinguish a fire

backdraft explosion a type of explosion caused by a sudden influx of air into a mixture of burning gases that have been heated to the ignition temperature of at least one of them

barrier obstacle; in communications, a barrier prevents the message from being understood by the receiver

benchmarks significant points in the emergency event usually marking the accomplishment of one of the three incident priorities: life safety, incident stabilization, or property conservation

branches level of supervision between divisions and groups and operations that is used when number of divisions or groups exceeds acceptable span of control

British thermal unit the amount of heat required to raise the temperature of one pound of water one degree Fahrenheit

budget a financial plan for an individual or organization

capital budget a financial plan to purchase high-dollar items that have a life expectancy of more than 1 year

certification document that attests that a person has demonstrated the knowledge and skills necessary to function in a particular craft or trade

chain of custody refers to the chronological documentation, and/or *paper trail*, showing the seizure, custody, control, transfer, analysis, and disposition of *evidence*

closed shop a term in labor relations denoting that union membership is a condition of employment

coach a person who helps another develop a skill

code a body of law or rules which is systematically arranged. *When and where to do, or not to do something*

codes systematic arrangement of a body of rules

command role a situation in which the first-arriving officer takes command until relieved by a senior officer

commanding third step in the management process, commanding, involves using the talents of others, giving them directions and setting them to work

communications unit prepares communication plan, distributes and maintains communication plan

community consequences an assessment of the consequences on the community, which includes the people, their property, and the environment

complaint an expression of discontent

confinement an activity required to prevent fire from extending to an uninvolved area or another structure

conflict a disagreement, quarrel, or struggle between two individuals or groups

consulting seeking advice or getting information from another; as a leadership style it implies that the leader seeks ideas and allows contributions to the decision-making process

controlling fifth step in the management process, controlling, involves monitoring the process to ensure that the work is accomplishing the intended goals and objectives, and taking corrective action when it is not

coordinating fourth step in the management process, coordinating, involves the manager's controlling the efforts of others

counseling one of several leadership tools that focuses on improving member performance

customer service service to people in the community

defensive mode actions intended to control a fire by limiting its spread to a defined area

delegate to grant to another a part of one's authority or power

delegating sharing work, authority, and responsibility with another; as a leadership style, delegating implies the most generous sharing of the officer's leadership role

delegation act of assigning duties to subordinates

demobilization unit ensures resources a released from the incident in an orderly, safe, and cost-effective manner

demotion reduction of a member to a lower grade

directing controlling a course of action; as a leadership style, it is characterized by an authoritarian approach

disciplinary action an administrative process whereby a member is punished for not conforming to the organizational rules or regulations

discipline system of rules and regulations

diversity a quality of being diverse, different, or not all alike

division tactical level management group that commands individual companies, task force, or strike teams; a geographical assignment versus a functional assignment

documentation unit maintains and archives all incident related documentation

emergency incident any situation to which a fire department or other emergency response organization responds to deliver emergency services

empowerment to give authority or power to another

energy the capacity to do work; in the fire triangle, heat represents energy

environmental factors factors like weather that impact on firefighting operations

Equal Employment Opportunity Commission (EEOC) federal government agency charged with administering laws related to nondiscrimination on the basis of race, color, religion, sex, age, or national origin

ethics system of values; a standard of conduct

expert power a recognition of authority by virtue of an individual's skill or knowledge

facilities unit sets up, maintains facilities, provides facility security

fact finding a collective bargaining process; the fact finder gathers information and makes recommendations

Fayol's bridge organizational principle that recognizes the practical necessity for horizontal as well as vertical communications within an organization

feedback reaction to a process that may alter or reinforce that process

finance section that part of the incident management system that is responsible for facilitating the procurement of resources and for tracking the costs associated with such procurement

fire behavior the science of the phenomena and consequences of fire

fire control activities associated with confining and extinguishing a fire

fire extension the movement of fire from one area to another

fire extinguishment activities associated with putting out all visible fires

fire resistance the resistance of a building to collapse or to total involvement in fire or the property of materials and their assemblies, which prevents or retards the passage of excessive heat, hot gases or flames under conditions of use

first-level supervisors the first supervisory rank in an organization

flashover a dramatic event in a room fire that rapidly leads to full involvement of all combustible materials present

floor plan a bird's-eye view of the structure with the roof removed showing walls, doors, stairs, and so on

food unit supplies food and potable water, operates food services unit

formal bid a formal request for cost of products and/or services that are over a specified dollar amount

free-burning phase second phase of fire growth, has sufficient fuel and oxygen to allow for continued fire growth

fuel load or fire load the total amount of combustible material within a fire area, expressed in terms of pounds per square foot

goal a target or other object by which achievement can be measured; in the context of management, a goal helps define purpose and mission

grievance a formal dispute between member and employer over some condition of work

grievance procedure formal process for handling disputed issues between member and employer; where a union contract is in place, the grievance procedure is part of the contract

gripe the least severe form of discontent

ground unit arranges for transportation of personnel, supplies, food, and equipment

group tactical level management group that commands individual companies, task force, or strike teams; a functional assignment versus a geographical assignment

group dynamics how individuals interact and influence one another

harassment to disturb, torment, or pester

health and safety officer a person assigned as the manager of the department's health and safety program

heat of combustion the amount of heat given off by a particular substance during the combustion process

high-achievement needs according to the needs theory of motivation, individuals with high-achievement needs accept challenges and work diligently

high-affiliation needs according to the needs theory of motivation, individuals with high-affiliation needs desire to be accepted by others

hygiene factors as used by Frederick Herzberg, hygiene factors keep people satisfied with their work environment

identification power a recognition of authority by virtue of the other individual's character or trust

ignition temperature the minimum temperature to which a substance must be heated to start combustion after an ignition source is introduced

incident command system a tool for dealing with emergencies of all kinds

incident commander (IC) the person in overall command of an incident

incident management system an organized system of roles, responsibilities, and standard operating procedures used to manage an emergency operation

incident safety officer (ISO) as part of the IMS, that person responsible for monitoring and assessing safety hazards and ensuring personnel safety

incipient phase first stage of fire growth, limited to the material originally ignited

initial report a vivid but brief description of the on-scene conditions relevant to the emergency

inspection model a method of setting inspection priorities based on a jurisdictions actual fire experience or identified hazards

inspection warrant an administrative warrant issued by a judicial officer to search or inspect a property, or perform a noncriminal investigation

internal customers members of the organization

interpersonal dynamics dynamic practices developed by groups that separate them from other individuals

ISO a family of standards for quality management systems; *ISO 9000* is maintained by ISO, the International Organization for Standardization, and is administered by accreditation and certification bodies

leadership the personal actions of managers and supervisors to get team members to carry out certain actions

legitimate power a recognition of authority derived from the government or other appointing agency

liaison officer as part of the incident management system, the contact between the incident commander and agencies not represented in the incident command structure

life risk factors the number of people in danger, the immediacy of their danger, and their ability to provide for their own safety

life safety the first priority during all emergency operations that addresses the safety of occupants and emergency responders

line authority characteristic of organizational structures denoting the relationship between supervisors and subordinates

line functions those activities that provide emergency services

line-item budgeting collecting similar items into a single account and presenting them on one line in a budget document

listed equipment or materials included in a document prepared by an approved testing agency indicating that the equipment or materials were tested in accordance with an approved test protocol and found suitable for a specific use

local ordinance law usually found in a municipal code. In the United States, these laws are enforced locally in addition to state law and federal law

lock box a locked container on the premises that can be opened by the fire department containing preplanning information, material safety data sheets, names and phone numbers of key personnel, and keys that allow emergency responders access to the property

logistics section that part of the incident management system that provides equipment, services, material, and other resources to support the response to an incident

management accomplishment of the organization's goals by utilizing the resources available

management by exception (MBE) management approach whereby attention is focused only on the exceptional situations when performance expectations are not being met

Maslow's Hierarchy of Needs a five-tiered representation of human needs developed by Abraham Maslow

mediation the process by which contract disputes are resolved with a facilitator

medical unit prepares medical plan, provides first aid and light medical treatment; mutual aid assistance provided by another fire department or agency

memorandum of understanding (MOU) a pre-established contract to provide goods or services from a vendor prior to an emergent event

message in communications; the message is the information being sent to another

mini-maxi code a code developed and adopted at the state level for either mandatory or optional enforcement by local governments; that cannot be amended by the local governments

mission statement a formal document indicating the focus and values for an organization

model representation or example of something

model code technical rules developed and maintained by organizations and made available for governments to adopt and enforce. They are generally developed through a consensus through the use of technical committees

motivators factors that are regarded as work incentives such as recognition and the opportunity to achieve personal goals

motive the goal or object of one's actions; the reason one sets a fire

National Integration Center (NIC) Incident Management Systems Integration Division established by the Secretary of Homeland Security to provide "strategic direction for and oversight of the National Incident Management System (NIMS)"

National Response Framework (NRF) defines the principles, roles, and structures that organize how we respond as a nation

noncombustible construction in broad terms, constructed of materials that will not burn

objective something that one's efforts are intended to accomplish

occupant services sector that part of the incident management system that focuses on the needs of the citizens who are directly or indirectly affected by an incident

offensive mode firefighting operations that make a direct attack on a fire for purposes of control and extinguishment

open shop a labor arrangement in which there is no requirement for the employee to join a union

operating budget a financial plan to acquire the goods and services needed to run an organization for a specific period of time, usually 1 year

operations section division in the incident management system that oversees the functions directly involved in rescue, fire suppression, or other activities within the mission of an organization

oral reprimand the first step in a formal disciplinary process

ordinance a law of an authorized subdivision of a state, such as a county, city or town

organization a group of people working together to accomplish a task

organizing second step in the management process, organizing, involves bringing together and arranging the essential resources to get a job done

overhaul searching the fire scene for possible hidden fires or sparks that may rekindle

oxidation a chemical reaction in which oxygen combines with other substances causing fire, explosions, and rust

performance code a prescriptive code that assigns an objective to be met and establishes criteria for determining compliance

performance-based design an alternative method for satisfying the fire protection and life safety intent of construction codes based on agreed-on performance goals and objectives, engineering analysis, and quantitative assessment of alternatives against the design goals and objectives

permit model a method of setting inspection priorities by inspecting those occupancies and processes that require fire code permits

personnel accountability the tracking of personnel as to location and activity during an emergency event

planning first step in the management process, planning, involves looking into the future and determining objectives

planning section that part of the incident management system that focuses on the collection, evaluation, dissemination, and use of information (information management) to support the incident command structure

plot plan a bird's-eye view of a property showing existing structures for the purpose of pre-incident planning, such as primary access points, barriers to access, utilities, water supply, and so on

policy formal statement that defines a course or method of action

power the command or control over others, status

power needs a recognized motivator to be in charge or have group affiliation

pre-incident planning preparing for operations at the scene of a given hazard or occupance

pre-incident survey the fact-finding part of the pre-incident planning process in which the facility is visited to gather information regarding the building and its contents property conservation the efforts to reduce primary and secondary (as a result of fire-fighting operations) damage

probable cause a reasonable cause for belief in the existence of facts

procedure defined course of action

program budget the expenses and possible income related to the delivery of a specific program within an organization

protected construction protected from the effects of fire by encasement; concrete, gypsum, and sprayed on fire-resistive coatings are all used to protect structural elements

public fire department part of local government

public information officer (PIO) as part of the IMS, that person who acts as the contact between the IC and the news media, responsible for gathering and releasing incident information

punishment power a recognition of authority by virtue of the supervisor's ability to administer punishment

Quick Access Prefire Plan a document that provides emergency responders with vital information pertaining to a particular occupancy

rapid intervention team (RIT) a team or company of emergency personnel kept immediately available for the potential rescue of other emergency responders

receiver in communications, the receiver is the intended recipient of the message

rehabilitation as applies to firefighting personnel, an opportunity to take a short break from firefighting duties to rest, cool off, and replenish liquids

request for bid a formal request for cost of products and/or services that are over a specified dollar amount

Request for Proposal (RFP) a solicitation used when you need design, consultant, products, or typically services

Request for Quote (RFQ) a non binding typically solicitation to determine cost for goods or services from pre-established vendors or contractors

resource unit conducts check-in activities and maintains status of all resources

responsibility being accountable for actions and activities; having a moral and perhaps legal obligation to carry out certain activities

reward power a recognition of authority by virtue of the supervisor's ability to give recognition

right-of-entry clause a code that permits the fire official to enter and inspect structures on a routine basis, exempt one and two-family dwellings and residential portions of multifamily structures

rollover reignition of gases that have risen and encountered fresh air, and thus a new supply of oxygen

scalar principle organizational concept that refers to the interrupted series of steps or layers within an organization

search warrant legal writ issued by a judge, magistrate, or other legal officer that directs certain law enforcement officers to conduct a search of certain property for certain things or persons, and if found, to bring them to court

sender part of the communications process, the sender transmits a thought or message to the receiver

shelter in place selecting a small, interior room, with no or few windows, and taking refuge there

situation unit collects and analyzes information on the current situation

size-up mental assessment of the situation; gathering and analyzing information that is critical to the outcome of an event

smoldering phase third stage of fire growth; once the oxygen has been reduced, visible fire diminishes

span of control organizational principle that addresses the number of personnel a supervisor can effectively manage

specific heat the heat-absorbing capacity of a substance

specification code a prescriptive code that specifies a type of construction or materials to be used

staff positions designated to aid the IC in fulfilling the IC's responsibilities, typically, an incident safety officer, an information officer, and a liaison officer

staff functions activities that support those providing emergency services

standard rule for measuring or model to be followed

standard operating procedure (SOP) an organized directive that establishes a standard course of action

statute a law enacted by a state or the federal legislature

strategy sets broad goals and outlines the overall plan to control the incident

strike team set number of like resources that operate with common communications and a single unit leader

structural factors an assessment of the age, condition, and structure type of a building, and the proximity of exposures

succession plan the formal or informal training of subordinates to assume your position within the organization

supply unit orders, receives, and distributes supplies

supporting as a leadership style, the supporting process involves open and continuous communications and a sharing in the decision-making process

suspension disciplinary action in which the member is relieved from duties, possibly with partial or complete loss of pay

tactics various maneuvers that can be used to achieve a strategy while fighting a fire or dealing with a similar emergency

target hazards locations where there are unusual hazards, or where an incident would likely overload the department's resources, or where there is a need for interagency cooperation to mitigate the hazard

task force mixed number of resources that operate with common communications and a single unit leader

tasks the duties and activities performed by individuals, companies, or teams that lead to successful accomplishments of assigned tactics

termination the final step in the disciplinary process or the incident command process

thermal lag the difference between the operating temperature of a fire detection device such a sprinkler head as and the actual air temperature when the device activates

theoretical fire flow the water flow requirements expressed in gallons per minute needed to control a fire in a given area

Theory X management style in which the manager believes that people dislike work and cannot be trusted

Theory Y management style in which the manager believes that people like work and can be trusted

Theory Z management style in which the manager believes that people not only like to work and can be trusted, but that they want to be collectively involved in the management process and recognized when successful

thermal lag the difference between the operating temperature of a fire detection device such a sprinkler head as and the actual air temperature when the device activates

thermal stratification rising of hotter gases in an enclosed space

total quality management focus of the organization on continuous improvement geared to customer satisfaction

transfer a step in the disciplinary process that provides the member a fresh start in another venue

transitional mode the critical process of shifting from the offensive mode or from the defensive to the offensive

union shop a term used to describe the situation in which an employee must agree to join the union after a specified period of time, usually 30 days after employment

unity of command organizational principle whereby there is only one boss

vented opened to the atmosphere by the fire burning through windows or walls through which heat and fire byproducts are released, and through which fresh air may enter

ventilation a systematic process to enhance the removal of smoke and fire by-products and the entry of cooler air to facilitate rescue and fire-fighting operations

vision an imaginary concept, usually favorable, of the result of an effort

warrant a legal writ issued by a judicial officer commanding an officer to arrest a person, seize property, or search a premises

written reprimand documents unsatisfactory performance and specifies the corrective action expected; usually follows an oral reprimand

Index